ENCYCLOPEDIA OF MATHEMATICS AND ITS APPLICATIONS

Edited by G.-C. ROTA
Volume 51

Handbook of Categorical Algebra 2

ENCYCLOPEDIA OF MATHEMATICS AND ITS APPLICATIONS

1. L A Santalo *Integral Geometry and Geometric Probability*
2. G E Andrews *The theory of partitions*
3. R J McEliece *The Theory of Information and Coding: A Mathematical Framework for Communication*
4. W Miller, Jr *Symmetry and separation of variables*
5. D Ruelle *Thermodynamic Formalism: The Mathematical Structures of Classical Equilibrium Statistical Mechanics*
6. H Minc *Permanents*
7. F S Roberts *Measurement Theory with Applications to Decisionmaking, Utility, and the Social Sciences*
8. L C Biedenharn and J D Louck *Angular Momentum in Quantum Physics: Theory and Applications*
9. L C Biedenharn and J D Louck *The Racah-Wigner Algebra in Quantum Theory*
10. J D Dollard and C N Friedman *Product Integration with Applications in Quantum Theory*
11. W B Jones and W J Thron *Continued Fractions: Analytic Theory and Applications*
12. N F G Martin and J W England *Mathematical Theory of Entropy*
13. G A Baker, Jr and P Graves-Morris *Padé Approximants, Part I, Basic Theory*
14. G A Baker, Jr and P Graves-Morris *Padé Approximants, Part II, Extensions and Applications*
15. E G Beltrametti and G Cassinelli *The Logic of Quantum Mechanics*
16. G D James and A Kerber *The Representation Theory of Symmetric Groups*
17. M Lothaire *Combinatorics on Words*
18. H O Fattorini *The Cauchy Problem*
19. G G Lorentz, K Jetter and S D Riemenschneider *Birkhoff Interpolation*
20. R Lidl and H Niederreiter *Finite Fields*
21. W T Tutte *Graph Theory*
22. J R Bastida *Field Extensions and Galois Theory*
23. J R Cannon *The One-Dimensional Heat Equation*
24. S Wagon *The Banach-Tarski Paradox*
25. A Salomaa *Computation and Automata*
26. N White (ed) *Theory of Matroids*
27. N H Bingham, C M Goldie and J L Teugels *Regular Variations*
28. P P Petrushev and V A Popov *Rational Approximation of Real Functions*
29. N White (ed) *Combinatorial Geometries*
30. M Phost and H Zassenhaus *Algorithmic Algebraic Number Theory*
31. J Aczel and J Dhombres *Functional Equations in Several Variables*
32. M Kuczma, B Choczewski and R Ger *Iterative Functional Equations*
33. R V Ambartzumian *Factorization Calculus and Geometric Probability*
34. G Gripenberg, S-O Londen and O Staffans *Volterra Integral and Functional Equations*
35. G Gasper and M Rahman *Basic Hypergeometric Series*
36. E Torgersen *Comparison of Statistical Experiments*
37. A Neumaier *Interval Methods for Systems of Equations*
38. N Korneichuk *Exact Constants in Approximation Theory*
39. R Brualdi and H Ryser *Combinatorial Matrix Theory*
40. N White (ed) *Matroid applications*
41. S Sakai *Operator Algebras in Dynamical Systems*
42. W Hodges *Model Theory*
43. H Stahl and V Totik *General Orthogonal Polynomials*
44. R Schneider *Convex Bodies*
45. G Da Prato and J Zabczyk *Stochastic Equations in Infinite Dimensions*
46. A Björner *et al* *Oriented Matroids*
47. G Edgar and L Sucheston *Stopping Times and Directed Processes*
48. C Sims *Computation with Finitely Presented Groups*
49. T W Palmer *C*-algebras I*
50. F Borceux *Handbook of Categorical Algebra 1, Basic Category Theory*
51. F Borceux *Handbook of Categorical Algebra 2, Categories and Structures*

ENCYCLOPEDIA OF MATHEMATICS AND ITS APPLICATIONS

Handbook of Categorical Algebra 2

Categories and Structures

Francis Borceux

Département de Mathématique

Université Catholique de Louvain

CAMBRIDGE UNIVERSITY PRESS
Cambridge, New York, Melbourne, Madrid, Cape Town, Singapore, São Paulo

Cambridge University Press
The Edinburgh Building, Cambridge CB2 8RU, UK

Published in the United States of America by Cambridge University Press, New York

www.cambridge.org
Information on this title: www.cambridge.org/9780521441797

First published 1994
This digitally printed version 2008

A catalogue record for this publication is available from the British Library

ISBN 978-0-521-44179-7 hardback
ISBN 978-0-521-06122-3 paperback

à René Lavendhomme,
mon maître

Contents

Preface to volume 2

This second volume of the *Handbook of categorical algebra* presents a selection of well-known specialized topics in category theory, with the exception of toposes which find their natural place in volume 3.

The first great achievement of category theory has certainly been the theory of abelian categories: these play an important role in homology and provide the correct setting for studying problems related to exact sequences. Entire books are devoted to abelian categories; in the first chapter of this volume we have selected some topics on abelian categories which appear to remain highly relevant in to-day's research in general category theory. The chapter starts by establishing the key "exactness" properties of limits and colimits in an abelian category, those properties being closely related to the existence of an "additive structure" on the sets of morphisms. The notion of exact sequence is then introduced together with the technique of "diagram chasing", used to prove the fundamental diagram lemmas. "Diagram chasing" is a technique for proving exactness properties in an abelian category, just by proving them in the categories of modules, where actual elements can be used. This is best achieved by applying the famous "embedding theorem" which asserts that every small abelian category can be fully and exactly embedded in a category of modules. This very difficult theorem is proved at the end of the chapter, using a Lubkin completion technique which turns out to be applicable in many other categorical situations. For those who do not want to enter these difficult matters, we give a very elementary "diagram chasing metatheorem" which is good enough for many purposes. We also study the localizations of abelian categories and their relations with torsion theories and universal closure operations.

Regular and exact categories are in a way those categories which recapture some essential exactness properties of abelian categories, but without any requirement or implication of additivity. As examples, one gets most "algebra-like" categories. Regular categories provide the correct setting for developing the theory of relations, in particular equivalence relations. Making the technique even harder, the proof we have given of the embedding theorem for abelian categories can be adapted to produce a full exact embedding of every small category in a topos of presheaves, i.e. in a category $\mathsf{Fun}(\mathscr{D}^{op}, \mathsf{Set})$ for some small category $\mathscr{D}$. We have preferred the much easier full exact embedding theorem in a topos of sheaves (see chapter 3, volume 3), which provides almost as good a "diagram chasing" metatheorem.

The next three chapters are devoted to various categorical approaches to the notion of "model of an algebraic theory". In chapter 3, we are interested in those cases where the algebraic structure is given by finitary everywhere defined operations, the axioms being expressed by equalities. The corresponding categories of models can be presented as categories of set-valued finite product preserving functors. We study the completeness and exactness properties of these "algebraic categories" and pay special attention to the case of free models. We prove a characterization theorem for algebraic categories and conclude the chapter with some special topics: commutative theories, tensor product of theories, and Morita equivalent theories.

The notion of "monad on a category" formalizes intuitively the idea of a theory defined by "operations of arbitrary arities". We show the close relations of this with the notion of adjoint functors and study the completeness and exactness properties of "monadic categories"; we prove also a corresponding characterization theorem. We pay some additional attention to the case of monads with finite rank (intuitively: the possible arities of operations are now bounded by some cardinal) and we conclude the chapter by exhibiting some relations with descent theory for modules.

In chapter 3, algebraic theories were defined using finite products. In chapter 5, we investigate those theories which can be defined by a "sketch", that is a set of small limits and colimits. The corresponding categories of models are the "accessible categories": they turn out to be exactly the categories of set-valued α-flat functors, for some regular cardinal α. When just limit cones are used, the corresponding accessible categories are "locally presentable" and coincide with the categories of set-valued α-left-exact functors, for some regular cardinal α.

Chapter 6 introduces some fundamental notions and results of enriched category theory. We are interested here in the case where the categories involved have an additional structure on their sets $\mathscr{C}(A,B)$ of morphisms; for example the categories of modules on a ring have abelian groups of morphisms: they are "enriched" in the category of abelian groups. We limit our investigations to the basic notions and questions concerning enriched limits, enriched adjunctions and enriched Kan extensions. Cartesian closed categories – i.e. those categories in which the cartesian product with every object has a right adjoint – constitute an important example of categories in which to enrich category theory: in particular, all toposes (see chapter 5, volume 3) are cartesian closed.

Unfortunately the category of topological spaces is not cartesian closed, i.e. given topological spaces Y, Z, there is no way to provide the set $\mathcal{C}(Y,Z)$ of continuous functions with a topology, in such a way that a mapping

$$X \longrightarrow \mathcal{C}(Y,Z)$$

is continuous if and only if the corresponding mapping

$$X \times Y \longrightarrow Z$$

is continuous, for every other space X. We pay some attention to "exponentiable spaces Y" (those for which the previous problem has a solution) and show that restricting one's attention to compactly generated spaces yields a cartesian closed category of topological spaces. Finally we introduce topological functors, i.e. those functors which satisfy axiomatically the conditions for the existence of topological-like initial structures.

The last chapter of this volume is devoted to the theory of fibred categories "à la Bénabou". Fibred categories formalize the idea of "families of objects and morphisms" indexed by an object in a base category. We study first the corresponding fibred notions of adjunction and completeness. We then pay special attention to "locally small" fibrations, which are those for which the formal "families of morphisms" are represented by objects in the base category. We conclude with the very crucial notion of "definability" which exhibits those classes of devices in a fibration which can be represented by objects in the base category.

Introduction to this handbook

My concern in writing the three volumes of this *Handbook of categorical algebra* has been to propose a directly accessible account of what – in my opinion – a Ph.D. student should ideally know of category theory before starting research on one precise topic in this domain. Of course, there are already many good books on category theory: general accounts of the state of the art as it was in the late sixties, or specialized books on more specific recent topics. If you add to this several famous original papers not covered by any book and some important but never published works, you get a mass of material which gives probably a deeper insight in the field than this *Handbook* can do. But the great number and the diversity of those excellent sources just act to convince me that an integrated presentation of the most relevant aspects of them remains a useful service to the mathematical community. This is the objective of these three volumes.

The first volume presents those basic aspects of category theory which are present as such in almost every topic of categorical algebra. This includes the general theory of limits, adjoint functors and Kan extensions, but also quite sophisticated methods (like categories of fractions or orthogonal subcategories) for constructing adjoint functors. Special attention is also devoted to some refinements of the standard notions, like Cauchy completeness, flat functors, distributors, 2-categories, bicategories, lax-functors, and so on.

The second volume presents a selection of the most famous classes of "structured categories", with the exception of toposes which appear in volume 3. The first historical example is that of abelian categories, which we follow by its natural non-additive generalizations: the regular and exact categories. Next we study various approaches to "categories of models of a theory": algebraic categories, monadic categories, locally presentable and accessible categories. We introduce also enriched category theory and devote some attention to topological categories. The volume ends with the theory of fibred categories "à la Bénabou".

The third volume is entirely devoted to the study of categories of sheaves: sheaves on a space, a locale, a site. This is the opportunity for developing the essential aspects of the theory of locales and introducing Grothendieck toposes. We relate this with the algebraic aspects of volume 2 by proving in this context the existence of a classifying topos for coherent theories. All these considerations lead naturally to the notion of an elementary topos. We study quite extensively the internal logic of toposes, including the law of excluded middle and the axiom of infinity. We conclude by showing how toposes are a natural context for defining sheaves.

Besides a technical development of the theory, many people appreciate historical notes explaining how the ideas appeared and grew. Let me tell you a story about that.

It was in July, I don't remember the year. I was participating in a summer meeting on category theory at the Isles of Thorns, in Sussex. Somebody was actually giving a talk on the history of Eilenberg and Mac Lane's collaboration in the forties, making clear what the exact contribution of the two authors was. At some point, somebody in the audience started to complain about the speaker giving credit to Eilenberg and Mac Lane for some basic aspect of their work which – he claimed – they borrowed from somebody else. A very sophisticated and animated discussion followed, which I was too ignorant to follow properly. The only things I can remember are the names of the two opponents: the speaker was Saunders Mac Lane and his opponent was Samuel Eilenberg. I was not born when they invented category theory. With my little story in mind, maybe you will forgive me for not having tried to give credit to anybody for the notions and results presented in this *Handbook*.

Let me conclude this introduction by thanking the various typists for their excellent job and my colleagues of the Louvain-la-Neuve category seminar for the fruitful discussions we had on various points of this *Handbook*. I want especially to acknowledge the numerous suggestions Enrico Vitale has made for improving the quality of my work.

Handbook of categorical algebra

Contents of the three volumes

Volume 1: Basic category theory

Volume 2: Categories and structures

Volume 3: Categories of sheaves

1
Abelian categories

1.1 Zero objects and kernels

In section 2.3, volume 1, we studied the notions of terminal and initial object. In the category of abelian groups, or any category of modules over a ring, both notions coincide and correspond to the group (or the module) reduced to $\{0\}$.

Definition 1.1.1 *By a zero object in a category $\mathscr{C}$, we mean an object $\mathbf{0}$ which is both an initial and a terminal object.*

It should be noticed that the notion of zero object is autodual!

Definition 1.1.2 *Consider a category $\mathscr{C}$ with a zero object $\mathbf{0}$. A morphism $f\colon A \longrightarrow B$ is called a zero morphism when it factors through the zero object $\mathbf{0}$.*

Proposition 1.1.3 *In a category $\mathscr{C}$ with a zero object $\mathbf{0}$, there is exactly one zero morphism from each object A to each object B.*

Proof This is just the composite of the unique morphisms $A \longrightarrow \mathbf{0}$, where $\mathbf{0}$ is considered as a terminal object, and $\mathbf{0} \longrightarrow B$, where $\mathbf{0}$ is considered as an initial object. □

Proposition 1.1.4 *In a category $\mathscr{C}$ with a zero object $\mathbf{0}$, the composite of a zero morphism with an arbitrary morphism is again a zero morphism.*

Proof Of course, the composite factors through $\mathbf{0}$. □

Definition 1.1.5 *In a category $\mathscr{C}$ with a zero object $\mathbf{0}$, the kernel of an arrow $f\colon A \longrightarrow B$ is – when it exists – the equalizer of f and the zero morphism $\mathbf{0}\colon A \longrightarrow B$. The cokernel of f is defined dually.*

Every kernel is, by definition, an equalizer and therefore a monomorphism (see 2.4.3, volume 1). But in a category with a zero object, a monomorphism or even an equalizer need not be a kernel, as is shown by example 1.1.9.a.

Let us make some trivial observations.

Proposition 1.1.6 *Let f be a monomorphism in a category with a zero object. If $f \circ g = 0$ for some morphism g, then $g = 0$.*

Proof $f \circ g = 0 = f \circ 0$ (see 1.1.4), thus $g = 0$. □

Proposition 1.1.7 *In a category with a zero object, the kernel of a monomorphism $f: A \longrightarrow B$ is just the zero arrow $\mathbf{0} \longrightarrow A$.*

Proof The composite $\mathbf{0} \longrightarrow A \xrightarrow{f} B$ is the zero morphism (see 1.1.4) and if another composite $f \circ g$ is zero,

$$X \underset{0}{\overset{g}{\rightrightarrows}} A \xrightarrow{\quad f \quad} B$$

then $f \circ g = f \circ 0$ and therefore $g = 0$ by 1.1.6, which means that g factors (uniquely) through the object $\mathbf{0}$ (see 1.1.3). □

Proposition 1.1.8 *In a category with a zero object, the kernel of a zero morphism $0: A \longrightarrow B$ is just, up to isomorphism, the identity on A.*

Proof By 1.1.4 $0 \circ 1_A = 0$. Now given $g: X \longrightarrow A$ with $0 \circ g = 0$ (of course!) there exists a unique factorization of g through 1_A: it is g itself! □

Examples 1.1.9

1.1.9.a In the category Gr of all groups and group homomorphisms, the zero group (0) is a zero object and the zero morphisms are precisely those morphisms which map every element onto the zero element. Given a group homomorphism $f: A \longrightarrow B$, its kernel is therefore

$$\operatorname{Ker} f = \{a \in A \mid f(a) = 0\}.$$

But we have immediately

$$f(a) = 0 \Rightarrow \forall a' \in A \;\; f(a' + a - a') = f(a') + f(a) - f(a') = 0$$

which proves that $\operatorname{Ker} f$ is a normal subgroup of A. Therefore given a subgroup $H \subseteq A$ which is not normal, we have already an example of a monomorphism (see 1.7.7.c, volume 1) which is not a kernel.

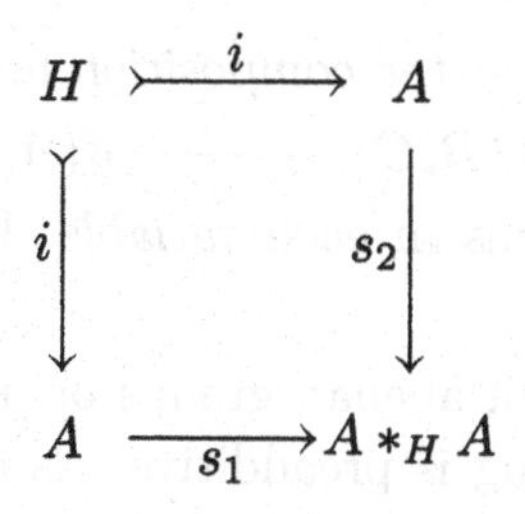

Diagram 1.1

Moreover, applying the amalgamation property for groups, described in 1.8.5.d, volume 1, we can compute the pushout of diagram 1.1 in Gr and conclude that s_1, s_2 are injective and the square is a pullback. In other words, the canonical inclusion i is the equalizer of s_1, s_2, so that every monomorphism of groups is an equalizer. Finally, not all equalizers of groups are kernels.

1.1.9.b Let R be a ring and Mod_R the category of right R-modules. The module (0) is clearly a zero object. For a R-linear mapping $f\colon A \longrightarrow B$, it is obvious that

$$\mathsf{Ker}\, f = \{a \in A \mid f(a) = 0\}.$$

On the other hand let us consider the submodule $f(A) \subseteq B$ and the corresponding quotient $B \longrightarrow B/f(A)$. It is clear that the composite

$$A \xrightarrow{\ f\ } B \xrightarrow{\ p\ } B/f(A)$$

is zero; but if a composite $A \xrightarrow{f} B \xrightarrow{g} C$ is zero, g maps every element of $f(A)$ onto 0, thus factors uniquely through p. Therefore

$$\mathsf{Coker}\, f = B/f(A).$$

Let us observe that given a submodule $M \subseteq B$, M is precisely the kernel of the projection $B \longrightarrow B/M$. Moreover, given a surjective linear mapping $p\colon B \longrightarrow C$, C is isomorphic to $B/\mathsf{Ker}\, p$, thus is the coequalizer of the inclusion $\mathsf{Ker}\, p \subseteq B$. So every monomorphism is a kernel and every epimorphism is a cokernel.

1.2 Additive categories and biproducts

Definition 1.2.1 *By a preadditive category we mean a category $\mathscr{C}$ together with an abelian group structure on each set $\mathscr{C}(A, B)$ of mor-*

phisms, in such a way that the composition mappings

$$c_{ABC}\colon \mathscr{C}(A,B) \times \mathscr{C}(B,C) \longrightarrow \mathscr{C}(A,C), \ (f,g) \mapsto g \circ f$$

are group homomorphisms in each variable. We shall write the group structure additively.

Clearly, the category of abelian groups or, more generally, any category of modules on a ring is preadditive. As a consequence, every full subcategory of a category of modules is preadditive as well.

Proposition 1.2.2 *The notion of preadditive category is autodual.* □

Proposition 1.2.3 *In a preadditive category $\mathscr{C}$, the following conditions are equivalent:*
(1) $\mathscr{C}$ has an initial object;
(2) $\mathscr{C}$ has a terminal object;
(3) $\mathscr{C}$ has a zero object.
In that case, the morphisms factoring through the zero object are exactly the identities for the group structure.

Proof (3) implies (1), (2) and, by duality, it suffices to prove that (1) implies (3). Let $\mathbf{0}$ be an initial object. The set $\mathscr{C}(\mathbf{0},\mathbf{0})$ has a single element, which proves that $1_{\mathbf{0}}$ is the zero element of the group $\mathscr{C}(\mathbf{0},\mathbf{0})$. Given an object C, $\mathscr{C}(C,\mathbf{0})$ has at least one element: the zero element of that group. But if $f\colon C \longrightarrow \mathbf{0}$ is any morphism, $f = 1_{\mathbf{0}} \circ f$ must be the zero element of $\mathscr{C}(C,\mathbf{0})$ since $1_{\mathbf{0}}$ is the zero element of $\mathscr{C}(\mathbf{0},\mathbf{0})$ ("bilinearity" of the composition). Thus $\mathbf{0}$ is a terminal object as well.

Given objects $C, D \in \mathscr{C}$, the groups $\mathscr{C}(C,\mathbf{0})$ and $\mathscr{C}(\mathbf{0},D)$ are thus reduced to their zero element, so that the composite of the two zero elements $C \longrightarrow \mathbf{0} \longrightarrow D$ is the zero element of $\mathscr{C}(C,D)$. □

Proposition 1.2.4 *Given two objects A, B in a preadditive category $\mathscr{C}$, the following conditions are equivalent:*
(1) the product (P, p_A, p_B) of A, B exists;
(2) the coproduct (P, s_A, s_B) of A, B exists;
(3) there exists an object P and morphisms

$$p_A\colon P \longrightarrow A, \ p_B\colon P \longrightarrow B, \ s_A\colon A \longrightarrow P, \ s_B\colon B \longrightarrow P$$

with the properties

$$p_A \circ s_A = 1_A, \ p_B \circ s_B = 1_B, \ p_A \circ s_B = 0, \ p_B \circ s_A = 0,$$

$$s_A \circ p_A + s_B \circ p_B = 1_P.$$

Moreover, under these conditions

$$s_A = \mathsf{Ker}\, p_B \ , \ s_B = \mathsf{Ker}\, p_A \ , \ p_A = \mathsf{Coker}\, s_B \ , \ p_B = \mathsf{Coker}\, s_A.$$

Proof By duality, it suffices to prove the equivalence of (1) and (3).

Given (1), define $s_A\colon A \longrightarrow P$ as the unique morphism with the properties $p_A \circ s_A = 1_A$, $p_B \circ s_A = 0$. In the same way $s_B\colon B \longrightarrow P$ is such that $p_A \circ s_B = 0$, $p_B \circ s_B = 1_B$. It is now easy to compute that

$$p_A \circ (s_A \circ p_A + s_B \circ p_B) = p_A + 0 = p_A,$$
$$p_B \circ (s_A \circ p_A + s_B \circ p_B) = 0 + p_B = p_B,$$

from which $s_A \circ p_A + s_B \circ p_B = 1_P$.

Given condition (3), consider $C \in \mathscr{C}$ and two morphisms $f\colon C \longrightarrow A$, $g\colon C \longrightarrow B$. Define $h\colon C \longrightarrow P$ as $h = s_A \circ f + s_B \circ g$. One has

$$p_A \circ h = p_A \circ s_A \circ f + p_A \circ s_B \circ g = f + 0 = f,$$
$$p_B \circ h = p_B \circ s_A \circ f + p_B \circ s_B \circ g = 0 + g = g.$$

Given $h'\colon C \longrightarrow P$ with the properties $p_A \circ h' = f$, $p_B \circ h' = g$, we deduce

$$\begin{aligned} h' &= 1_P \circ h' = (s_A \circ p_A + s_B \circ p_B) \circ h' \\ &= s_A \circ p_A \circ h' + s_B \circ p_B \circ h' = s_A \circ f + s_B \circ g \\ &= h. \end{aligned}$$

Now assuming conditions (1) to (3), let us prove that $s_A = \mathsf{Ker}\, p_B$. We have already $p_B \circ s_A = 0$. Choose $x\colon X \longrightarrow P$ such that $p_B \circ x = 0$. The composite $p_A \circ x\colon X \longrightarrow A$ is the required factorization since the relations

$$p_A \circ s_A \circ p_A \circ x = p_A \circ x,$$
$$p_B \circ s_A \circ p_A \circ x = 0 \circ p_A \circ x = 0 = p_B \circ x$$

imply $s_A \circ p_A \circ x = x$. The factorization is unique because $p_A \circ s_A = 1_A$ and thus s_A is a monomorphism.

The relation $s_B = \mathsf{Ker}\, p_A$ is true by analogy and the relations $p_A = \mathsf{Coker}\, s_B$, $p_B = \mathsf{Coker}\, s_A$ hold by duality. □

Definition 1.2.5 *Given two objects A, B in a preadditive category, a quintuple (P, p_A, p_B, s_A, s_B) as in 1.2.4.(3) is called a "biproduct" of A and B. The object P will generally be written $A \oplus B$.*

Definition 1.2.6 *By an additive category we mean a preadditive category with a zero object and binary biproducts.*

Clearly, the notion of additive category is again autodual. The following result is somewhat amazing.

Proposition 1.2.7 *On a category $\mathscr{C}$, any two additive structures are necessarily isomorphic.*

Proof Given an object $C \in \mathscr{C}$, we consider the diagonal

$$\Delta_C \colon C \longrightarrow C \oplus C$$

characterized by $p_1 \circ \Delta_C = 1_C$, $p_2 \circ \Delta_C = 1_C$. On the other hand we consider the difference $\sigma_C = p_1 - p_2 \colon C \oplus C \longrightarrow C$ of the two projections. It is immediate that

$$\sigma_C \circ \Delta_C = (p_1 - p_2) \circ \Delta_C = (p_1 \circ \Delta_C) - (p_2 \circ \Delta_C) = 1_C - 1_C = 0.$$

We shall prove that $\sigma_C = \mathsf{Coker}\, \Delta_C$.

First of all observe that $\Delta_C = s_1 + s_2$. Indeed one has

$$p_1 \circ (s_1 + s_2) = p_1 \circ s_1 = 1_C = p_1 \circ \Delta_C,$$
$$p_2 \circ (s_1 + s_2) = p_2 \circ s_2 = 1_C = p_2 \circ \Delta_C.$$

Therefore if $f \colon C \oplus C \longrightarrow D$ is such that $f \circ \Delta_C = 0$, one has

$$f \circ s_1 + f \circ s_2 = f \circ (s_1 + s_2) = f \circ \Delta_C = 0.$$

Therefore defining $g \colon C \longrightarrow D$ to be $f \circ s_1$, we have indeed

$$\begin{aligned} g \circ \sigma_C &= f \circ s_1 \circ (p_1 - p_2) = f \circ s_1 \circ p_1 - f \circ s_1 \circ p_2 \\ &= f \circ s_1 \circ p_1 + f \circ s_2 \circ p_2 = f \circ (s_1 \circ p_1 + s_2 \circ p_2) \\ &= f. \end{aligned}$$

Conversely if $g' \colon C \longrightarrow D$ is such that $g' \circ \sigma_C = f$, one has

$$g' = (g' \circ 1_C) - (g' \circ 0) = (g' \circ p_1 \circ s_1) - (g' \circ p_2 \circ s_1) = g' \circ \sigma_C \circ s_1 = f \circ s_1.$$

Thus we have proved that $\mathsf{Coker}\, \Delta_C = p_1 - p_2$, which proves that the difference $p_1 - p_2$ is characterized, up to isomorphism, by the limit–colimit structure of $\mathscr{C}$.

But now given two morphisms $a, b \colon A \rightrightarrows C$, we have a unique factorization $c \colon A \longrightarrow C \oplus C$ such that $p_1 \circ c = a$, $p_2 \circ c = b$. Therefore

$$a - b = (p_1 \circ c) - (p_2 \circ c) = (p_1 - p_2) \circ c,$$

which proves that the difference $a - b$ is characterized by the limit–colimit properties of $\mathscr{C}$. And finally $a + b$ can be written $a - (0 - b)$, which concludes the proof. □

If products and coproducts have very special properties in a preadditive category (see 1.2.4), not much can be said about arbitrary equalizers and coequalizers. Nevertheless, we have the following.

Proposition 1.2.8 *Let $f, g\colon A \rightrightarrows B$ be two morphisms in a preadditive category. The following conditions are equivalent:*

(1) the equalizer $\mathsf{Ker}\,(f,g)$ exists;

(2) the kernel $\mathsf{Ker}\,(f-g)$ exists;

(3) the kernel $\mathsf{Ker}\,(g-f)$ exists.

When this is the case, those three objects are isomorphic.

Proof Since in any case $\mathsf{Ker}\,(f,g) = \mathsf{Ker}\,(g,f)$, it suffices to prove that (1) $\Leftrightarrow$ (2). Given a morphism $x\colon X \longrightarrow A$, $f \circ x = g \circ x$ is equivalent to $(f-g)(x) = 0$, from which the result follows. □

Let us conclude this section with a point of notation. If A_1, A_2, B_1, B_2 are four objects in an additive category $\mathscr{C}$, a morphism

$$f\colon A_1 \oplus A_2 \longrightarrow B_1 \oplus B_2$$

is completely characterized by the four morphisms

$$\begin{aligned} f_{11} &= p_1 \circ f \circ s_1\colon A_1 \longrightarrow B_1,\\ f_{12} &= p_1 \circ f \circ s_2\colon A_2 \longrightarrow B_1,\\ f_{21} &= p_2 \circ f \circ s_1\colon A_1 \longrightarrow B_2,\\ f_{22} &= p_2 \circ f \circ s_2\colon A_2 \longrightarrow B_2, \end{aligned}$$

so that it makes sense to use the notation

$$\begin{pmatrix} f_{11} & f_{12} \\ f_{21} & f_{22} \end{pmatrix}$$

to denote f. It is routine to verify that given another morphism

$$g\colon B_1 \oplus B_2 \longrightarrow C_1 \oplus C_2,$$

the composite $g \circ f$ is precisely represented by the product of the two individual matrices. That notation extends obviously to the case of n-ary products. In particular given two morphisms $f\colon A_1 \longrightarrow B$, $g\colon A_2 \longrightarrow B$ we write $(f,g)\colon A_1 \oplus A_2 \longrightarrow B$ for the corresponding factorization. In an analogous way given two morphisms $f\colon A \longrightarrow B_1$ and $g\colon A \longrightarrow B_2$, we write $\binom{f}{g}\colon A \longrightarrow B_1 \oplus B_2$ for the induced factorization.

Examples 1.2.9

1.2.9.a Every category of modules on a ring is additive.

1.2.9.b The category Gr of all groups and group homomorphisms is not additive. Indeed, the coproduct of two groups is not their cartesian product (see 2.2.4.e, volume 1).

1.2.9.c The category Ban_1 of Banach spaces and linear contractions is not additive. Indeed, isomorphisms are isometric bijections (see 1.9.6.f) while the product of two objects and their coproduct are not isometric (see 2.1.7.d and 2.2.4.i, volume 1).

1.2.9.d The category Ban_∞ of Banach spaces and bounded linear mappings is additive: the sum of two bounded linear mappings is again such a mapping. Notice that binary products and coproducts are computed as in Ban_1, but they are now isomorphic since in Ban_∞ isomorphisms are just the bounded linear bijections (see 1.9.6.e, volume 1).

1.2.9.e The category of abelian groups is obviously additive. In particular finite products coincide with finite coproducts (see 1.2.4, volume 1). But infinite products do not coincide with infinite coproducts (see 2.1.7.c and 2.2.4.f, volume 1).

1.3 Additive functors

Definition 1.3.1 *Given two preadditive categories $\mathscr{A}$ and $\mathscr{B}$, a functor $F\colon \mathscr{A} \longrightarrow \mathscr{B}$ is additive when, for all objects A, A' in $\mathscr{A}$, the mapping*

$$F_{AA'}\colon \mathscr{A}(A, A') \longrightarrow \mathscr{B}(F(A), F(A')), \quad f \mapsto F(f)$$

is a group homomorphism.

It is clear that a composite of additive functors is again additive.

Proposition 1.3.2 *Given two preadditive categories $\mathscr{A}$, $\mathscr{B}$ with $\mathscr{A}$ small, the category $\mathsf{Add}(\mathscr{A}, \mathscr{B})$ of additive functors from $\mathscr{A}$ to $\mathscr{B}$ and natural transformations between them is again preadditive. Moreover the preadditive structure of $\mathsf{Add}(\mathscr{A}, \mathscr{B})$ is defined pointwise.*

Proof Given two natural transformations $\alpha : F \Rightarrow G, \beta\colon F \Rightarrow G$, define $\alpha + \beta\colon F \Rightarrow G$ by $(\alpha + \beta)_A = \alpha_A + \beta_A$. The details of the proof are straightforward. □

Proposition 1.3.3 *Consider an additive category $\mathscr{B}$ and a small preadditive category $\mathscr{A}$. In that case the category $\mathsf{Add}(\mathscr{A}, \mathscr{B})$ is additive and biproducts in it are computed pointwise. Moreover if $\mathscr{B}$ is finitely complete, so is $\mathsf{Add}(\mathscr{A}, \mathscr{B})$ and finite limits are computed pointwise.*

Proof The constant functor on the zero object of $\mathscr{B}$ is obviously additive and it is also the zero object in the category $\mathsf{Fun}(\mathscr{A}, \mathscr{B})$ of all functors (see 2.15.1, volume 1). Thus it is a zero object in $\mathsf{Add}(\mathscr{A}, \mathscr{B})$.

Given two functors $F, G\colon \mathscr{A} \rightrightarrows \mathscr{B}$, we consider their pointwise product $F \times G\colon \mathscr{A} \longrightarrow \mathscr{B}$ in the category $\mathsf{Fun}(\mathscr{A}, \mathscr{B})$ of all functors from $\mathscr{A}$ to

$\mathscr{B}$. An object $A \in \mathscr{A}$ is thus mapped onto $F(A) \oplus G(A)$ and a morphism $f\colon A \longrightarrow A'$ onto the morphism

$$\begin{pmatrix} F(f) & 0 \\ 0 & G(f) \end{pmatrix}.$$

From this description it follows immediately that $F \times G$ is an additive functor and is the biproduct of F and G in $\mathsf{Add}(\mathscr{A}, \mathscr{B})$.

If $\mathscr{B}$ has kernels, consider a natural transformation $\alpha\colon F \Rightarrow G$ and its pointwise kernel $\gamma\colon K \Rightarrow F$, let us prove that K is additive. Given $f, g\colon A \rightrightarrows A'$ in $\mathscr{A}$ one has

$$\begin{aligned} \gamma_{A'} \circ K(f-g) &= F(f-g) \circ \gamma_A = \big(F(f) - F(g)\big) \circ \gamma_A \\ &= \big(F(f) \circ \gamma_A\big) - \big(F(g) \circ \gamma_A\big) \\ &= \big(\gamma_{A'} \circ K(f)\big) - \big(\gamma_{A'} \circ K(g)\big) \\ &= \gamma_{A'} \circ \big(K(f) - K(g)\big) \end{aligned}$$

from which $K(f-g) = K(f) - K(g)$, since $\gamma_{A'}$ is a monomorphism. In the same way, given the zero morphism $0\colon A \longrightarrow A'$,

$$\gamma_{A'} \circ k(0) = f(0) \circ \gamma_A = 0 \circ \gamma_A = 0$$

from which $K(0) = 0$ since $\gamma_{A'}$ is a monomorphism. Thus $\mathsf{Add}(\mathscr{A}, \mathscr{B})$ has equalizers (see 1.2.8) and we get the conclusion by 2.8.1, volume 1. □

The next criterion for additivity is particularly interesting.

Proposition 1.3.4 *For a functor $F\colon \mathscr{A} \longrightarrow \mathscr{B}$ between additive categories, the following conditions are equivalent:*
(1) F is additive;
(2) F preserves biproducts;
(3) F preserves finite products;
(4) F preserves finite coproducts.

Proof (3) and (4) are equivalent by duality. (3), (4) imply in particular the preservation of both binary products and binary coproducts, thus the preservation of biproducts (see 1.2.4). To prove the converse implication (2) $\Rightarrow$ (3), it remains just to prove the preservation of the product of the empty family, i.e. of the terminal object $\mathbf{0}$. For every object $B \in \mathscr{B}$ there is at least the zero morphism $B \longrightarrow F(\mathbf{0})$, because $\mathscr{B}$ is preadditive. Now since $\mathbf{0} \in \mathscr{A}$ is a terminal object, the two projections

$$p_1, p_2\colon \mathbf{0} \oplus \mathbf{0} \rightrightarrows \mathbf{0}$$

are equal. By assumption, $F(\mathbf{0} \oplus \mathbf{0}) \cong F(\mathbf{0}) \oplus F(\mathbf{0})$ with $F(p_1) = F(p_2)$ as projections of the product. Therefore given two morphisms $f, g\colon B \rightrightarrows F(\mathbf{0})$ and the factorization $\binom{f}{g}\colon B \longrightarrow F(\mathbf{0}) \oplus F(\mathbf{0})$ through the product, the considerations of 1.2.7 show that

$$f - g = \sigma_{F(\mathbf{0})} \circ \begin{pmatrix} f \\ g \end{pmatrix} = (F(p_1) - F(p_2)) \circ \begin{pmatrix} f \\ g \end{pmatrix} = 0 \circ \begin{pmatrix} f \\ g \end{pmatrix} = 0,$$

thus $f = g$. So $F(\mathbf{0})$ is the terminal, thus the zero object of $\mathscr{B}$ (see 1.2.3).

(1) $\Rightarrow$ (2) is obvious and it remains to prove (2) $\Rightarrow$ (1). Assuming (2), we have already seen that F preserves the zero object, thus the zero morphisms. It remains to prove that F preserves the difference of two morphisms. Applying again the considerations of 1.2.7, it suffices to prove that F preserves the difference $p_1 - p_2$, for every biproduct $A \oplus A$ in $\mathscr{A}$. Since Fs_1, Fs_2 are the canonical injections of the biproduct $FA \oplus FA$, it suffices to apply the relations

$$F(p_1 - p_2) \circ F(s_1) = F\big((p_1 - p_2) \circ s_1\big) = F(1_A) = 1_{F(A)}$$
$$\big(F(p_1) - F(p_2)\big) \circ F(s_1) = 1_{F(A)}$$

and correspondingly with $F(s_2)$. □

Let us now make explicit a canonical example of a category $\mathsf{Add}(\mathscr{A}, \mathscr{B})$.

Consider a ring R with unit. We view it as a preadditive category $\mathscr{R}$ in the following way:

- $\mathscr{R}$ has a single object $*$;
- $\mathscr{R}(*, *) = R$;
- the composition $r \circ s$ of two arrows is their product rs as elements of the ring R;
- the sum $r + s$ of two arrows is their sum as elements of the ring R.

On the other hand consider the (additive) category Ab of abelian groups.

Proposition 1.3.5 *Given a ring R with unit, the category Mod_R of left R-modules is isomorphic to the category $\mathsf{Add}(\mathscr{R}, \mathsf{Ab})$ of additive functors, where $\mathscr{R}$ is just the ring R viewed as a preadditive category.*

Proof Given a left R-module M, define a functor $F\colon \mathscr{R} \longrightarrow \mathsf{Ab}$ by the following data:

- $F(*) = (M, +)$;
- $\forall r \in R \;\; F(r)\colon M \longrightarrow M, \;\; x \mapsto rx.$

It is immediate from the axioms of modules that F is an additive functor.

Now given an R-linear mapping $f\colon M \longrightarrow N$ to another left R-module N, consider the additive functor G associated with N. We get a natural transformation $\varphi\colon F \Rightarrow G$ by defining $\varphi_*\colon M \longrightarrow N$ to be just f.

Conversely an additive functor $F\colon \mathscr{R} \longrightarrow \mathsf{Ab}$ gives in particular an abelian group $F(*)$ and, for every element $r \in R$, a group homomorphism $F(r)\colon F(*) \longrightarrow F(*)$. Let us provide $F(*)$ with a scalar multiplication

$$R \times F(*) \longrightarrow F(*), \quad (r, x) \mapsto rx = F(r)(x).$$

The fact that $F(r)$ is a group homomorhism implies

$$r(x + y) = rx + ry$$

and the fact that F is a functor means

$$(rs)x = r(sx),$$

$$1x = x.$$

Finally the additivity of F implies

$$(r + s)x = rx + sx,$$

so that $F(*)$ has been provided with the structure of a left R-module.

If $G\colon \mathscr{R} \longrightarrow \mathsf{Ab}$ is another additive functor and $\varphi\colon F \Rightarrow G$ is a natural transformation, the group homomorphism $\varphi_*\colon F(*) \longrightarrow G(*)$ satisfies the relation $\varphi_* \circ F(r) = G(r) \circ \varphi_*$ or, in other words, $\varphi_*(rx) = r\varphi_*(x)$ for every element $x \in F(*)$. Therefore φ_* is R-linear.

It is obvious that we have constructed reciprocal isomorphisms. □

Clearly, a right R-module is just an additive functor $\mathscr{R}^* \longrightarrow \mathsf{Ab}$, or in other words a contravariant additive functor $\mathscr{R} \longrightarrow \mathsf{Ab}$.

Let us finally replace $\mathscr{R}$ by an arbitrary preadditive category $\mathscr{A}$. The additive functors $\mathscr{A} \longrightarrow \mathsf{Ab}$ have many of the properties of ordinary functors with values in the category of sets. Let us point out three examples. In fact those examples (as many other results of this section) are just special cases of a more general theory developed in chapter 6.

Proposition 1.3.6 *If $\mathscr{A}$ is a preadditive category and $A \in \mathscr{A}$, the "representable functor"*

$$\mathscr{A}(A, -)\colon \mathscr{A} \longrightarrow \mathsf{Ab}, \quad \mathscr{A}(A, -)(B) = \mathscr{A}(A, B)$$

is additive.

Proof Given $f, g\colon X \rightrightarrows Y$ in $\mathscr{A}$ consider

$$\mathscr{A}(A, -)(f - g)\colon \mathscr{A}(A, X) \longrightarrow \mathscr{A}(A, Y).$$

One has

$$\mathscr{A}(A,f-g)(h)=(f-g)\circ h=f\circ h-g\circ h=\mathscr{A}(A,f)(h)-\mathscr{A}(A,g)(h),$$

from which $\mathscr{A}(A,f-g)=\mathscr{A}(A,f)-\mathscr{A}(A,g)$. In the same way given the zero morphism $0\colon X\longrightarrow Y$,

$$\mathscr{A}(A,-)(0)(h)=0\circ h=0.$$

□

Proposition 1.3.7 (Additive Yoneda lemma)
If $\mathscr{A}$ is a preadditive category, $A\in\mathscr{A}$ and $F\colon\mathscr{A}\longrightarrow\mathsf{Ab}$ is an additive functor, there exist isomorphisms of abelian groups

$$\mathsf{Nat}\big(\mathscr{A}(A,-),F\big)\cong_{\theta_{F,A}}F(A)$$

where $\mathscr{A}(A,-)$ stands for the additive representable functor of 1.3.6. These isomorphisms are natural both in A and in F.

Proof Of course $\mathscr{A}(A,-)$ is the Ab-valued representable functor described in 1.3.6.

Given $\alpha\colon\mathscr{A}(A,-)\Rightarrow F$, we define $\theta_{F,A}(\alpha)=\alpha_A(1_A)$. Conversely given $a\in F(A)$, we consider for every object $B\in\mathscr{A}$ the mapping

$$\tau(a)_B\colon\mathscr{A}(A,B)\longrightarrow F(B),\ \ \tau(a)_B(f)=F(f)(a).$$

Since F is additive, $\tau(a)_B$ is a group homomorphism. Therefore the proof of 1.3.3, volume 1, applies. □

Proposition 1.3.8 *Consider a small preadditive category $\mathscr{C}$ and an additive functor $F\colon\mathscr{C}\longrightarrow\mathsf{Ab}$. In the category $\mathsf{Add}(\mathscr{C},\mathsf{Ab})$ of additive functors, F can be presented as the colimit of a diagram just constituted of Ab-valued representable functors and natural transformations between them.*

Proof It suffices to consider the composite

$$\mathsf{Elts}(F)\xrightarrow{\Phi_F}\mathscr{C}\xrightarrow{Y^*}\mathsf{Add}(\mathscr{C},\mathsf{Ab})$$

where $\mathsf{Elts}(F)$ is the category of elements of F (see 1.6.3) and $Y^*(C)=\mathscr{C}(C,-)$. The proof of 2.15.6, volume 1, applies without any change since colimits in $\mathsf{Add}(\mathscr{C},\mathsf{Ab})$ are computed pointwise (see 1.3.3), thus as in $\mathsf{Fun}(\mathscr{A},\mathsf{Ab})$ (see 2.15.2, volume 1), and the additive Yoneda lemma holds (see 1.3.7). □

1.4 Abelian categories

Historically, the study of abelian categories played a very important role in the development of category theory. Algebraic topology and homological algebra make wide use of them.

Definition 1.4.1 *A category $\mathscr{C}$ is abelian when it satisfies the following properties:*

(1) $\mathscr{C}$ has a zero object;

(2) every pair of objects of $\mathscr{C}$ has a product and a coproduct;

(3) every arrow of $\mathscr{C}$ has a kernel and a cokernel;

(4) every monomorphism of $\mathscr{C}$ is a kernel; every epimorphism of $\mathscr{C}$ is a cokernel.

Observing that the dual of each of those four axioms is the axiom itself, we conclude that

Proposition 1.4.2 (Abelian duality principle)
The dual notion of "abelian category" is again "abelian category". □

The previous proposition is fundamental since it implies that when a property is proved for abelian categories, so is the dual property.

To provide examples of abelian categories, we now use freely several results which will be proved in the following sections.

Proposition 1.4.3 *If a category $\mathscr{C}$ is abelian, each localization, and each colocalization, of $\mathscr{C}$ is again abelian.*

Proof An abelian category is finitely complete and finitely cocomplete (see 1.5.3). By duality (see 1.4.2), it suffices to prove the assertion concerning localizations (see 3.5.5, volume 1). We consider a localization $l \dashv i\colon \mathscr{L} \underset{\longrightarrow}{\longleftarrow} \mathscr{C}$; as a full subcategory of a preadditive category, $\mathscr{L}$ is obviously preadditive.

If $\mathbf{0}$ is the zero object of $\mathscr{C}$, $l(\mathbf{0})$ is both initial and final in $\mathscr{L}$ (l preserves colimits by 3.2.2, volume 1, and finite limits by 3.5.5, volume 1) and $il(\mathbf{0})$ is terminal in $\mathscr{C}$ (i preserves limits by 3.2.2, volume 1); in other words, viewing i as a full embedding, we conclude that $\mathbf{0} \in \mathscr{L}$ and $\mathbf{0}$ is also the zero object of $\mathscr{L}$.

Applying the arguments of 3.5.3,4, volume 1, to the special cases of those limits and colimits involved in definition 1.3.1, we conclude that $\mathscr{L}$ has binary products, binary coproducts, kernels and cokernels.

Now given a monomorphism $f\colon A \longrightarrow B$ in $\mathscr{L}$, it is still a monomorphism as an arrow of $\mathscr{C}$ (see 2.9.3, volume 1). Therefore $i(f) = \operatorname{Ker} g$

with g some arrow of $\mathscr{C}$, and thus $f = li(f) = \mathsf{Ker}\, l(g)$ since l preserves kernels.

The case of epimorphisms is more subtle. Consider an epimorphism $f\colon A \longrightarrow B$ in $\mathscr{L}$. Viewed as an arrow of $\mathscr{C}$, it has no reason still to be an epimorphism (and it is not in general). But in $\mathscr{C}$, f admits a factorization $i(f) = k \circ p$ where $p = \mathsf{Coker}\,(\mathsf{Ker}\, i(f))$ and $k = \mathsf{Ker}\,(\mathsf{Coker}\, i(f))$ (see 1.5.5). Applying l we get $f = li(f) = l(k) \circ l(p)$ with, in $\mathscr{L}$, $l(p) = \mathsf{Coker}\,(\mathsf{Ker}\, f)$ and $l(k) = \mathsf{Ker}\,(\mathsf{Coker}\, f)$ since l preserves kernels and cokernels and $li(f)$ is isomorphic to f. But since f is an epimorphism in $\mathscr{L}$, $\mathsf{Coker}\, f = 0$ in $\mathscr{L}$ (see 1.1.7) and therefore $\mathsf{Ker}\,(\mathsf{Coker}\, f)$ is an isomorphism in $\mathscr{L}$ (see 1.1.8). Thus f is isomorphic to $\mathsf{Coker}\,(\mathsf{Ker}\, f)$ in $\mathscr{L}$ and is a cokernel. □

Proposition 1.4.4 *Let $\mathscr{C}$ be an abelian category and $\mathscr{A}$ a small category. In that case the category of all functors and natural transformations $\mathsf{Fun}(\mathscr{A}, \mathscr{C})$ is again abelian.*

Proof By 2.15.1, volume 1, $\mathsf{Fun}(\mathscr{A}, \mathscr{C})$ has a zero object, binary products and coproducts, kernels and cokernels; all those limits and colimits are computed pointwise.

Choose now a natural transformation $\alpha\colon F \Rightarrow G$ which is a monomorphism in $\mathsf{Fun}(\mathscr{A}, \mathscr{C})$. The kernel pair of α is just $(1_F, 1_F)$ (see 2.5.4, volume 1) and since $\mathscr{C}$ has kernel pairs (see 1.5.3), the kernel pair of each morphism $\alpha_A\colon F(A) \longrightarrow G(A)$ is just $1_{F(A)}$ (see 2.15.1, volume 1). This means that each α_A is a monomorphism in $\mathscr{C}$. Computing the cokernel $\beta\colon G \Rightarrow H$ of α in $\mathsf{Fun}(\mathscr{A}, \mathscr{C})$, we know that each $\beta_A\colon G(A) \longrightarrow H(A)$ is the cokernel of the monomorphism α_A, thus α_A is the kernel of β_A (see 1.5.7). This proves that α is the kernel of β.

The fact that each epimorphism of $\mathsf{Fun}(\mathscr{A}, \mathscr{C})$ is a cokernel follows by duality. □

Proposition 1.4.5 *Let $\mathscr{C}$ be an abelian category and $\mathscr{A}$ a small additive category. In that case the category of additive functors and natural transformations $\mathsf{Add}(\mathscr{A}, \mathscr{C})$ is again abelian.*

Proof By 1.4.4, 1.3.3 and 1.5.3, $\mathsf{Add}(\mathscr{A}, \mathscr{C})$ has finite limits and colimits computed pointwise as in $\mathsf{Fun}(\mathscr{A}, \mathscr{C})$ (see 2.15.2, volume 1). In particular it has a zero object, products, coproducts, kernels and cokernels. Now given a monomorphism $\alpha\colon F \Rightarrow G$ in $\mathsf{Add}(\mathscr{A}, \mathscr{C})$, consider its cokernel $\beta\colon G \Rightarrow H$ in $\mathsf{Add}(\mathscr{A}, \mathscr{C})$. Since kernel pairs and cokernels in $\mathsf{Add}(\mathscr{A}, \mathscr{C})$ are computed as in $\mathsf{Fun}(\mathscr{A}, \mathscr{C})$, α is still a monomorphism in $\mathsf{Fun}(\mathscr{A}, \mathscr{C})$

(see 2.5.4) and $\beta = \mathsf{Coker}\,\alpha$ in $\mathsf{Fun}(\mathscr{A},\mathscr{C})$. Therefore $\alpha = \mathsf{Ker}\,\beta$ in $\mathsf{Fun}(\mathscr{A},\mathscr{C})$ and thus a fortiori in $\mathsf{Add}(\mathscr{A},\mathscr{C})$. □

Examples 1.4.6

1.4.6.a If R is a ring, the category of right (respectively, left) modules over R admits (0) as a zero object and is complete and cocomplete. Moreover each monomorphism is a kernel and each epimorphism is a cokernel (see 1.1.9.b). So Mod_R is abelian.

1.4.6.b If R is a ring and $\mathscr{C}$ is a small category, the category of presheaves of R-modules on $\mathscr{C}$ (see 4.1.7, volume 3) is again abelian (see 1.4.4). Indeed, writing $\mathscr{P}$ for the full subcategory of of Mod_R given by the finitely presented modules (see 5.1.1) and applying 5.2.7, we are interested in the category $\mathsf{Lex}(\mathscr{P}^*, \mathsf{Fun}(\mathscr{C}^*, \mathsf{Set}))$ of left exact functors from $\mathscr{P}^*$ to the category $\mathsf{Fun}(\mathscr{C}^*, \mathsf{Set})$. This category is obviously equivalent to the category of those functors $\mathscr{P}^* \times \mathscr{C}^* \longrightarrow \mathsf{Set}$ which are left exact in the first variable, i.e. finally to the category $\mathsf{Fun}(\mathscr{C}^*, \mathsf{Lex}(\mathscr{P}^*, \mathsf{Set}))$, which is just $\mathsf{Fun}(\mathscr{C}^*, \mathsf{Mod}_R)$ by 5.2.7. Since Mod_R is abelian, this last category is abelian as well (see 1.4.6).

1.4.6.c If R is a ring and $(\mathscr{C}, \mathscr{T})$ is a site (see 3.2.4, volume 3), we shall prove that the category of sheaves of R-modules on $(\mathscr{C}, \mathscr{T})$ is a localization of the corresponding category of presheaves of R-modules on $\mathscr{C}$, thus is abelian (see 1.4.6.a and 1.4.3). We write $\mathsf{Sh}(\mathscr{C},\mathcal{T})$ for the category of sheaves on $(\mathscr{C}, \mathscr{T})$; this is a localization of the category $\mathsf{Fun}(\mathscr{C}^*, \mathsf{Set})$ of presheaves and we write i, a for the canonical inclusion and the associated sheaf functor (see 3.3.12, volume 3). As in the previous example, we write $\mathscr{P}$ for the category of finitely presented R-modules and we are interested in exhibiting a localization

$$l \dashv j\colon \mathsf{Lex}(\mathcal{P}^*, \mathsf{Sh}(\mathscr{C},\mathcal{T})) \rightleftarrows \mathsf{Lex}(\mathscr{P}^*, \mathsf{Fun}(\mathscr{C}^*, \mathsf{Set})).$$

It suffices to take for j the composite with i, and for l the composite with a; this makes sense since both i and a are left exact. Applying 3.2.4, volume 1, we get immediately, for $F \in \mathsf{Lex}(\mathscr{P}^*, \mathsf{Sh}(\mathscr{C},\mathcal{T}))$ and $G \in \mathsf{Lex}(\mathscr{P}^*, \mathsf{Fun}(\mathscr{C}^*, \mathsf{Set}))$,

$$\mathsf{Nat}(l(G), F) \cong \mathsf{Nat}(a \circ G, F) \cong \mathsf{Nat}(G, i \circ F) \cong \mathsf{Nat}(G, j(F)),$$

from which the adjunction $l \dashv j$. l is left exact because finite limits are computed pointwise both in $\mathsf{Fun}(\mathscr{C}^*, \mathsf{Set})$ and in $\mathsf{Sh}(\mathscr{C},\mathcal{T})$ (see 2.15.2, volume 1, and 3.4.3, volume 3) and are preserved by a.

1.4.6.d We have seen in 1.3.6 that a ring can be seen as a particular preadditive category $\mathcal{R}$, while an R-module is just an additive functor $\mathcal{R} \longrightarrow \mathsf{Ab}$. More generally if $\mathcal{R}$ is any small preadditive category, the category $\mathsf{Add}(\mathcal{R}, \mathsf{Ab})$ of additive functors is abelian (see 1.4.5); an additive functor $\mathcal{R} \longrightarrow \mathsf{Ab}$ is sometimes called a "module on $\mathcal{R}$".

1.5 Exactness properties of abelian categories

By "exactness properties" we mean essentially properties related with finite limits and finite colimits.

Proposition 1.5.1 *In an abelian category, the following conditions are equivalent:*
(1) f is an isomorphism;
(2) f is both a monomorphism and an epimorphism.

Proof (1) $\Rightarrow$ (2) by 1.9.2, volume 1. Conversely $f = \mathsf{Ker}\, g$ for some morphism g. But from $g \circ f = 0$ we deduce $g = 0$ since f is an epimorphism (1.1.6) and therefore f is an isomorphism (1.1.8). □

Proposition 1.5.2 *In an abelian category $\mathscr{C}$, the intersection of two subobjects always exists.*

Proof Consider two monomorphisms $a\colon A \rightarrowtail C$ and $b\colon B \rightarrowtail C$. There exist morphisms $f\colon C \longrightarrow D$ and $g\colon C \longrightarrow E$ with $a = \mathsf{Ker}\, f$, $b = \mathsf{Ker}\, g$. Let us consider the factorization $\binom{f}{g}\colon C \longrightarrow D \times E$.

In the case $\mathscr{C} = \mathsf{Ab}$, one could write

$$A = \{c \in C \mid f(c) = 0\},$$
$$B = \{c \in C \mid g(c) = 0\},$$
$$A \cap B = \left\{ c \in C \,\middle|\, \begin{pmatrix} f \\ g \end{pmatrix}(c) = 0 \right\},$$

so that $A \cap B$ would just be $\mathsf{Ker}\,\binom{f}{g}$. Let us generalize this to the case of an arbitrary abelian category $\mathscr{C}$. We refer to diagram 1.2.

We consider $k = \mathsf{Ker}\,\binom{f}{g}$. Since $f \circ k = p_D \circ \binom{f}{g} \circ k = p_D \circ 0 = 0$, k factors through a as $k = a \circ a'$; in the same way, $k = b \circ b'$. Let us prove that $\left(\mathsf{Ker}\,\binom{f}{g}, a', b'\right)$ is the pullback of the pair (a, b). Given u, v such that $a \circ u = b \circ v$, we have $f \circ b \circ v = f \circ a \circ u = 0 \circ u = 0$; since $a = \mathsf{Ker}\, f$, there is a unique w such that $k \circ w = b \circ v$. So $b \circ v = k \circ w = b \circ b' \circ w$ and $b' \circ w = v$ since b is a monomorphism. Moreover $a \circ u = b \circ v = k \circ w = a \circ a' \circ w$, from which $u = a' \circ w$ since a is a monomorphism. This factorization w is unique since a, b are monomorphisms (see 1.7.2, volume 1). □

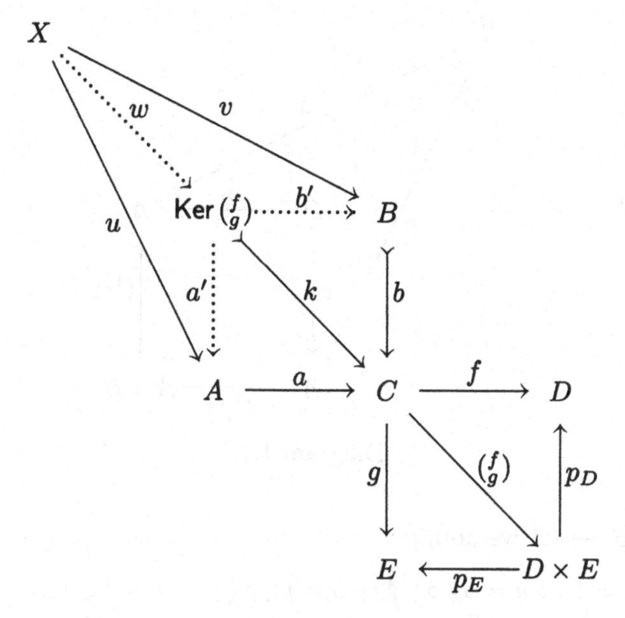

Diagram 1.2

Proposition 1.5.3 *An abelian category is finitely complete and finitely cocomplete.*

Proof By duality, it suffices to prove the finite completeness. Since finite products exist by definition, it suffices to prove the existence of equalizers (see 2.8.2, volume 1). We refer to diagram 1.3. Let us consider two morphisms $f, g: A \rightrightarrows B$. Observe that $\binom{1_A}{f}$ and $\binom{1_A}{g}$ are monomorphisms, since their composite with the projection $p_A: A \times B \longrightarrow A$ is just the identity on A. The intersection (P, u, v) of those two monomorphisms exists by 1.5.2.

In the case of the category of abelian groups, one would have

$$\begin{aligned} A &\cong \{(a, f(a)) \mid a \in A\}, \\ A &\cong \{(a, g(a)) \mid a \in A\}, \\ P &\cong \{(a, a') \mid a \in A\,,\ a' \in A\,,\ a = a'\,,\ f(a) = g(a')\} \\ &\cong \{a \mid a \in A\,,\ f(a) = g(a)\}, \end{aligned}$$

and therefore P would be the equalizer of f and g. Let us generalize this fact to the case of an arbitrary abelian category $\mathscr{C}$.

We refer still to diagram 1.3. Composing it with the first projection

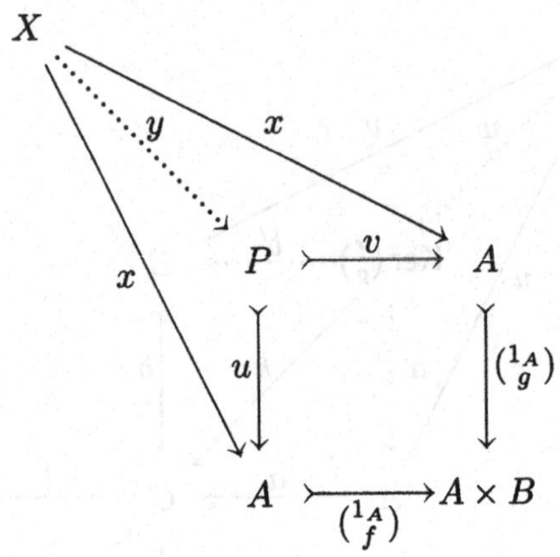

Diagram 1.3

$p_A \colon A \times B \longrightarrow A$ we obtain

$$u = 1_A \circ u = p_A \circ \binom{1_A}{f} \circ u = p_A \circ \binom{1_A}{g} \circ v = 1_A \circ v = v,$$

while composing with the projection $p_B \colon A \times B \longrightarrow B$ yields

$$f \circ u = p_B \circ \binom{1_A}{f} \circ u = p_B \circ \binom{1_A}{g} \circ v = g \circ v.$$

Putting these equalities together, we get $f \circ u = g \circ u$. Now if $x \colon X \longrightarrow A$ is another morphism such that $f \circ x = g \circ x$, then clearly $\binom{1_A}{f} \circ x = \binom{1_A}{g} \circ x$ and we find a unique factorization y such that $u \circ y = x = v \circ y$. □

Proposition 1.5.4 *For a morphism $f \colon A \longrightarrow B$ in an abelian category $\mathscr{C}$, the following conditions are equivalent:*
(1) f is a monomorphism;
(2) $\mathsf{Ker}\, f = 0$;
(3) $\forall C \in \mathscr{C}\ \forall g \colon C \longrightarrow A\ \ f \circ g = 0 \Rightarrow g = 0$.

Proof (1) ⇒ (2) and (1) ⇒ (3) have been proved in 1.1.7 and 1.1.6. To prove (2) ⇒ (1), we refer to diagram 1.4. If $\mathsf{Ker}\, f = 0$, consider two morphisms u, v such that $f \circ u = f \circ v$ and their coequalizer $q = \mathsf{Coker}\,(u, v)$ (which exists by 1.5.3). Since q is an epimorphism, $q = \mathsf{Coker}\, w$ for some w. Since $f \circ u = f \circ v$, f factors through q as $f = m \circ q$. From $f \circ w = m \circ q \circ w = m \circ 0 = 0$ we deduce that w factors through the kernel k of f as $w = k \circ n$. But since $\mathsf{Ker}\, f = 0$, $w = 0$ and q is an isomorphism (see 1.1.8). In particular q is a monomorphism and since $q \circ u = q \circ v$, one concludes that $u = v$. So f is a monomorphism.

Finally let us prove (3) ⇒ (2). Clearly the composite $0 \longrightarrow A \xrightarrow{f} B$ is the zero morphism. Now if the composite $C \xrightarrow{g} A \xrightarrow{f} B$ is the zero

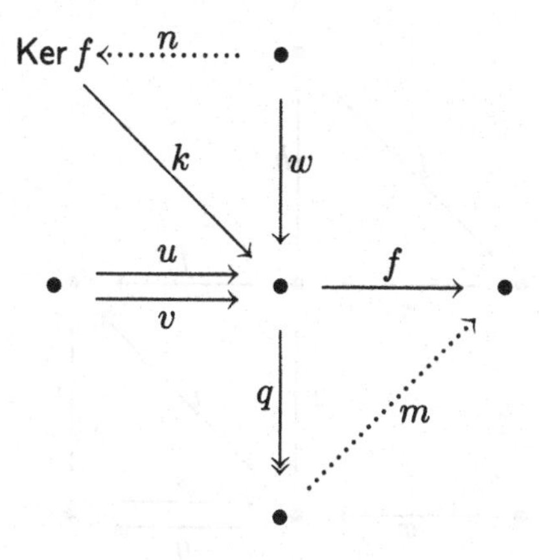

Diagram 1.4

morphism, g factors through $\mathbf{0}$ by assumption and this factorization is unique since $\mathbf{0}$ is terminal. Thus $0 = \mathsf{Ker}\, f$. □

Theorem 1.5.5 *Every morphism f in an abelian category can be factored uniquely (up to isomorphism) as $f = i \circ p$, where i is a monomorphism and p is an epimorphism. Moreover, $i = \mathsf{Ker}\,(\mathsf{Coker}\, f)$ and $p = \mathsf{Coker}\,(\mathsf{Ker}\, f)$.*

The factorization of f refered to in theorem 1.5.5 is called, for obvious reasons, the "image factorization" of f. Observe that in the case of a homomorphism $f\colon A \longrightarrow B$ of abelian groups, the image $f(A) \subseteq B$ is such that $f\colon A \longrightarrow f(A)$ is surjective. Therefore $f(A)$ is the quotient of A by the kernel of f, i.e. the cokernel of the kernel of f.

Proof We refer to diagram 1.5 where the various morphisms are introduced as follows. First, f is the given morphism, $k = \mathsf{Ker}\, f$ and $p = \mathsf{Coker}\, k$. Since $f \circ k = 0$ ($k = \mathsf{Ker}\, f$), f factors through p as $f = i \circ p$. We shall prove first that i is a monomorphism. To do this we apply 1.5.4. It suffices to choose x such that $i \circ x = 0$ and prove that $x = 0$. Since $i \circ x = 0$, i factors uniquely as $i = r \circ q$, where $q = \mathsf{Coker}\, x$. Now $q \circ p$ is an epimorphism as composite of two epimorphisms, thus there exists some h such that $q \circ p = \mathsf{Coker}\, h$ (see 1.4.1). Since $f \circ h = r \circ q \circ p \circ h = r \circ 0 = 0$, h factors uniquely as $h = k \circ l$ ($k = \mathsf{Ker}\, f$). Finally $p \circ h = p \circ k \circ l = 0 \circ l = 0$, thus p factors as $p = s \circ (q \circ p)$ since $q \circ p = \mathsf{Coker}\, h$. But since p is an epimorphism this last relation implies $s \circ q = 1$. So q is a monomorphism from which $q \circ x = 0$ implies $x = 0$. Thus we have already proved that $f = i \circ p$, with i a monomorphism and $p = \mathsf{Coker}\,(\mathsf{Ker}\, f)$.

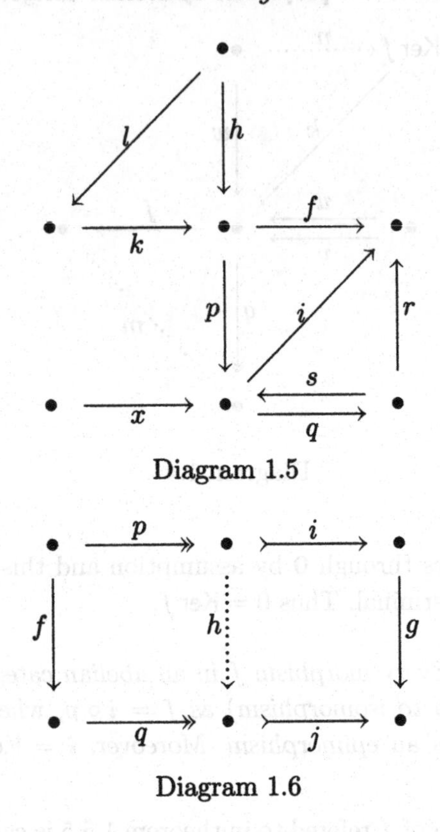

Diagram 1.5

Diagram 1.6

The uniqueness of such a factorization is attested by 4.4.5, volume 1, since regular epimorphisms are strong (see 4.3.6, volume 1).

By duality, f factors as $f = i' \circ p'$ with $i' = \mathsf{Ker}\,(\mathsf{Coker}\, f)$ and p' an epimorphism. By uniqueness of the mono–epi factorization, the factorizations $f = i \circ p$ and $f = i' \circ p'$ are isomorphic, which concludes the proof. □

Proposition 1.5.6 *In an abelian category, consider diagram 1.6 where the outer rectangle is commutative, p, q are epimorphisms and i, j are monomorphisms. There exists a unique morphism h making the completed diagram commutative.*

Proof Every epimorphism is regular (see 1.4.1), thus strong (see 4.3.6, volume 1), so the result follows from 4.4.5, volume 1. □

The previous construction is generally referred to as the "naturality" of the image construction.

Proposition 1.5.7 *In an abelian category*

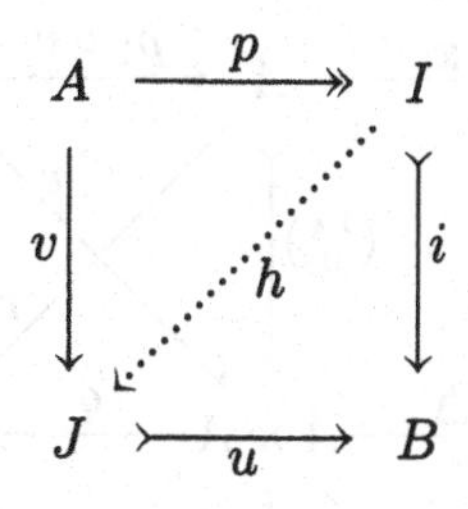

Diagram 1.7

(1) every monomorphism is the kernel of its cokernel,
(2) every epimorphism is the cokernel of its kernel.

Proof By 1.5.5, an epimorphism f can be factored as $f = i \circ p$, where i is a monomorphism and $p = \mathsf{Coker}(\mathsf{Ker}\, f)$. But since f is an epimorphism, so is i and therefore i is an isomorphism (see 1.5.1); finally f is isomorphic to $p \cong \mathsf{Coker}(\mathsf{Ker}\, f)$. The first assertion follows by duality. □

Proposition 1.5.8 *Consider a morphism $a: A \longrightarrow B$ in an abelian category, with image factorization $a = i \circ p$. Given another factorization $a = u \circ v$, with u a monomorphism, there exists a unique morphism h such that $u \circ h = i$, $h \circ p = v$ (see diagram 1.7).*

Proof Apply 1.5.6 with $f = v$, $q = 1_J$, $j = u$ and $g = 1_B$. □

Observe that the previous statement is just a particularization of the definition of a strong epimorphism p (see 4.3.5, volume 1).

1.6 Additivity of abelian categories

Having in mind the considerations of 1.2.7, we shall now introduce an additive structure on every abelian category.

Lemma 1.6.1 *Consider an object A of an abelian category, the diagonal morphism $\Delta: A \longrightarrow A \times A$ and its cokernel $q: A \times A \longrightarrow Q$. The object Q is isomorphic to A.*

Proof We refer to diagram 1.8 where p_1, p_2 denote the two projections of the product $A \times A$. We define $r: A \longrightarrow Q$ to be the composite $q \circ \binom{1_A}{0}$ and we shall prove it is an isomorphism. Since Δ is a monomorphism, $\Delta = \mathsf{Ker}(\mathsf{Coker}\,\Delta) = \mathsf{Ker}\, q$ (see 1.5.7). Since $p_1 \circ \binom{1_A}{0} = 1_A$, p_1 is an epimorphism and $\binom{1_A}{0}$ is a monomorphism; in the same way p_2 is an epimorphism and $\binom{0}{1_A}$ is a monomorphism. By definition $p_2 \circ \binom{1_A}{0} = 0$ and if $p_2 \circ v = 0$, $\binom{1_A}{0}(p_1 \circ v) = v$ as checked immediately by composing

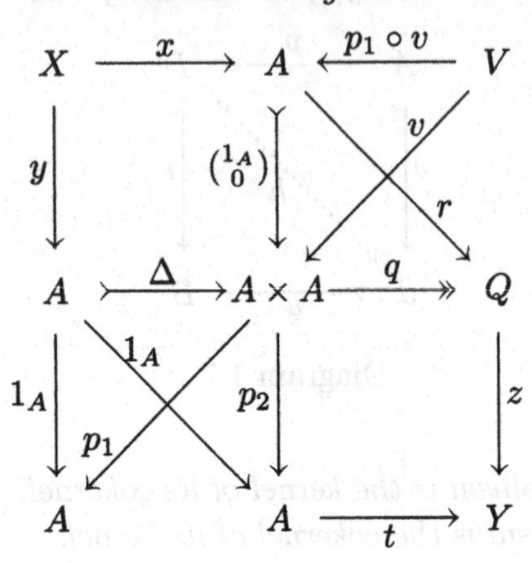

Diagram 1.8

with both projections p_1, p_2. This factorization $p_1 \circ v$ is unique since $\binom{1_A}{0}$ is a monomorphism, thus $\binom{1_A}{0} = \operatorname{Ker} p_2$. In the same way $\binom{0}{1_A} = \operatorname{Ker} p_1$. But since p_2 is an epimorphism, $p_2 = \operatorname{Coker}(\operatorname{Ker} p_2) = \operatorname{Coker}\binom{1_A}{0}$. In the same way $p_1 = \operatorname{Coker}\binom{0}{1_A}$.

To prove that r is a monomorphism, choose x such that $r \circ x = 0$. From $q \circ \binom{1_A}{0} \circ x = r \circ x = 0$ and $\Delta = \operatorname{Ker} q$, we deduce a factorization $\binom{1_A}{0} \circ x = \Delta \circ y$. Therefore $y = p_2 \circ \Delta \circ y = p_2 \circ \binom{1_A}{0} \circ x = 0 \circ x = 0$. It follows that $\Delta \circ y = \binom{1_A}{0} \circ x = 0$ and since $\binom{1_A}{0}$ is a monomorphism, then $x = 0$. By 1.5.4, r is a monomorphism.

We have still to prove that r is an epimorphism (see 1.5.1). Choose z such that $z \circ r = 0$. From $z \circ q \circ \binom{1_A}{0} = z \circ r = 0$ and $p_2 = \operatorname{Coker}\binom{1_A}{0}$, we deduce a factorization $z \circ q = t \circ p_2$. Therefore

$$t = t \circ p_2 \circ \Delta = z \circ q \circ \Delta = z \circ 0 = 0.$$

It follows that $z \circ q = t \circ p_2 = 0$ and thus $z = 0$ since p_2 is an epimorphism. So r is an epimorphism as well (see 1.5.4). □

Definition 1.6.2 *Consider two arrows $f, g \colon B \rightrightarrows A$ in an abelian category $\mathscr{C}$. With the notation of 1.6.1, we write σ_A for the composite*

$$A \times A \xrightarrow{q} Q \xrightarrow{r^{-1}} A$$

and call it "substraction on A". We write $f - g$ for the composite

$$B \xrightarrow{\binom{f}{g}} A \times A \xrightarrow{\sigma_A} A$$

and define $f + g = f - (0 - g)$.

We shall now prove that definition 1.6.2 introduces a (pre)additive structure on the abelian category $\mathscr{C}$. The following lemma will be crucial.

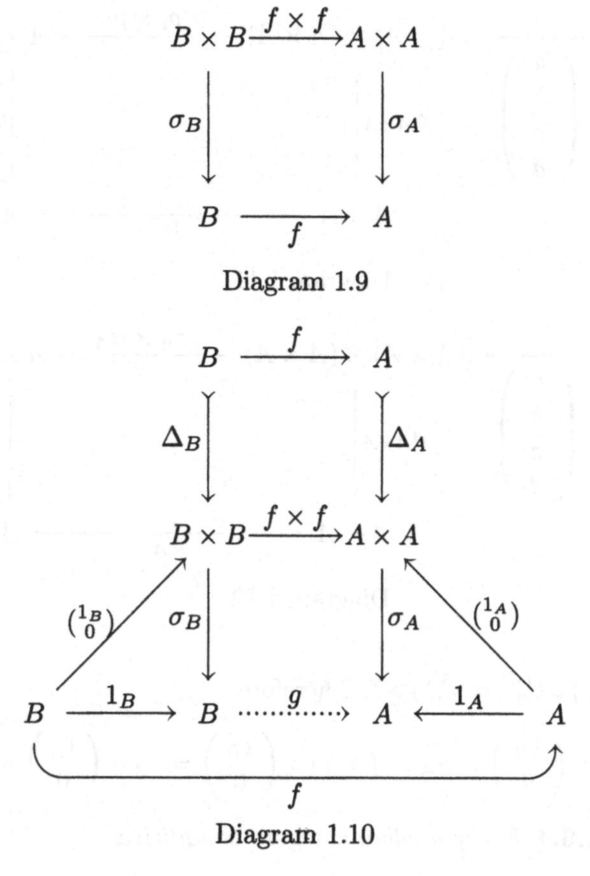

Diagram 1.9

Diagram 1.10

Lemma 1.6.3 *Given a morphism $f\colon B \longrightarrow A$ in an abelian category and with the notation of 1.6.2, $f \circ \sigma_B = \sigma_A \circ (f \times f)$; see diagram 1.9.*

Proof We consider diagram 1.10 where the relation

$$\sigma_A \circ (f \times f) \circ \Delta_B = \sigma_A \circ \Delta_A \circ f = 0 \circ f = 0$$

implies the existence of a unique g such that $g \circ \sigma_B = \sigma_A \circ (f \times f)$, just because by construction, $\sigma_B = \operatorname{Coker} \Delta_B$. We must prove that $f = g$.

With the notation of 1.6.1, we have also $q \circ \binom{1_A}{0} = r$, from which $\sigma_A \circ \binom{1_A}{0} = r^{-1} \circ q \circ \binom{1_A}{0} = 1_A$. In the same way $\sigma_B \circ \binom{1_B}{0} = 1_B$. On the other hand the relations

$$p_1 \circ (f \times f) \circ \begin{pmatrix} 1_B \\ 0 \end{pmatrix} = f = p_1 \circ \begin{pmatrix} 1_A \\ 0 \end{pmatrix} \circ f,$$

$$p_2 \circ (f \times f) \circ \begin{pmatrix} 1_B \\ 0 \end{pmatrix} = 0 = p_2 \circ \begin{pmatrix} 1_A \\ 0 \end{pmatrix} \circ f$$

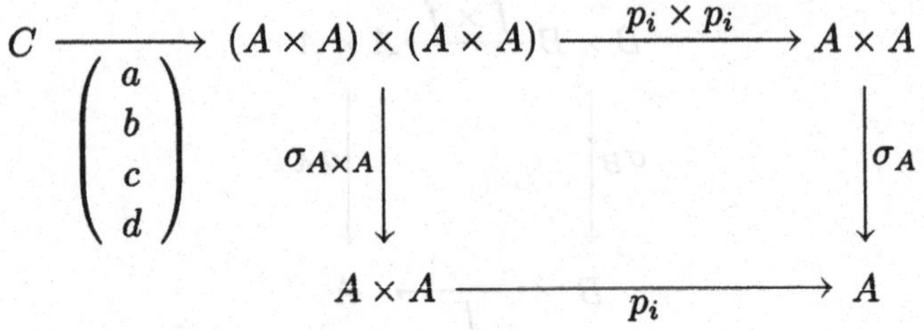

$$\begin{array}{ccc} C \xrightarrow{\begin{pmatrix} a \\ b \\ c \\ d \end{pmatrix}} (A \times A) \times (A \times A) & \xrightarrow{p_i \times p_i} & A \times A \\ \downarrow \sigma_{A \times A} & & \downarrow \sigma_A \\ A \times A & \xrightarrow{p_i} & A \end{array}$$

Diagram 1.11

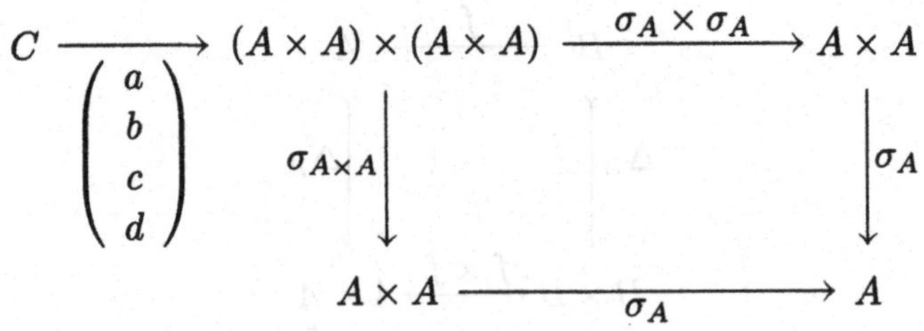

$$\begin{array}{ccc} C \xrightarrow{\begin{pmatrix} a \\ b \\ c \\ d \end{pmatrix}} (A \times A) \times (A \times A) & \xrightarrow{\sigma_A \times \sigma_A} & A \times A \\ \downarrow \sigma_{A \times A} & & \downarrow \sigma_A \\ A \times A & \xrightarrow{\sigma_A} & A \end{array}$$

Diagram 1.12

imply $(f \times f) \circ \binom{1_B}{0} = \binom{1_A}{0} \circ f$. Therefore

$$g = g \circ \sigma_B \circ \begin{pmatrix} 1_B \\ 0 \end{pmatrix} = \sigma_A \circ (f \times f) \circ \begin{pmatrix} 1_B \\ 0 \end{pmatrix} = \sigma_A \circ \begin{pmatrix} 1_A \\ 0 \end{pmatrix} \circ f = f. \square$$

Theorem 1.6.4 *Every abelian category is additive.*

Proof Let us first apply lemma 1.6.3 with $B = A \times A$ and f the i-th ($i = 1, 2$) projection. Given an arbitrary object C and four morphisms a, b, c, d: $C \longrightarrow A$, consideration of diagram 1.11, where $i = 1, 2$, shows the validity of the formula

$$\begin{pmatrix} a \\ b \end{pmatrix} - \begin{pmatrix} c \\ d \end{pmatrix} = \begin{pmatrix} a - c \\ b - d \end{pmatrix}.$$

Next let us apply lemma 1.6.3 with again $B = A \times A$, but $f = \sigma_A$. Given an arbitrary object C and four morphisms a, b, c, d: $C \longrightarrow A$, the consideration of diagram 1.12 together with the previous formula, shows the validity of the relation

$$(a - c) - (b - d) = (a - b) - (c - d).$$

On the other hand we have noticed in the proof of 1.6.3 that $\sigma_A \circ \binom{1_A}{0} = 1_A$, which implies that given a morphism a: $C \longrightarrow A$, one has $a - 0 = a$. In the same way the relation $\sigma_A \circ \Delta_A = 0$ implies $a - a = 0$.

It is now rather straightforward to prove the theorem. Applying the previous relations, one proves successively that

$$\begin{aligned}
&(0-b)-c=(0-b)-(c-0)=(0-c)-(b-0)=(0-c)-b,\\
&0-(0-d)=(d-d)-(0-d)=(d-0)-(d-d)=(d-0)-0=d,\\
&\quad b+c=b-(0-c)=\big(0-(0-b)\big)-(0-c)\\
&\qquad=(0-0)-\big((0-b)-c\big)=(0-0)-\big((0-c)-b\big)\\
&\qquad=\big(0-(0-c)\big)-(0-b)=c-(0-b)=c+b,\\
&b+(0-c)=b-\big((0-(0-c)\big)=b-c,\\
&b+(0-b)=b-b=0,\\
&0-(c-d)=(0-0)-(c-d)=(0-c)-(0-d)=(0-c)+d,\\
&0-(c+d)=0-\big(c-(0-d)\big)=(0-c)+(0-d)=(0-c)-d,\\
&(a-b)+d=(a-b)-(0-d)=(a-0)-(b-d)=a-(b-d),\\
&(a+b)+d=\big(a-(0-b)\big)+d=a-\big((0-b)-d\big)=a-\big(0-(b+d)\big)\\
&\qquad=a+(b+d),
\end{aligned}$$

which shows already that $\mathscr{C}(C,A)$ has been provided with the structure of an abelian group.

Now given an arrow $x\colon X\longrightarrow C$,

$$(a-b)\circ x=\sigma_A\circ\begin{pmatrix}a\\ b\end{pmatrix}\circ x=\sigma_A\circ\begin{pmatrix}a\circ x\\ b\circ x\end{pmatrix}=a\circ x-b\circ x.$$

On the other hand given an arrow $y\colon A\longrightarrow Y$, we can apply lemma 1.6.3 with $f=y$, yielding

$$y\circ(a-b)=y\circ\sigma_A\circ\begin{pmatrix}a\\ b\end{pmatrix}=\sigma_Y\circ(y\times y)\circ\begin{pmatrix}a\\ b\end{pmatrix}=\sigma_Y\circ\begin{pmatrix}y\circ a\\ y\circ b\end{pmatrix}=(y\circ a)-(y\circ b).$$

The conclusion follows at once. □

Corollary 1.6.5 *Let A,B be two objects of an abelian category, with product $(A\times B,p_A,p_B)$ and coproduct $(AamalgB,s_A,s_B)$. There exists an isomorphism $A\times B\cong A\amalg B$ and, via this isomorphism,*

$$s_A=\mathsf{Ker}\,p_B,\quad s_B=\mathsf{Ker}\,p_A,\quad p_A=\mathsf{Coker}\,s_B,\quad p_B=\mathsf{Coker}\,s_A,$$

$$p_A\circ s_A=1_A,\quad p_B\circ s_B=1_B,\quad s_A\circ p_A+s_B\circ p_B=1_{A\oplus B},$$

where we have written $A\oplus B$ to denote the object $A\times B\cong A\amalg B$.

Proof See 1.2.4. □

1.7 Union of subobjects

First of all, we deduce from the considerations of 4.2, volume 1:

Proposition 1.7.1 *In an abelian category, the intersection and the union of a finite family of subobjects always exists.*

Proof The intersection of two subobjects exists by 1.5.2. The union of two subobjects exists by 4.2.6, volume 1, because an abelian category has coproducts (see 1.5.3), epi–mono factorizations (see 1.5.5), and every epimorphisms is regular (see 1.4.1), thus strong (see 4.3.6, volume 1).

Moreover every object A of an abelian category has a biggest subobject (i.e. the intersection of the empty family), namely A itself. Now the unique morphism $\mathbf{0} \longrightarrow A$ is a monomorphism because $\mathbf{0}$ is terminal; it represents the smallest subobject of A because $\mathbf{0}$ is initial. □

In other words:

Corollary 1.7.2 *In an abelian category, the subobjects of every object constitute a lattice, with top and bottom elements.* □

In the category of sets, the distributive formulae

$$R \cap \left(\bigcup_{i\in I} S_i\right) = \bigcup_{i\in I} (R \cap S_i), \quad R \cup \left(\bigcap_{i\in I} S_i\right) = \bigcap_{i\in I} (R \cup S_i)$$

are well-known to be valid for subsets $R \rightarrowtail A$, $S_i \rightarrowtail A$ of a set A. Nothing analogous holds for abelian categories, i.e. the lattices of subobjects are no longer distributive. For example, in the category $\mathsf{Vect}_{\mathbb{R}}$ of real vector spaces, consider the following subobjects:

$$s_1\colon \mathbb{R} \rightarrowtail \mathbb{R}^2, \quad s_2\colon \mathbb{R} \rightarrowtail \mathbb{R}^2, \quad \Delta\colon \mathbb{R} \rightarrowtail \mathbb{R}^2,$$

where s_1 and s_2 are the two canonical axes and Δ is the diagonal. Clearly $s_1 \vee s_2 = \mathbb{R}^2$, $s_1 \wedge \Delta = (0)$ and $s_2 \wedge \Delta = (0)$. Therefore

$$(s_1 \vee s_2) \wedge \Delta = \mathbb{R}^2 \wedge \Delta = \Delta,$$
$$(s_1 \wedge \Delta) \vee (s_2 \wedge \Delta) = (0) \vee (0) = (0),$$

which shows that in general

$$(s_1 \vee s_2) \wedge \Delta \neq (s_1 \wedge \Delta) \vee (s_2 \wedge \Delta).$$

But unions in an abelian category share another important property of unions in the category of sets: namely, effectiveness.

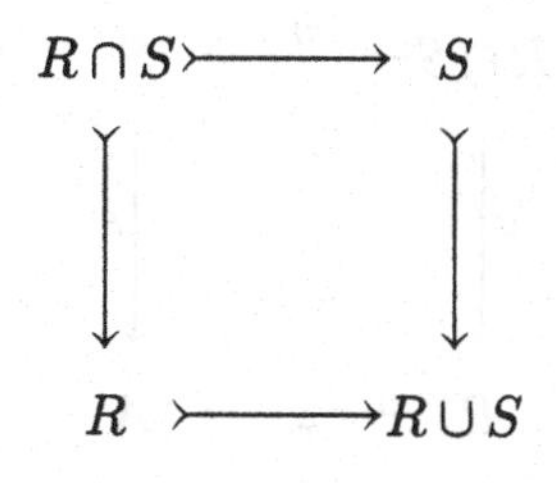

Diagram 1.13

Definition 1.7.3 *Consider a category $\mathscr{C}$ with binary intersections and binary unions of subobjects. The union of two subobjects $r\colon R \rightarrowtail A$ and $s\colon S \rightarrowtail A$ is effective when $R \cup S$ is the pushout of R, S over their common subobject $R \cap S$. In other words, the square in diagram 1.13 is both a pullback and a pushout.*

Rephrasing the previous definition, the union $R \cup S$ is effective when, in defining a morphism $R \cup S \longrightarrow B$, it suffices to find morphisms $R \longrightarrow B$, $S \longrightarrow B$ which agree on $R \cap S$. See exercise 1.15.6 for an example of a non-effective union.

Proposition 1.7.4 *In an abelian category, binary unions are effective.*

Proof By 4.2.6, volume 1, we know that $R \cup S$ is obtained as the image factorization

$$R \oplus S \xrightarrow{\;p\;} R \cup S \overset{i}{\rightarrowtail} A$$

of the morphism $(r, s)\colon R \oplus S \longrightarrow A$. By 1.5.5, $p = \mathsf{Coker}\,(\mathsf{Ker}\,(r, s))$.

Let us first compute $\mathsf{Ker}\,(r, s)$. Giving a morphism $x\colon X \longrightarrow R \oplus S$ is equivalent to giving a pair $\binom{x_1}{x_2}$ of morphisms, with $x_1 = p_R \circ x\colon X \longrightarrow R$, $x_2 = p_S \circ x\colon X \longrightarrow S$. The relation $(r, s) \circ x = 0$ is just $(r, s) \circ \binom{x_1}{x_2} = 0$, or more explicitly $(r \circ x_1) + (s \circ x_2) = 0$, which we can write $r \circ x_1 = (-s) \circ x_2$. Observe that $-1_S\colon S \longrightarrow S$ is an isomorphism (it is its own inverse!) so that $(S, -s)$ is still a subobject of A (for a while, avoiding ambiguity in mentioning subobjects requires writing the monomorphisms explicitly, not just their sources). Since intersections of subobjects are just pullbacks (see 4.2.3, volume 1), giving a pair $x_1\colon X \longrightarrow R$, $x_2\colon X \longrightarrow S$ such that $r \circ x_1 = (-s) \circ x_2$ is equivalent to giving a morphism $x'\colon X \longrightarrow (R, r) \cap (S, -s)$. All this proves that the kernel of

$$(r, s)\colon R \oplus S \longrightarrow A$$

is exactly $(R, r) \cap (S, -s)$ or, to make notation easier, the morphism

$$k\colon R \cap S \longrightarrow R \oplus S,$$

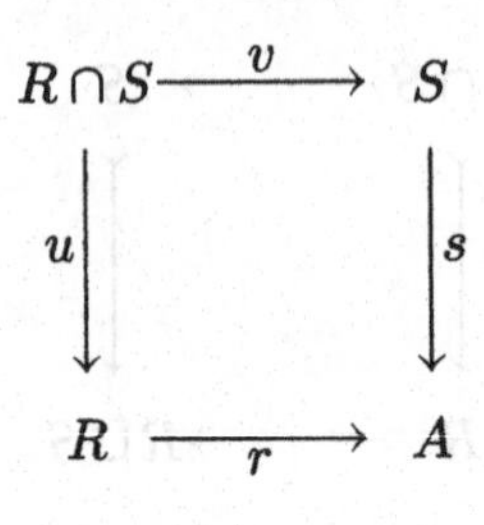

Diagram 1.14

where $R \cap S$ stands now for the usual intersection of R, S as in diagram 1.14 and k is the factorization of the two morphisms

$$R \cap S \rightarrowtail^{u} R, \quad R \cap S \rightarrowtail^{v} S \xrightarrow{-1_S} S$$

through the biproduct $R \oplus S$.

Now we must prove that for two morphisms $\alpha: R \longrightarrow Y$, $\beta: S \longrightarrow Y$ which agree on $R \cap S$ (i.e. $\alpha \circ u = \beta \circ v$), there exists a common and unique extension $\gamma: R \cup S \longrightarrow Y$. Considering the factorization

$$(\alpha, \beta): R \oplus S \longrightarrow Y$$

and the relation $p = \mathsf{Coker}\,(\mathsf{Ker}\,(r,s))$, it suffices to prove the relation $(\alpha, \beta) \circ \mathsf{Ker}\,(r,s) = 0$. Using the description of $k = \mathsf{Ker}\,(r,s)$ we have just given, we have indeed

$$(\alpha, \beta) \circ k = (\alpha, \beta) \circ \begin{pmatrix} u \\ -v \end{pmatrix} = (\alpha \circ u) - (\beta \circ v) = 0. \qquad \square$$

Using the "quotient notation" $p: A \longrightarrow A/R$ for the coequalizer of a monomorphism $R \rightarrowtail A$, we obtain the following corollary:

Corollary 1.7.5 *If $r: R \rightarrowtail A$ and $s: S \rightarrowtail A$ are subobjects in an abelian category, $A/(R \cup S)$ is the pushout of A/R and A/S under A. In other words the square in diagram 1.15 is a pushout.*

Proof We refer to diagram 1.16 where r, s, m, n, l are the original monomorphisms, $p = \mathsf{Coker}\, r$, $q = \mathsf{Coker}\, s$ and $t = \mathsf{Coker}\, l$. From $t \circ r = t \circ l \circ m = 0$ we find a factorization u yielding $t = u \circ p$ and in the same way we find v such that $t = v \circ q$.

Now consider x and y such that $x \circ p = y \circ q$. One has $x \circ p \circ l \circ m = x \circ p \circ r = 0$ and $x \circ p \circ l \circ n = y \circ q \circ l \circ n = y \circ q \circ s = 0$, from which $x \circ p \circ l = 0$ since m, n are the canonical morphisms of a pushout (see 1.7.4). Therefore we find a factorization z with the property $z \circ t = x \circ p$.

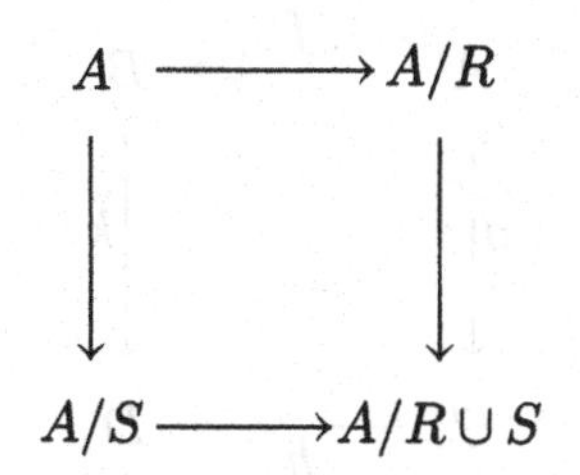

Diagram 1.15

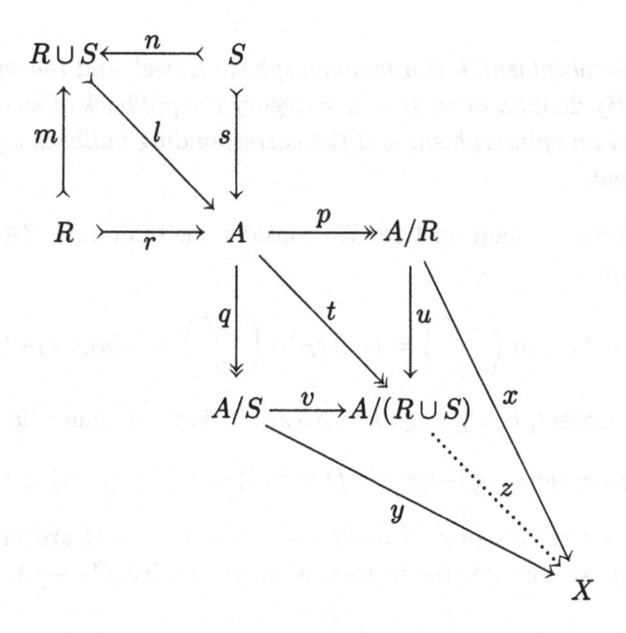

Diagram 1.16

This implies $z \circ u \circ p = z \circ t = x \circ p$ and thus $z \circ u = x$, since p is an epimorphism. In the same way $z \circ v \circ q = z \circ t = x \circ p = y \circ q$ and $z \circ v = y$ since q is an epimorphism. Since t is an epimorphism, such a factorization z is necessarily unique. □

Let us now prove a property which generalizes somewhat the existence of effective unions (see 1.7.3). It should be observed that proposition 1.7.6 and its dual do not hold in general for arbitrary categories (see exercise 1.15.7).

Proposition 1.7.6 *In an abelian category $\mathscr{C}$, the pushout of a monomorphism is still a monomorphism and the pushout square is also a pullback. More precisely if the square in diagram 1.17 is a pushout and*

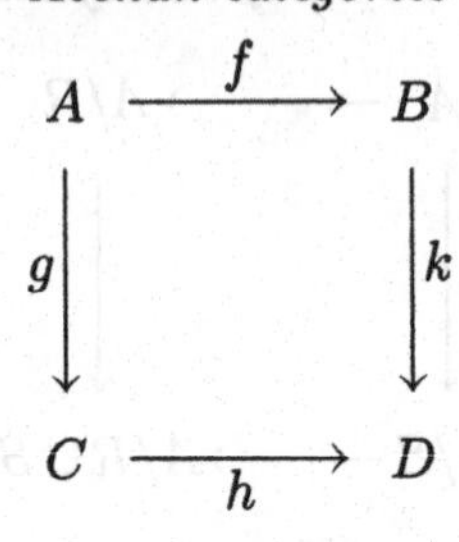

Diagram 1.17

if g is a monomorphism, k is a monomorphism as well and the square is a pullback. By duality, in an abelian category the pullback of an epimorphism is still an epimorphism and the corresponding pullback square is also a pushout.

Proof We refer to diagram 1.18. We consider the biproduct $B \oplus C$ and the morphism

$$l = \begin{pmatrix} -f \\ g \end{pmatrix} = 1_{B\oplus C} \circ \begin{pmatrix} -f \\ g \end{pmatrix} = (s_B, s_C) \circ \begin{pmatrix} -f \\ g \end{pmatrix} = -(s_B \circ f) + (s_C \circ g).$$

We put $p = \mathsf{Coker}\, l$, $h = p \circ s_C$, $k = p \circ s_B$. This yields immediately

$$(h \circ g) - (k \circ f) = (p \circ s_C \circ g) - (p \circ s_B \circ f) = p \circ \big((s_C \circ g) - (s_B \circ f)\big) = p \circ l = 0,$$

thus $h \circ g = k \circ f$. Moreover, if $m: B \longrightarrow M$, $n: C \longrightarrow M$ are such that $m \circ f = n \circ g$, we consider the factorization $(m, n): B \oplus C \longrightarrow M$, which yields

$$(m, n) \circ l = (m, n) \circ \begin{pmatrix} -f \\ g \end{pmatrix} = -(m \circ f) + (n \circ g) = 0,$$

from which we get a unique factorization $d: D \longrightarrow M$ through the cokernel of l, yielding $d \circ p = (m, n)$. This implies immediately

$$d \circ k = z \circ p \circ s_B = (m, n) \circ s_B = m$$

$$d \circ h = z \circ p \circ s_C = (m, n) \circ s_C = m$$

and the uniqueness of such a factorization follows at once from the uniqueness conditions in the definitions of (m, n) and d. So the previous constructions describe the pushout (h, k) of the pair (g, f); compare with 2.17.1, volume 1.

Let us prove now that l is a monomorphism. Choosing $x: X \longrightarrow A$ such that $l \circ x = 0$, it suffices indeed to compose on the left with p_C to

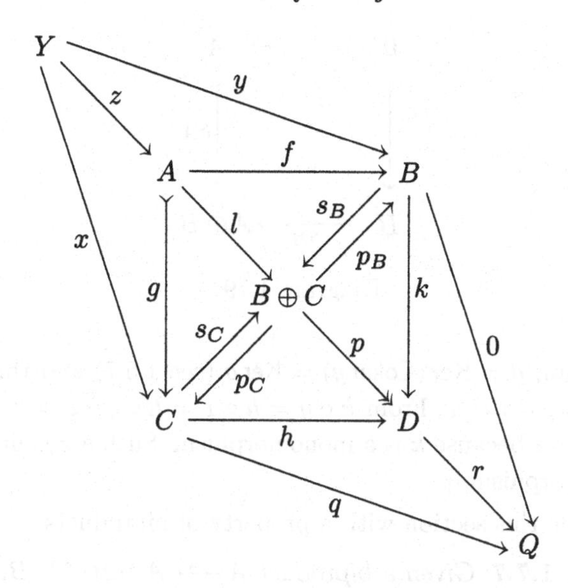

Diagram 1.18

get

$$0 = p_C \circ 0 = p_C \circ l \circ x = p_C \circ \big((s_C \circ g) - (s_B \circ f)\big) \circ x = g \circ x,$$

from which $x = 0$ since g is a monomorphism. Thus l is indeed a monomorphism (see 1.5.4) and therefore $l = \mathsf{Ker}\,(\mathsf{Coker}\, l)$ by 1.5.7; in other words, $l = \mathsf{Ker}\, p$.

Now let us choose $y\colon Y \longrightarrow B$ such that $k \circ y = 0$, as in diagram 1.18. One has

$$p \circ s_B \circ y = k \circ y = 0$$

from which one gets a factorization $z\colon Y \longrightarrow A$ of y through the kernel l of p, yielding $l \circ z = s_B \circ y$. Let us observe that

$$g \circ z = p_C \circ \begin{pmatrix} -f \\ g \end{pmatrix} \circ z = p_C \circ l \circ z = p_C \circ s_B \circ y = 0 \circ y = 0$$

and thus $z = 0$, since g is a monomorphism. Finally $s_B \circ y = l \circ z = 0$ and $y = 0$ because s_B is a monomorphism. By 1.5.4 again, this proves that k is a monomorphism.

We must still prove that the pushout square is a pullback. Let us consider diagram 1.18 again, where now $h \circ x = k \circ y$ and $q = \mathsf{Coker}\, g$. Since $q \circ g = 0 = 0 \circ f$, we find a unique r such that $r \circ h = q$, $r \circ k = 0$. Therefore $q \circ x = r \circ h \circ x = r \circ k \circ y = 0 \circ y = 0$. But since g is a

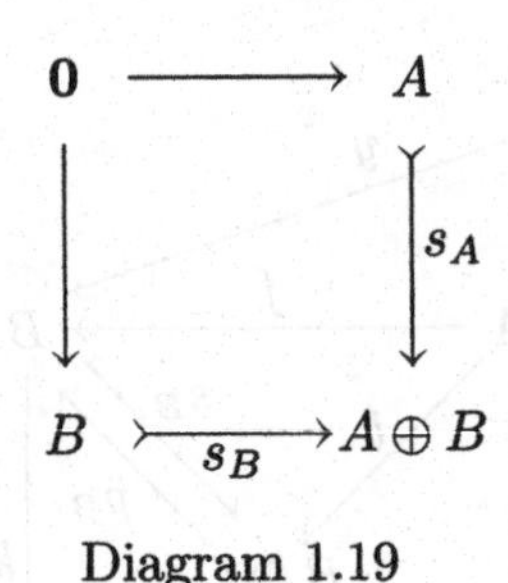

Diagram 1.19

monomorphism, $g = \mathsf{Ker}\,(\mathsf{Coker}\, g) = \mathsf{Ker}\, q$ (see 1.5.7) and thus there is a z such that $g \circ z = x$. From $k \circ y = h \circ x = h \circ g \circ z = k \circ f \circ z$, we deduce $y = f \circ z$ because k is a monomorphism. Such a z is unique since g is a monomorphism. □

We conclude this section with a property of biproducts.

Proposition 1.7.7 *Given a biproduct $A \overset{s_A}{\rightarrowtail} A \oplus B \overset{s_B}{\leftarrowtail} B$, the intersection of the subobjects s_A, s_B is the zero object and their union is $A \oplus B$.*

Proof Given morphisms $x\colon X \longrightarrow A$, $y\colon X \longrightarrow B$ such that $s_A \circ x = s_B \circ y$, one has $x = p_A \circ s_A \circ x = p_A \circ s_B \circ y = 0 \circ y = 0$. In the same way $y = 0$, so that x and y factor (uniquely!) through $X \longrightarrow \mathbf{0}$. So the square of diagram 1.19 is a pullback, but it is also a pushout since $\mathbf{0}$ is initial. Applying 1.7.4, we conclude that $A \oplus B$ is the union of s_A, s_B. □

1.8 Exact sequences

Several aspects of this section are valid for additive or even for preadditive categories. But for the sake of simplicity, we shall restrict our attention to the case of abelian categories.

Definition 1.8.1 *In an abelian category $\mathscr{C}$, a composable pair of morphisms*

$$A \xrightarrow{\ f\ } B \xrightarrow{\ g\ } C$$

is called an exact sequence when the image of f coincides with the kernel of g.

To be precise, in 1.8.1 we consider the image factorization

$$A \xrightarrow{\ p\ } I \rightarrowtail^{\ i\ } B$$

of f; the requirement is thus that (I, i) is the kernel of g.

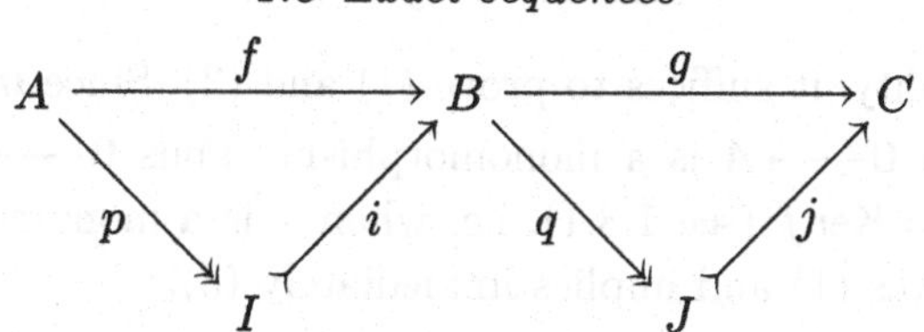

Diagram 1.20

Proposition 1.8.2 *Consider diagram 1.20 in an abelian category $\mathscr{C}$, where both triangles are commutative while (i,p) and (j,q) are the image factorizations of f and g. The following conditions are equivalent:*

(1) (f,g) is an exact sequence;

(2) (i,g) is an exact sequence;

(3) (f,q) is an exact sequence;

(4) (i,q) is an exact sequence.

Proof It suffices to observe that $\mathsf{Ker}\, g = \mathsf{Ker}\, q$, since j is a monomorphism. □

Corollary 1.8.3 *In abelian categories, the notion of exact sequence is autodual.*

Proof With the notation of 1.8.2, the sequence (f,g) is coexact when, in the dual category $\mathscr{C}^*$, (g^*, f^*) is an exact sequence. By 1.8.2 this is equivalent to $q^* = \mathsf{Ker}\, i^*$ in $\mathscr{C}^*$, thus to $q = \mathsf{Coker}\, i$ in $\mathscr{C}$. But since i is a monomorphism, $i = \mathsf{Ker}\,(\mathsf{Coker}\, i) = \mathsf{Ker}\, q$ (see 1.5.7) and the sequence (f,g) is exact. By duality, exactness implies coexactness. □

Definition 1.8.4 *A finite or infinite sequence of morphisms*

$$\cdots \longrightarrow A_n \xrightarrow{f_n} A_{n+1} \xrightarrow{f_{n+1}} A_{n+2} \longrightarrow \cdots$$

in an abelian category is called exact when each pair of consecutive morphisms is an exact sequence in the sense of 1.8.1.

Proposition 1.8.5 *In an abelian category, the following equivalences hold:*

(1) $\mathbf{0} \longrightarrow A \xrightarrow{f} B$ is an exact sequence iff f is a monomorphism;

(2) $B \xrightarrow{f} A \longrightarrow \mathbf{0}$ is an exact sequence iff f is an epimorphism;

(3) $\mathbf{0} \longrightarrow A \xrightarrow{f} B \xrightarrow{g} C$ is an exact sequence iff $f = \mathsf{Ker}\, g$;

(4) $C \xrightarrow{g} B \xrightarrow{f} A \longrightarrow \mathbf{0}$ is an exact sequence iff $f = \mathsf{Coker}\, g$.

Proof By duality, it suffices to prove (1) and (3). Since $\mathbf{0}$ is terminal, each morphism $\mathbf{0} \longrightarrow A$ is a monomorphism. Thus $\mathbf{0} \longrightarrow A \xrightarrow{f} B$ is exact when $\mathbf{0} = \mathsf{Ker}\, f$ (see 1.8.1), i.e. when f is a monomorphism (see 1.5.4); this proves (1) and implies immediately (3). □

Definition 1.8.6 *By a short exact sequence in an abelian category is meant an exact sequence of the form*

$$\mathbf{0} \longrightarrow A \xrightarrow{f} B \xrightarrow{g} C \longrightarrow \mathbf{0}.$$

Proposition 1.8.7 *Consider a short exact sequence*

$$\mathbf{0} \longrightarrow A \xrightarrow{f} B \xrightarrow{g} C \longrightarrow \mathbf{0}$$

in an abelian category. The following conditions are equivalent:

(1) there exists a morphism $s\colon C \longrightarrow B$ such that $g \circ s = 1_C$;
(2) thre exists a morphism $r\colon B \longrightarrow A$ such that $r \circ f = 1_A$;
(3) there exist morphisms $s\colon C \longrightarrow B$, $r\colon B \longrightarrow A$ such that the quintuple (B, r, g, f, s) is the biproduct of A and C.

Proof By duality, it suffices to prove (1) $\Leftrightarrow$ (3). But (3) $\Rightarrow$ (1) is obvious (see 1.6.5). Given (1), consider the morphism $1_B - s \circ g\colon B \longrightarrow B$. One has $g \circ (1_B - s \circ g) = g - g \circ s \circ g = g - g = 0$, from which one gets a unique morphism $r\colon B \longrightarrow A$ such that $f \circ r = 1_B - s \circ g$, since $f = \mathsf{Ker}\, g$. We have already $f \circ r + s \circ g = 1_B$, $g \circ f = 0$ and $g \circ s = 1_C$. We have also $f \circ r \circ f = (1_B - s \circ g) f = f - s \circ g \circ f = f - 0 = f$, thus $r \circ f = 1_A$ since f is a monomorphism, see 1.8.5(1). Finally $f \circ r \circ s = (1_B - s \circ g) s = s - s \circ g \circ s = s - s = 0$, thus $r \circ s = 0$ since f is a monomorphism. This concludes the proof (see 1.2.4). □

Definition 1.8.8 *In an abelian category, a split exact sequence is a short exact sequence which satisfies the conditions of 1.8.7.*

Let us recall that given a biproduct as in 1.2.4, the sequence

$$\mathbf{0} \longrightarrow A \xrightarrow{s_A} A \oplus B \xrightarrow{p_B} B \longrightarrow \mathbf{0}$$

is always exact since $s_A = \mathsf{Ker}\, p_B$, $p_B = \mathsf{Coker}\, s_A$; see 1.8.5. It is a split exact sequence by 1.8.7.3. And 1.8.7 asserts that all split exact sequences are of this type up to isomorphism.

1.9 Diagram chasing

In any abelian category Mod_R of modules over a ring R, monomorphisms are just injections, and epimorphisms are just surjections, as attested by 1.7.7.e and 1.8.5.e, volume 1. Therefore

$f\colon A \longrightarrow B$ is a monomorphism
 iff $\forall a \in A \quad f(a) = 0 \;\Rightarrow\; a = 0$
 iff $\forall a, a' \in A \quad f(a) = f(a') \;\Rightarrow\; a = a'$,

$f\colon A \longrightarrow B$ is an epimorphism
 iff $\forall b \in B \quad \exists a \in A \;\; f(a) = b$,

$A \xrightarrow{f} B \xrightarrow{g} C$ is an exact sequence
 iff $\forall a \in A \;\; g\big(f(a)\big) = 0$,
 $\forall b \in B \;\; g(b) = 0 \;\Rightarrow\; \exists a \in A \;\; b = f(a)$.

In the case of the exact sequence, the first condition says exactly that $\mathsf{Im}\, f \subseteq \mathsf{Ker}\, g$, while the second condition means $\mathsf{Ker}\, g \subseteq \mathsf{Im}\, f$.

In the case of a category of modules, proofs concerning exact sequences, thus in particular monomorphisms, epimorphisms, kernels and cokernels, can be performed quite easily using the characterization in terms of elements. This technique is called "diagram chasing". We shall prove that, amazingly enough, this same technique of "diagram chasing" can be applied to every abelian category.

The final result concerning diagram chasing in an abelian category is obtained by proving that every abelian category $\mathscr{A}$ can be fully embedded in a category of modules over a ring, in a way which preserves and reflects exact sequences (see 1.14.9). We shall present here a weakened version of the diagram chasing theorem, very easy to prove and good enough for many purposes.

Definition 1.9.1 *In an abelian category $\mathscr{C}$, consider an object A and a morphism $f\colon A \longrightarrow B$.*

(1) A pseudo-element of A is an arrow $\bullet \xrightarrow{a} A$ with codomain A; we shall write simply $a \in^* A$;

(2) two pseudo-elements $X \xrightarrow{a} A$ and $X' \xrightarrow{a'} A$ are pseudo-equal when there exist epimorphisms $Y \overset{p}{\twoheadrightarrow} X$, $Y \overset{p'}{\twoheadrightarrow} X'$ with the property $a \circ p = a' \circ p'$; we shall write simply $a =^* a'$;

(3) the pseudo-image under $f\colon A \longrightarrow B$ of a pseudo-element $\bullet \xrightarrow{a} A$ of A is the composite $f \circ a$; we shall write simply $f(a)$.

Proposition 1.9.2 *Consider two morphisms $f\colon A \longrightarrow B$, $g\colon B \longrightarrow C$ of an abelian category. The following properties hold:*

(1) pseudo-equality is an equivalence relation on the pseudo-elements of A;

(2) for pseudo-elements $a \in^* A$, $a' \in^* A$

$$a =^* a' \;\Rightarrow\; f(a) =^* f(a');$$

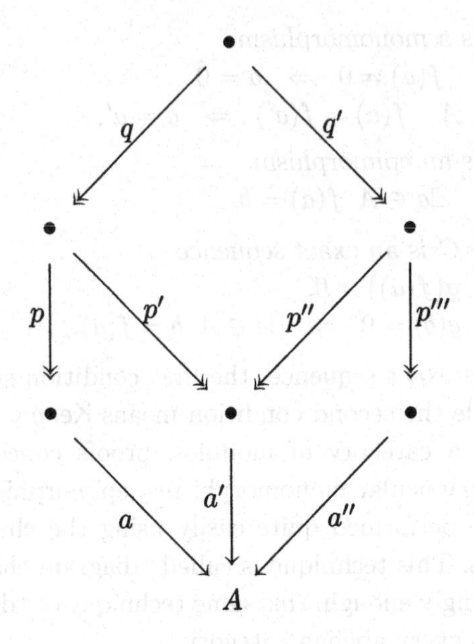

Diagram 1.21

(3) *for a pseudo-element* $a \in^* A$
$$f\big(g(a)\big) =^* (f \circ g)(a).$$

Proof The reflexivity and the symmetry of pseudo-equality are obvious; let us prove the transitivity. We consider $a, a', a'' \in^* A$ with $a =^* a'$, $a' =^* a''$. There exist epimorphisms p, p', p'', p''' making all parts of diagram 1.21 commutative; we obtain new epimorphisms q, q' by performing the pullback of p', p'' (see dual of 1.7.6). The equality $a \circ (p \circ q) = a'' \circ (p''' \circ q')$ allows us to conclude the proof. The validity of the two other statements is obvious. □

Lemma 1.9.3 *Let A be an object of an abelian category. There exists an equivalence class for pseudo-equality on pseudo-elements of A, consisting exactly of all the zero morphisms with codomain A.*

Proof Consider $a\colon X \longrightarrow A$ pseudo-equal to $0\colon Y \longrightarrow A$. There exist epimorphisms p, q such that $a \circ p = 0 \circ q = 0$; thus $a = 0$. Conversely if $a = 0$, it suffices to consider the epimorphisms $p_X\colon X \oplus Y \longrightarrow X$ and $p_Y\colon X \oplus Y \longrightarrow Y$ (see 1.2.4) to get $a \circ p_X = 0 = 0 \circ p_Y$. □

The previous lemma allows us to refer freely to "the" zero pseudo-element of a given object. The interest of the notion of pseudo-element is attested by the following result:

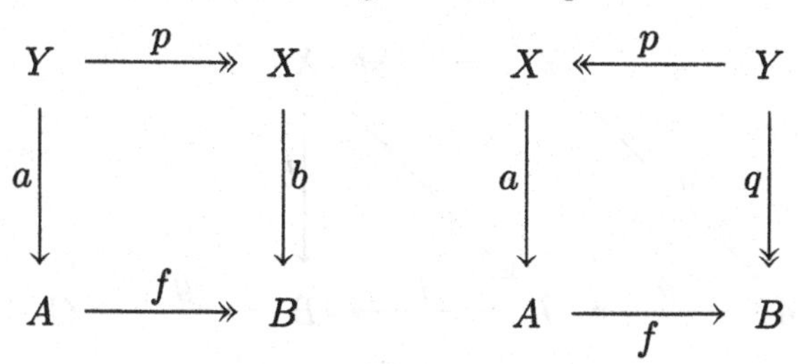

Diagram 1.22

Proposition 1.9.4 *In an abelian category $\mathscr{C}$, the following equivalences hold:*

(1) $f\colon A \longrightarrow B$ is the zero morphism
iff $\forall a \in^* A \;\; f(a) =^* 0$;

(2) $f\colon A \longrightarrow B$ is a monomorphism
iff $\forall a \in^* A \;\; f(a) =^* 0 \;\Rightarrow\; a =^* 0$
iff $\forall a, a' \in^* A \;\; f(a) =^* f(a') \;\Rightarrow\; a =^* a'$;

(3) $f\colon A \longrightarrow B$ is an epimorphism
iff $\forall b \in^* B \;\exists a \in^* A \;\; f(a) =^* b$;

(4) $A \xrightarrow{f} B \xrightarrow{g} C$ is an exact sequence
iff $\forall a \in^* A \;\; g\big(f(a)\big) =^* 0$,
$\forall b \in^* B \;\; g(b) =^* 0 \;\Rightarrow\; \exists a \in^* A \;\; f(a) =^* b$;

(5) if $f\colon A \longrightarrow B$ and $a, a' \in^* A$ with $f(a) =^* f(a')$
then there exists $a'' \in^* A$ such that $f(a'') =^* 0$ and
$\forall g\colon A \longrightarrow C \;\; g(a') =^* 0 \;\Rightarrow\; g(a'') =^* g(a)$.

Proof (1) If $f = 0$, then $f(a) = f \circ a = 0$ and one gets the conclusion by 1.9.3. Conversely, $f = f(1_A) =^* 0$ and thus $f = 0$, again by 1.9.3.

(2) If f is a monomorphism and $f(a) =^* f(a')$, there are epimorphisms p, q with $f \circ a \circ p = f \circ a' \circ q$ and thus $a \circ p = a' \circ q$. So the second characterization holds and it implies the first one just by putting $a' = 0$. This first characterization is just condition 1.5.4.(3) for being a monomorphism.

(3) If f is an epimorphism and $b \in^* B$, construct the pullback of f and b, getting $a \in^* A$ and an epimorphism p, as in diagram 1.22. From $f \circ a \circ 1_Y = b \circ p$ we deduce $f(a) =^* b$. Conversely since $1_B \in^* B$, we find $a \in^* A$ and epimorphisms p, q such that $f \circ a \circ p = q$ (see the second square in diagram 1.22). Since q is an epimorphism, so is f.

(4) Suppose first that (f, g) is an exact sequence. Since $g \circ f = 0$, we already have the first assertion (see first part of the proof). Now choose $b \in^* B$ such that $g(b) =^* 0$ and consider the image factorization

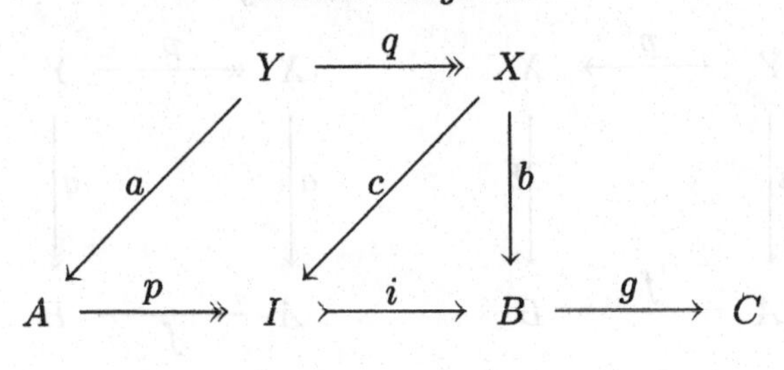

Diagram 1.23

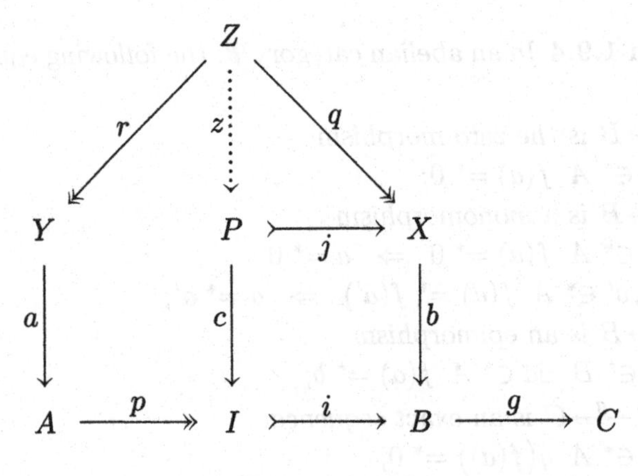

Diagram 1.24

$f = i \circ p$ of f as in diagram 1.23. Since $g \circ b = 0$, there is a factorization c, yielding $b = i \circ c$, through the kernel $i = \mathsf{Ker}\, g$. Computing the pullback of (p, c), we get $a \in^* A$ and an epimorphism q (see 1.7.6). The relation $f \circ a \circ 1_Y = i \circ p \circ a = b \circ q$ implies $f(a) =^* b$.

Conversely suppose condition (4) is satisfied and consider again the image factorization $f = i \circ p$ of f, as in diagram 1.24. We must prove that $i = \mathsf{Ker}\, g$. From (a) and the first part of the proof we deduce that $g \circ i \circ p = g \circ f = 0$, thus $g \circ i = 0$ since p is an epimorphism. Now if b is such that $g \circ b = 0$, (b) implies the existence of $a \in^* A$ and epimorphisms q, r such that $i \circ p \circ a \circ r = b \circ q$. Computing the pullback (c, j) of (i, b), we find a factorization z such that $j \circ z = q$ and $c \circ z = p \circ a \circ r$. The arrow j is a monomorphism since i is (see 2.5.3, volume 1) and an epimorphism since q is (see 1.8.2, volume 1). Thus j is an isomorphism (see 1.5.1) and b factors through i as $b = i \circ c \circ j^{-1}$. The factorization is unique since i is a monomorphism.

(5) We have the pseudo-elements $a \colon X \longrightarrow A$, $a' \colon X' \longrightarrow A$ and two epimorphisms $p \colon Y \longrightarrow X$, $p' \colon Y \longrightarrow X'$ such that $f \circ p \circ a = f \circ p' \circ a'$.

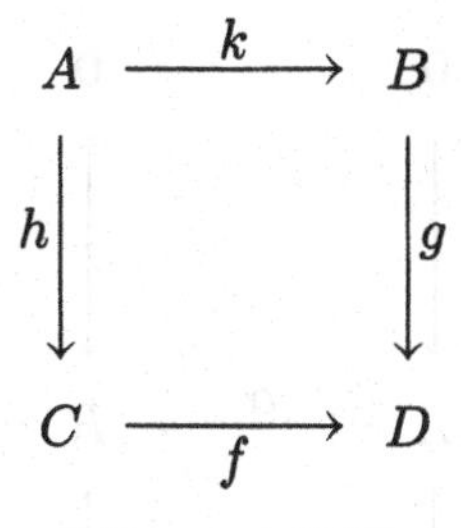

Diagram 1.25

It suffices to define $a'' = p \circ a - p' \circ a'$. □

The proof of 1.9.4.(5) indicates that $a'' \in^* A$ should be thought of, intuitively, as the difference $a - a' \in^* A$. Now let us be careful: this does not provide the pseudo-elements of A with the structure of an abelian group. For example, given a pseudo-element $a: X \longrightarrow A$ of A, the relation $a \circ (-1_X) = (-a) \circ 1_X$ implies that $a =^* -a \ldots$ while in general one does not have $a + a =^* 0$. This same example shows the weakness of this notion of pseudo-elements: it cannot be used to prove the equality of two morphisms. In other words given $f, g: A \rightrightarrows B$

$$\forall a \in^* A \quad f(a) =^* g(a)$$

does not imply $f = g$. The previous remarks show indeed that putting $g = -f$ will give (in general) a counterexample.

Here is another (useful) example of a proposition which is just an implication, not an equivalence. Again this is due to the fact that our present notion of pseudo-element is too weak.

Proposition 1.9.5 *In an abelian category $\mathscr{A}$ consider the pullback in diagram 1.25. Given two pseudo-elements $c \in^* C$ and $b \in^* B$ such that $f(c) =^* g(b)$, there exists a pseudo-unique pseudo-element $a \in^* A$ such that $h(a) =^* c$, $k(a) =^* b$.*

Proof If $f(c) =^* g(b)$, there are epimorphisms p, q such that $f \circ c \circ p = g \circ b \circ q$. By definition of a pullback, this implies the existence of some $a \in^* A$ such that $h \circ a = c \circ p$, $k \circ a = b \circ q$. In particular $h(a) =^* c$ and $k(a) =^* b$.

Consider now $a' \in^* A$ such that $h(a') =^* c$ and $k(a') =^* b$. There are epimorphisms p', q', p'', q'' such that $h \circ a' \circ p' = c \circ q'$ and $k \circ a' \circ p'' = b \circ q''$. All the epimorphisms p, p', p'', q, q', q'' can, by successive pullbacks, be replaced by epimorphisms with the same domain, from which $a =^* a'$. □

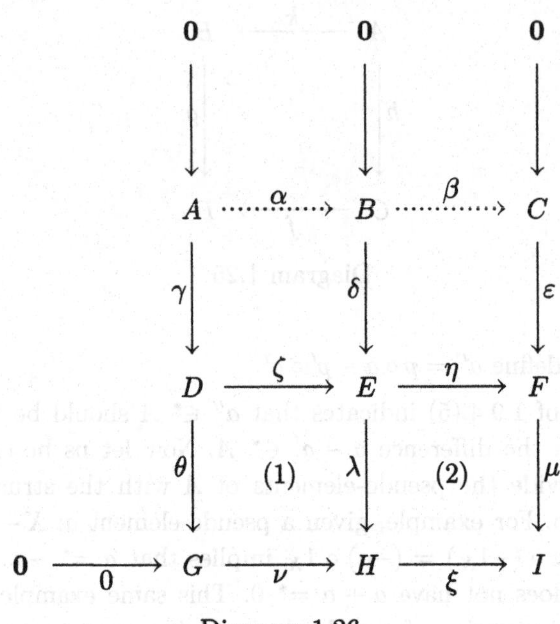

Diagram 1.26

To get a better insight into what a pseudo-element $a\colon X \longrightarrow A$ is, let us consider its image factorization $X \xrightarrow{p} I \xrightarrow{i} A$. From the relation $a \circ 1_X = i \circ p$ we deduce $a =^* i$, so that each pseudo-element can be represented by a monomorphism. Now if two monomorphisms $i\colon I \rightarrowtail A$ and $j\colon J \rightarrowtail A$ are pseudo-equal, we have epimorphisms q, r such that $i \circ q = j \circ r$. The uniqueness of the image factorization (see 1.5.5) implies the isomorphism of i and j. In conclusion, the equivalence classes of pseudo-elements of A are in bijection with the subobjects of A (see 4.1.1, volume 1).

1.10 Some diagram lemmas

In this section, we freely use 1.8.5 and 1.9.4 without referring to them every time.

Lemma 1.10.1 (The kernels' lemma) *Consider diagram 1.26 in an abelian category, with commutative squares (1) and (2) and exact rows (ζ, η), $(0, \nu, \xi)$. Put $\gamma = \mathsf{Ker}\,\theta$, $\delta = \mathsf{Ker}\,\lambda$, $\varepsilon = \mathsf{Ker}\,\mu$. There exist unique morphisms α, β making the diagram commutative. Moreover, (α, β) is an exact sequence.*

$$\begin{array}{ccccccc}
A & \xrightarrow{\alpha} & B & \xrightarrow{\beta} & C & \longrightarrow & 0 \\
\gamma\downarrow & (1) & \downarrow\delta & & \downarrow\varepsilon & & \\
D & \xrightarrow[\zeta]{} & E & \xrightarrow[\eta]{} & F & &
\end{array}$$

Diagram 1.27

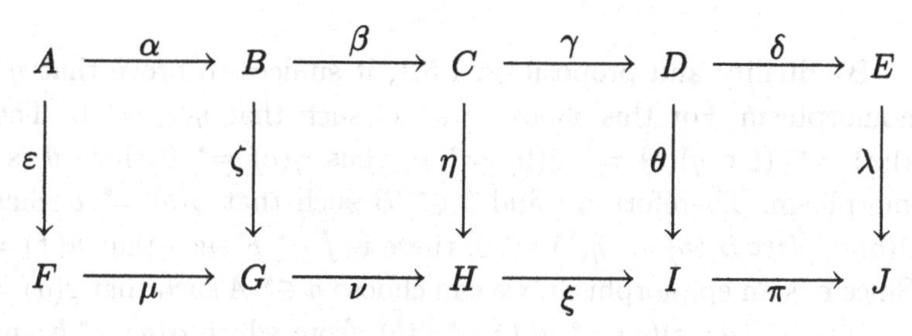

Diagram 1.28

Proof From $\lambda \circ \zeta \circ \gamma = \nu \circ \theta \circ \gamma = \nu \circ 0 = 0$ and $\delta = \mathsf{Ker}\,\lambda$, we deduce the existence and uniqueness of α. In an analogous way, we construct β. Since $\varepsilon \circ \beta \circ \alpha = \eta \circ \zeta \circ \gamma = 0 \circ \gamma = 0$ and ε is a monomorphism, $\beta \circ \alpha = 0$. Now choose $b \in^* B$ such that $\beta(b) =^* 0$. Since $(\eta \circ \delta)(b) =^* (\varepsilon \circ \beta)(b) =^* \varepsilon(0) =^* 0$, we get some $d \in^* D$ with the property $\zeta(d) =^* \delta(b)$. From $(\nu \circ \theta)(d) =^* (\lambda \circ \zeta)(d) =^* (\lambda \circ \delta)(b) =^* 0$, we deduce $\theta(d) =^* 0$, since ν is a monomorphism. Therefore we have $a \in^* A$ such that $\gamma(a) =^* d$. Then $(\delta \circ \alpha)(a) =^* \zeta\gamma(a) =^* \zeta(d) =^* \delta(b)$, from which $\alpha(a) =^* b$ since δ is a monomorphism. □

Lemma 1.10.2 *In an abelian category, consider diagram 1.27, where the two rows are exact, the square (1) is a pullback and the other square is commutative. Then ε is a monomorphism.*

Proof Choose $c \in^* C$ such that $\varepsilon(c) =^* 0$. Since β is an epimorphism, there exists $b \in^* B$ such that $\beta(b) =^* c$. From $(\eta \circ \delta)(b) =^* (\varepsilon \circ \beta)(b) =^* \varepsilon(c) = 0$, we deduce the existence of $d \in^* D$ such that $\zeta(d) =^* \delta(b)$. Since (1) is a pullback, by 1.9.5 there is some $a \in^* A$ with the properties $\gamma(a) =^* d$ and $\alpha(a) =^* b$. Finally $c =^* \beta(b) =^* (\beta \circ \alpha)(a) =^* 0$. □

Lemma 1.10.3 (The five lemma) *In an abelian category, consider diagram 1.28 which is commutative with exact rows. If $\varepsilon, \zeta, \theta, \lambda$ are isomorphisms, η is an isomorphism as well.*

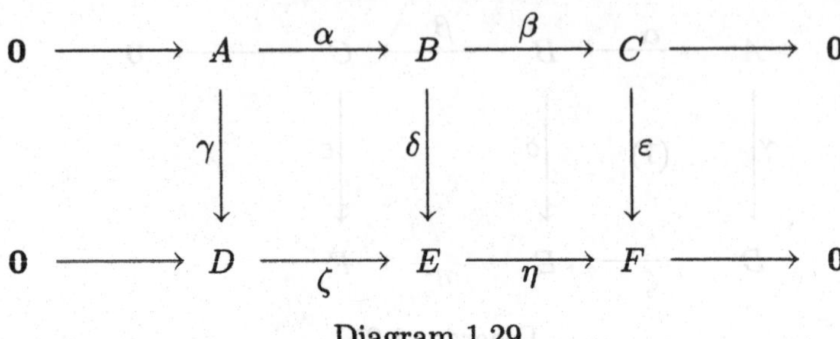

Diagram 1.29

Proof By duality and proposition 1.5.1, it suffices to prove that η is a monomorphism. For this choose $c \in^* C$ such that $\eta(c) =^* 0$. Then $(\theta \circ \gamma)(c) =^* (\xi \circ \eta)(c) =^* \xi(0) =^* 0$, thus $\gamma(c) =^* 0$ since θ is a monomorphism. Therefore we find $b \in^* B$ such that $\beta(b) =^* c$. Since $(\nu \circ \zeta)(b) =^* (\eta \circ \beta)(b) =^* \eta(c) =^* 0$, there is $f \in^* F$ such that $\mu(f) =^* \zeta(b)$. Since ε is an epimorphism, we can choose $a \in^* A$ such that $\varepsilon(a) =^* f$. $(\zeta \circ \alpha)(a) =^* (\mu \circ \varepsilon)(a) =^* \mu(f) =^* \zeta(b)$, from which $\alpha(a) =^* b$ since ζ is a monomorphism. Finally $c =^* \beta(b) =^* (\beta \circ \alpha)(a) =^* 0$. □

Lemma 1.10.4 (The short five lemma) *In an abelian category consider diagram 1.29, where both squares are commutative and both rows are exact. If γ and ε are monomorphisms, δ is a monomorphism as well.*

Proof Choose a pseudo-element $b \in^* B$ such that $\delta(b) =^* 0$. From the relation $(\varepsilon \circ \beta)(b) =^* (\eta \circ \delta)(b) =^* \eta(0) =^* 0$ we deduce that $\beta(b) =^* 0$, since ε is a monomorphism. Therefore we find $a \in^* A$ such that $\alpha(a) =^* b$. From $(\zeta \circ \gamma)(a) =^* (\delta \circ \alpha)(a) =^* \delta(b) =^* 0$, we deduce $a =^* 0$ since both ζ and γ are monomorphisms. Finally $b =^* \alpha(a) =^* \alpha(0) =^* 0$. □

Lemma 1.10.5 (The nine lemma) *In an abelian category, consider diagram 1.30, where all squares are commutative. Suppose five rows and columns are exact, including the central row and the central column. Then the remaining row or column is exact as well. Moreover the square (1) is a pullback and the square (2) is a pushout.*

Proof Just by symmetry, the roles of rows and columns can be interchanged. So let us suppose that the three columns are exact as well as the central row. The roles of the first and the last row can now be interchanged by duality, so that we can assume the last row to be exact. The arrow α is a monomorphism since γ and ζ are. Applying lemma 1.10.1, it remains to prove that β is an epimorphism.

Choose $c \in^* C$. Then $\varepsilon(c) \in^* F$ and, since η is an epimorphism, $\varepsilon(c) =^* \eta(e)$ for some $e \in^* E$. So $(\xi \circ \lambda)(e) =^* (\mu \circ \eta)(e) =^* (\mu \circ \varepsilon)(c) =^* 0$

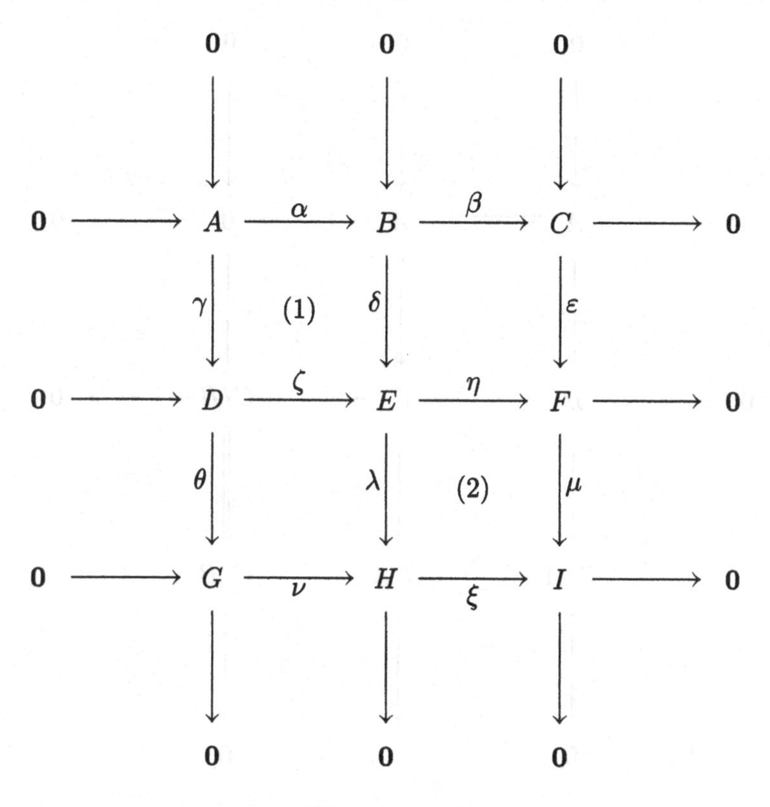

Diagram 1.30

and therefore $\lambda(e) =^* \nu(g)$ for some $g \in^* G$. Since θ is an epimorphism, there exists $d \in^* D$ such that $\theta(d) =^* g$. We have thus $(\lambda \circ \zeta)(d) =^* (\nu \circ \theta)(d) =^* \nu(g) =^* \lambda(e)$. Applying 1.9.4(5), there is a pseudo-element writtten formally $e - \zeta(d) \in^* E$ such that $\lambda\big(e - \zeta(d)\big) =^* 0$ and, since $(\eta \circ \zeta)(d) =^* 0$, $\eta\big(e - \zeta(d)\big) =^* \eta(e)$. From $\lambda\big(e - \zeta(d)\big) =^* 0$ we deduce the existence of $b \in^* B$ such that $\delta(b) =^* e - \zeta(d)$. Finally $(\varepsilon \circ \beta)(b) =^* (\eta \circ \delta)(b) =^* \eta\big(e - \zeta(d)\big) =^* \eta(e) =^* \varepsilon(c)$ and since ε is a monomorphism, $\beta(b) =^* c$.

Let us prove now that (1) is a pullback; by duality, (2) will be a pushout. Given morphisms $x\colon X \longrightarrow D$, $y\colon X \longrightarrow B$ such that $\zeta \circ x = \delta \circ y$, one has $\varepsilon \circ \beta \circ y = \eta \circ \delta \circ y = \eta \circ \zeta \circ x = 0$, and thus $\beta \circ y = 0$, since ε is a monomorphism. From $\beta \circ y = 0$ and $\alpha = \mathsf{Ker}\,\beta$ we get $z\colon X \longrightarrow A$ such that $\alpha \circ z = y$. Then $\zeta \circ \gamma \circ z = \delta \circ \alpha \circ z = \delta \circ y = \zeta \circ x$ and $\gamma \circ z = x$ since ζ is a monomorphism. Such a z is necessarily unique since α and γ are monomorphisms. $\square$

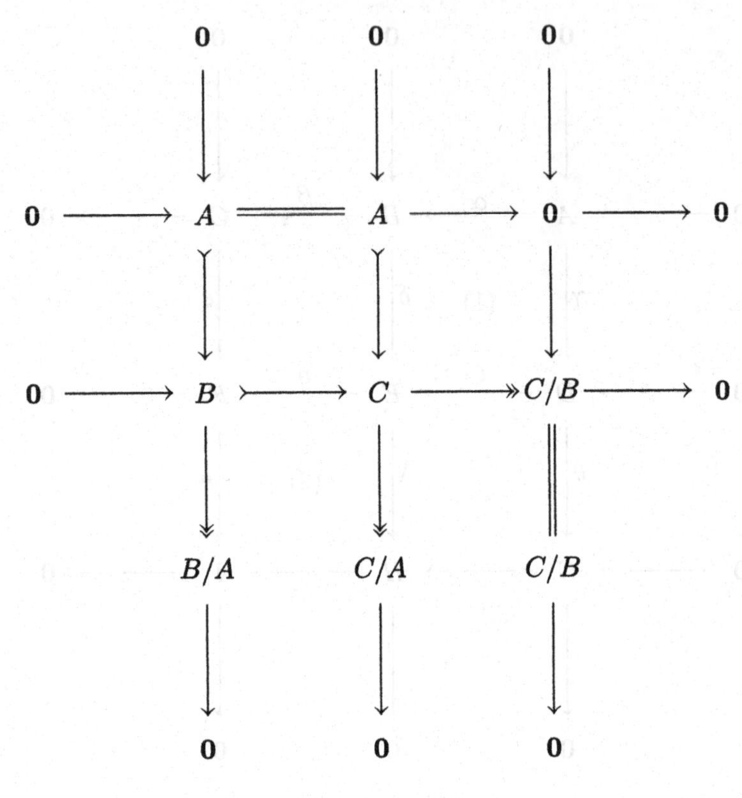

Diagram 1.31

When $f\colon A \rightarrowtail B$ is a monomorphism, let us again use the notation $p\colon B \twoheadrightarrow B/A$ for the cokernel.

Lemma 1.10.6 (First Noether isomorphism theorem)
In an abelian category, consider subobjects $A \rightarrowtail B \rightarrowtail C$. In this case B/A is a subobject of C/A and $(C/A)/(B/A)$ is isomorphic to C/B.

Proof It suffices to apply the nine lemma to diagram 1.31, where the existence of a last row follows from 1.10.1. □

Lemma 1.10.7 (Second Noether isomorphism theorem)
Consider two subobjects $R \rightarrowtail A$ and $S \rightarrowtail A$ in an abelian category. The following isomorphism holds:

$$S/(R \cap S) \cong (R \cup S)/R.$$

Proof Consider diagram 1.32, where the two first rows and the two first columns are exact. Since (1) is a pullback, the last row is exact as well

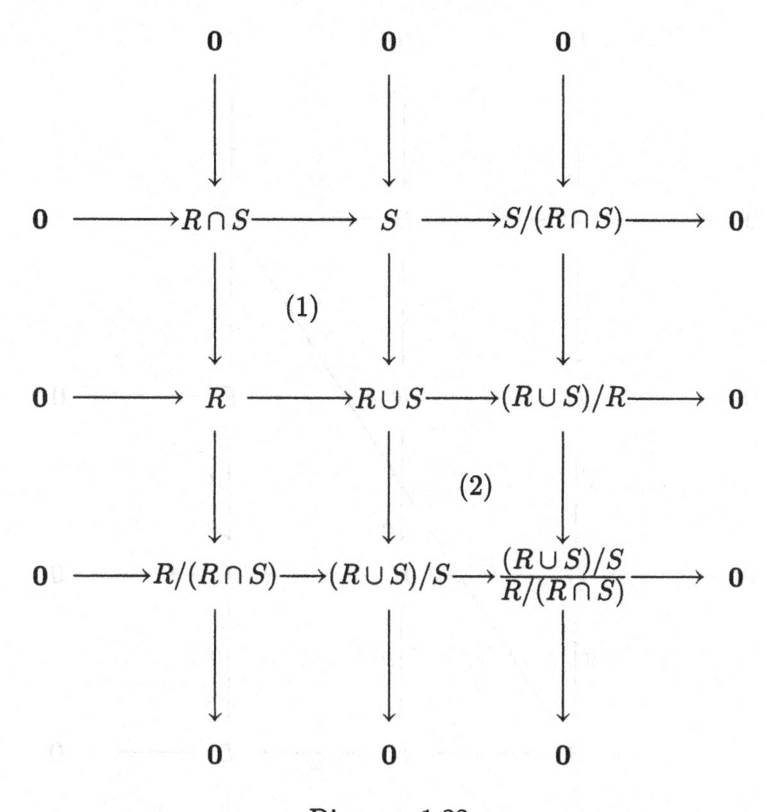

Diagram 1.32

(see 1.10.2). Applying 1.10.1 we construct the last column and lemma 1.10.5 implies its exactness.

Applying 1.10.5 again, we know that (2) is a pushout square. But by 1.7.5 the pushout of $(R \cup S)/R$ and $(R \cup S)/S$ is just $(R \cup S)/(R \cup S)$, i.e. the zero object. Therefore we have an exact sequence

$$\mathbf{0} \longrightarrow S/(R \cap S) \longrightarrow (R \cup S)/R \longrightarrow \mathbf{0},$$

which implies the required isomorphism (see 1.5.1). □

Lemma 1.10.8 (The restricted snake lemma) *In an abelian category, consider diagram 1.33, where all squares are commutative and where all rows and columns are exact. In those conditions there exists a diagonal morphism ω which makes the sequence (β, ω, τ) an exact sequence.*

Proof To construct ω, we consider diagram 1.34 where Γ, Δ are obtained as pullback of ε, η and Λ, Ξ as pushout of π, ν. We also define

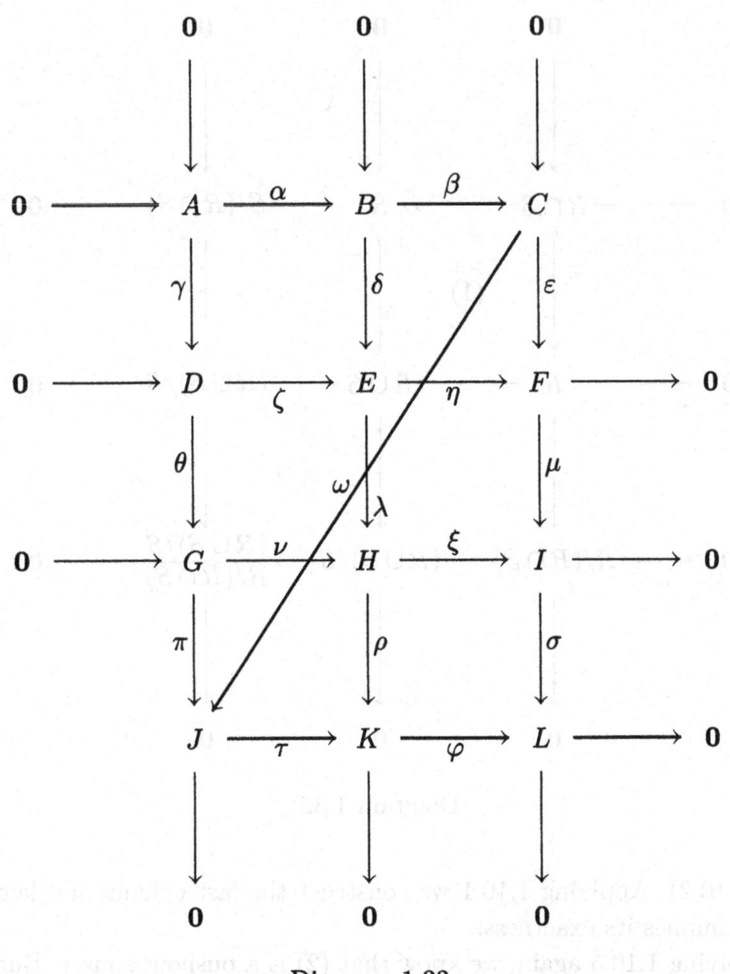

Diagram 1.33

$\Sigma = \mathsf{Ker}\,\Delta$ and $\Upsilon = \mathsf{Coker}\,\Xi$. By lemma 1.10.1 and its dual, there are morphisms Ψ and Ω making the diagram commutative and the two outer columns exact. Since $\Sigma = \mathsf{Ker}\,\Delta$ and Δ is an epimorphism (dual of 1.7.6), $\Delta = \mathsf{Coker}\,(\mathsf{Ker}\,\Delta) = \mathsf{Coker}\,\Sigma$ (see 1.5.7); in the same way $\Xi = \mathsf{Ker}\,\Upsilon$. So in diagram 1.34, just the central column is not exact.

Since $\Lambda \circ \lambda \circ \Gamma \circ \Sigma = \Xi \circ \pi \circ \theta \circ \Psi = 0$ and $\Delta = \mathsf{Coker}\,\Sigma$, there is a unique factorization χ such that $\chi \circ \Delta = \Lambda \circ \lambda \circ \Gamma$. Now $\Upsilon \circ \chi \circ \Delta = \Upsilon \circ \Lambda \circ \lambda \circ \Gamma = \Omega \circ \mu \circ \varepsilon \circ \Delta = 0$, so that $\Upsilon \circ \chi = 0$ because Δ is an epimorphism. From $\Upsilon \circ \chi = 0$ and $\Xi = \mathsf{Ker}\,\Upsilon$, we obtain the required factorization ω such that $\Xi \circ \omega = \chi$.

Let us now study the action of ω on the pseudo-elements. Given $c \in^*$

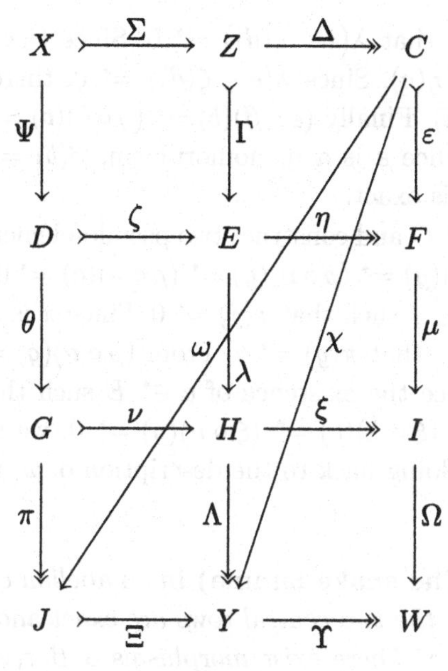

Diagram 1.34

C, we have $\varepsilon(c) \in^* F$ and since η is an epimorphism, there is $e \in^* E$ such that $\eta(e) =^* \varepsilon(c)$. Now $(\xi \circ \lambda)(e) =^* (\mu \circ \eta)(e) =^* (\mu \circ \varepsilon)(c) =^* 0$, so that we find $g \in^* G$ such that $\nu(g) =^* \lambda(e)$. We shall prove that $\pi(g) =^* \omega(c)$.

By 1.9.5, the relation $\eta(e) =^* \varepsilon(c)$ implies the existence of $z \in^* Z$ such that $\Delta(z) =^* c$ and $\Gamma(z) =^* e$. Now

$$\begin{aligned}(\Xi \circ \omega)(c) &=^* \chi(c) =^* (\chi \circ \Delta)(z) =^* (\Lambda \circ \lambda \circ \Gamma)(z) \\ &=^* (\Lambda \circ \lambda)(e) =^* (\Lambda \circ \nu)(g) =^* \Xi\pi(g),\end{aligned}$$

from which $\omega(c) =^* \pi(g)$ since Ξ is a monomorphism.

We can now prove the exactness of the sequence (β, ω, τ) by chasing on diagram 1.33. First of all choose $b \in^* B$. To prove that $(\omega\circ\beta)(b) =^* 0$, put $c =^* \beta(b)$ and $e =^* \delta(b)$ in the previous description of ω; one has indeed $\eta(e) =^* (\eta \circ \delta)(b) =^* (\varepsilon \circ \beta)(b) =^* \varepsilon(c)$. As before, we choose $g \in^* G$ such that $\nu(g) =^* \lambda(e) =^* (\lambda \circ \delta)(b) =^* 0$. This implies $g =^* 0$ since ν is a monomorphism and thus $\omega(c) =^* \pi(g) =^* 0$.

Now choose $c \in^* C$ and construct $e \in^* E$, $g \in^* G$ as indicated in the description of ω in terms of pseudo-elements. Suppose $\pi(g) =^* 0$. This implies the existence of $d \in^* D$ such that $\theta(d) =^* g$. Since $(\lambda \circ \zeta)(d) =^* (\nu \circ \theta)(d) =^* \nu(g) =^* \lambda(e)$, we can consider an element written formally

$e - \zeta(d) \in^* E$ such that $\lambda(e - \zeta(d)) =^* 0$. Since $(\eta \circ \zeta)(d) =^* 0$, one has $\eta(e - \zeta(d)) =^* \eta(e)$. Since $\lambda(e - \zeta(d)) =^* 0$, there is $b \in^* B$ such that $\delta(b) =^* e - \zeta(d)$. Finally $(\varepsilon \circ \beta)(b) =^* (\eta \circ \delta)(b) =^* \eta(e - \zeta(d)) =^* \eta(e) =^* \varepsilon(c)$ and since ε is a monomorphism, $\beta(b) =^* c$. This proves already that (β, ω) is exact.

Choose again $c \in^* C$ and construct two pseudo-elements e, g as before. $(\tau \circ \omega)(c) =^* (\tau \circ \pi)(g) =^* (\rho \circ \nu)(g) =^* (\rho \circ \lambda)(e) =^* 0$.

Finally choose $j \in J$ such that $\tau(j) =^* 0$. Since π is an epimorphism, we find $g \in^* G$ such that $\pi(g) =^* j$. From $(\rho \circ \nu)(g) =^* (\tau \circ \pi)(g) =^* \tau(j) =^* 0$, we deduce the existence of $e \in^* E$ such that $\lambda(e) =^* \nu(g)$. From $(\mu \circ \eta)(e) =^* (\xi \circ \lambda)(e) =^* (\xi \circ \nu)(g) =^* 0$, we find $c \in^* C$ such that $\varepsilon(c) =^* \eta(e)$. Going back to the description of ω, we conclude that $j =^* \pi(g) =^* \omega(c)$. □

Lemma 1.10.9 (The snake lemma) *In an abelian category, consider diagram 1.35 where the two central rows are exact and the squares (1), (2) are commutative. There exist morphisms $\alpha, \beta, \tau, \varphi$ making the diagram commutative as well as a diagonal morphism ω such that the sequence $(\alpha, \beta, \omega, \tau, \varphi)$ is exact.*

Proof The existence of the exact sequences (α, β) and (τ, φ) is attested by 1.10.1. Let us consider diagram 1.36 where $\zeta = \zeta_2 \circ \zeta_1$ and $\xi = \xi_2 \circ \xi_1$ are image factorizations. Applying 1.5.6, we find morphisms Γ, Δ keeping the diagram commutative. We define $\Lambda = \mathsf{Ker}\,\Gamma$, $\Xi = \mathsf{Ker}\,\Delta$, $\Sigma = \mathsf{Coker}\,\Gamma$, $\Upsilon = \mathsf{Coker}\,\Delta$. There are obvious factorizations $\alpha_i, \beta_i, \tau_i, \varphi_i$ through the kernels and cokernels, with $\alpha = \alpha_2 \circ \alpha_1$, $\beta = \beta_2 \circ \beta_1$, $\tau = \tau_2 \circ \tau_1$ and $\varphi = \varphi_2 \circ \varphi_1$. α_2 and β_2 are obviously monomorphisms, and in the same way τ_1, φ_1 are epimorphisms. Observe that by definition, all the columns are exact.

Let us prove that β_2 is also an epimorphism, from which it will be an isomorphism (see 1.5.1). Given $c \in^* C$, $(\xi_2 \circ \Delta \circ \varepsilon)(c) =^* (\mu \circ \varepsilon)(c) =^* 0$ so that $(\Delta \circ \varepsilon)(c) =^* 0$ since ξ_2 is a monomorphism. This implies the existence of $u \in^* U$ such that $\Xi(u) =^* \varepsilon(c)$. Finally $(\varepsilon \circ \beta_2)(u) =^* \Xi(u) =^* \varepsilon(c)$, and thus $\beta_2(u) =^* c$ since ε is a monomorphism. By duality, τ_1 is an isomorphism as well.

Applying proposition 1.8.2, we conclude that (ζ_2, η) and (ν, ξ_1) are exact sequences, so that lemma 1.10.8 can be applied to the central part of the diagram, giving a morphism χ such that the sequence (β_1, χ, τ_2) is exact. It remains to define $\omega = \tau_1^{-1} \circ \chi \circ \beta_2^{-1}$ and, since τ_1 and β_2 are isomorphisms, (β, ω, τ) is an exact sequence as well. □

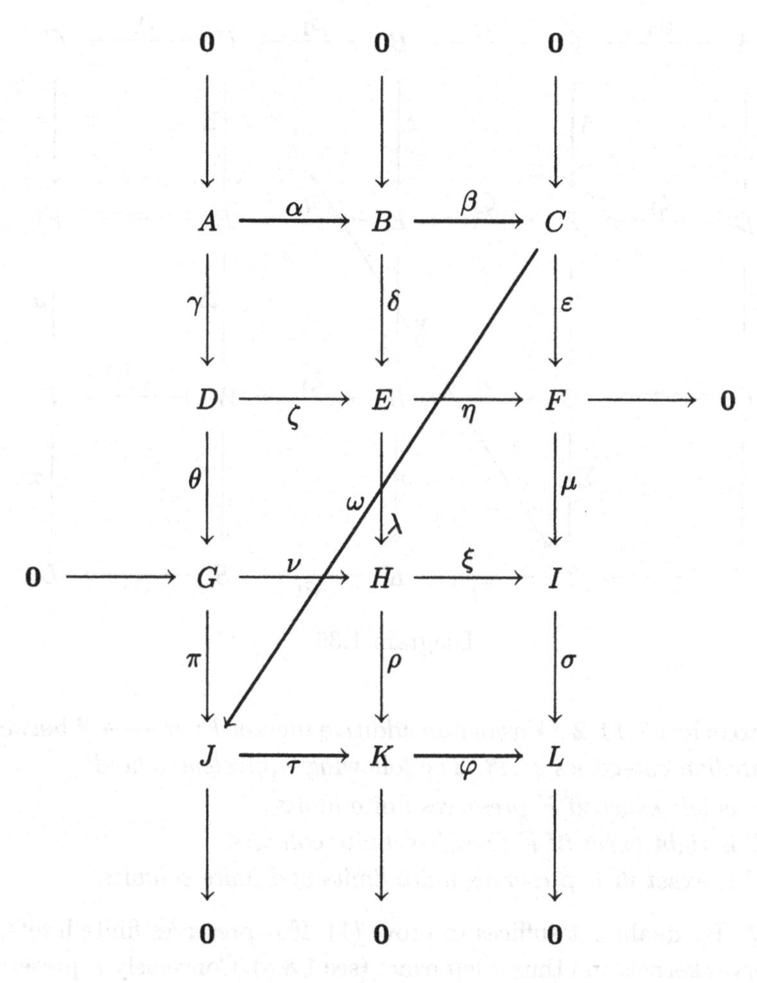

Diagram 1.35

1.11 Exact functors

Exact functors are those which preserve exact sequences. More precisely:

Definition 1.11.1 *Consider an additive functor $F\colon \mathscr{A} \longrightarrow \mathscr{B}$ between two abelian categories $\mathscr{A}, \mathscr{B}$. We say:*

(1) F is left exact when it preserves exact sequences of the form

$$0 \longrightarrow A \longrightarrow B \longrightarrow C;$$

(2) F is right exact when it preserves exact sequences of the form

$$A \longrightarrow B \longrightarrow C \longrightarrow 0;$$

(3) F is exact when it preserves exact sequences of the form

$$0 \longrightarrow A \longrightarrow B \longrightarrow C \longrightarrow 0.$$

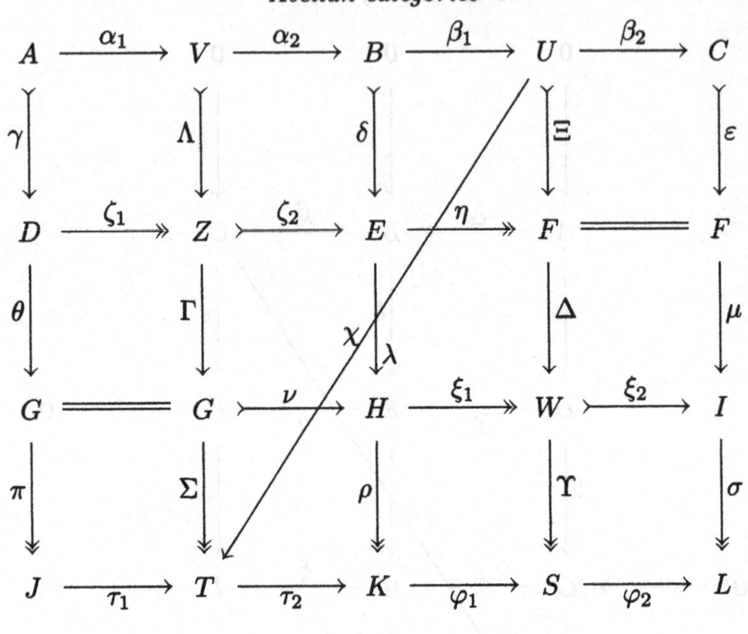

Diagram 1.36

Proposition 1.11.2 *Consider an additive functor $F: \mathscr{A} \longrightarrow \mathscr{B}$ between two abelian categories $\mathscr{A}, \mathscr{B}$. The following equivalences hold:*

(1) F is left exact iff F preserves finite limits;

(2) F is right exact iff F preserves finite colimits;

(3) F is exact iff F preserves finite limits and finite colimits.

Proof By duality, it suffices to prove (1). If F preserves finite limits, it preserves kernels and thus is left exact (see 1.8.5). Conversely F preserves biproducts since it is additive (see 1.3.4) and preserves kernels since it is left exact (see 1.8.5). Applying 2.8.2, volume 1, it remains to prove that F preserves equalizers. But the equalizer of a pair $f, g: A \rightrightarrows B$ is just the kernel of $f - g$. □

Proposition 1.11.3 *Consider an additive functor $F: \mathscr{A} \longrightarrow \mathscr{B}$ between two abelian categories $\mathscr{A}, \mathscr{B}$. The following conditions are equivalent:*

(1) F is exact;

(2) F preserves all exact sequences.

Proof (2) $\Rightarrow$ (1) is obvious. Conversely, we consider an exact sequence (f, g) and the image factorizations $f = i \circ p$, $g = j \circ q$ as in diagram 1.37. By 1.11.2, F preserves monomorphisms and epimorphisms. But the sequence (i, q) is exact (see 1.8.2) and, by definition, is preserved by F.

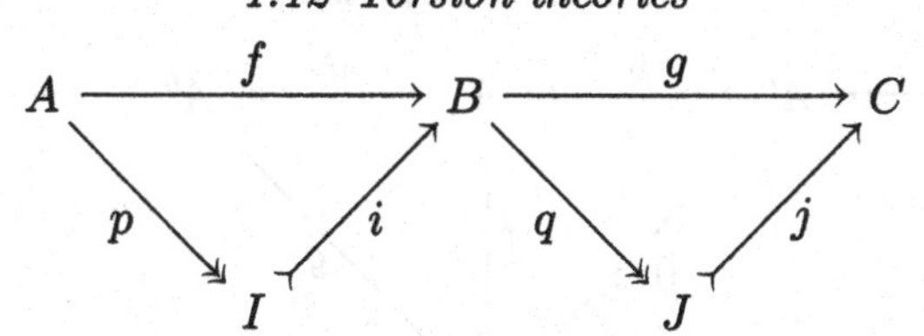

Diagram 1.37

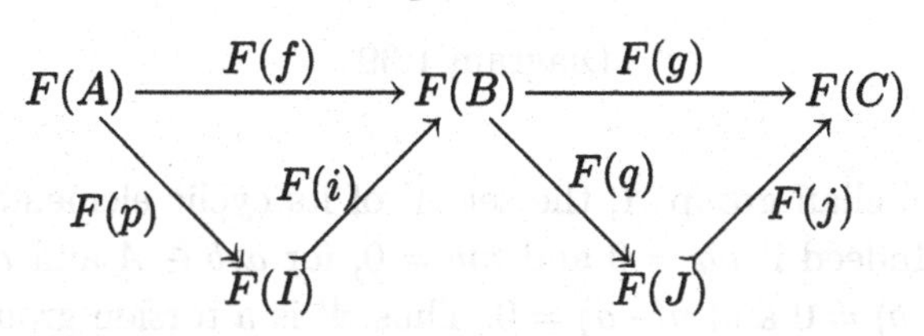

Diagram 1.38

Thus we obtain diagram 1.38 in $\mathscr{B}$ with $(F(i), F(q))$ exact. Therefore $(F(f), F(g))$ is exact (see 1.8.2). □

Proposition 1.11.4 *Consider a left exact functor $F\colon \mathscr{A} \longrightarrow \mathscr{B}$ between abelian categories. The following conditions are equivalent:*

(1) F is exact;
(2) F preserves epimorphisms.

Proof (1) ⇒ (2) is already proved (see 1.11.2). So assume (2) and consider a short exact sequence

$$0 \longrightarrow A \xrightarrow{\alpha} B \xrightarrow{\beta} C \longrightarrow 0$$

in $\mathscr{A}$. Applying F we get an exact sequence

$$0 \longrightarrow F(A) \xrightarrow{F(\alpha)} F(B) \xrightarrow{F(\beta)} F(C).$$

Since β is an epimorphism, so is $F(\beta)$ and the second sequence is in fact a short exact sequence (see 1.8.5). □

1.12 Torsion theories

An element a of an abelian group A is cyclic (or has a torsion) when there exists a non-zero natural number $n \in \mathbb{N}$ such that $na = 0$. An abelian group A is torsion free when it does not contain any non-zero cyclic element; on the other hand A is a torsion group when all its elements are cyclic.

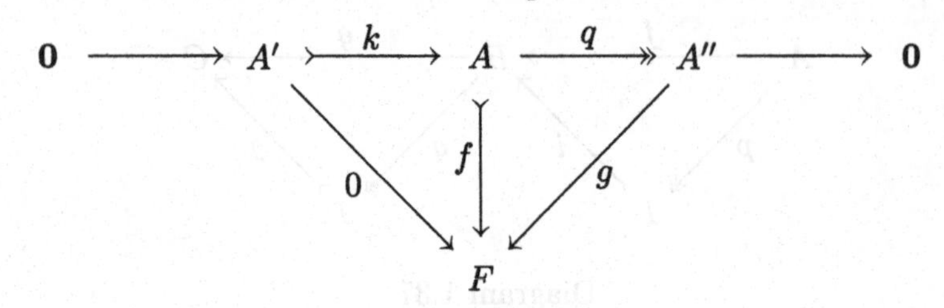

Diagram 1.39

Given an abelian group A, the set A' of its cyclic elements is a subgroup of A. Indeed if $na = 0$ and $mb = 0$, for $a, b \in A$ and $n, m \in \mathbb{N}^*$, then $nm(a+b) = 0$ and $n(-a) = 0$. Thus A' is a torsion group. On the other hand the quotient A/A' is obviously torsion free, which yields an exact sequence

$$\mathbf{0} \longrightarrow A' \longrightarrow A \longrightarrow A/A' \longrightarrow \mathbf{0}.$$

Observe also that given a group homorphism $f\colon A \longrightarrow B$, if $na = 0$ in A, then $nf(a) = 0$ in B. Thus if A is a torsion group and B is torsion free, f must be the zero homomorphism, since $\mathbf{0}$ is the only cyclic element in B.

This example is the leading one for understanding the definition of a torsion theory. We recall that a full subcategory is replete when it is closed under isomorphisms (see 3.5.1, volume 1).

Definition 1.12.1 *A torsion theory on an abelian category $\mathscr{B}$ is a pair $(\mathcal{T}, \mathcal{F})$ of full replete subcategories of $\mathscr{B}$ such that:*

(1) every morphism $f\colon T \longrightarrow F$, with $T \in \mathcal{T}$ and $F \in \mathcal{F}$, is the zero morphism;

(2) for every $B \in \mathscr{B}$, there exist $T \in \mathcal{T}$ and $F \in \mathcal{F}$ together with an exact sequence

$$\mathbf{0} \longrightarrow T \longrightarrow B \longrightarrow F \longrightarrow \mathbf{0}.$$

Observe that the previous definition is autodual.

Proposition 1.12.2 *Consider an abelian category $\mathscr{B}$ provided with a torsion theory $(\mathcal{T}, \mathcal{F})$. Under these conditions:*

(1) if $f\colon A \rightarrowtail F$ is a monomorphism with $F \in \mathcal{F}$, then $A \in \mathcal{F}$;

(2) if $g\colon T \longrightarrow C$ is an epimorphism with $T \in \mathcal{T}$, then $C \in \mathcal{T}$.

Proof By duality, it suffices to prove the first statement. Consider an exact sequence (k, q) with $A' \in \mathcal{T}$, $A'' \in \mathcal{F}$ as in diagram 1.39. Since $A' \in \mathcal{T}, F \in \mathcal{F}$, $f \circ k = 0$. Since $q = \operatorname{Coker} k$, one gets g such that $g \circ q = f$.

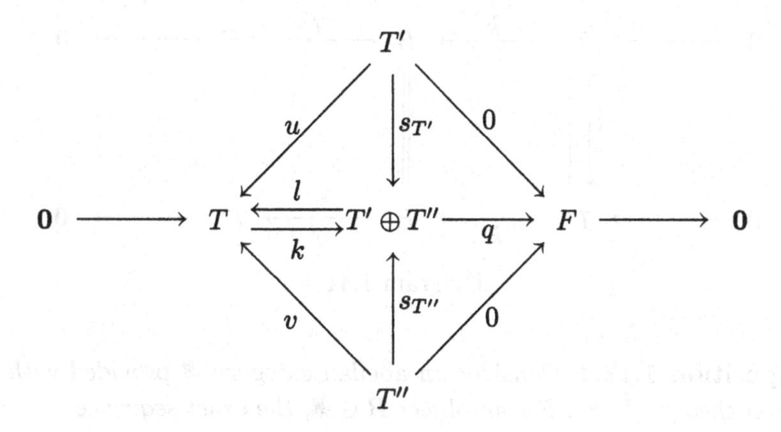

Diagram 1.40

Since f is a monomorphism, so is q. Since q is both a monomorphism and an epimorphism, it is an isomorphism (see 1.5.1). Since $A'' \in \mathcal{F}$ and $\mathcal{F}$ is replete, $A \in \mathcal{F}$. □

Proposition 1.12.3 *Consider an abelian category $\mathscr{B}$ provided with a torsion theory $(\mathcal{T}, \mathcal{F})$. Then:*

(1) $\mathcal{T}$ and $\mathcal{F}$ contain the zero object;

(2) $\mathcal{T}$ and $\mathcal{F}$ are closed under biproducts.

Proof By 1.12.1 there exists an exact sequence

$$\mathbf{0} \longrightarrow T \longrightarrow \mathbf{0} \longrightarrow F \longrightarrow \mathbf{0}$$

with $T \in \mathcal{T}$, $F \in \mathcal{F}$. Therefore $T \longrightarrow \mathbf{0}$ is a monomorphism (see 1.8.5); but it is also an epimorphism since $\mathbf{0}$ is initial. Thus $T \longrightarrow \mathbf{0}$ is an isomorphism (see 1.4.1) and $\mathbf{0} \in \mathcal{T}$ since $\mathcal{T}$ is replete. By duality, $\mathbf{0} \in \mathcal{F}$.

Next consider $T', T'' \in \mathcal{T}$. By 1.12.1 there exists a short exact sequence (k, q) with $T \in \mathcal{T}$, $F \in \mathcal{F}$; see diagram 1.40. Considering the canonical morphisms $s_{T'}$, $s_{T''}$ of the biproduct, one has $q \circ s_{T'} = 0$, $q \circ s_{T''} = 0$ since $T', T'' \in \mathcal{T}$ and $F \in \mathcal{F}$. But $k = \operatorname{Ker} q$ (see 1.8.5), thus there exist u, v such that $k \circ u = s_{T'}$, $k \circ v = s_{T''}$. Since $T' \oplus T''$ is a coproduct, we get l such that $l \circ s_{T'} = u$, $l \circ s_{T''} = v$. The relations

$$k \circ l \circ s_{T'} = k \circ u = s_{T'}, \quad k \circ l \circ s_{T''} = k \circ v = s_{T''}$$

imply $k \circ l = 1_{T' \oplus T''}$. Thus $k \circ l \circ k = k$ and since k is a monomorphism, $l \circ k = 1_T$. So k is an isomorphism and since $T \in \mathcal{T}$, one gets $T' \oplus T'' \in \mathcal{T}$.

The stability of $\mathcal{F}$ under biproducts follows by duality. □

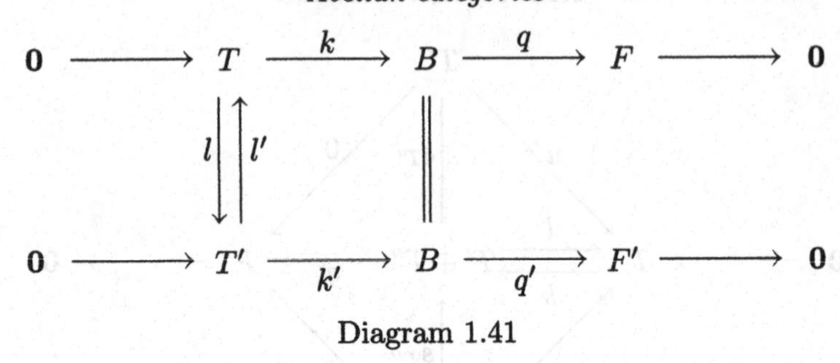

Diagram 1.41

Proposition 1.12.4 *Consider an abelian category $\mathscr{B}$ provided with a torsion theory $(\mathcal{T},\mathcal{F})$. For an object $B\in\mathscr{B}$, the exact sequence*

$$\mathbf{0}\longrightarrow T\longrightarrow B\longrightarrow F\longrightarrow \mathbf{0}$$

as in 1.12.1 is unique up to isomorphism.

Proof Suppose we are given two such sequences (k,q) and (k',q') as in diagram 1.41. Since $T\in\mathcal{T}$ and $F'\in\mathcal{F}$, $q'\circ k=0$ and we get a factorization l such that $k'\circ l=k$, because $k'=\mathsf{Ker}\,q'$. Analogously we get l' such that $k\circ l'=k'$. From $k\circ l'\circ l=k'\circ l=k$ and $k'\circ l\circ l'=k\circ l'=k'$, we deduce $l'\circ l=1_T$ and $l\circ l'=1_{T'}$ since k,k' are monomorphisms. So l,l' are indeed isomorphisms making the diagram commutative. By duality, one gets the corresponding result for F,F'. □

Proposition 1.12.5 *Consider an abelian category $\mathscr{B}$ provided with a torsion theory $(\mathcal{T},\mathcal{F})$. Given an object $B\in\mathscr{B}$,*
(1) $B\in\mathcal{T}$ iff $\forall F\in\mathcal{F}$ $\mathscr{B}(B,F)=\{0\}$,
(2) $B\in\mathcal{F}$ iff $\forall T\in\mathcal{T}$ $\mathscr{B}(T,F)=\{0\}$.

Proof By duality, it suffices to prove the first statement. If $B\in\mathcal{T}$, $\mathscr{B}(B,F)=\{0\}$ for every $F\in\mathcal{F}$, just by definition of a torsion theory.

Conversely suppose $\mathscr{B}(B,F)=\{0\}$ for every $F\in\mathcal{F}$ and consider the following exact sequence, with $T\in\mathcal{T}$ and $F\in\mathcal{F}$ (see 1.12.1):

$$\mathbf{0}\longrightarrow T\xrightarrow{\;k\;}B\xrightarrow{\;q\;}F\longrightarrow \mathbf{0}.$$

By assumption, $q=0$ so that its image is just $\mathbf{0}$. By 1.8.5, q is already an epimorphism, so that F is isomorphic to the image $\mathbf{0}$ of q. By 1.8.5 again, k is both a monomorphism and an epimorphism, so it is an isomorphism by 1.5.1. Since $T\in\mathcal{T}$ with $\mathcal{T}$ replete, one concludes that $B\in\mathcal{T}$. □

In section 5.7, volume 1, we studied universal closure operations. We shall now prove an interesting relation between universal closure operations and torsion theories.

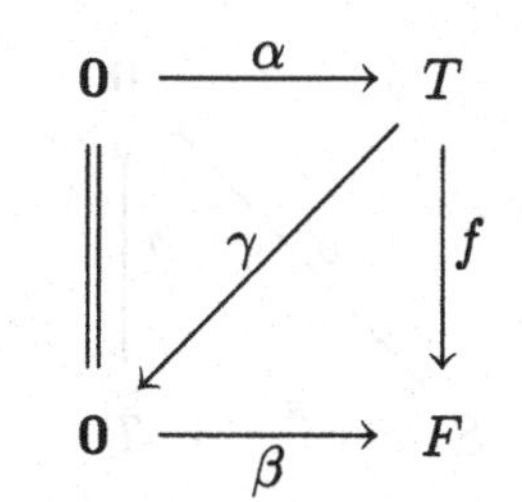

Diagram 1.42

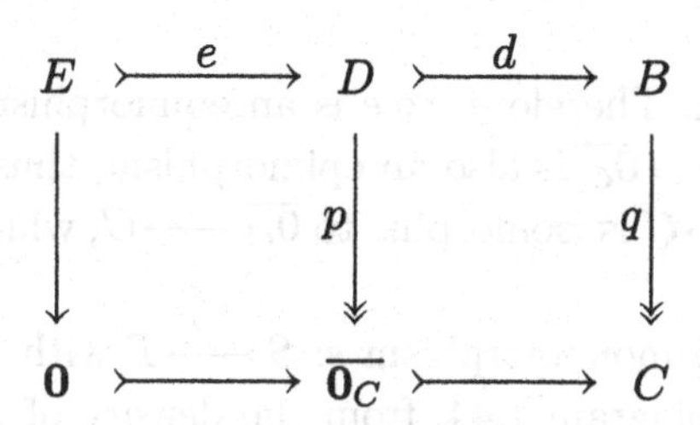

Diagram 1.43

Proposition 1.12.6 *Consider an abelian category $\mathscr{B}$ provided with a universal closure operation. Define two full subcategories $\mathcal{T}, \mathcal{F}$ of $\mathscr{B}$ by*
(1) $T \in \mathcal{T}$ if $\mathbf{0} \longrightarrow T$ is dense,
(2) $F \in \mathcal{F}$ if $\mathbf{0} \longrightarrow F$ is closed.
Under these conditions, the pair $(\mathcal{T}, \mathcal{F})$ is a torsion theory on $\mathscr{B}$. Moreover this torsion theory satisfies the additional property:

If $s\colon S \rightarrowtail T$ is a monomorphism and $T \in \mathcal{T}$, then $S \in \mathcal{T}$.

Proof Obviously $\mathcal{T}$ and $\mathcal{F}$ are replete. Moreover if $T \in \mathcal{T}$, $F \in \mathcal{F}$ and $f\colon T \longrightarrow F$, consider diagram 1.42. Since α is dense and β is closed, 5.7.10 of volume 1 implies the existence of a unique γ making the diagram commutative. In other words, $f = 0$.

Now given an object $B \in \mathscr{B}$, consider the subobject $\mathbf{0} \longrightarrow B$, its closure $\overline{\mathbf{0}_B} \rightarrowtail B$ and the coequalizer $B \twoheadrightarrow C$ of $\overline{\mathbf{0}_B} \rightarrowtail B$. By 5.7.5 of volume 1, $\mathbf{0} \rightarrowtail \overline{\mathbf{0}_B}$ is dense and thus $\overline{\mathbf{0}_B} \in \mathcal{T}$. We have an exact sequence

$$\mathbf{0} \longrightarrow \overline{\mathbf{0}_B} \xrightarrow{\ k\ } B \xrightarrow{\ q\ } C \longrightarrow \mathbf{0}$$

with $\overline{\mathbf{0}_B} \in \mathcal{T}$. We shall prove that $C \in \mathcal{F}$. Consider diagram 1.43 where both squares are pullbacks and $\overline{\mathbf{0}_C}$ is the closure of $\mathbf{0}$ in C. One has $E \cong q^{-1}(\mathbf{0})$ thus $E \cong \operatorname{Ker} q$. But $\operatorname{Ker} q = k$ by construction, so that $E \cong \overline{\mathbf{0}_B}$. On the other hand $D \cong \overline{E}$; see 5.7.1, volume 1. Thus $D \cong \overline{\overline{\mathbf{0}_B}} \cong \overline{\mathbf{0}_B}$ and

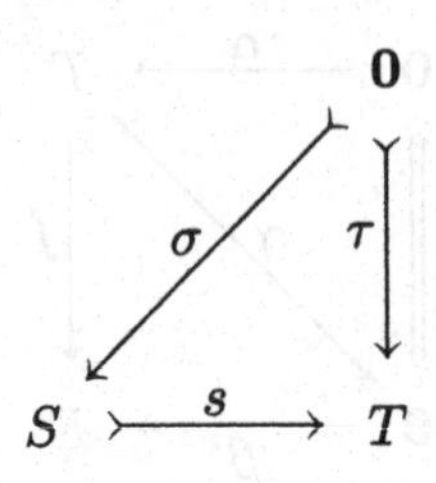

Diagram 1.44

e is an isomorphism. Therefore $p \circ e$ is an epimorphism and finally the monomorphism $\mathbf{0} \rightarrowtail \overline{\mathbf{0}_C}$ is also an epimorphism, thus an isomorphism (see 1.5.1). So $\mathbf{0} \rightarrowtail C$ is isomorphic to $\overline{\mathbf{0}_C} \rightarrowtail C$, which is closed. Thus $C \in \mathcal{F}$.

Finally consider a monomorphism $s\colon S \rightarrowtail T$ with $T \in \mathcal{T}$. Considering the triangle of diagram 1.44, from the density of τ we deduce that of σ; see 5.7.7, volume 1. □

Definition 1.12.7 *Consider an abelian category $\mathscr{B}$ provided with a torsion theory $(\mathcal{T}, \mathcal{F})$. This torsion theory is hereditary when it satisfies the additional condition:*

If $s\colon S \rightarrowtail T$ is a monomorphism and $T \in \mathcal{T}$, then $S \in \mathcal{T}$.

Theorem 1.12.8 *Consider an abelian category $\mathscr{B}$. There exists a bijection between*

(1) the universal closure operations on $\mathscr{B}$ and

(2) the hereditary torsion theories on $\mathscr{B}$.

Proof In 1.12.6 we have constructed an hereditary torsion theory from a given universal closure operation. Conversely, let us start with an hereditary torsion theory $(\mathcal{T}, \mathcal{F})$ of $\mathscr{B}$. Given a subobject $A \rightarrowtail B$, we consider diagram 1.45 where the vertical and the horizontal sequence are exact and $T \in \mathcal{T}$, $F \in \mathcal{F}$; see 1.12.1. The pair (c, d) is the pullback of (k, b). Let us define the closure of $a\colon A \rightarrowtail B$ as being $d\colon \overline{A} \rightarrowtail B$. Since $b \circ a = 0 = k \circ 0$, there exists e such that $d \circ e = a$, $c \circ e = 0$. This proves already that $A \subseteq \overline{A}$.

Choose now $A \subseteq A' \subseteq B$. Write $u\colon A \rightarrowtail A'$ for the inclusion; we shall prove that $\overline{A} \subseteq \overline{A'}$. For this consider diagram 1.46 where k, q, a, b, c, d are defined as in diagram 1.45 and k', q', a', b', c', d' are defined analogously from a'. Since $b' \circ a = b' \circ a' \circ u = 0$, we get a factorization v through $b = \mathsf{Coker}\, a$, yielding $w \circ c = c' \circ x$. Since $T \in \mathcal{T}$ and $F' \in \mathcal{F}$, $q' \circ v \circ k = 0$

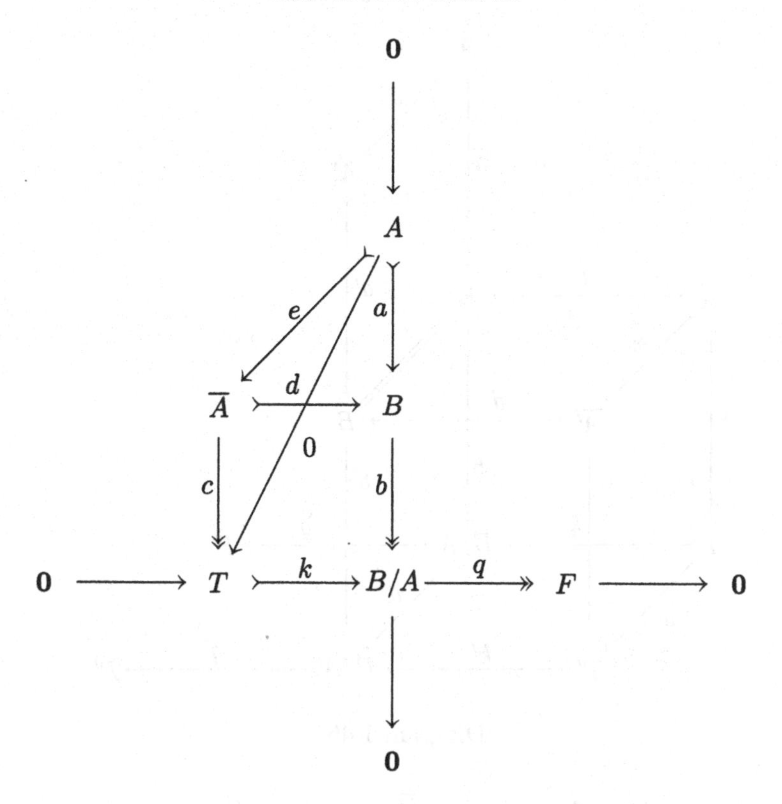

Diagram 1.45

from which we get the factorization w through $k' = \mathsf{Ker}\, q'$, yielding $v \circ k = k' \circ w$. Finally $k' \circ w \circ c = v \circ k \circ c = v \circ b \circ d = b' \circ d$, from which we get the factorization x through the pullback (c', d'), yielding $w \circ c = c' \circ x$, $d = d' \circ x$. The existence of x indicates precisely that $\overline{A} \subseteq \overline{A'}$.

Consider again $A \subseteq B$ as in diagram 1.45. We shall prove that $\overline{A} = \overline{\overline{A}}$. For that consider diagram 1.47 where $z = \mathsf{Coker}\, d$. From the equality $q \circ b \circ d = q \circ k \circ c = 0$, we obtain a factorization p with $p \circ z = q \circ b$. The arrow p is an epimorphism since q and b are. But the square $k \circ c = b \circ d$ is a pullback, by definition of $\overline{A}$; by lemma 1.10.2, p is then a monomorphism, thus an isomorphism; see 1.5.1. Therefore the kernel of p is $\mathbf{0} \rightarrowtail B/\overline{A}$; see 1.1.7. Now $\mathbf{0} \in \mathcal{T}$ by 1.12.3, so that diagram 1.48 defines $\overline{\overline{A}}$ via its pullback square. But $\overline{A} = z^{-1}(\mathbf{0})$ is precisely the kernel of z, i.e. d: $\overline{A} \longrightarrow B$. Thus $\overline{\overline{A}} = \overline{A}$.

Finally let us consider a subobject a: $A \rightarrowtail B$ and an arbitrary mor-

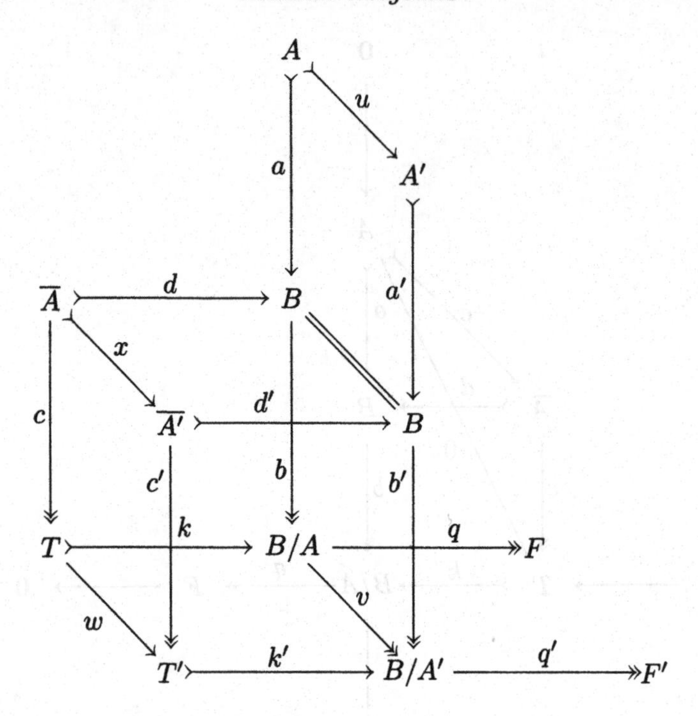

Diagram 1.46

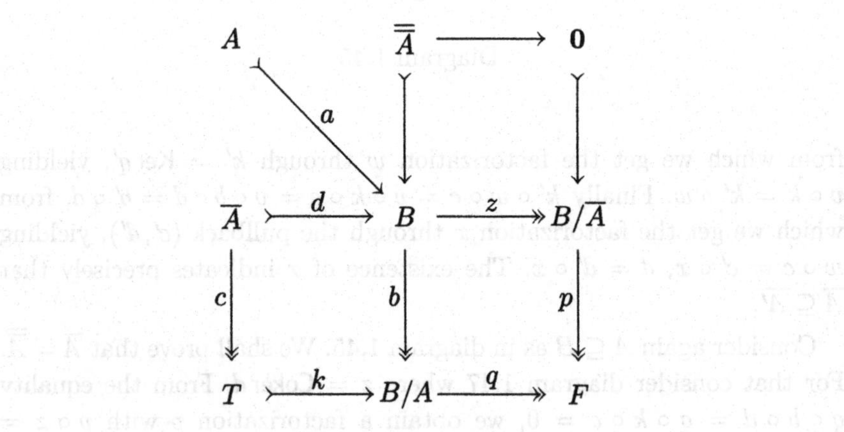

Diagram 1.47

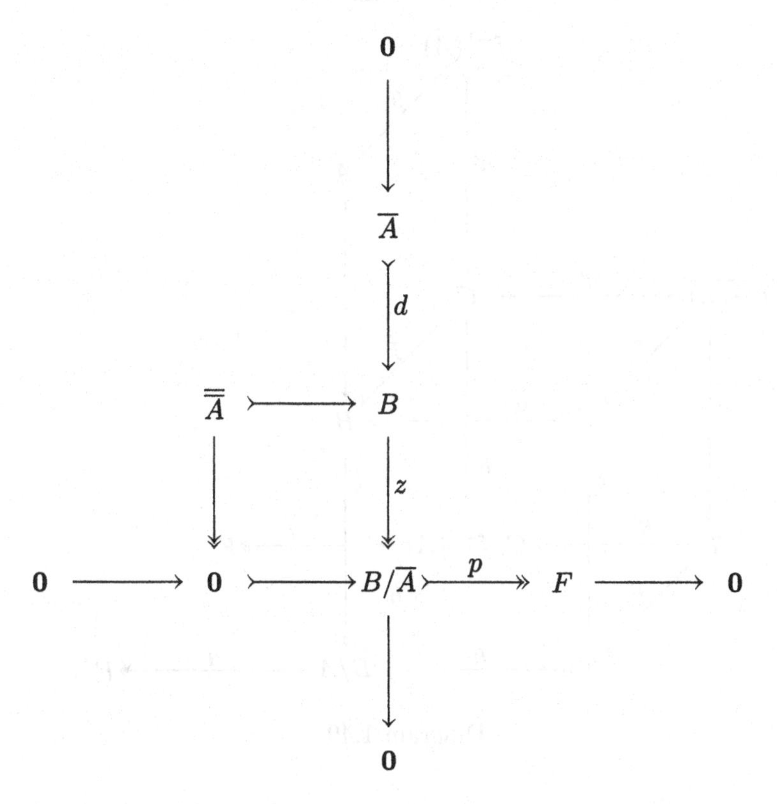

Diagram 1.48

phism $f\colon C \longrightarrow B$. We consider diagram 1.49 where (a', g) is the pullback of (f, a), a, b, c, d, k, q are defined as in diagram 1.45 and a', b', c', d', k', q' are defined analogously from the subobject $f^{-1}(A) \rightarrowtail C$. From the relation $b \circ f \circ a' = b \circ a \circ g = 0$ we get the factorization h such that $b \circ f = h \circ b'$, because $b' = \mathsf{Coker}\, a'$. Since the square $f \circ a' = a \circ g$ is a pullback, h is a monomorphism (see lemma 1.10.2). Since $T' \in \mathcal{T}$ and $F \in \mathcal{F}$, $q \circ h \circ k' = 0$ from which we get the factorization t through $k = \mathsf{Ker}\, q$, yielding $h \circ k' = k \circ t$. The arrow t is a monomorphism since h and k' are. From $b \circ f \circ d' = h \circ b' \circ d' = h \circ k' \circ c'$ we obtain the factorization s through the pullback (c, d) of (k, b), yielding $f \circ d' = d \circ s$, $t \circ c' = c \circ s$. This completes the diagram. Let us now consider diagram 1.50, where the square is a pullback. Since $h \circ k' = k \circ t$, we get a factorization l making the diagram commutative, i.e. $k' = n \circ l$ and $t = m \circ l$. Since $T \in \mathcal{T}$ and the torsion theory is hereditary, $S \in \mathcal{T}$. But since $S \in \mathcal{T}$ and $F' \in \mathcal{F}$, $q' \circ n = 0$. So we get a factorization r through $k' = \mathsf{Ker}\, q'$,

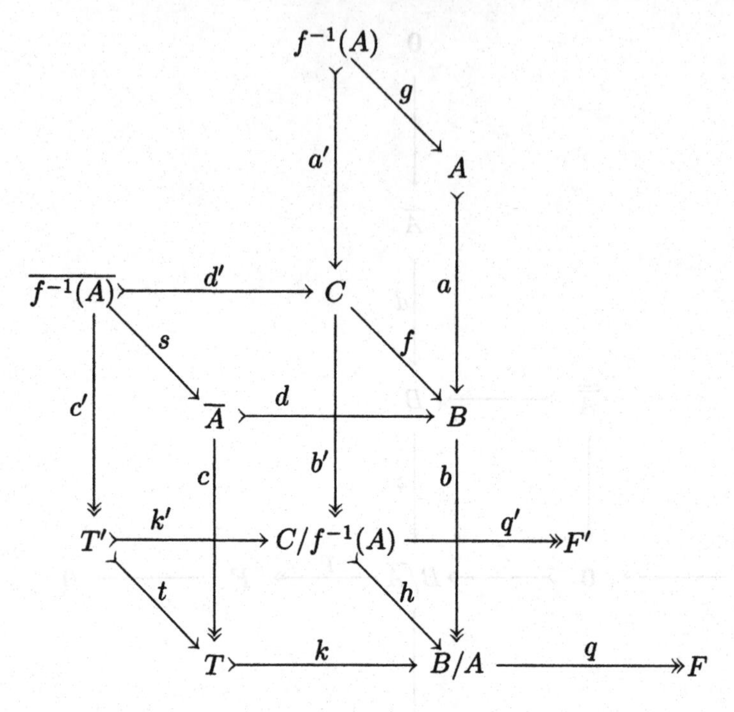

Diagram 1.49

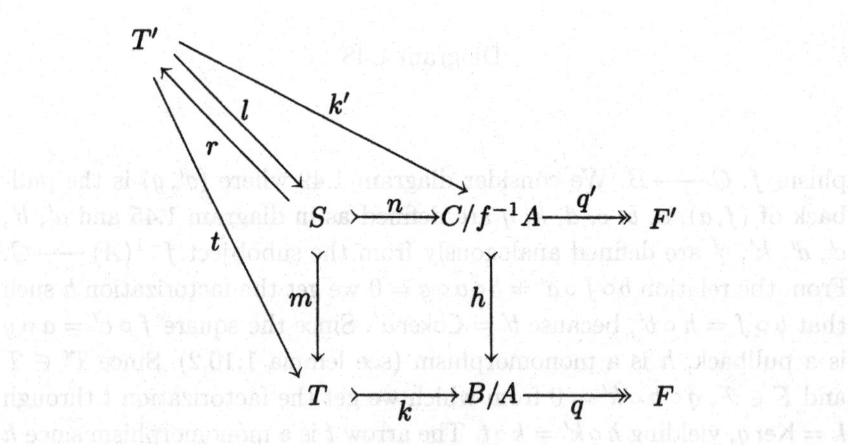

Diagram 1.50

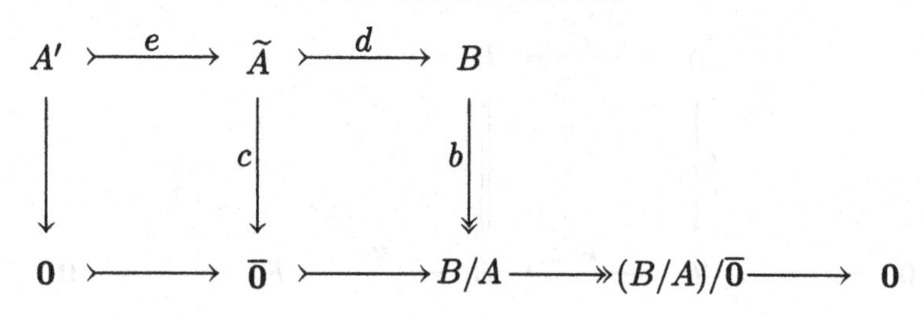

Diagram 1.51

yielding $k' \circ r = n$. Therefore $k' \circ r \circ l = n \circ l = k'$, from which $r \circ l = 1_T$, since k' is a monomorphism. In the same way $n \circ l \circ r = k' \circ r = n$ and $l \circ r = 1_S$ since n is a monomorphism. So l, r are isomorphisms and the square $k \circ t = h \circ k'$ is a pullback. But the squares $k' \circ c' = b' \circ d'$ and $k \circ c = b \circ d$ are also pullbacks by definition of the closure operation. Applying 2.5.9 of volume 1, (associativity of pullbacks), we deduce that $d \circ s = f \circ d'$ is a pullback, or in other words $\overline{f^{-1}(A)} = f^{-1}(\overline{A})$.

It remains to prove that we have defined a bijective correspondence. Let us start with the universal closure operation and consider the pair $(\mathcal{T}, \mathcal{F})$ defined in 1.12.6. Write $\overline{(-)}$ for the original closure operation and $\widetilde{(-)}$ for that deduced from the pair $(\mathcal{T}, \mathcal{F})$, as in diagram 1.22. Given a subobject $A \rightarrowtail B$, consider diagram 1.51 where the bottom sequence is exact with $\overline{0} \in \mathcal{T}$ and $(B/A)/\overline{0} \in \mathcal{F}$; see proof of 1.12.5. The pair (c, d) is the pullback defining $\widetilde{A}$ and the left-hand square is defined as a pullback, i.e. $e = \mathsf{Ker}\, c$. Since $b^{-1}(0) = \mathsf{Ker}\, b = A$, one has in fact $A' = A$. By universality of the original closure operation (see 5.7.1.(4), volume 1), one obtains $\widetilde{A} = \overline{A'} = \overline{A}$.

Conversely start with an hereditary torsion theory $(\mathcal{T}, \mathcal{F})$, construct the associated closure operation $\overline{(-)}$ and the corresponding pair $(\mathcal{T}', \mathcal{F}')$, as in 1.12.6. Given $B \in \mathcal{B}$ consider the closure $\overline{0} \rightarrowtail B$ of $0 \rightarrowtail B$ as constructed in diagram 1.45. Since the cokernel of $0 \rightarrowtail B$ is just the identity on B (see 1.1.8), this closure is given by the pullback square of diagram 1.52 where the bottom sequence is exact, with $T \in \mathcal{T}$ and $F \in \mathcal{F}$; see 1.12.1. Since the pullback of an identity is an identity, $c = 1_T$ and thus $\overline{0} = T$. Since 0 is dense in $\overline{0}$, $\overline{0} \in \mathcal{T}'$; see 1.12.6.

If $B \in \mathcal{T}'$, then $\overline{0} = B$; see 1.12.6. Thus $B = \overline{0} = T \in \mathcal{T}$. Conversely if $B \in \mathcal{T}$, the sequence

$$0 \longrightarrow B \xrightarrow{1_B} B \longrightarrow 0 \longrightarrow 0$$

is exact with $B \in \mathcal{T}$ and $0 \in \mathcal{F}$; see 1.12.3 and 1.8.5. Thus $B = T$ by

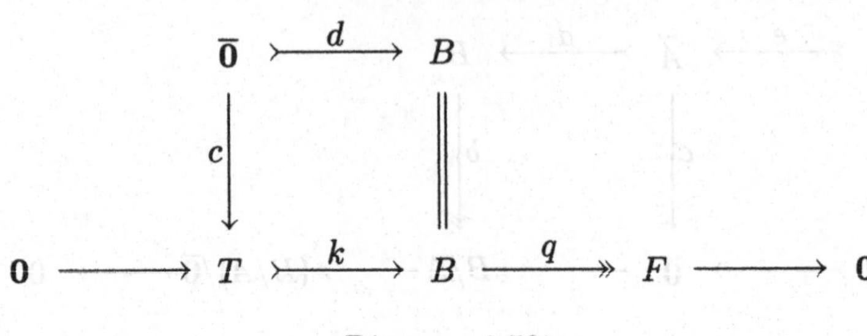

Diagram 1.52

uniqueness of such a sequence (see 1.12.4). But then $B = T = \overline{0} \in \mathcal{T}'$.

Since in a torsion theory $(\mathcal{T}, \mathcal{F})$, the class $\mathcal{F}$ is completely determined by $\mathcal{T}$ (see 1.12.5), $\mathcal{T} = \mathcal{T}'$ implies $\mathcal{F} = \mathcal{F}'$. □

It should be observed that in the construction of the universal closure operation associated with an hereditary torsion theory, the hereditary condition has been used just to prove the axiom $\overline{f^{-1}(A)} = f^{-1}(\overline{A})$. So an arbitrary torsion theory induces a closure operation satisfying the three axioms:

(1) $S \subseteq \overline{S}$;
(2) $S \subseteq T \;\Rightarrow\; \overline{S} \subseteq \overline{T}$;
(3) $\overline{\overline{S}} = \overline{S}$.

But it is false in general that such a closure operation is induced by a torsion theory (see exercise 1.15.15).

1.13 Localizations of abelian categories

For "good" abelian categories $\mathcal{B}$, the localizations of $\mathcal{B}$ (see 3.5.5, volume 1) are exactly described by the hereditary torsion theories on $\mathcal{B}$ or, equivalently, by the universal closure operations on $\mathcal{B}$ (see 1.12.8). In this section, "good" will mean "locally finitely presentable". The reader should refer to chapter 5 for the theory of locally presentable categories. In fact this assumption will just be used in this section through the references to chapter 5, volume 1.

In an abelian category, the bidense morphisms (see 5.8.1, volume 1) with respect to a universal closure operation can be described in a somehow simpler way.

Proposition 1.13.1 *Consider an abelian category $\mathcal{B}$ provided with a universal closure operation. Given a morphism f of $\mathcal{B}$, the following conditions are equivalent:*

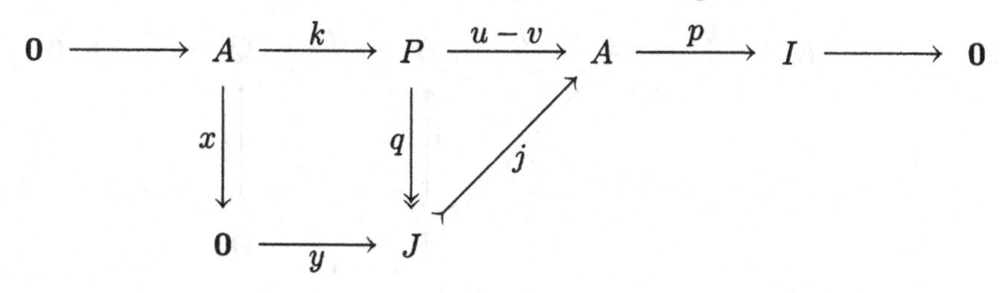

Diagram 1.53

(1) *f is bidense;*
(2) *(a) the image of f is dense,*
(b) the monomorphism $\mathbf{0} \longrightarrow \mathsf{Ker}\, f$ is dense.

Proof Consider the image factorization $f = i \circ p$ of f. In both cases i is required to be dense.

Since i is a monomorphism, the kernel pair (u, v) of f coincides with that of p. But since p is an epimorphism and every epimorphism is regular (see 1.4.1), $p = \mathsf{Coker}\,(u, v)$ (see 2.5.7, volume 1) or in other words $p = \mathsf{Coker}\,(u - v)$; see 1.2.8. On the other hand if $k = \mathsf{Ker}\,(u, v)$, $k = \mathsf{Ker}\,(u-v)$; see 1.2.8. Therefore the upper exact sequence in diagram 1.53 is exact.

Considering the image J of $u - v$, we know that $j = \mathsf{Ker}\, p$; (see 1.8.1) and $q = \mathsf{Coker}\, k$; see 1.8.3. Since i is a monomorphism, the kernel of f is also that of p, thus it is j. Since $q = \mathsf{Coker}\, k$ and k is a monomorphism, $k = \mathsf{Ker}\, q$; see 1.5.7. Thus $k = q^{-1}(\mathbf{0})$ and the square is a pullback. If y is dense, so is k (see 5.7.4, volume 1) and if k is dense so is y; see 5.7.8, volume 1. □

Proposition 1.13.2 *Consider an abelian category $\mathscr{B}$ provided with a universal closure operation. Write $(\mathcal{T}, \mathcal{F})$ for the corresponding hereditary torsion theory (see 1.12.6). Given a morphism $f \in \mathscr{B}$, the following conditions are equivalent:*

(1) *f is bidense;*
(2) *the objects $\mathsf{Ker}\, f$ and $\mathsf{Coker}\, f$ belong to $\mathcal{T}$.*

Proof For a morphism $f \colon A \longrightarrow B$, consider the situation of diagram 1.54, where $k = \mathsf{Ker}\, f$, $q = \mathsf{Coker}\, f$ and $f = i \circ p$ is the image factorization. Observe that $I \longrightarrow\!\!\!\!\!\rightarrow \mathbf{0}$ is an epimorphism since $\mathbf{0}$ is initial and dually, $\mathbf{0} \longrightarrow Q$ is a monomorphism since $\mathbf{0}$ is final. We know that $p = \mathsf{Coker}\, k$ and dually $i = \mathsf{Ker}\, q$; see 1.5.5. But $q^{-1}(\mathbf{0}) = \mathsf{Ker}\, q = i$, so that the square is a pullback.

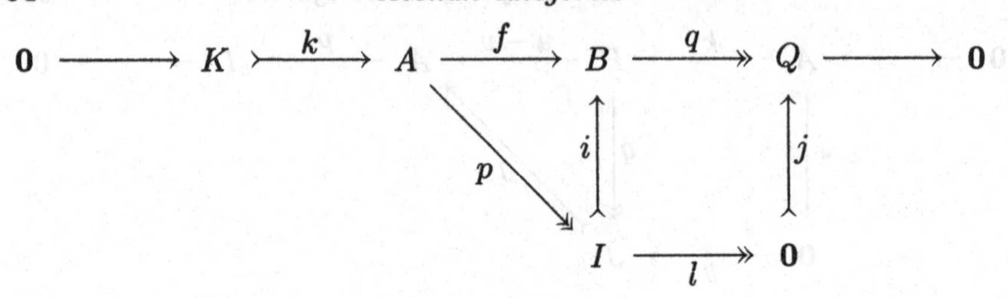

Diagram 1.54

If i is dense, so is j because q is a (regular) epimorphism (see 5.7.8, volume 1) and conversely if j is dense, so is i; see 5.7.4, volume 1. But f is bidense iff $\mathbf{0} \longrightarrow K$ and i are dense (see 1.13.1) thus iff $\mathbf{0} \longrightarrow K$ and $\mathbf{0} \longrightarrow Q$ are dense, i.e. iff K and Q are in $\mathcal{T}$; see 1.12.6. □

Definition 1.13.3 *Consider an abelian category $\mathscr{B}$ and a full replete subcategory $\mathcal{T} \subseteq \mathscr{B}$. Define a class $\Sigma \subseteq \mathscr{B}$ of morphisms by*

$$f \in \Sigma \text{ iff } \mathsf{Ker}\, f \in \mathcal{T}, \ \mathsf{Coker}\, f \in \mathcal{T}.$$

The subcategory $\mathcal{T}$ is called "localizing" when Σ is the class of those morphisms of $\mathscr{B}$ inverted by a localization $r \dashv i\colon \mathscr{A} \rightleftarrows \mathscr{B}$.

Let us prove now that for a locally finitely presentable abelian category $\mathscr{B}$, if $(\mathcal{T}, \mathcal{F})$ is an hereditary torsion theory, $\mathcal{T}$ is a localizing subcategory.

In order to apply theorem 5.8.8, volume 1, we first prove the following result.

Proposition 1.13.4 *If $\mathscr{B}$ is an abelian category provided with a universal closure operation, the class of bidense morphisms is stable under finite colimits (see 5.4.9, volume 1).*

Proof We consider a finite category $\mathscr{D}$, two functors $F, G\colon \mathscr{D} \rightrightarrows \mathscr{B}$ and a natural transformation $\alpha\colon F \Rightarrow G$ such that, for each object $D \in \mathscr{D}$, $\alpha_D\colon F(D) \longrightarrow G(D)$ is bidense. We must prove that the factorization

$$\operatorname{colim} \alpha\colon \operatorname{colim} F \longrightarrow \operatorname{colim} G$$

is bidense.

A finite colimit can be computed from finite coproducts and coequalizers (see 2.8.1, volume 1), thus from finite coproducts and cokernels (see 1.2.8).

The case where $\mathscr{D}$ is empty is obvious, since both colimits are the zero object and the factorization is just the identity on it.

The case of a finite coproduct follows at once from the case of a binary coproduct, just by associativity of coproducts (see 2.2.3, vol-

$$
\begin{array}{ccccccc}
0 & \longrightarrow & K_i & \xrightarrow{k_i} & A_i & \xrightarrow{\alpha_i} & B_i \\
 & & & & \Big\downarrow p_i & \nearrow j_i & \\
 & & & & I_i & &
\end{array}
$$

Diagram 1.55

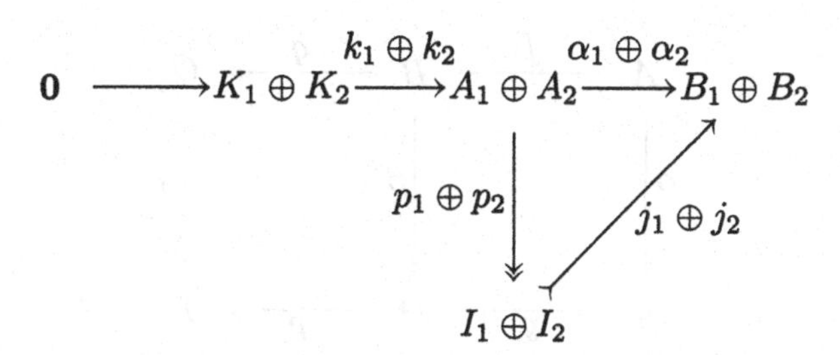

Diagram 1.56

ume 1). Coproducts are biproducts (see 1.6.5). Therefore given two bidense morphisms α_1, α_2 as in diagram 1.55, their images $\alpha_i = j_i \circ p_i$ and their kernels k_i, let us compute the various biproducts to get diagram 1.56. By interchange of products and kernels (see 2.12.1, volume 1), $k_1 \oplus k_2 = \mathsf{Ker}\,(\alpha_1 \oplus \alpha_2)$ and by interchange of coproducts and cokernels, $p_1 \oplus p_2 = \mathsf{Coker}\,(k_1 \oplus k_2)$. In particular $j_1 \oplus j_2$ is the image of $\alpha_1 \oplus \alpha_2$. To conclude the proof in this case, it suffices to prove that the monomorphisms $j_1 \oplus j_2$ and $\mathbf{0} \oplus \mathbf{0}$ are dense.

It is indeed a general fact that the biproduct of two dense monomorphisms j_1, j_2 is again dense. It suffices to consider diagram 1.57 where both quadrilaterals are pullbacks. Since the bottom monomorphisms are dense, the top ones are dense as well (see 5.7.4, volume 1) and thus also their composite (see 5.7.7, volume 1).

It remains to prove that bidense morphisms are stable under cokernels. For this consider diagram 1.58 where $g \circ \alpha = \beta \circ f$, $q = \mathsf{Coker}\, f$ and $p = \mathsf{Coker}\, g$. From $p \circ \beta \circ f = p \circ g \circ \alpha = 0$ we find γ such that $\gamma \circ q = p \circ \beta$. Assuming α, β bidense, we must prove that γ is too.

First of all, we reduce the problem to the case where f, g are monomorphisms. For this we factor f and g through their images $f = f_2 \circ f_1$, $g = g_2 \circ g_1$ and we consider the corresponding factorization $\delta\colon I \longrightarrow J$ given by 4.4.5, volume 1, and making the left-hand squares of dia-

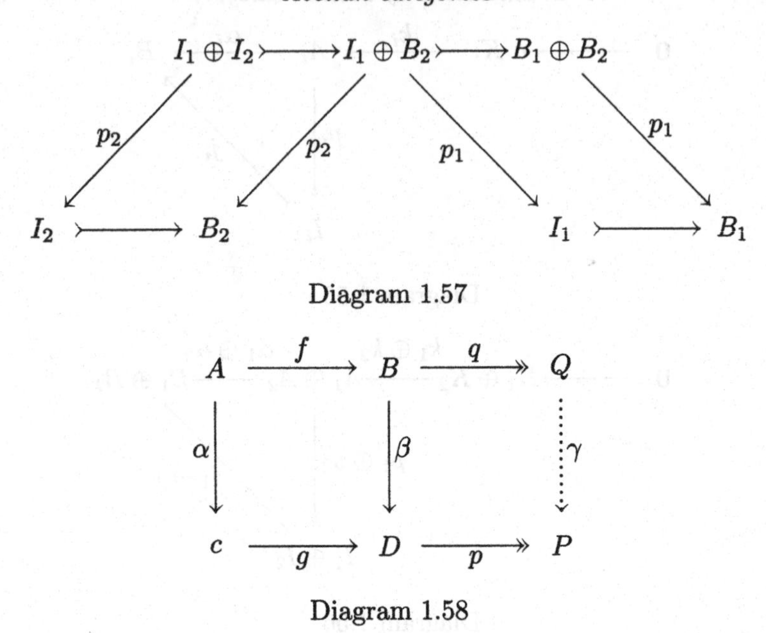

Diagram 1.57

Diagram 1.58

gram 1.59 commutative. Since f_1 and g_1 are epimorphisms, we still have $q = \mathsf{Coker}\, f_2$, $p = \mathsf{Coker}\, g_2$ with, now, f_2, g_2 monomorphisms and β bidense; we want to prove that δ is bidense as well. Considering the image factorizations $\alpha = \alpha_2 \circ \alpha_1$, $\beta = \beta_2 \circ \beta_1$, $\delta = \delta_2 \circ \delta_1$ of α, β, δ, we find again by 4.4.5, volume 1, factorizations h_1, h_2 making the corresponding squares of diagram 1.60 commutative. Since g_1 is an epimorphism and α_2 is dense, δ_2 is dense by 5.7.8, volume 1. Next we compute the kernels $\alpha_0, \beta_0, \delta_0$ of α, β, δ, which are also those of $\alpha_1, \beta_1, \gamma_1$ since $\alpha_2, \beta_2, \gamma_2$ are monomorphisms. Obvious diagram chasing (like in 1.10.1) shows the existence of morphisms k_1, k_2 making the corresponding squares of diagram 1.60 commutative. $\mathbf{0} \longrightarrow I_0$ is a monomorphism because $\mathbf{0}$ is final while k_2 is a monomorphism because so are δ_0 and f_2. $\mathbf{0} \longrightarrow B_0$ is dense because β is bidense (see 5.8.1, volume 1), thus $\mathbf{0} \longrightarrow I_0$ is dense as well (see 5.7.7, volume 1). This concludes the proof that δ is bidense.

Next we split the problem into two parts: the case where δ, β are both epimorphisms and the case where they are both monomorphisms. For this we refer further to diagram 1.60 and consider $r = \mathsf{Coker}\, h_2$. Notice that h_2 is still a monomorphism since so are g_2, δ_2. Obvious diagram chasing produces γ_1, γ_2 making the diagram commmutative and from

$$\gamma_2 \circ \gamma_1 \circ q = \gamma_2 \circ r \circ \beta_1 = p \circ \beta_2 \circ \beta_1 = p \circ \beta = \gamma \circ q$$

we deduce $\gamma = \gamma_2 \circ \gamma_1$ since q is an epimorphism. By 5.8.4, volume 1,

$$\begin{array}{ccccccc}
A & \overset{f_1}{\twoheadrightarrow} & I & \overset{f_2}{\rightarrowtail} & B & \overset{q}{\twoheadrightarrow} & Q \\
\downarrow{\scriptstyle\alpha} & & \downarrow{\scriptstyle\delta} & & \downarrow{\scriptstyle\beta} & & \downarrow{\scriptstyle\gamma} \\
C & \underset{g_1}{\twoheadrightarrow} & J & \underset{g_2}{\rightarrowtail} & D & \underset{p}{\twoheadrightarrow} & P
\end{array}$$

Diagram 1.59

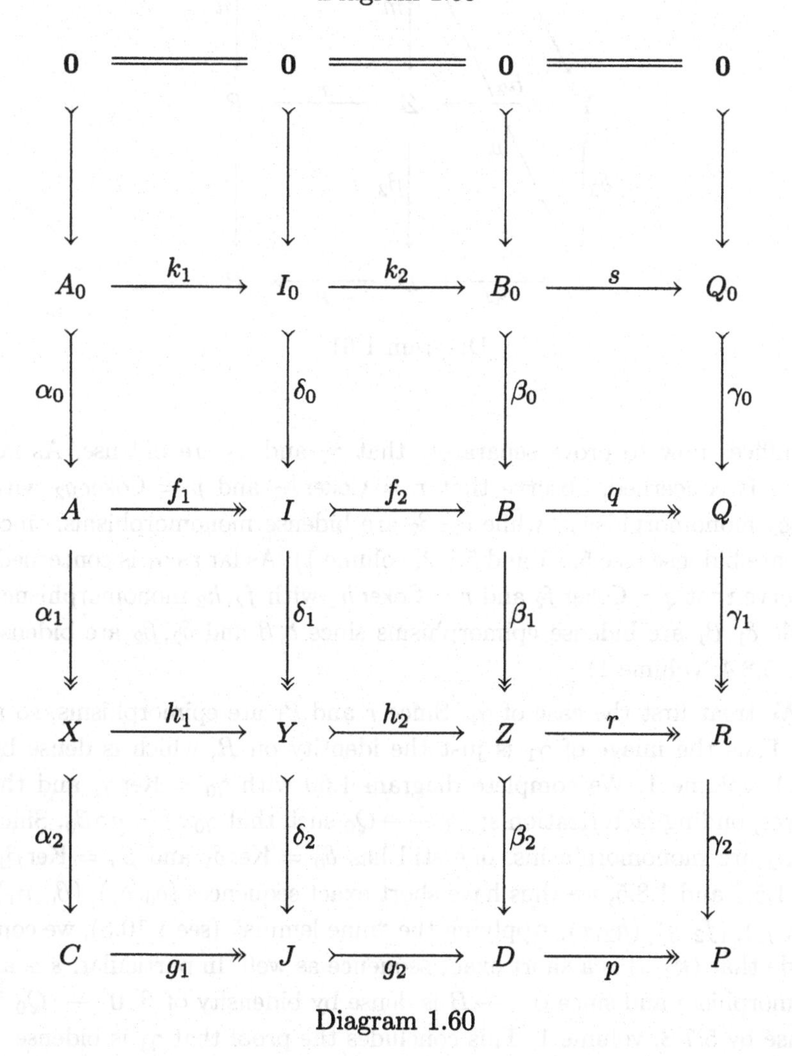

Diagram 1.60

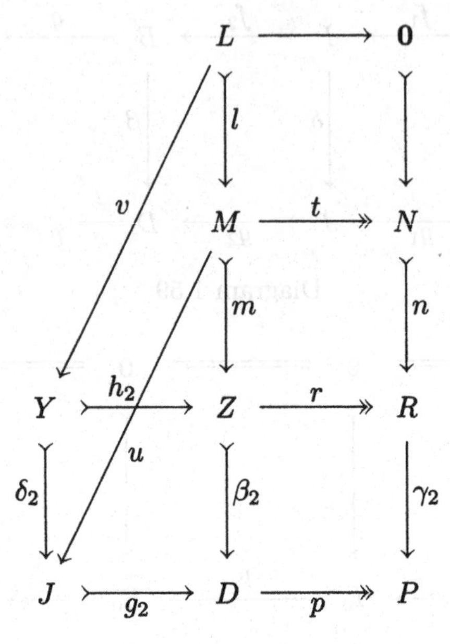

Diagram 1.61

it suffices now to prove separately that γ_1 and γ_2 are bidense. As far as γ_2 is concerned, observe that $r = \mathsf{Coker}\, h_2$ and $p = \mathsf{Coker}\, g_2$ with h_2, g_2 monomorphisms, while δ_2, β_2 are bidense monomorphisms, since δ, β are bidense (see 5.8.1 and 5.8.2, volume 1). As far as γ_1 is concerned, observe that $q = \mathsf{Coker}\, f_2$ and $r = \mathsf{Coker}\, h_2$ with f_2, h_2 monomorphisms, while δ_1, β_1 are bidense epimorphisms since δ, β and δ_2, β_2 are bidense (see 5.8.4, volume 1).

We treat first the case of γ_1. Since r and β_1 are epimorphisms, so is γ_1. Thus the image of γ_1 is just the identity on R, which is dense by 5.8.4, volume 1. We complete diagram 1.60 with $\gamma_0 = \mathsf{Ker}\, \gamma_1$ and the corresponding factorization $s \colon B_0 \longrightarrow Q_0$ such that $\gamma_0 \circ s = q \circ \beta_0$. Since δ_2, β_2 are monomorphisms, one still has $\delta_0 = \mathsf{Ker}\, \delta_1$ and $\beta_0 = \mathsf{Ker}\, \beta_1$. By 1.5.7 and 1.8.5, we thus have short exact sequences (δ_0, δ_1), (β_0, β_1), (γ_0, γ_1), (f_2, q), (h_2, r). Applying the "nine lemma" (see 1.10.5), we conclude that (k_2, s) is a short exact sequence as well. In particular, s is an epimorphism and since $\mathbf{0} \longrightarrow B$ is dense by bidensity of β, $\mathbf{0} \longrightarrow Q_0$ is dense by 5.7.8, volume 1. This concludes the proof that γ_1 is bidense.

To prove that γ_2 is bidense, we observe first that since β_2 is a dense monomorphism and p is an epimorphism, the image of γ_2 is dense by 5.7.8, volume 1. Next we construct diagram 1.61 where $n = \mathsf{Ker}\, \gamma_2$ and

(m, t) is the pullback of (r, n). The arrow t is an epimorphism since r is (see 1.7.6). From

$$p \circ \beta_2 \circ m = \gamma_2 \circ r \circ m = \gamma_2 \circ n \circ t = 0$$

we get a unique $u: M \longrightarrow J$ such that $g_2 \circ u = \beta_2 \circ m$, because $g_2 = \mathsf{Ker}(\mathsf{Coker}\, g_2) = \mathsf{Ker}\, p$; see 1.5.7. The pair (l, v) is defined as the pullback of (δ_2, u), so that l is a dense monomorphism because so is δ_2; see 5.7.4. Observe next that

$$\beta_2 \circ m \circ l = g_2 \circ u \circ l = g_2 \circ \delta_2 \circ v = \beta_2 \circ h_2 \circ v$$

from which $m \circ l = h_2 \circ v$, since β_2 is a monomorphism. Therefore

$$n \circ t \circ l = r \circ m \circ l = r \circ h_2 \circ v = 0$$

since $r = \mathsf{Coker}\, h_2$. This implies $t \circ l = 0$ because n is a monomorphism. So the upper square in diagram 1.61 is commutative, with l a dense monomorphism and t an epimorphism. Again by 5.7.8, volume 1, $\mathbf{0} \longrightarrow N$ is dense and finally, γ_2 is bidense. □

Theorem 1.13.5 *Consider a locally finitely presentable abelian category $\mathscr{B}$. There are bijections between:*
(1) the localizations of $\mathscr{B}$;
(2) the universal closure operations on $\mathscr{B}$;
(3) the hereditary torsion theories on $\mathscr{B}$;
(4) the localizing subcategories of $\mathscr{B}$.

Proof By 5.8.8, volume 1, and 1.13.4, this volume, we have a bijection between the localizations and the universal closure operations. By 1.12.8, there exists also a bijection between the universal closure operations and the hereditary torsion theories.

Every localization corresponds to an hereditary torsion theory $(\mathcal{T}, \mathcal{F})$ and the morphisms f inverted by the reflection are precisely those with $\mathsf{Ker}\, f \in \mathcal{T}$, $\mathsf{Coker}\, f \in \mathcal{T}$; see 1.13.2, this volume, and 5.8.8, volume 1. Thus $\mathcal{T}$ is a localizing subcategory. Observe that two different localizations induce different localizing subcategories, just because each one of the classes $\mathcal{T}, \mathcal{F}$ characterizes the other one (see 1.12.5). We have thus defined an injection from the class of localizations to that of localizing subcategories.

But this is also a surjection. Indeed every localizing subcategory $\mathcal{T}$, by definition, induces a localization. Considering the corresponding universal closure operation, a monomorphism $\mathbf{0} \longrightarrow B$ is inverted when it is bidense (see 5.8.8, volume 1), thus when it is dense (see 5.8.2, volume 1).

On the other hand this monomorphism is inverted iff its kernel and its cokernel are in $\mathcal{T}$; see 1.13.3. The kernel of the monomorphism $\mathbf{0} \longrightarrow B$ is $\mathbf{0} = \mathbf{0}$ (see 1.1.7) and the cokernel of the zero morphism $\mathbf{0} \longrightarrow B$ is $B = B$ (see 1.1.8). Since $\mathbf{0} = \mathbf{0}$ is inverted by the localization, its kernel and its cokernel are in $\mathcal{T}$, thus $\mathbf{0} \in \mathcal{T}$. So finally $\mathbf{0} \longrightarrow B$ is dense iff $B \in \mathcal{T}$. Applying 1.12.6, we conclude that the hereditary torsion theory associated with the localization has indeed the form $(\mathcal{T}, \mathcal{F})$, for $\mathcal{T}$ the original localizing category. □

Counterexample 1.13.6
An abelian group T is divisible when for each $a \in T$ and each $n \in \mathbb{N}^*$, there exists $b \in T$ such that $nb = a$. An abelian group F is reduced when the only divisible subgroup of F is (0). Write $\mathcal{T}$ for the full subcategory of divisible groups and $\mathcal{F}$ for the full subcategory of reduced groups. We show $(\mathcal{T}, \mathcal{F})$ is a torsion theory, but not an hereditary torsion theory.

First of all observe that given a divisible abelian group T and a morphism of abelian groups $f\colon T \longrightarrow F$, the image $f(T)$ is still a divisible group. When F is reduced, $f(T) = \mathbf{0}$, proving that $f = 0$.

Next observe that given a family of divisible subgroups $T_i \subseteq A$, $i \in I$, their sum $+_{i\in I} T_i \subseteq A$ is still a divisible subgroup. Indeed an element $t \in +_{i\in I} T_i$ has the form $t_1 + \ldots + t_j$ with $t_k \in T_{i_k}$; given $n \in \mathbb{N}^*$, choose $s_k \in T_{i_k}$ such that $ns_k = t_k$. It follows that $n(s_1 + \ldots + s_j) = t$. Given an abelian group A, write T for the biggest divisible subgroup of A, i.e. the sum of all the divisible subgroups of A. Consider the exact sequence

$$\mathbf{0} \longrightarrow T \rightarrowtail^{k} A \xrightarrow{q} A/T \longrightarrow \mathbf{0}.$$

Let us prove that the quotient A/T is reduced. If $X \subseteq A/T$ is a divisible subgroup, then $q^{-1}(X)$ is divisible since, given $a \in q^{-1}(X)$ and $n \in \mathbb{N}^*$, we find $[a] \in X$ and thus $[b] \in X$ with $n[b] = [a]$. This means that $b \in q^{-1}(X)$ with $nb - a \in T$. Since T is divisible, choose $t \in T$ such that $nb - a = nt$. This indicates that $a = nb - nt = n(b - t)$. But $q(b-t) = q(b) - q(t) = q(b) \in X$, thus $b - t \in q^{-1}(X)$. This proves that $q^{-1}(X)$ is divisible, and thus contained in T. As a consequence, $X = (0)$ and A/T is reduced.

Thus $(\mathcal{T}, \mathcal{F})$ is a torsion theory. It is not hereditary because the group $\mathbb{Q}$ of rational numbers is divisible, while the subgroup $\mathbb{Z}$ of integers is not.

1.14 The embedding theorem

This section is devoted to the proof that every small abelian category has an exact, full and faithful embedding in a category of modules. Let us first observe that:

Proposition 1.14.1 *Given a small abelian category and a filtered diagram of left exact additive functors* $F_i\colon \mathscr{A} \longrightarrow \mathsf{Ab}$*, the pointwise colimit* $\operatorname{colim} F_i$ *is again left exact and additive.*

Proof By 2.12.1, volume 1, an arbitrary colimit of functors preserving finite coproducts is again a functor preserving finite coproducts. Therefore $\operatorname{colim} F_i$ is additive (see 1.3.4).

By 2.13.6, volume 1, a filtered colimit of functors $F_i\colon \mathscr{A} \longrightarrow \mathsf{Ab}$ preserving finite limits is again a functor preserving finite limits. Thus $\operatorname{colim} F_i$ is left exact (see 1.11.2). □

Given a small abelian category $\mathscr{A}$, we construct first a faithful and exact embedding $U\colon \mathscr{A} \longrightarrow \mathsf{Ab}$ in the category of abelian groups. By 1.3.8, U can in any case be described as a colimit $U = \operatorname{colim} \mathscr{A}(A_i, -)$ of additive Ab-valued representable functors. We shall thus define U as the colimit of a composite

$$\mathcal{D} \xrightarrow{\ \phi\ } \mathscr{A} \xrightarrow{\ Y\ } \mathsf{Add}(\mathscr{A}, \mathsf{Ab})$$

where

(1) Y is the contravariant Yoneda embedding defined on the objects by $Y(A) = \mathscr{A}(A, -)$,

(2) $\mathcal{D}$ and ϕ are chosen in a way which ensures the exactness and faithfulness of U.

Lemma 1.14.2 *With the previous notation,* U *is left exact as long as* $\mathcal{D}$ *is cofiltered.*

Proof Each representable functor $\mathscr{A}(A, -)\colon \mathscr{A} \longrightarrow \mathsf{Ab}$ is left exact, because the representable functor $\mathscr{A}(A, -)\colon \mathscr{A} \longrightarrow \mathsf{Set}$ is (see 2.9.5, volume 1) and the forgetful functor $\mathsf{Ab} \longrightarrow \mathsf{Set}$ reflects limits (see 2.9.8.d, volume 1).

Since Y is contravariant, the cofilteredness of $\mathcal{D}$ implies that U is a filtered colimit of representable, thus left exact, functors. So U is left exact by 1.14.1. □

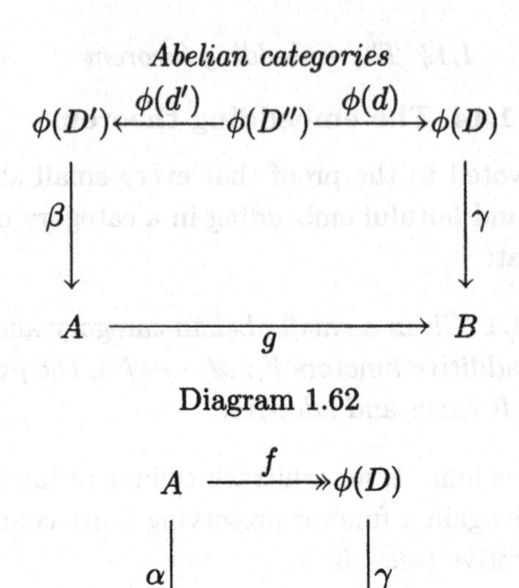

Diagram 1.62

Diagram 1.63

Lemma 1.14.3 *With the previous notation and supposing that the category $\mathcal{D}$ is cofiltered, U preserves epimorphisms as soon as every epimorphism of the form $f\colon A \longrightarrow \phi(D)$ in $\mathscr{A}$ can be written as $f = \phi(d)$, with $d\colon D' \longrightarrow D$ in $\mathcal{D}$.*

Proof Consider an epimorphism $g\colon B \longrightarrow C$ in $\mathscr{A}$; $U(g)$ is an epimorphism precisely if it is a surjection (see 1.8.5.e, volume 1). Recall that filtered colimits in **Ab** are computed as in **Set** (see 2.13.6 and 2.13.3, volume 1).

Choose now $x \in U(C) \cong \operatorname{colim} \mathscr{A}\big(\phi(D), C\big)$. The element x is thus the class of some arrow $\gamma\colon \phi(D) \longrightarrow C$. For $U(g)$ being surjective, we must find an element $y \in U(B) \cong \operatorname{colim} \mathscr{A}\big(\phi(D), B\big)$ mapped to x. The element y is the equivalence class of some arrow $\beta\colon \phi(D') \longrightarrow B$ and the relation $U(g)(y) = x$ means that $g \circ \beta$ and γ are equal "at some further level D''". More precisely, $U(g)(y) = x$ if one can find $D'' \in \mathcal{D}$ and $d\colon D'' \longrightarrow D$, $d'\colon D'' \longrightarrow D'$ such that $g \circ \beta \circ \phi(d') = \gamma \circ \phi(d)$, as in diagram 1.62.

Let us consider the pullback of diagram 1.63. By 1.7.6 f is an epimorphism and, by assumption, can be written as $f = \phi(\overline{d})$, with $\overline{d}\colon \overline{D} \longrightarrow D$ in $\mathcal{D}$. Putting $D' = \overline{D}$, $\beta = \alpha$, $D'' = \overline{D}$, $d' = 1_{\overline{D}}$ and $d = \overline{d}$, the required conditions for having $U(g)(y) = x$ are now satisfied. □

Lemma 1.14.4 *With the previous notation and supposing $\mathcal{D}$ cofiltered,*

U is faithful as long as

(1) every object $A \in \mathscr{A}$ has the form $A = \phi(D)$ for some $D \in \mathcal{D}$,

(2) for every morphism $d \in \mathcal{D}$, $\phi(d)$ is an epimorphism in $\mathscr{A}$.

Proof Consider two morphisms $f, g\colon A \rightrightarrows B$ in $\mathscr{A}$ such that $U(f) = U(g)$. Since $U(A) \cong \operatorname{colim} \mathscr{A}(\phi(D), A)$, an element $a \in U(A)$ is the equivalence class of some arrow $\alpha\colon \phi(D) \longrightarrow A$. The equality $U(f)(a) = U(g)(a)$ means that $f \circ \alpha$ and $g \circ \alpha$ are equal "at some further level D'". More precisely, $U(f)(a) = U(g)(a)$ means the existence of an arrow $d\colon D' \longrightarrow D$ in $\mathcal{D}$ such that $f \circ \alpha \circ \phi(d) = g \circ \alpha \circ \phi(d)$.

Now assume $u(f) = U(g)$. If A is of the form $\phi(\overline{D})$ for $\overline{D} \in \mathcal{D}$, $1_A\colon \phi(\overline{D}) \longrightarrow A$ corresponds to some element $\overline{a} \in U(A)$. By the first part of the proof we can find $d \in \mathcal{D}$ such that $f \circ 1_A \circ \phi(d) = g \circ 1_A \circ \phi(d)$, which implies $f = g$ if $\phi(d)$ is an epimorphism. □

Theorem 1.14.5 (The faithful embedding theorem)
Every small abelian category admits a faithful and exact embedding in the category of abelian groups.

Proof Using the previous notation, it remains to construct a small category $\mathcal{D}$ and a functor $\phi\colon \mathcal{D} \longrightarrow \mathscr{A}$ which satisfies the conditions of the three previous lemmas. We shall do this by constructing a sequence (indexed by the natural numbers) of posets

$$\mathcal{D}_0 \subseteq \mathcal{D}_1 \subseteq \ldots \subseteq \mathcal{D}_n \subseteq \ldots$$

and a corresponding sequence of functors $\phi_n\colon \mathcal{D}_n \longrightarrow \mathscr{A}$, satisfying the following conditions:

(1) each $\mathcal{D}_n$ is a $\wedge$-semi-lattice;

(2) if $n \leq m$, ϕ_n and ϕ_m coincide on $\mathcal{D}_n$;

(3) $\forall n \in \mathbb{N}\ \forall d \in \mathcal{D}_n\ \ \phi_n(d)$ is an epimorphism;

(4) $\forall A \in \mathscr{A}\ \exists D \in \mathcal{D}_1\ \ \phi_1(D) = A$;

(5) $\forall A \in \mathscr{A}\ \forall D \in \mathcal{D}_n\ \forall f\colon A \twoheadrightarrow \phi_n(D)$
f epimorphism $\Rightarrow \exists d \in \mathcal{D}_{n+1},\ \ d\colon D' \longrightarrow D,\ \ \phi_{n+1}(d) = f$.

Defining $\mathcal{D} = \bigcup_{n \in \mathbb{N}} \mathcal{D}_n$, one gets at once an extension $\phi\colon \mathcal{D} \longrightarrow \mathscr{A}$ of the various ϕ_n and it is then obvious that the pair $(\mathcal{D}, \phi)$ fulfils the requirements.

We define $\mathcal{D}_0$ to be the one-point poset $\{*\}$; $\phi_0(*)$ is just the zero object of $\mathscr{A}$. The poset $\mathcal{D}_0$ is obviously an $\wedge$-semi-lattice and, since its only arrow is 1_*, its image under ϕ is clearly an epimorphism.

Suppose $\mathcal{D}_0, \ldots, \mathcal{D}_n$ have been defined and satisfy conditions (1), (2), (3), (5). Consider all the pairs (D, f) where $D \in \mathcal{D}_n$ and $f \in \mathscr{A}$ is an epimorphism with codomain $\phi_n(D)$. We index those pairs by the successive successor ordinals. We shall now construct a sequence of posets, indexed by the ordinals, up to the supremum of the ordinals used for the indexation,

$$\mathcal{D}_n^0 \subseteq \mathcal{D}_n^1 \subseteq \ldots \subseteq \mathcal{D}_n^\alpha \subseteq \ldots,$$

and a corresponding sequence of functors $\phi_n^\alpha \colon \mathcal{D}_n^\alpha \longrightarrow \mathscr{A}$, satisfying the following conditions:

(a) each $\mathcal{D}_n^\alpha$ is a $\wedge$-semi-lattice;
(b) if $\beta < \alpha$, ϕ_n^β and ϕ_n^α coincide on $\mathcal{D}_n^\beta$;
(c) $\forall \alpha \ \forall d \in \mathcal{D}_n^\alpha$, $\phi_n^\alpha(d)$ is an epimorphism;
(d) if the ordinal α indexes the pair (D, f) with $f \colon A \longrightarrow \phi_n(D)$, there exists $d \colon D' \longrightarrow D$ in $\mathcal{D}_n^\alpha$ such that $\phi_n^\alpha(d) = f$.

Defining $\mathcal{D}_{n+1} = \bigcup_\alpha \mathcal{D}_n^\alpha$, we get an obvious extension $d_{n+1} \colon \mathcal{D}_{n+1} \longrightarrow \mathscr{A}$ of the various ϕ_n^α and it is then clear that the pair $(\mathcal{D}_{n+1}, \phi_{n+1})$ satisfies conditions (1), (2), (3), (5).

We define $(\mathcal{D}_n^0, \phi_n^0)$ to be $(\mathcal{D}_n, \phi_n)$; $(\mathcal{D}_n^0, \phi_n^0)$ satisfies conditions (a), (b), (c) because $(\mathcal{D}_n, \phi_n)$ satisfies conditions (1), (3). For a limit ordinal α, $\mathcal{D}_n^\alpha = \bigcup_{\beta<\alpha} \mathcal{D}_n^\beta$ and ϕ_n^α is the obvious extension of the various ϕ_n^β. It is immediate that $\mathscr{D}_n^\alpha$ satisfies conditions (a), (b), (c) as soon as they are satisfied for every $\beta < \alpha$. Now suppose $(\mathcal{D}_n^\alpha, \phi_n^\alpha)$ is defined and $\alpha + 1$ indexes the pair (D, f). We consider

$$\downarrow D = \{D' \in \mathcal{D}_n^\alpha \mid D' \leq D\}$$

and perform the disjoint union $\mathcal{D}_n^\alpha \amalg \downarrow D$. For simplicity, when $D' \leq D$ in $\mathcal{D}_n^\alpha$, we write D'^* for the corresponding copy of D' in the second component $\downarrow D$ of the disjoint union. On $\mathcal{D}_n^\alpha \amalg \downarrow D$, we provide $\mathcal{D}_n^0$ and $\downarrow D$ with their original ordering and moreover we impose the relations $D'^* \leq D'$ for all $D' \leq D$. The poset structure generated in this way is given by the relations valid in $\mathcal{D}_n^\alpha$ and $\downarrow D$ together with the relation $D'^* \leq D''$ for all elements $D' \leq D$, $D' \leq D''$ in $\mathcal{D}_n^\alpha$. We choose $\mathcal{D}_n^\alpha \amalg \downarrow D$ with that poset structure as $\mathcal{D}_n^{\alpha+1}$. Since $\mathcal{D}_n^\alpha$ is a $\wedge$-semi-lattice, $\downarrow D$ is a $\wedge$-semi-lattice as well. Now choosing $D_1 \in \downarrow D$ and $D_2 \in \mathcal{D}_n^\alpha$, we compute the infimum $D_1 \wedge D_2$ in $\mathcal{D}_n^\alpha$ and $(D_1 \wedge D_2)^* = D_1^* \wedge D_2$ in $\mathcal{D}_n^{\alpha+1}$.

Let us define $\phi_n^{\alpha+1}$, which coincides with ϕ_n^α on $\mathcal{D}_n^\alpha$. We put

$$\phi_n^{\alpha+1}(D^*) = A \quad \text{and} \quad \phi_n^{\alpha+1}(D^* \leq D) = f.$$

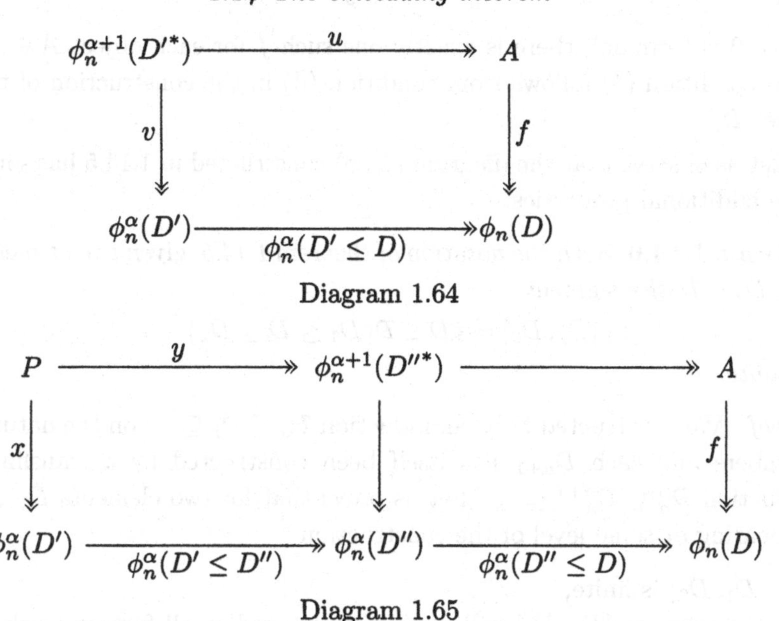

Diagram 1.64

Diagram 1.65

If $D' \leq D$ in $\mathcal{D}_n^\alpha$, we define $\phi_n^{\alpha+1}(D'^*)$ by the pullback of diagram 1.64, which takes place in $\mathscr{A}$. Notice that $\phi_n(D) = \phi_n^\alpha(D)$, thus considering $\phi_n^\alpha(D' \leq D)$ makes sense; since this morphism and f are in fact epimorphisms, u and v are epimorphisms as well (see 1.7.6). Since $A = \phi_n^\alpha(D^*)$, we put $u = \phi_n^{\alpha+1}(D'^* \leq D^*)$. Since $\phi_n^\alpha(D') = \phi_n^{\alpha+1}(D')$, we put $v = \phi_n^{\alpha+1}(D'^* \leq D')$.

If $D' \leq D'' \leq D$ in $\mathcal{D}_n^\alpha$, let us consider diagram 1.65 where both squares are pullbacks and thus all morphisms are epimorphisms. Since the outer square is a pullback and the lower composite is $\phi_n^\alpha(D' \leq D)$, one gets $P = \phi_n^{\alpha+1}(D'^*)$ and $x = \phi_n^{\alpha+1}(D'^* \leq D')$. We put $y = \phi_n^{\alpha+1}(D'^* \leq D''^*)$.

Finally if $D' \leq D$ and $D' \leq D''$ in $\mathcal{D}_n^\alpha$, let us define

$$\phi_n^{\alpha+1}(D'^* \leq D'') = \phi_n^{\alpha+1}(D' \leq D'') \circ \phi_n^{\alpha+1}(D'^* \leq D').$$

Just by associativity of pullbacks, $\phi_n^{\alpha+1}$ is a functor. By definition, each $\phi_n^{\alpha+1}(d)$ is an epimorphism and $\phi_n^{\alpha+1}$ extends ϕ_n^α, thus every ϕ_n^β with $\beta \leq \alpha$. And also by definition we have the relation $D^* \leq D$ in $\mathcal{D}_n^{\alpha+1}$ with $\phi_n^{\alpha+1}(D^* \leq D) = f$, which takes care of condition (d).

It remains to verify condition (4). The construction of $\mathcal{D}_1$ is indexed by all the pairs $(*, f)$, where $f\colon A \longrightarrow \phi_0(*) = 0$ is an epimorphism. Since $\mathbf{0}$ is initial, every morphism to $\mathbf{0}$ is an epimorphism and the construction is finally indexed by all the pairs $(*, f)$, where $f\colon A \longrightarrow \mathbf{0}$ is any morphism.

Since $\mathbf{0}$ is terminal, there is exactly one such f for each object $A \in \mathscr{A}$, thus condition (4) follows from condition (d) in the construction of the poset $\mathcal{D}_1$. □

Let us observe that the diagram $(\mathcal{D}, \phi)$ constructed in 1.14.5 has some nice additional properties.

Lemma 1.14.6 *With the notation of theorem 1.14.5, given two elements $D_1, D_2 \in \mathcal{D}$ the segment*

$$[D_1, D_2] = \{D \in \mathcal{D} | \, D_1 \leq D \leq D_2\}$$

is finite.

Proof We constructed $\mathcal{D}$ by an induction $\mathcal{D}_0 \subseteq \mathcal{D}_1 \subseteq \ldots$ on the natural numbers and each $\mathcal{D}_{n+1}$ has itself been constructed by a transfinite induction $\mathcal{D}_n^\alpha \subseteq \mathcal{D}_n^{\alpha+1} \subseteq \ldots$. Let us prove that for two elements D_1, D_2 appearing at some level of the construction

(1) $[D_1, D_2]$ is finite,
(2) the segment $[D_1, D_2]$ will remain unchanged at all further levels of the construction.

Condition (1) is certainly satisfied by $\mathcal{D}_0 = \{*\}$. Consider D_1, D_2 in $\mathcal{D}_n^{\alpha+1}$ and suppose $\mathcal{D}_n^\alpha$ satisfies condition (1). Let us observe that the construction of $\mathcal{D}_n^{\alpha+1}$ implies that one has never $D' \leq D^*$, for a new element D^* and an element $D' \in \mathcal{D}_n^\alpha$. Therefore if $D_1, D_2 \in \mathcal{D}_n^\alpha$, $[D_1, D_2]$ in $\mathcal{D}_n^{\alpha+1}$ is just $[D_1, D_2]$ in $\mathcal{D}_n^\alpha$; thus this segment is finite by induction and condition (2) has been verified for the passage from $\mathcal{D}_n^\alpha$ to $\mathcal{D}_n^{\alpha+1}$. If $D_1, D_2 \notin \mathcal{D}_n^\alpha$, then $D_1 = D^*$ and $D_2 = D'^*$ for $D, D' \in \mathcal{D}_n^\alpha$; $[D_1, D_2]$ in $\mathcal{D}_n^{\alpha+1}$ is isomorphic to $[D, D']$ in $\mathcal{D}_n^\alpha$ and thus is finite. If $D_1 \notin \mathcal{D}_n^\alpha$ and $D_2 \in \mathcal{D}_n^\alpha$, then $D_1 = D^*$ and $[D_1, D_2] = [D^*, D_2^*] \cup [D, D_2]$ where both subsegments are finite. The case $D_1 \leq D_2$ with $D_1 \in \mathcal{D}_n^\alpha$ and $D_2 \notin \mathcal{D}_n^\alpha$ is impossible as we have seen.

Thus conditions (1), (2) remain satisfied at each successor step. Because of that, they remain obviously satisfied at a limit step. □

Lemma 1.14.7 *With the notation of 1.14.5, given elements $D_1 \leq D_2$ in $\mathcal{D}$, the canonical factorization*

$$\phi(D_1) \longrightarrow \lim_{D_1 < D \leq D_2} \phi(D)$$

is an epimorphism.

Proof Again we prove this by induction on the level at which the object D_1 has been introduced. Since at a given level of the induction one never introduces elements greater than already existing elements, when D_1 is

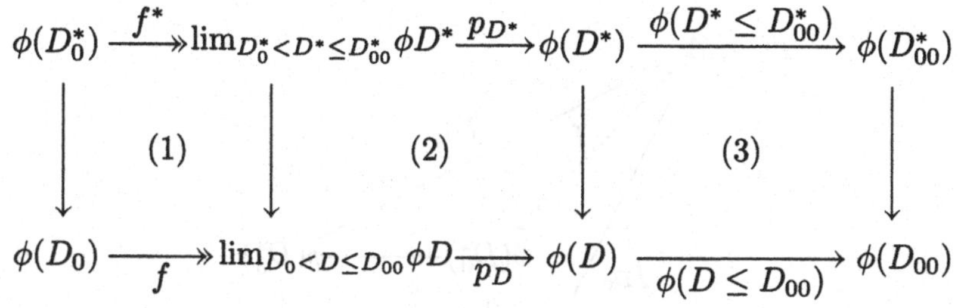

$$\begin{array}{ccccccc}
\phi(D_0^*) & \xrightarrow{f^*} & \lim_{D_0^* < D^* \leq D_{00}^*} \phi D^* & \xrightarrow{p_{D^*}} & \phi(D^*) & \xrightarrow{\phi(D^* \leq D_{00}^*)} & \phi(D_{00}^*) \\
\downarrow & (1) & \downarrow & (2) & \downarrow & (3) & \downarrow \\
\phi(D_0) & \xrightarrow[f]{} & \lim_{D_0 < D \leq D_{00}} \phi D & \xrightarrow[p_D]{} & \phi(D) & \xrightarrow[\phi(D \leq D_{00})]{} & \phi(D_{00})
\end{array}$$

Diagram 1.66

introduced, D_2 will be present as well and thus also all the D's such that $D_1 < D \leq D_2$.

If $D_1 \in \mathcal{D}_0$, $D_1 = * = D_2$ and the limit is the empty one. The factorization indicated is thus the identity on the zero morphism, which is indeed an epimorphism.

An object can never be introduced at a limit level in the construction. So it remains to consider the case where D_1 is introduced at the level $\mathcal{D}_n^{\alpha+1}$. There is some $D_0 \in \mathcal{D}_n^\alpha$ such that $D_1 = D_0^*$. There are two possibilities for D_2: namely, $D_2 \in \mathcal{D}_n^\alpha$ or $D_2 \notin \mathcal{D}_n^\alpha$.

If $D_1 = D_2 \notin \mathcal{D}_n^\alpha$, the limit is that of the empty diagram and the factorization is the canonical morphism $\phi(D_1) \longrightarrow \mathbf{0}$, which is an epimorphism since $\mathbf{0}$ is initial.

If $D_1 \neq D_2 \notin \mathcal{D}_n^\alpha$, $D_2 = D_{00}^*$ for some $D_{00} \in \mathcal{D}_n^\alpha$. Consider diagram 1.66. By construction of $(\mathcal{D}, \phi)$ each square (3) is a pullback and, by definition of a limit, $\phi(D \leq D_{00}) \circ p_D = p_{D_{00}}$ and $\phi(D^* \leq D_{00}^*) \circ p_{D^*} = p_{D_{00}^*}$. So there is one single composite rectangle (2)–(3), independent of the choice of D. This rectangle is a pullback since it is just the limit of the pullback rectangles (3) (see 2.12.1, volume 2). But since the outer rectangle is also a pullback by definition of $(\mathcal{D}, \phi)$, we conclude by 2.5.9, volume 1, that rectangle (1) is a pullback as well. By the induction hypothesis, f is an epimorphism and thus f^* is an epimorphism as well (see 1.7.6).

There remains the case $D_1 = D_0^*$, $D_2 \in \mathcal{D}_n^\alpha$. Suppose the index $\alpha + 1$ corresponds to the pair $(\overline{D}, \overline{f})$ (in the notation of 1.14.5). If $D_0 = \overline{D}$,

$$\{D \mid D_1 < D \leq D_2\} = [\overline{D}, D_2]$$

has an initial object $\overline{D}$, so the corresponding limit is just $\phi(\overline{D})$ (see 2.11.4, volume 1) and the factorization of the statement is just the epimorphism $\overline{f}$.

Finally let us suppose $D_1 = D_0^* < \overline{D}^*$ and $D_2 \in \mathcal{D}_n^\alpha$. We shall prove

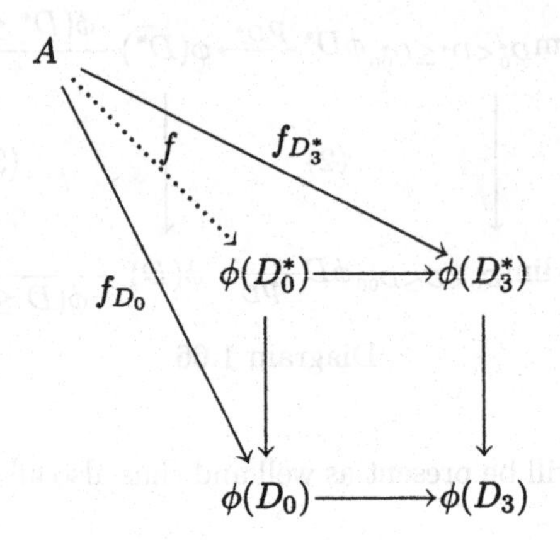

Diagram 1.67

that in this case the limit indicated is just $\phi(D_1)$ and the required factorization is the identity. Put $D_3 = D_2 \wedge \overline{D}$, which is an element of $\mathcal{D}_n^\alpha$. If $D_1 < D \leq D_2$ then, since $D_1 < \overline{D}$, $D_1 < D \wedge \overline{D} \leq D_2 \wedge \overline{D}$. This proves that

$$\{D \mid D_1 < D \leq D_3\} \text{ is initial in } \{D \mid D_1 < D < D_2\}$$

so that it suffices to compute the limit on this initial part (see 2.11.2, volume 1). We want thus to prove that the cone

$$\left(\phi(D_1), \phi(D_1 \leq D)_{D_1 < D \leq D_3}\right)$$

is a limit cone. To do this, take another cone

$$\Big(A, \big(f_D \colon A \longrightarrow \phi(D)\big)_{D_1 < D \leq D_3}\Big).$$

In diagram 1.67, the square is a pullback by definition. Since $D_0^* < D_3$, one also has $D_0^* < D_0 \leq D_3$ and $D_0^* < D_3^* \leq D_3$ by construction of the poset $\mathcal{D}$. Thus the outer part of diagram 1.67 commutes, from which we get a unique factorization $f \colon A \longrightarrow \phi(D^*)$ making the whole diagram commutative. Let us prove that this (unique) factorization works for every index D of the given cone. If $D_0^* < D \leq D_3$, one can have $D \in \mathcal{D}_n^\alpha$ or $D \notin \mathcal{D}_n^\alpha$. If $D \in \mathcal{D}_n^\alpha$, then $D_0 \leq D$ and one has

$$\phi(D_0^* \leq D) \circ f = \phi(D_0 \leq D) \circ \phi(D_0^* \leq D_0) \circ f = \phi(D_0 \leq D) \circ f_{D_0} = f_D$$

since D_0 and D are indices in the original cone. If $D \notin \mathcal{D}_n^\alpha$, then $D = D'^*$ with $D' \in \mathcal{D}_n^\alpha$ and $D_0^* < D_0 \leq D' \leq D_3$. We can then consider diagram 1.68 where both squares are pullbacks by definition. Since $D_0, D', D'^*, D_3, D_3^*$ are indices in the original cone,

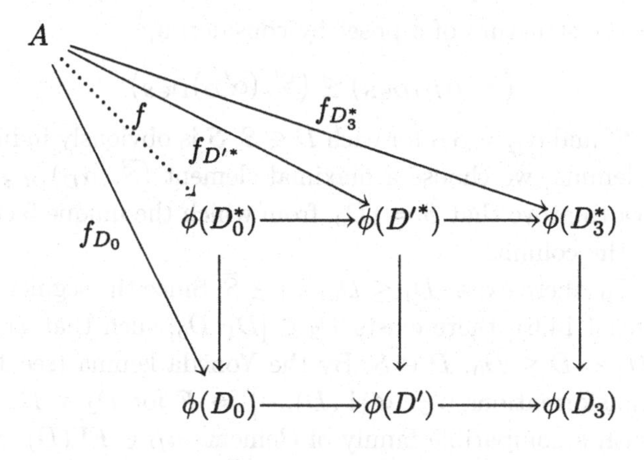

Diagram 1.68

$$\begin{aligned}\phi(D'^* \leq D_3^*) \circ \phi(D_0^* \leq D'^*) \circ f &= \phi(D'^* \leq D_3^*) \circ f \\ &= f_{D_3^*} = \phi(D'^* \leq D_3^*) \circ f_{D'},\\ \phi(D'^* \leq D') \circ \phi(D_0^* \leq D'^*) \circ f &= \phi(D_0 \leq D') \circ \phi(D_0^* \leq D_0) \circ f \\ &= \phi(D_0 \leq D') \circ f_{D_0} = f_{D'} = \phi(D'^* \leq D') \circ f_{D'^*}.\end{aligned}$$

Since the right-hand square in diagram 1.68 is a pullback, those relations imply $\phi(D_0^* \leq D'^*) \circ f = f_{D'^*}$ as required. □

Lemma 1.14.8 *With the notation of 1.14.5, consider an object $D_0 \in \mathcal{D}$ and a functor $\Gamma\colon \downarrow D_0 \longrightarrow \mathscr{A}$ such that for every $D_1 \leq D_2 \leq D_0$ the factorization*

$$\Gamma(D_1) \longrightarrow \lim_{D_1 < D \leq D_2} \Gamma(D)$$

is an epimorphism. In that case, given an exact functor $F\colon \mathscr{A} \longrightarrow \mathsf{Ab}$ and a natural transformation

$$\alpha_{D_0}\colon \mathscr{A}(\Gamma D_0, -) \Rightarrow F,$$

α_{D_0} can be factored as

$$\mathscr{A}\big(\Gamma(D_0), -\big) \xrightarrow{\ s_{D_0}\ } \operatorname{colim}_{D \leq D_0} \mathscr{A}\big(\Gamma(D), -\big) \xrightarrow{\ \alpha\ } F$$

for some α, with s_{D_0} the canonical morphism of the colimit.

Proof Let $\downarrow D_0$ indicate the set of elements $D \leq D_0$. By a final segment of $\downarrow D_0$ we mean a subset $S \subseteq \downarrow D_0$ such that $D \in S$ and $D \leq D' \leq D_0$ imply $D' \in S$.

Let us consider the set $\mathcal{S}$ of all pairs $\big(S, (\alpha_D)_{D \in S}\big)$ where

(1) S is a final segment of $\downarrow D_0$,
(2) $\alpha_D\colon \mathscr{A}\big(\Gamma(D), -\big) \Rightarrow F$, for $D \in S$, constitute a cocone.

We give S the structure of a poset by considering

$$\big(S,(\alpha_D)_{D\in S}\big) \leq \big(S',(\alpha'_D)_{D\in S}\big)$$

when $S \subseteq S'$ and $\alpha'_D = \alpha_D$ for each $D \in S$. S is obviously inductive and by Zorn's lemma, we choose a maximal element $\big(\overline{S},(\alpha_D)_{D\in S}\big)$ in it. It suffices now to prove that $\overline{S} = \downarrow D_0$, from which the unique factorization α through the colimit.

If $\overline{S} \neq \downarrow D_0$, there exists $D_1 \leq D_0$, $D_1 \notin \overline{S}$. Since the segment $[D_1, D_0]$ is finite (see 1.14.6), there exists $D_2 \in [D_1, D_0]$ such that $D_2 \notin \overline{S}$ but for each $D_2 < D \leq D_0$, $D \in \overline{S}$. By the Yoneda lemma (see 1.3.8), the natural transformations $\alpha_D\colon \mathscr{A}\big(\Gamma(D),-\big) \Rightarrow F$ for $D_2 < D \leq D_0$ correspond with a compatible family of elements $a_D \in F\Gamma(D)$. Since F is left exact and there are finitely many such D's, this compatible family corresponds with an element $a \in F\big(\lim \Gamma(D)\big)$. Since the factorization $\beta\colon \Gamma(D_2) \longrightarrow \lim \Gamma(D)$ is an epimorphism by assumption and since the functor F is exact, the morphism $F(\beta)\colon F\Gamma(D_2) \longrightarrow F\big(\lim \Gamma(D)\big)$ is surjective. Therefore we find $a_{D_2} \in F\Gamma(D_2)$ such that $F(a_{D_2}) = a$. In particular, composing with the projections, we find $F\Gamma(D_2 \leq D)(a_{D_2}) = a_D$ for each $D_2 < D \leq D_0$. But a_{D_2} corresponds by the Yoneda lemma with a natural transformation $\alpha_{D_2}\colon \mathscr{A}(\Gamma D_2, -) \Rightarrow F$ such that

$$\alpha_D = \alpha_{D_2} \circ \mathscr{A}\big(\Gamma(D_2 \leq D), -\big).$$

It remains to observe that, due to the choice of D_2, $\overline{S} \cup \{D_2\}$ is final in $\downarrow D_0$. Thus $\Big(\overline{S} \cup \{D_2\}, (\alpha_D)_{D\in \overline{S}\cup\{D_2\}}\Big)$ extends strictly $\big(\overline{S},(\alpha_D)_{D\in\overline{S}}\big)$, which contradicts its maximality. □

We are now ready to prove the final result of this chapter.

Theorem 1.14.9 (The full and faithful embedding theorem) *Every small abelian category $\mathscr{A}$ has a full, faithful, and exact embedding in a category Mod_R of modules over a ring R.*

Proof We use the notation of 1.14.5.

Step 1: construction of the embedding

We consider the embedding $U\colon \mathscr{A} \longrightarrow \mathsf{Ab}$ of 1.14.5 and the abelian group $\mathsf{Nat}(U,U)$ of natural transformations on U (see 1.3.2). Taking the composition of natural transformations as a multiplication, the group $R = \mathsf{Nat}(U,U)$ becomes a ring, by preadditivity of $\mathscr{A}$.

Given an object $A \in \mathscr{A}$, $U(A)$ is naturally provided with a scalar multiplication

$$R \times U(A) \longrightarrow U(A), \ \ (r,x) \mapsto r_A(x).$$

$$\begin{array}{ccccc}
\pi(D) & \overset{u_D}{\underset{v_D}{\rightrightarrows}} & \phi(D) & \xrightarrow{\phi(D \leq \delta_A)} & A \\
\Big\downarrow{\scriptstyle \pi(D \leq D')} & & \Big\downarrow{\scriptstyle \phi(D \leq D')} & & \Big\downarrow{\scriptstyle 1_A} \\
\pi(D') & \overset{u_{D'}}{\underset{v_{D'}}{\rightrightarrows}} & \phi(D') & \xrightarrow[\phi(D' \leq \delta_A)]{} & A
\end{array}$$

Diagram 1.69

Since r_A is a group homomorphism and the addition on $R = \mathsf{Nat}(U, U)$ is defined pointwise, this is an R-module structure.

Given an arrow $f\colon A \longrightarrow B$, the R-linearity of $U(f)$ reduces to the naturality of $r \in R$:

$$r_B\big(U(f)\big)(x) = \big(U(f)\big)r_A(x)$$

where again $x \in U(A)$.

Therefore U factors through Mod_R via a functor $V\colon \mathscr{A} \longrightarrow \mathsf{Mod}_R$. Since U is faithful, V is faithful. Since limits and colimits are computed in the same way in Mod_R and in Ab, the exactness of U implies that of V (see 1.11.12). It remains to prove that V is full.

Step 2: canonical presentation of representable functors

Given an object $A \in \mathscr{A}$ and the corresponding terminal epimorphism $p_A\colon A \longrightarrow \mathbf{0}$, we consider the unique element $* \in D_0$ and the index $\alpha + 1$ corresponding with the pair $(*, p_A)$ in the construction of $\mathcal{D}_1$ (see 1.14.5). We obtain $\mathcal{D}_0^{\alpha+1}$ by "duplicating" the initial segment $\downarrow\! * \subseteq \mathcal{D}_0^{\alpha}$, which is $\mathcal{D}_0^{\alpha}$ itself. Let us write δ_A for the new copy of $* \in \mathcal{D}_0$ which is so introduced. One has $\phi(\delta_A) = A$ and $\phi(\delta_A \leq *) = p_A$.

For every $D \leq \delta_A$ consider diagram 1.69 where (u_D, v_D) is the kernel pair of the morphism $\phi(D \leq \delta_A)$. If $D \leq D' \leq \delta_A$, consider in the same way the kernel pair $(u_{D'}, v_{D'})$ of $\phi(D' \leq \delta_A)$. The right-hand square of diagram 1.69 is commutative, from which we deduce the existence of a factorization $\pi(D \leq D')$ making $\pi\colon \downarrow\!\delta_A \longrightarrow \mathscr{A}$ a functor.

Since $\phi(D \leq \delta_A)$ is an epimorphism, it is a cokernel (see 1.4.1) and thus the coequalizer of its kernel pair (u_D, v_D); see 2.5.7, volume 1. But the contravariant Yoneda embedding transforms coequalizers in equalizers (see 2.9.5 and the fact that equalizers in Ab are computed as in

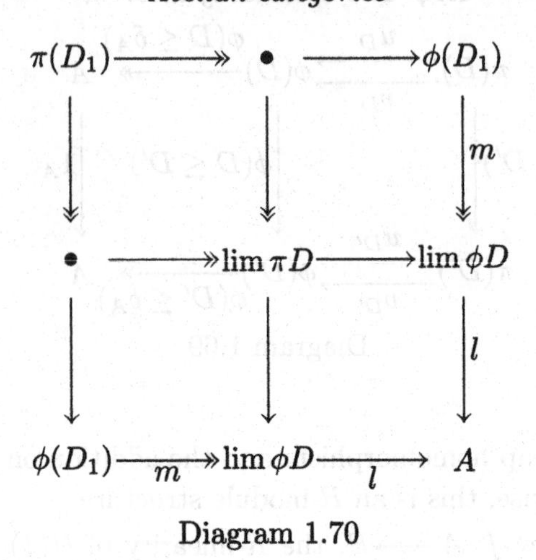

Diagram 1.70

Set). Thus we have an equalizer diagram

$$\mathscr{A}(\pi D,-) \overset{\mathscr{A}(u_D,-)}{\underset{\mathscr{A}(v_D,-)}{\leftleftarrows}} \mathscr{A}(\phi D,-) \xleftarrow{\mathscr{A}(\phi(D\leq\delta_A),-)} \mathscr{A}(A,-).$$

Let us compute the colimit on $D \leq \delta_A$ of each piece of this diagram:

$$P \overset{u}{\underset{v}{\leftleftarrows}} U \xleftarrow{w} \mathscr{A}(A,-).$$

The colimit of the central terms is indeed U, since $\downarrow\delta_A$ is an initial part of $\mathcal{D}$; see 2.11.2, volume 1. The colimit of the right-hand terms is the filtered colimit of the constant diagram on $\mathscr{A}(A,-)$, so it is obviously $\mathscr{A}(A,-)$. Since filtered colimits commute with equalizers in Ab, one has $w = \mathsf{Ker}\,(u,v)$; see 2.13.6, volume 1.

Step 3: the functor π satisfies the assumptions of 1.14.8

Consider $D_1 \leq D_2 \leq \delta_A$. By interchange of limits (see 2.12.1, volume 2) the following diagram is a kernel pair:

$$\lim_{D_1<D\leq D_2}\pi(D) \overset{\lim u_D}{\underset{\lim v_D}{\rightrightarrows}} \lim_{D_1<D\leq D_2}\phi(D) \xrightarrow{l} A.$$

Once again, we have used the obvious fact that the limit of the constant diagram on A is just A, because the indexing diagram is connected (see 2.6.7.e, volume 1).

Now it suffices to consider diagram 1.70 where all the squares are pullbacks and the factorization $m\colon \phi(D_1) \longrightarrow \lim_{D_1<D\leq D_2}\phi(D)$ is epimorphic by 1.14.7. The composite $l \circ m$ is just $\phi(D_1 \leq \delta_A)$ so that the outer pullback in diagram 1.70 is the one defining $\pi(D_1)$. Since m is

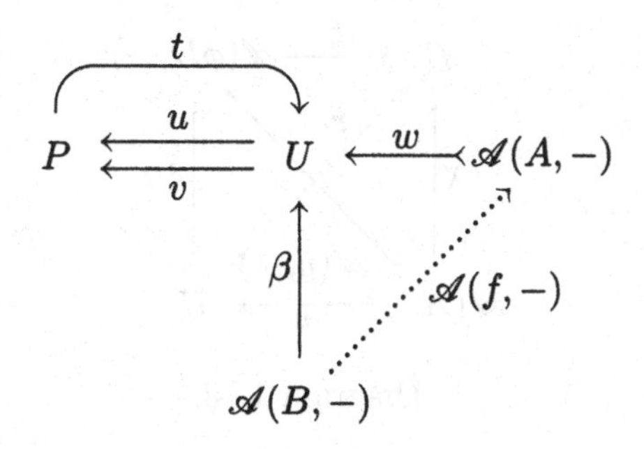

Diagram 1.71

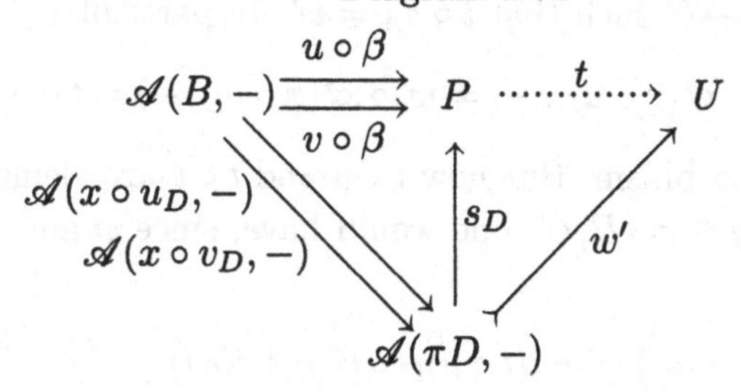

Diagram 1.72

an epimorphism, so are the four sides of the left upper square, which concludes the argument.

Step 4: V is full

Choose now $A, B \in \mathscr{A}$ and a group homomorphism $\varphi\colon U(A) \longrightarrow U(B)$ which is R-linear for the structures defined in step 1. The monomorphism $w\colon \mathscr{A}(A,-) \longrightarrow U$ of step 2 corresponds by the Yoneda lemma to some "generic" element $a = w_A(1_A) \in U(A)$. Let us consider the element $\varphi(a) \in U(B)$ which, by the Yoneda lemma, corresponds to a natural transformation $\beta\colon \mathscr{A}(B,-) \Rightarrow U$ defined by $\beta_C(g) = U(g)\big(\varphi(a)\big)$. Refering to diagram 1.71, we shall prove that $u \circ \beta = v \circ \beta$, from which there exists a unique factorization through $w = \mathsf{Ker}\,(u,v)$. By the Yoneda lemma, this factorization will have the form $\mathscr{A}(f,-)$ for some $f\colon A \longrightarrow B$.

Let us suppose $u \circ \beta \neq v \circ \beta$. β corresponds to $\varphi(a) \in U(B)$ via the Yoneda lemma and $\varphi(a) \in U(B) = \operatorname{colim}_{D \leq \delta_B} \mathscr{A}(\phi D, B)$ can be represented by some morphism $x\colon \phi(D) \longrightarrow B$ with $D \leq \delta_B$. If $u \circ \beta \neq v \circ \beta$, one must have $x \circ u_D \neq x \circ v_D$. In diagram 1.72, consider the monomorphism $w'\colon \mathscr{A}\big(\pi(D),-\big) \rightarrowtail U$ obtained in step 2, starting with $\pi(D)$ as object A. Because of step 3 and lemma 1.14.8, there exists a

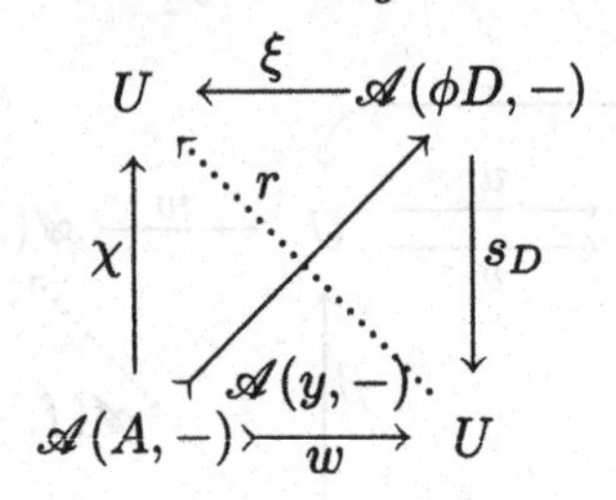

Diagram 1.73

factorization $t\colon P \longrightarrow U$ such that $t \circ s_A = w'$. In particular

$$t \circ u \circ \beta = w' \circ \mathscr{A}(x \circ u_D, -) \neq w' \circ \mathscr{A}(x \circ v_D, -) = t \circ v \circ \beta$$

since w' is a monomorphism. But now $t \circ u$ and $t \circ v$ are elements of R and given $C \in \mathscr{A}$, $g \in \mathscr{A}(B, C)$ one would have, since φ and $U(g)$ are R-linear,

$$\begin{aligned}
(t_C \circ u_C \circ \beta_C)(g) &= (t \circ u)_C \circ U(g)\big(\varphi(a)\big) \\
&= U(g) \circ \varphi \circ (t \circ u)_A(a) \\
&= U(g) \circ \varphi \circ t_A \circ u_A \circ w_A(1_A) \\
&= U(g) \circ \varphi \circ t_A \circ v_A \circ w_A(1_A) \\
&= U(g) \circ \varphi \circ (t \circ v)_A(a) \\
&= (t \circ v)_C \circ U(g)\big(\varphi(a)\big) \\
&= (t_C \circ v_C \circ \beta_C)(g).
\end{aligned}$$

Finally one would have $t \circ u \circ \beta = t \circ v \circ \beta$, which is a contradiction.

So we have indeed obtained a morphism $f\colon A \longrightarrow B$ and it remains to prove that $U(f) = \varphi$. Given any element $x \in U(A)$, we consider in diagram 1.73 the corresponding natural transformation $\chi\colon \mathscr{A}(A, -) \longrightarrow U$ obtained from the Yoneda lemma. Let us consider also the monomorphism w of step 1, which corresponds to some element of $U(A)$ represented by a morphism $y\colon \phi(D) \longrightarrow \mathscr{A}$. Since $s_D \circ \mathscr{A}(y, -) = w$ and w is a monomorphism, $\mathscr{A}(y, -)$ is a monomorphism as well. This implies that given $p, q\colon A \rightrightarrows B$ in $\mathscr{A}$,

$$\mathscr{A}(y, -) \circ \mathscr{A}(p, -) = \mathscr{A}(y, -) \circ \mathscr{A}(q, -) \Rightarrow p = q$$

or in other words $p \circ y = q \circ y \Rightarrow p = q$. Thus y is an epimorphism in $\mathscr{A}$ and, since U is exact, $U(y)\colon U\big(\phi(D)\big) \longrightarrow U(A)$ is an epimorphism in Ab, thus a surjection. Choose $z \in U\big(\phi(D)\big)$ such that $U(y)(z) = x$ and consider the corresponding natural transformation $\xi\colon \mathscr{A}\big(\phi(D), -\big) \Rightarrow$

U given by the Yoneda lemma. Since $U(y)(z) = x$, one has also $\xi \circ \mathscr{A}(y, -) = \chi$. By lemma 1.14.7, U satisfies the conditions of lemma 1.14.8 so that we obtain $r: U \Rightarrow U$ such that $r \circ s_D = \xi$. One has also

$$r \circ w = r \circ s_D \circ \mathscr{A}(y, -) = \xi \circ \mathscr{A}(y, -) = \chi.$$

But since φ is R-linear and $r \in R$, we obtain

$$\begin{aligned}
\varphi(x) &= \varphi(\chi_A(1_A)) \\
&= \varphi(r_A \circ w_A(1_A)) \\
&= r_B \circ \varphi(w_A(1_A)) \\
&= r_B \circ \varphi(a) \\
&= r_B \circ \beta_B(1_B) \\
&= r_B \circ w_B \circ \mathscr{A}(f, B)(1_B) \\
&= \chi_B(f) \\
&= U(f)(x).
\end{aligned}$$

This concludes the proof. □

Let us observe that since the embedding $V: \mathscr{A} \longrightarrow \mathsf{Mod}_R$ constructed in 1.14.9 is full and faithful, it reflects isomorphisms (see 1.9.5, volume 1). Since V is exact, V reflects finite limits and finite colimits (see 2.9.7, volume 1) thus in particular monomorphisms, epimorphisms and as a consequence the construction of images. Therefore V reflects exact sequences. Thus to prove a statement about exact sequences for all small abelian categories $\mathscr{A}$, it suffices to prove it for all categories of modules.

The case of arbitrary abelian categories $\mathscr{A}$ can generally be reduced to the case of small ones. Indeed if $\mathcal{B}$ is a diagram of objects and arrows in $\mathscr{A}$, let us consider the sequence

$$\mathscr{B}_0 \subseteq \mathscr{B}_1 \subseteq \cdots \subseteq \mathscr{B}_n \subseteq \cdots$$

where $\mathscr{B}_0$ is the full subcategory of $\mathscr{A}$ generated by $\mathcal{B}$ and $\mathscr{B}_{n+1}$ is the full subcategory of $\mathscr{A}$ generated by the limits and colimits of all finite diagrams in $\mathscr{B}_n$ (choose one limit and one colimit for each finite diagram). Putting $\mathscr{B} = \bigcup_{n \in \mathbb{N}} \mathscr{B}_n$, $\mathscr{B}$ is stable in $\mathscr{A}$ under finite limits and finite colimits, thus in particular $\mathscr{B}$ is an abelian subcategory of $\mathscr{A}$ (see 1.5.7). And when the diagram $\mathcal{B}$ is small (which is generally the case in the applications), the abelian category $\mathscr{B}$ is still small and so theorem 1.14.9 can be applied to $\mathscr{B}$, in which the whole problem takes place.

Theorem 1.14.9 can also be seen as a way to introduce diagram chasing on a small abelian category $\mathscr{A}$ (see section 1.9). Indeed, given an object $A \in \mathscr{A}$, define a pseudo-element $a \in^* A$ to be an actual element $a \in V(A)$. The action of a morphism $f\colon A \longrightarrow B$ on pseudo-elements is just given by $f(a) =_{\text{def}} V(f)(a)$. Since V preserves and reflects exact sequences, proposition 1.9.4 is certainly valid for this new notion of pseudo-element. But moreover, since V is full and faithful:

(1) it is now possible to prove the equality of two morphisms using pseudo-elements:

$$\forall f, g\colon A \rightrightarrows B \quad f = g \quad \text{iff} \quad \forall a \in^* A \;\; f(a) =^* g(a);$$

(2) it is now possible to construct a morphism $f\colon A \longrightarrow B$ by describing its action on pseudo-elements.

Moreover let us observe that since V preserves and reflects pullbacks, 1.9.5 now becomes an equivalence.

1.15 Exercises

1.15.1 In the category Gr of groups, the zero group is a zero object. Every epimorphism is the cokernel of its kernel. Every monomorphism is an equalizer, but only the normal subgroups are kernels.

1.15.2 Consider the category Set_b of sets with a base point: an object is a pair (A, a), with A a set and $a \in A$; a morphism $f\colon (A, a) \longrightarrow (B, b)$ is a mapping $f\colon A \longrightarrow B$ such that $f(a) = b$. The singleton is a zero object. Epimorphisms are just surjections. $f\colon (A, a) \longrightarrow (B, b)$ is a cokernel if and only if $f\colon A \setminus f^{-1}(b) \longrightarrow B \setminus \{b\}$ is bijective. Prove that proposition 1.2.3 holds in Set_b.

1.15.3 Consider the matrix notation described in section 1.2. Prove that the sum of two morphisms corresponds with the sum of matrices and the composite of two morphisms with the product of their matrices.

1.15.4 Consider a field K and an infinite-dimensional K-vector-space V. The ring $R = \mathsf{End}_{\mathbf{K}}(V, V)$ of $\mathbf{K}$-linear-endomorphisms of V, seen as preadditive category with a single object, has biproducts but no zero object.

1.15.5 Given an abelian category $\mathscr{A}$, consider the category $\mathsf{Ar}(\mathscr{A})$ of arrows of $\mathscr{A}$; see 1.2.7.c, volume 1. Observe that $\mathsf{Ar}(\mathscr{A})$ is abelian and prove the existence of an exact functor $I\colon \mathsf{Ar}(\mathscr{A}) \longrightarrow \mathscr{A}$ mapping an object $f\colon A \longrightarrow B$ of $\mathsf{Ar}(\mathscr{A})$ to its image factorization.

1.15.6 In the category Mon of abelian monoids, consider the monoid $(\mathbb{Z},+)$ of integers and its two subobjects $(\mathbb{N},+)$ and $(-\mathbb{N},+)$ of natural numbers and negative numbers. The union of those two subobjects is $(\mathbb{Z},+)$, but it is not effective.

1.15.7 By applying 1.8.5.c, prove that epimorphisms are not stable under pullback in the category Haus of Hausdorff spaces and continuous mappings.

1.15.8 By applying 2.2.4.h, prove that monomorphisms are not stable under pushouts in the category Rng of rings.

1.15.9 Applying the adjoint functor theorem (see 3.3.3, volume 1), prove that for a small abelian category $\mathscr{A}$, the category $\mathsf{Lex}(\mathscr{A},\mathsf{Ab})$ of left exact functors is reflective in the category $\mathsf{Add}(\mathscr{A},\mathsf{Ab})$ of additive functors. (In fact, this reflection is a localization, but this is much harder to prove.)

1.15.10 Produce a counterexample showing that the following "nine lemma" is false: suppose five rows and columns are exact in diagram 1.14, then the remaining row or column is exact as well.

1.15.11 Given $f = g \circ h$ in an abelian category, prove the existence of an exact sequence

$$0 \longrightarrow \mathsf{Ker}\, h \longrightarrow \mathsf{Ker}\, f \longrightarrow \mathsf{Ker}\, g \cdots$$
$$\cdots \longrightarrow \mathsf{Coker}\, h \longrightarrow \mathsf{Coker}\, f \longrightarrow \mathsf{Coker}\, g \longrightarrow 0.$$

1.15.12 Prove the following version of the nine lemma. Consider diagram 1.14 where the middle row and the middle column are exact, the square (1) is a pullback, the square (2) is a pushout and $\varepsilon \circ \beta$, $\nu \circ \theta$ are image factorizations. Then all the rows and columns are exact.

1.15.13 Consider an abelian category $\mathscr{A}$ and the category $\mathsf{Sex}(\mathscr{A})$ of short exact sequences of $\mathscr{A}$: an object is an exact sequence and a morphism is a triple f, g, h of morphisms making the squares of diagram 1.74 commutative. Prove that this category has biproducts, kernels and cokernels. Prove that when g and h are monomorphisms in $\mathscr{A}$, (f,g,h) is a monomorphism in $\mathsf{Sex}(\mathscr{A})$. Prove that when (f,g,h) is a kernel in $\mathsf{Sex}(\mathscr{A})$, f, g, h are monomorphisms in $\mathscr{A}$ and the left-hand square is a pullback. Produce a counterexample showing that in general $\mathsf{Sex}(\mathscr{A})$ is not abelian. [Hint: construct the kernel of (f,g,h) as the following exact sequence:

$$0 \longrightarrow \mathsf{Ker}\, f \longrightarrow \mathsf{Ker}\, g \longrightarrow X \longrightarrow 0.]$$

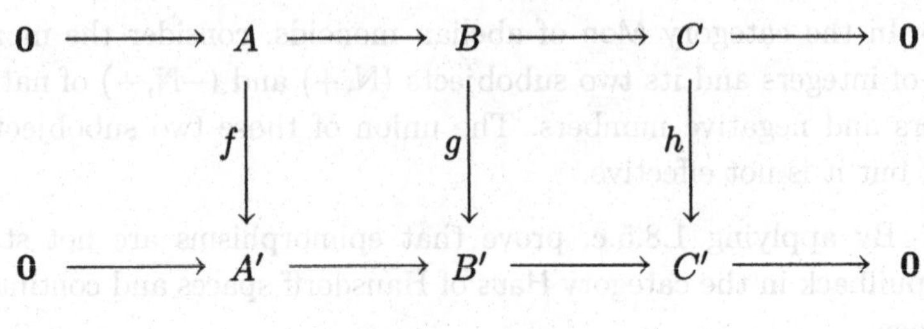

Diagram 1.74

1.15.14 Prove the snake lemma using the full embedding theorem 1.12.9, i.e., prove the existence of the diagonal morphism by just working with "elements".

1.15.15 Consider the category of abelian groups and the torsion theory defined at the beginning of section 1.12. Prove that the corresponding closure operation is given by

$$\overline{S} = \{a \in A \mid \exists n \in \mathbb{N},\ n \neq 0,\ na \in S\}$$

for a subgroup $S \subseteq A$.

1.15.16 On the category of abelian groups, consider the closure operation defined as in 1.15.15 for all subgroups $S \subseteq A$ with $A \neq \mathbb{Z}$; for a subgroup $S \subseteq \mathbb{Z}$, define $\overline{S} = S$. Prove that this closure operation satisfies the three axioms indicated at the end of section 1.12, but it is not induced by a torsion theory.

2
Regular categories

Abelian categories are additive (see 1.6.4), which excludes many interesting situations in which, nevertheless, several "exactness properties" of abelian categories are still valid, like the existence of images (see 1.5.5). The notion of a regular category recaptures many "exactness properties" of abelian categories, but avoids requiring additivity. For example, the category of sets and most "algebraic-like" categories are regular.

Convention: as a matter of convention, in the present chapter, the symbol $\longrightarrow\!\!\!\!\!\rightarrow$ will denote a regular epimorphism; no special notation will be used for ordinary epimorphisms.

2.1 Exactness properties of regular categories

We recall that an epimorphism is regular when it can be written as the coequalizer of some pair of morphisms (see 4.3.1, volume 1). Regular epimorphisms and kernel pairs are closely related via the properties:

- if a regular epimorphism has a kernel pair, it is the coequalizer of that kernel pair (see 2.5.7, volume 1);
- if a kernel pair has a coequalizer, it is the kernel pair of that regular epimorphism (see 2.5.8, volume 1).

The reader will also remember from 4.3.6 of volume 1 that

- every regular epimorphism is strong.

The properties of strong epimorphisms have been studied in section 4.3, volume 1.

There exist in the literature many different definitions of regular categories, which are all equivalent under the assumptions that finite limits

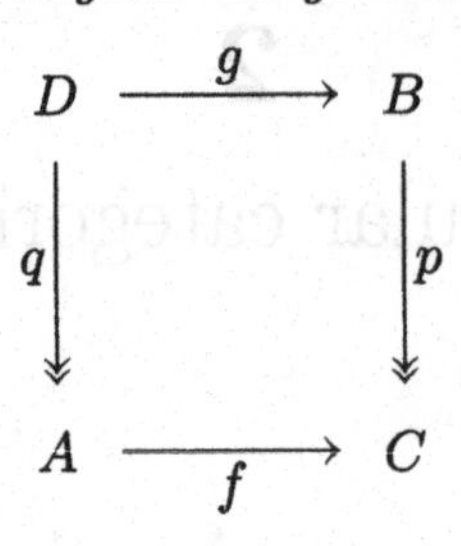

Diagram 2.1

and coequalizers exist. We choose here a somehow "weakest" possible definition. Requesting the existence of more finite limits in the definition of a regular category is essentially a matter of personal taste or convenience; requiring the existence of all coequalizers is less innocuous: for example arbitrary coequalizers are not preserved by exact functors (see 2.3.5 and 2.7.4), by pulling back in a regular category (see remark after 2.1.1), by forgetful functors in algebraic situations (see 3.9.1 and 4.4.4), and so on.

Definition 2.1.1 *A category $\mathscr{C}$ is regular when it satisfies the following conditions:*

(1) every arrow has a kernel pair;

(2) every kernel pair has a coequalizer;

(3) the pullback of a regular epimorphism along any morphism exists and is again a regular epimorphism.

The third axiom means that in diagram 2.1, if f and p are given with p a regular epimorphism, then the pullback (q, g) of (f, p) exists and q is a regular epimorphism.

It is probably useful to emphasize the fact that the third axiom in 2.1.1 requires the preservation of regular epimorphisms under pullbacks, not the preservation of coequalizers (see exercise 2.9.1).

Lemma 2.1.2 *Consider a regular epimorphism $f: A \twoheadrightarrow B$ and an arbitrary morphism $g: B \longrightarrow C$. In these conditions the factorization*

$$f \times_C f: A \times_C A \longrightarrow B \times_C B$$

exists and is an epimorphism.

Proof The pullback $B \times_C B$ of the pair (g, g) is just the kernel pair of g, thus it exists. Since f is a regular epimorphism, the three other partial pullbacks involved in diagram 2.2 exist, yielding regular epimorphisms d, e, i, j. Then $f \times_C f = d \circ i = e \circ j$, which is an epimorphism as composite of two (regular) epimorphisms (see 1.8.2, volume 1). □

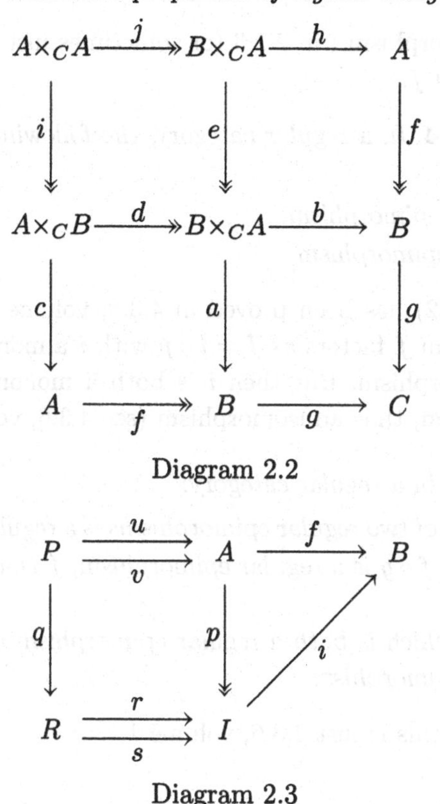

Diagram 2.2

Diagram 2.3

Theorem 2.1.3 *In a regular category, every morphism factors as a regular epimorphism followed by a monomorphism and this factorization is unique up to isomorphism.*

Proof Consider a morphism f, its kernel pair (u, v) and the coequalizer $p = \mathsf{Coker}\,(u, v)$ of that kernel pair, as in diagram 2.3. Since $f \circ u = f \circ v$, we get a unique factorization i through the coequalizer, yielding $f = i \circ p$. It remains to prove that i is a monomorphism.

Let (r, s) be the kernel pair of i. Since $i \circ p \circ u = i \circ p \circ v$, there exists a unique morphism q such that $r \circ q = p \circ u$, $s \circ q = p \circ v$. Applying 2.1.2 to the regular epimorphism p and the morphism i, we observe that

$$P = A \times_B A, \quad R = I \times_B I, \quad q = p \times_B p,$$

thus q is an epimorphism. Then $r \circ q = p \circ u = p \circ v = s \circ q$ implies $r = s$ and i is indeed a monomorphism (see 2.5.6, volume 1). The uniqueness of the factorization is attested by 4.4.5, volume 1. □

As usual, the morphism i in 2.1.3 (or sometimes just its domain I) is called the *image* of f.

Proposition 2.1.4 *In a regular category, the following conditions are equivalent:*

(1) f is a regular epimorphism;
(2) f is a strong epimorphism.

Proof (1) $\Rightarrow$ (2) has been proved in 4.3.6, volume 1. Conversely a strong epimorphism f factors as $f = i \circ p$ with i a monomorphism and p a regular epimorphism. But then i is both a monomorphism and a strong epimorphism, thus an isomorphism (see 4.3.6, volume 1). □

Corollary 2.1.5 *In a regular category:*

(1) the composite of two regular epimorphisms is a regular epimorphism;
(2) if a composite $f \circ g$ is a regular epimorphism, f is a regular epimorphism;
(3) a morphism which is both a regular epimorphism and a monomorphism is an isomorphism.

Proof Via 2.1.4, this is just 4.3.6, volume 1. □

2.2 Definition in terms of strong epimorphisms

In a regular category, regular epimorphisms coincide with strong epimorphisms (see 2.1.4). Therefore one can expect a definition of a regular category in terms of strong epimorphisms.

In definition 2.1.1, the second axiom asserts the existence of "enough" regular epimorphisms, while the third axiom is an "exactness" property of regular epimorphisms. If the third axiom makes perfectly good sense with "regular" replaced by "strong", the second axiom must be put into an equivalent form to allow an analogous translation.

Proposition 2.2.1 *A category $\mathscr{C}$ is regular when it satisfies the following conditions:*

(1) every arrow has a kernel pair;
(2) every arrow f can be factored as $f = i \circ p$ with i a monomorphism and p a regular epimorphism;
(3) the pullback of a regular epimorphism along any morphism exists and is a regular epimorphism.

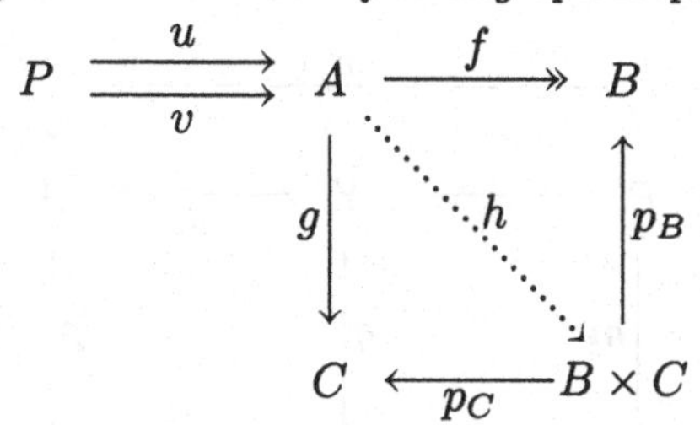

Diagram 2.4

Proof Theorem 2.1.3 proves the necessary condition. Conversely, consider an arrow f, its kernel pair (u, v) and its mono–regular-epi factorization $f = i \circ p$. Since i is a monomorphism, (u, v) is still the kernel pair of p and since p is a regular epimorphism, p is the coequalizer of (u, v); see 2.5.7, volume 1. □

The conditions of 2.2.1 can now be stated for strong monomorphisms, yielding the following result.

Proposition 2.2.2 *Let $\mathscr{C}$ be a category with finite limits. The category $\mathscr{C}$ is regular if and only if it satisfies the following conditions:*

(1) every arrow f can be factored as $f = i \circ p$ with i a monomorphism and p a strong epimorphism;

(2) the pullback of a strong epimorphism along any morphism is again a strong epimorphism.

Proof The necessity of conditions (1) and (2) follows from 2.1.5 and 2.2.1. Conversely, assuming the conditions of 2.2.2, it suffices by 2.2.1 to prove the coincidence between strong and regular epimorphisms. To do this, let us consider a strong epimorphism $f: A \longrightarrow B$, the kernel pair (u, v) of f and a morphism g such that $g \circ u = g \circ v$; we shall prove the existence of $w: B \longrightarrow C$ such that $g = w \circ f$ (such a w is necessarily unique since f is an epimorphism). We consider the product of B, C and the unique factorization $h: A \longrightarrow B \times C$ as in diagram 2.4, such that $p_B \circ h = f$, $p_C \circ h = g$. The morphism h can be factored as $h = i \circ p$ with i a monomorphism and p a strong epimorphism. We shall prove that $p_B \circ i$ is an isomorphism and $w = p_C \circ i \circ (p_B \circ i)^{-1}$ is the required factorization.

Let us consider diagram 2.5, where all the individual squares are pullbacks. Since $p_B \circ i \circ p = p_B \circ h = f$, we can identify the global pullback with the kernel pair of f, yielding $P' = P$, $x \circ m = u$, $y \circ n = v$. Since p is a strong epimorphism, so are t, q, m, n; r, s are strong epimorphisms

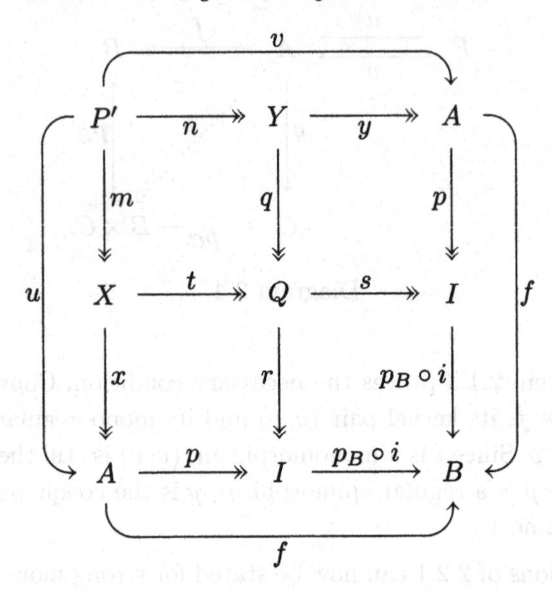

Diagram 2.5

as well as parts of a kernel pair, but this does not play any role here. By commutativity of the diagram, we have immediately

$$p_B \circ i \circ r \circ t \circ m = p_B \circ i \circ s \circ t \circ m.$$

On the other hand

$$\begin{aligned} p_C \circ i \circ r \circ t \circ m &= p_C \circ i \circ p \circ x \circ m \\ &= p_C \circ h \circ x \circ m \\ &= g \circ u \\ &= g \circ v \\ &= p_C \circ h \circ y \circ n \\ &= p_C \circ i \circ p \circ y \circ n \\ &= p_C \circ i \circ s \circ q \circ n \\ &= p_C \circ i \circ s \circ t \circ m. \end{aligned}$$

By definition of a product, this yields $i \circ r \circ t \circ m = i \circ s \circ t \circ m$, thus $r \circ t \circ m = s \circ t \circ m$ since i is a monomorphism and $r = s$ since t, m are epimorphisms. This already proves that $p_B \circ i$ is a monomorphism (see 2.5.6, volume 1). But since $f = (p_B \circ i) \circ p$ and f is a strong epimorphism, $p_B \circ i$ is a strong epimorphism as well and finally $p_B \circ i$ is an isomorphism (see 4.3.6, volume 1).

So we put $w = p_C \circ i \circ (p_B \circ i)^{-1}$. It is straightforward to observe that

$$\begin{aligned} w \circ f &= p_C \circ i \circ (p_B \circ i)^{-1} \circ f \\ &= p_C \circ i \circ (p_B \circ i)^{-1} \circ p_B \circ i \circ p \\ &= p_C \circ i \circ p \\ &= g \end{aligned}$$

On the other hand we have noticed already that such a factorization w is unique since f is an epimorphism. This proves that $f = \mathsf{Coker}\,(u, v)$. In particular, every strong epimorphism is regular and thus regular epimorphisms coincide with strong epimorphisms (see 4.3.6, volume 1). □

2.3 Exact sequences

Definition 2.3.1 *By an exact sequence in a regular category we mean a diagram*

$$P \underset{v}{\overset{u}{\rightrightarrows}} A \xrightarrow{f} B$$

where (u, v) is the kernel pair of f and f is the coequalizer of (u, v).

Observe that in definition 2.3.1, u, v are regular epimorphisms since (for example) f is a regular epimorphism.

We shall observe in 2.4.1 that an abelian category is always regular. The relation of 2.3.1 with the notion of exact sequence in an abelian category is given by the following proposition.

Proposition 2.3.2 *Let $\mathscr{C}$ be an abelian category. The following conditions are equivalent:*

(1) $P \underset{v}{\overset{u}{\rightrightarrows}} A \xrightarrow{f} B$ is an exact sequence in the sense of 2.3.1;

(2) the sequence

$$0 \longrightarrow P \xrightarrow{\binom{u}{v}} A \oplus A \xrightarrow{(f, -f)} B \longrightarrow 0$$

is a short exact sequence in the sense of 1.8.6.

Proof For morphisms $x, y \colon X \rightrightarrows A$, the relation $f \circ x = f \circ y$ is equivalent to $(f, -f) \circ \binom{x}{y} = 0$; this proves that (u, v) is the kernel pair of f iff $\binom{u}{v}$ is the kernel of $(f, -f)$. Now assuming (1), f is an epimorphism and the sequence in (2) is a short exact sequence (see 1.8.5). Conversely assuming (2), $(f, -f) = f \circ (1_A, -1_A)$ so that f is an epimorphism, thus a regular epimorphism since $\mathscr{C}$ is abelian; one concludes the proof by 2.5.7, volume 1. □

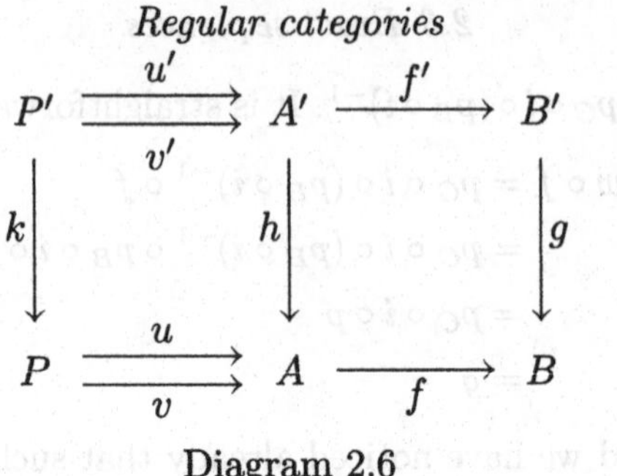

Diagram 2.6

Proposition 2.3.3 *In a regular category, pulling back along any arrow preserves exact sequences.*

Proof We consider the situation of diagram 2.6, where all the individual squares are pullbacks and $(f; u, v)$ is an exact sequence. Observe that since $f \circ u = f \circ v$, their pullbacks with g are the same and by associativity of pullbacks (see 2.5.9, volume 1); this means the existence of a single morphism k such that (u', k) is the pullback of (u, h) and (v', k) is the pullback of (v, h). We must prove that $(f'; u', v')$ is an exact sequence.

An easy "diagram chasing argument" shows that (u', v') is the kernel pair of f'. Indeed $f' \circ x = f' \circ y$ implies $f \circ h \circ x = g \circ f' \circ x = g \circ f' \circ y = f \circ h \circ y$, from which there is a unique w such that $u \circ w = h \circ x$, $v \circ w = h \circ y$. This yields z_1, z_2 such that $u' \circ z_1 = x$, $k \circ z_1 = w$ and $v' \circ z_2 = y$, $k \circ z_2 = w$. The relations $k \circ z_1 = w = k \circ z_2$ and $f' \circ u' \circ z_1 = f' \circ x = f \circ' y = f' \circ v' \circ z_2 = f' \circ u' \circ z_2$ imply $z_1 = z_2$, since the global diagram is a pullback. This yields a morphism $z = z_1 = z_2$ such that $u' \circ z = x$ and $v' \circ z = y$. The uniqueness of such a z is proved in the same way.

Now f' is regular since f is. The pair (u', v') is the kernel pair of f' and f' is regular: this implies that f' is the coequalizer of (u', v'); see 2.5.7, volume 1. □

Proposition 2.3.4 *Let $\mathscr{C}$ be a regular category with binary products. Given two exact sequences*

$$P \underset{v}{\overset{u}{\rightrightarrows}} A \xrightarrow{f} B, \quad P' \underset{v'}{\overset{u'}{\rightrightarrows}} A' \xrightarrow{f'} B',$$

the product sequence

$$P \times P' \underset{v \times v'}{\overset{u \times u'}{\rightrightarrows}} A \times A' \xrightarrow{f \times f'} B \times B'$$

is again exact.

Proof It is just an obvious exercise on pullbacks and products to check that $(u \times u', v \times v')$ is the kernel pair of $f \times f'$. By 2.5.7, volume 1, it

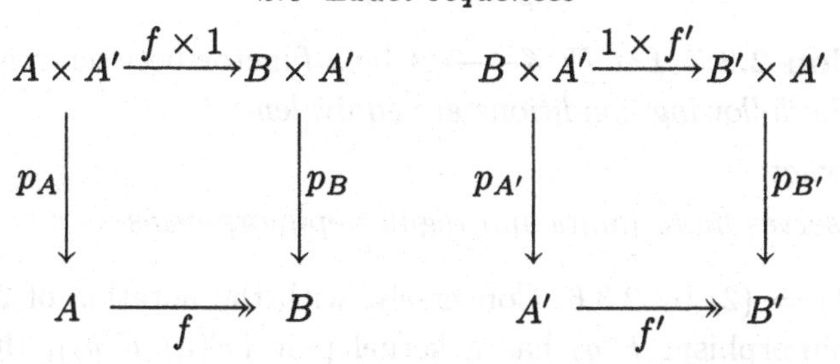

Diagram 2.7

remains to prove that $f \times f'$ is a regular epimorphism. Observing that the squares of diagram 2.7 are pullbacks one concludes that $f \times 1$ and $1 \times f'$ are regular epimorphisms. This implies that $f \times f' = (f \times 1) \circ (1 \times f')$ is also a regular epimorphism (see 2.1.5). □

In 1.11.2, we observed that for an additive functor between abelian categories, preservation of short exact sequences did imply preservation of finite limits and colimits. This was essentially due to the equational characterization of biproducts, given in 1.2.4, and the close relation between kernels and equalizers. No such phenomena appear for regular categories, so that the definition which is generally admitted is the following one.

Definition 2.3.5 *Let $F\colon \mathscr{C} \longrightarrow \mathscr{D}$ be a functor between regular categories $\mathscr{C}, \mathscr{D}$. The functor F is exact when it preserves:*

(1) all finite limits which happen to exist in $\mathscr{C}$;
(2) exact sequences.

In most applications, the regular categories one considers have finite limits. When this is not the case, it is not clear if condition (1) in definition 2.3.5 is pertinent. One could imagine replacing it by a weaker condition, like the preservation of just those finite limits whose existence is required in the definition of a regular category, or by a stronger condition, like being flat (see 6.7.5).

Proposition 2.3.6 *Let $F\colon \mathscr{C} \longrightarrow \mathscr{D}$ be an exact functor between regular categories $\mathscr{C}, \mathscr{D}$. The functor F preserves:*

(1) regular epimorphisms;
(2) kernel pairs;
(3) coequalizers of kernel pairs;
(4) mono–regular-epi factorizations. □

Proposition 2.3.7 *Let $F\colon \mathscr{C} \longrightarrow \mathscr{D}$ be a functor between regular categories. The following conditions are equivalent:*

(1) F is exact;

(2) F preserves finite limits and regular epimorphisms.

Proof (1) $\Rightarrow$ (2) by 2.3.6. Conversely, with the notation of 2.3.1, the regular epimorphism $F(q)$ has a kernel pair $\big(F(u), F(v)\big)$, thus is its coequalizer (see 2.5.7, volume 1). □

Proposition 2.3.8 *Let $F\colon \mathscr{C} \longrightarrow \mathscr{D}$ be an additive functor between abelian categories $\mathscr{C}, \mathscr{D}$. The following conditions are equivalent:*

(1) F is exact in the sense of definition 2.3.5;

(2) F is exact in the sense of definition 1.11.1.

Proof (2) $\Rightarrow$ (1) since by 1.11.2, F preserves finite limits and finite colimits.

Conversely, assuming the conditions of 2.3.5, F preserves kernels, since it preserves finite limits. It also preserves (regular) epimorphisms by 2.3.6. Thus F is exact by 1.11.4. □

2.4 Examples

Example 2.4.1

An abelian category is finitely complete and cocomplete (see 1.5.3) and every epimorphism is regular (see 1.5.7). By the dual of 1.7.6, every abelian category is regular.

Example 2.4.2

The category of sets is finitely complete and cocomplete (see 2.8.6, volume 1). All epimorphisms are strong (see 4.3.10, volume 1): they are the surjections (see 1.8.5.a, volume 1); monomorphisms are just injections (see 1.7.8.a, volume 1). Every mapping factors as a surjection followed by an injection. Moreover the pullback of a surjection is obviously a surjection. Indeed if diagram 2.8 is a pullback of sets with g a surjection,

$$A \times_C B = \big\{(a,b) \big| \, a \in A,; \;\; b \in B, \;\; f(a) = g(b)\big\}.$$

Given $a \in A$, there exists $b \in B$ such that $f(a) = g(b)$, just because g is a surjection. Therefore $(a,b) \in A \times_C B$ and $p_A(a,b) = a$. This proves that p_A is surjective. So the category of sets is regular by 2.2.2.

$$\begin{array}{ccc} A\times_C B & \xrightarrow{p_B} & B \\ {\scriptstyle p_A}\downarrow & & \downarrow{\scriptstyle g} \\ A & \xrightarrow[f]{} & C \end{array}$$

Diagram 2.8

Example 2.4.3
The category of groups is finitely complete and cocomplete (see 2.8.6, volume 1). All epimorphisms are strong (see 4.3.10, volume 1) and coincide with the surjections (see 1.8.5.d, volume 1); monomorphisms are just injections (see 1.7.8.c, volume 1). The pullback of a surjection is a surjection and every homomorphism factors as a surjective one followed by an injective one. By 2.2.2, the category of groups is regular.

Example 2.4.4
The kernel pair of a monomorphism is just the identity pair (see 2.5.6, volume 1) and, of course, the coequalizer of the identity pair is just the identity. With 2.8.7, volume 1, in mind, this proves that in a category $\mathscr{C}$ where every arrow is a monomorphism, the regular epimorphisms are exactly the isomorphisms. Therefore a category where every arrow is a monomorphism is necessarily regular. In particular, every poset, seen as a category, is regular (see 1.2.6.b, volume 1). The category of fields is also regular, since every homomorphism of fields is injective.

Counterexample 2.4.5
The category Top of topological spaces and continuous mappings is not regular. Indeed, the strong epimorphisms are just the quotient maps $f\colon A \longrightarrow B$, i.e. the surjections f where B is provided with the corresponding quotient topology (see 4.3.10.b, volume 1). But quotient maps are not stable under pullbacks, so that Top is not regular. Here is an elementary counterexample. Let us put

$$\begin{array}{ll} A = \{a,b,c,d\} & \text{with } \{a,b\} \text{ open,} \\ B = \{l,m,n\} & \text{with } \{l,m\} \text{ open,} \\ C = \{x,y,z\} & \text{with the indiscrete topology.} \end{array}$$

We define $f\colon A \longrightarrow C$, $g\colon B \longrightarrow C$ by

$$f(a) = x, \quad f(b) = y = f(c), \quad f(d) = z,$$

$$g(l) = x, \;\; g(m) = z = g(n).$$

Now f is surjective and no subset of C has $\{a,b\}$ as inverse image; thus f is a quotient map. The product $A \times B$ has a single non-trivial open subset, namely $\{a,b\} \times \{l,m\}$. The pullback of f, g is thus given by

$$P = \{(a,l),(d,m),(d,n)\} \quad \text{with } \{(a,l)\} \text{ open.}$$

The projection $p_C\colon P \longrightarrow C$ is not a quotient map since $p_C^{-1}(l) = \{(a,l)\}$ is open while $\{l\}$ is not.

Counterexample 2.4.6
The category **Cat** of small categories and functors is not regular. Indeed, strong and regular epimorphisms do not coincide in this category **Cat** (see 4.5.17.h, volume 1 and 2.1.4).

Example 2.4.7
Consider a regular category $\mathscr{C}$ and a small category $\mathscr{D}$. The category of functors and natural trasformations $\mathsf{Fun}(\mathscr{D},\mathscr{C})$ is again regular. Indeed, the considerations of 2.15.1, volume 1, indicate that kernel pairs and their coequalizers can be constructed pointwise in $\mathsf{Fun}(\mathscr{D},\mathscr{C})$ since they exist in $\mathscr{C}$. Now if $\alpha\colon F \Rightarrow G$ is a regular epimorphism in $\mathsf{Fun}(\mathscr{D},\mathscr{C})$, it is the coequalizer in $\mathsf{Fun}(\mathscr{D},\mathscr{C})$ of its kernel pair (see 2.5.7, volume 1); therefore for every object $D \in \mathscr{D}$, α_D is in $\mathscr{C}$ the coequalizer of its kernel pair, yielding an exact sequence in $\mathscr{C}$. Since exact sequences in $\mathscr{C}$ are preserved under pulling back, one concludes immediately that pulling back α in $\mathsf{Fun}(\mathscr{D},\mathscr{C})$ yields another regular epimorphism.

Example 2.4.8
Choosing a small category $\mathscr{D}$, the category $\mathsf{Fun}(\mathscr{D},\mathsf{Set})$ of set valued functors on $\mathscr{D}$ is regular (see 2.4.2 and 2.4.7). This example is a "generic" one in the sense that every small regular category $\mathscr{C}$ admits a full exact embedding

$$\mathscr{C} \longrightarrow \mathsf{Fun}\,(\mathscr{D},\mathsf{Set})$$

in some category of set-valued functors. The proof of this theorem is essentially a more sophisticated version of the method used in section 1.14 (see **Barr**, *Exact categories*). It yields as a corollary a metatheorem asserting that to prove, in a regular category, a property of exact sequences, it suffices to prove it in a category of set valued functors. Now since exact sequences in $\mathsf{Fun}(\mathscr{D},\mathsf{Set})$ are constructed pointwise, this often reduces the problem to giving a proof in the category of sets. In section

2.7 we shall prove an alternative metatheorem using the techniques of topos theory (see chapter 6, volume 3).

Example 2.4.9
If $\mathscr{C}$ is a regular category, for every object $C \in \mathscr{C}$ the category $\mathscr{C}/C$ (see 1.2.7.a, volume 1) is regular. Indeed pullbacks and coequalizers in $\mathscr{C}/C$ are easily seen to be computed as in $\mathscr{C}$ (see 2.16.3, volume 1), from which the result follows immediately.

Observe that when $\mathscr{C}$ is abelian (see 2.4.1), $\mathscr{C}/C$ does not need to be so since it is in general not additive (see 1.6.4). Indeed given two morphisms $f, g\colon (X,x) \rightrightarrows (Y,y)$ in $\mathscr{C}/C$, one has $y \circ f = x$, $y \circ g = x$ from which $y \circ (f+g) = x + x \ldots$ and $x + x \neq x$ except when $x = 0$.

Example 2.4.10
In chapters 3, 4, we shall prove that the algebraic and monadic categories over Set are regular (see 3.5.4 and 4.3.5).

2.5 Equivalence relations

A relation on a set A is a subset $R \subseteq A \times A$. But giving the embedding $r\colon R \rightarrowtail A \times A$ is equivalent to giving the two composites

$$p_1 \circ r, p_2 \circ r\colon R \rightarrowtail A \times A \rightrightarrows A.$$

We prefer this second formulation since it does not refer to the existence of cartesian products. The injectivity of r can immediately be translated by the fact that $(p_1 \circ r, p_2 \circ r)$ is a monomorphic family (see 4.8.5, volume 1), i.e. that two arrows $x, y\colon X \rightrightarrows R$ are equal iff $p_i \circ r \circ x = p_i \circ r \circ y$ $(i = 1, 2)$.

Definition 2.5.1 *By a relation on an object A of a category $\mathscr{C}$, we mean an object $R \in \mathscr{C}$ together with a monomorphic pair of arrows*

$$r_1, r_2\colon R \rightrightarrows A$$

(i.e. given arrows $x, y\colon X \rightrightarrows R$, $x = y$ iff $r_1 \circ x = r_1 \circ y$ and $r_2 \circ x = r_2 \circ y$). For every object $X \in \mathscr{C}$ we write

$$R_X = \{(r_1 \circ x, r_2 \circ x) \mid x \in \mathscr{C}(X, R)\}$$

for the corresponding relation (in the usual sense) generated by R on the set $\mathscr{C}(X, A)$.

Given a relation (R, r_1, r_2) on an object $A \in \mathscr{C}$ of some category, it is now possible to require classical properties on the various relations R_X on the sets $\mathscr{C}(X, A)$.

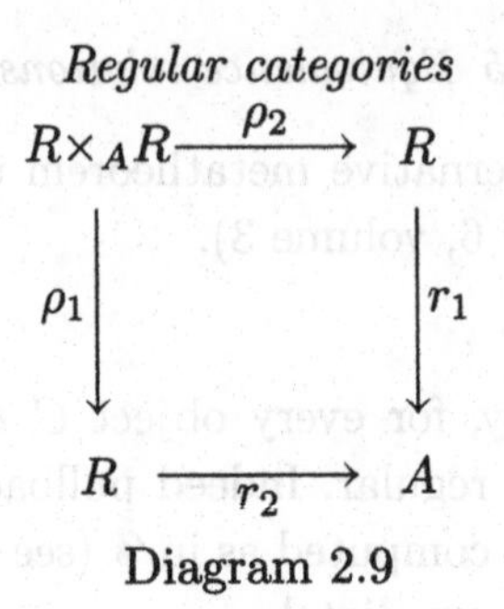

Diagram 2.9

Definition 2.5.2 *By an equivalence relation on an object A of a category $\mathscr{C}$, we mean a relation (R, r_1, r_2) on A such that, for every object $X \in \mathscr{C}$, the corresponding relation R_X on the set $\mathscr{C}(X, A)$ is an equivalence relation. More generally, the relation R is reflexive (respectively, transitive, symmetric, antisymmetric, ...) when each relation R_X is.*

In the category of sets, given an equivalence relation $R \subseteq A \times A$, one can perform the quotient of A by R, yielding a diagram

$$R \underset{r_2}{\overset{r_1}{\rightrightarrows}} A \xrightarrow{\;q\;} A/R.$$

The coequalizer of r_1, r_2 is the quotient of A by the equivalence relation generated by the pairs $(r_1(x), r_2(x)), x \in R, \ldots$, i.e. the quotient of A by R, which is q. On the other hand $q(a) = q(a')$ iff $(a, a') \in R$, which indicates that (r_1, r_2) is the kernel pair of q.

Definition 2.5.3 *An equivalence relation (R, r_1, r_2) on an object A of a category $\mathscr{C}$ is effective when the coequalizer q of (r_1, r_2) exists and (r_1, r_2) is the kernel pair of q.*

Proposition 2.5.4 *Let $\mathscr{C}$ be a category admitting pullbacks of strong epimorphisms. A relation (R, r_1, r_2) on an object $A \in \mathscr{C}$ is an equivalence relation precisely when there exist:*

(1) a morphism $\delta\colon A \longrightarrow R$ such that $r_1 \circ \delta = 1_A$, $r_2 \circ \delta = 1_A$;

(2) a morphism $\sigma\colon R \longrightarrow R$ such that $r_1 \circ \sigma = r_2$, $r_2 \circ \sigma = r_1$;

(3) a morphism $\tau\colon R \times_A R \longrightarrow R$ such that $r_1 \circ \tau = r_1 \circ \rho_1$, $r_2 \circ \tau = r_2 \circ \rho_2$, where the pullback is that of diagram 2.9.

Such morphisms δ, σ, τ are necessarily unique.

Proof The reflexivity of the relation (R, r_1, r_2) implies that given the pair $(1_A, 1_A)\colon A \rightrightarrows A$, there exists a morphism $\delta\colon A \longrightarrow R$ such that $r_1 \circ \delta = 1_A = r_2 \circ \delta$. Conversely, if given the relation (R, r_1, r_2) such a morphism δ exists, then for every arrow $x\colon X \longrightarrow A$ one has

$$x = r_1 \circ \delta \circ x, \quad x = r_2 \circ \delta \circ x,$$

which proves that $(x,x) \in R_X$ and thus R is reflexive.

The pair $(r_1, r_2)\colon R \rightrightarrows A$ is obviously in R_R, since $r_1 = r_1 \circ 1_R$ and $r_2 = r_2 \circ 1_R$. By symmetry of (R, r_1, r_2), the pair $(r_2, r_1)\colon R \rightrightarrows A$ is in R_R, yielding a morphism $\sigma\colon R \longrightarrow R$ such that $r_1 \circ \sigma = r_2$, $r_2 \circ \sigma = r_1$. Conversely, if given the relation (R, r_1, r_2) such a morphism σ exists, then for every pair of arrows $x, y\colon X \rightrightarrows A$ in R_X, one has a morphism $z\colon X \longrightarrow R$ such that $r_1 \circ z = x$, $r_2 \circ z = y$; as a consequence,

$$r_1 \circ \sigma \circ z = r_2 \circ z = y, \quad r_2 \circ \sigma \circ z = r_1 \circ z = x,$$

and the pair (y, x) is in R_X as well, proving the symmetry of R.

Observe that the definitions of δ and σ were independent of the existence of finite limits. This is no longer true for the morphism τ "representing" the transitivity of (R, r_1, r_2). Let us consider the pullback of diagram 2.9, which exists since the relations $r_1 \circ \delta = 1_A$, $r_2 \circ \delta = 1_A$ imply that r_1, r_2 are strong epimorphisms (see 4.3.6, volume 1). Intuitively, $R \times_A R$ represents the pairs $((a_1, a_2), (a_2, a_3))$ where $a_1 \approx a_2$ and $a_2 \approx a_3$. Considering the diagrams

$$R \times_A R \xrightarrow{\rho_1} R \underset{r_2}{\overset{r_1}{\rightrightarrows}} A, \quad R \times_A R \xrightarrow{\rho_2} R \underset{r_2}{\overset{r_1}{\rightrightarrows}} A.$$

We conclude that $(r_1 \circ \rho_1, r_2 \circ \rho_1)$ and $(r_1 \circ \rho_2, r_2 \circ \rho_2)$ are in $R_{(R \times_A R)}$. Since $r_2 \circ \rho_1 = r_1 \circ \rho_2$, this implies that $(r_1 \circ \rho_1, r_2 \circ \rho_2)$ is in $R_{(R \times_A R)}$, yielding an arrow $\tau\colon R \times_A R \longrightarrow R$ such that $r_1 \circ \tau = r_1 \circ \rho_1$ and $r_2 \circ \tau = r_2 \circ \rho_2$. Conversely, suppose we are given a relation (R, r_1, r_2) with the property that such a morphism τ exists. Given three arrows $x, y, z\colon X \longrightarrow A$ with $(x, y) \in R_X$ and $(y, z) \in R_X$, we get two morphisms $u, v\colon X \rightrightarrows R$ such that $r_1 \circ u = x$, $r_2 \circ u = y$, $r_1 \circ v = y$, $r_2 \circ v = z$. From the relation $r_1 \circ v = r_2 \circ u$ we get a morphism $w\colon X \longrightarrow R \times_A R$ such that $\rho_1 \circ w = u$, $\rho_2 \circ w = v$. Finally one has

$$x = r_1 \circ u = r_1 \circ \rho_1 \circ w = r_1 \circ \tau \circ w,$$

$$z = r_2 \circ v = r_2 \circ \rho_2 \circ w = r_2 \circ \tau \circ w,$$

so that $(x, z) \in R_X$ and R is transitive.

It remains to observe that the morphisms δ, σ, τ with the indicated properties are unique, because (r_1, r_2) is monomorphic. □

Finally when products exist, equivalence relations admit a more usual description.

Proposition 2.5.5 *Let $\mathscr{C}$ be a category with finite limits. An equivalence relation on an object A of $\mathscr{A}$ can equivalently be defined as a subobject $r\colon R \rightarrowtail A \times A$ such that:*

(1) the diagonal $\Delta_A \colon A \rightarrowtail A \times A$ factors through r;
(2) with $\Sigma_A \colon A \times A \longrightarrow A \times A$ as the symmetry on A, i.e. the unique morphism such that $p_1 \circ \Sigma_A = p_2$ and $p_2 \circ \Sigma_A = p_1$, then $\Sigma_A \circ r$ factors through r;
(3) consider $r_1 = p_1 \circ r$, $r_2 = p_2 \circ r$ and the pullback of diagram 2.9; the morphism

$$\begin{pmatrix} r_1 \circ \rho_1 \\ r_2 \circ \rho_2 \end{pmatrix} \colon R \times_A R \longrightarrow A \times A$$

factors through r.

Proof Given an equivalence relation $r_1, r_2 \colon R \rightrightarrows A$ on an object A, we get a corresponding factorization

$$r \colon R \longrightarrow A \times A, \quad p_1 \circ r = r_1, \quad p_2 \circ r = r_2,$$

which is a monomorphism, since the pair (r_1, r_2) is monomorphic. Conversely given a monomorphism $r \colon R \rightarrowtail A \times A$, one defines r_1, r_2 as in the statement and this family is monomorphic because r is.

Conditions (1), (2), (3) of 2.5.4 for r_1, r_2 are just conditions (1), (2), (3) of 2.5.5 for r. □

Examples 2.5.6

2.5.6.a As we observed before giving definition 2.5.3, equivalence relations are effective in the category of sets.

2.5.6.b If $\mathscr{D}$ is a small category, equivalence relations are effective in the category of set valued functors $\mathsf{Fun}(\mathscr{D}, \mathsf{Set})$. Indeed, coequalizers and kernel pairs are constructed pointwise and the result holds in Set.

2.5.6.c In the category Gr of groups, equivalence relations are effective. Indeed, an equivalence relation $r_1, r_2 \colon R \rightrightarrows G$ on a group G is just a congruence on G and the coequalizer is performed exactly as in the case of Set (see 2.4.6.d, volume 1).

2.5.6.d In an abelian category, equivalence relations are effective. Indeed, given an equivalence relation $r_1, r_2 \colon R \rightrightarrows A$, we can compute its coequalizer $q \colon A \longrightarrow Q$ (see 1.5.3) and it remains to prove that (r_1, r_2) is the kernel pair of q. But being an equivalence relation is a property which can be expressed entirely in term of finite limits (see 2.5.4); thus by the embedding theorem for abelian categories (see 1.14.9) it suffices to prove the result in a category of modules. And this is obvious since the coequalizer of r_1, r_2 in a category of modules is just the set theoretical quotient of A by the congruence generated by the pairs $(r_1(x), r_2(x))$, i.e. exactly the set theoretical quotient of A by the equivalence relation

R, which is already a congruence (which means an equivalence relation which is a submodule of $A \times A$).

2.5.6.e In every category $\mathscr{C}$, the pair $(1_A, 1_A)\colon A \rightrightarrows A$ is obviously an effective equivalence relation on the object A. When $\mathscr{C}$ has finite limits, it corresponds with the diagonal subobject of $\Delta_A\colon A \rightarrowtail A \times A$.

2.5.6.f In the category Top of topological spaces and continuous mappings, equivalence relations are not effective. Indeed given a topological space A and an equivalence relation R on A, one gets an equivalence relation $R \rightarrowtail A \times A$ in $A \times A$ by providing the set R with any topology stronger than the topology induced by the product topology on $A \times A$. On the other hand, the kernel pair of a morphism $f\colon A \longrightarrow B$ has always the topology induced by that of $A \times A$.

2.5.6.g In every category $\mathscr{C}$ with coequalizers, a kernel pair is always an effective equivalence relation. The fact it is an equivalence relation is just obvious and the effectiveness follows from 2.5.8, volume 1.

Let us conclude this section with a warning. In 2.5.5.d we used the embedding theorem of 1.14.9 to prove the effectiveness of equivalence relations in an abelian category. By 2.5.5.b, equivalence relations are effective in every category $\mathsf{Fun}(\mathscr{D}, \mathsf{Set})$ and, as mentioned in 2.4.8, there exists for every small regular category $\mathscr{C}$ a full exact embedding of the form $\mathscr{C} \longrightarrow \mathsf{Fun}(\mathscr{D}, \mathsf{Set})$. Nevertheless, one cannot conclude that equivalence relations are effective in every regular category (an explicit counterexample will be produced in 2.6.12). The difference with the abelian case is that an exact functor in the sense of 2.3.5 preserves coequalizers of kernel pairs, not arbitrary coequalizers (compare with 1.11.2). And in the construction developed in 2.5.5.d, the whole problem is precisely to prove that q (supposing it exists) admits (r_1, r_2) as a kernel pair.

2.6 Exact categories

Definition 2.6.1 *An exact category is a regular category in which equivalence relations are effective.*

Putting together the considerations of 2.4 and 2.5.6, we get the following examples of exact categories: every abelian category, the category of sets, the category of groups, every category $\mathsf{Fun}(\mathscr{D}, \mathsf{Set})$ of set-valued functors on a small category $\mathscr{D}$. In chapters 3 and 4 we shall prove that algebraic and monadic categories over Set are exact (see 3.5.4 and 4.3.5).

In this section, we want to emphasize the fact that exact categories are exactly the "non-additive version of abelian categories". More precisely, we want to show that an exact additive category is necessarily abelian.

Lemma 2.6.2 *In a non-empty and preadditive regular category $\mathscr{C}$, the biproduct $A \oplus A$ exists for every object A.*

Proof Consider an arbitrary object $A \in \mathscr{C}$, the zero map $0: A \longrightarrow A$ and its kernel pair $u, v: P \rightrightarrows A$. Given arbitrary morphisms $x, y: X \rightrightarrows A$, one has $0 \circ x = 0 \circ y$, from which there is a unique morphism $z: X \longrightarrow P$ such that $u \circ z = x$, $v \circ z = y$. This proves that (P, u, v) is the product $A \times A$. One derives the conclusion by 1.2.4. ☐

Lemma 2.6.3 *A non-empty and preadditive regular category $\mathscr{C}$ has a zero object.*

Proof Choose an arbitrary object A. By lemma 2.6.2, the zero map $0: A \longrightarrow A$ admits $p_1, p_2: A \oplus A \rightrightarrows A$ as kernel pair. Let $q: A \longrightarrow Q$ be the coequalizer of p_1, p_2. Given a morphism $x: Q \longrightarrow X$ one has

$$x \circ q = x \circ q \circ p_1 \circ s_1 = x \circ q \circ p_2 \circ s_1 = x \circ q \circ 0 = 0$$

(see 1.2.4), from which $x = 0$, since q is an epimorphism. Thus 0 is the unique morphism from Q to X and Q is an initial object. One derives the conclusion by 1.2.3. ☐

Lemma 2.6.4 *A non-empty and preadditive regular category $\mathscr{C}$ has biproducts.*

Proof Given objects A, B, the morphisms $A \longrightarrow \mathbf{0}$, $B \longrightarrow \mathbf{0}$ to the zero object (see 2.6.3) are retractions, with zero as a section. Therefore they are regular epimorphisms (see 6.5.4, volume 1) and the pullback of those two arrows exists, yielding the product $A \times B$; see 2.8.2, volume 1. One derives the conclusion by 1.2.4. ☐

Lemma 2.6.5 *A non-empty and preadditive regular category $\mathscr{C}$ has kernels.*

Proof Take a morphism $f: A \longrightarrow B$ and its kernel pair $u, v: P \rightrightarrows A$. The morphism $u - v: P \longrightarrow A$ can be factored as $u - v = i \circ p$ with i a monomorphism and p a regular epimorphism. We shall prove that i is the kernel of f.

First $f \circ i \circ p = f \circ (u - v) = (f \circ u) - (f \circ v) = 0$, which proves $f \circ i = 0$ since p is an epimorphism. Next if $x: X \longrightarrow A$ is such that

$f \circ x = 0 = f \circ 0$, we get a factorization $y\colon X \longrightarrow P$ such that $u \circ y = x$, $v \circ y = 0$. This yields

$$i \circ p \circ y = (u - v) \circ y = (u \circ y) - (v \circ y) = x - 0 = x$$

and $p \circ y$ is a factorization of x through i. This factorization is unique since i is a monomorphism. □

Lemma 2.6.6 *A non-empty preadditive regular category $\mathscr{C}$ is finitely complete.*

Proof By 1.2.8 and 2.6.5, $\mathscr{C}$ has equalizers. One derives the conclusion by 2.6.4, this volume and 2.8.1, volume 1. □

Lemma 2.6.7 *In a preadditive category $\mathscr{C}$, every reflexive relation is necessarily an equivalence relation.*

Proof Consider a reflexive relation $s_1, s_2\colon S \rightrightarrows A$. Given an object $X \in \mathscr{C}$, the relation

$$S_X = \{(s_1 \circ x, s_2 \circ x) \mid x \in \mathscr{C}(X, S)\}$$

on the abelian group $\mathscr{C}(X, A)$ contains the diagonal, just by assumption. Since S_X is obviously a subgroup of $\mathscr{C}(X, A) \times \mathscr{C}(X, A)$, this reduces the problem to proving the lemma in the category of abelian groups.

Supposing $\mathscr{C} = \mathsf{Ab}$, we know already that all the pairs (a, a) belong to S, thus S is reflexive. Next if $(a, b) \in S$, using the reflexivity we get

$$(b, a) = (a, a) - (a, b) + (b, b) \in S$$

proving the symmetry of S. Finally if $(a, b) \in S$ and $(b, c) \in S$, again using the reflexivity we obtain

$$(a, c) = (a, b) - (b, b) + (b, c) \in S,$$

which proves the transitivity of S. □

Lemma 2.6.8 *In an additive exact category $\mathscr{C}$, every monomorphism has a cokernel and is the kernel of its cokernel.*

Proof Let $f\colon A \rightarrowtail B$ be a monomorphism. Applying 2.6.4 and using the matrix notations of 1.2, let us consider the morphism

$$r \equiv \begin{pmatrix} f & 1_B \\ 0 & 1_B \end{pmatrix} : A \oplus B \longrightarrow B \oplus B.$$

This is a monomorphism since given morphisms

$$a, a' \colon X \rightrightarrows A, \ \ b, b' \colon X \rightrightarrows B,$$

$$\begin{pmatrix} f & 1_B \\ 0 & 1_B \end{pmatrix} \circ \begin{pmatrix} a \\ b \end{pmatrix} = \begin{pmatrix} (f \circ a) + b \\ b \end{pmatrix},$$

so that from the relation

$$\begin{pmatrix} f & 1_B \\ 0 & 1_B \end{pmatrix} \circ \begin{pmatrix} a \\ b \end{pmatrix} = \begin{pmatrix} f & 1_B \\ 0 & 1_B \end{pmatrix} \begin{pmatrix} a' \\ b' \end{pmatrix}$$

we deduce $b = b'$ and $(f \circ a) + b = (f \circ a') + b'$, thus $f \circ a = f \circ a'$ and finally $a = a'$ since f is a monomorphism. Observing moreover that

$$\begin{pmatrix} f & 1_B \\ 0 & 1_B \end{pmatrix} \circ \begin{pmatrix} 0 \\ 1_B \end{pmatrix} = \begin{pmatrix} 1_B \\ 1_B \end{pmatrix} = \Delta_B,$$

we conclude that the monomorphism r, seen as a relation on B, contains the diagonal Δ_B. Therefore r is an equivalence relation by 2.6.7 and thus an effective equivalence relation, because $\mathscr{C}$ is an exact category (see 2.6.1).

Writing q for the coequalizer of the effective equivalence relation r on B, we thus have an exact sequence in $\mathscr{C}$

$$A \oplus B \underset{(0, 1_B)}{\overset{(f, 1_B)}{\rightrightarrows}} B \xrightarrow{\ q\ } Q.$$

In particular

$$q \circ f = q \circ (f, 1_B) \circ s_A = q \circ (0, 1_B) \circ s_A = q \circ 0 = 0.$$

Now given a morphism $x \colon B \longrightarrow X$ such that $x \circ f = 0$

$$x \circ (f, 1_B) = (x \circ f, x) = (0, x) = x \circ (0, 1_B)$$

from which we get a unique factorization $z \colon Q \longrightarrow X$ such that $z \circ q = x$. This proves that $q = \mathsf{Coker}\, f$.

Thus f admits q as a cokernel. It remains to prove that f is the kernel of q. Given $y \colon Y \longrightarrow B$ such that $q \circ y = 0 = q \circ 0$, we find a unique $z \colon Y \longrightarrow A \oplus B$ such that $(f, 1_B) \circ z = y$, $(0, 1_B) \circ z = 0$. The arrow z has the form $\binom{u}{v}$ for some morphisms $u \colon Y \longrightarrow A$, $v \colon Y \longrightarrow B$. The arrow u is the required factorization since

$$0 = (0, 1_B) \circ \begin{pmatrix} u \\ v \end{pmatrix} = (0 \circ u) + (1_B \circ v) = v,$$

$$y = (f, 1_B) \circ \begin{pmatrix} u \\ v \end{pmatrix} = (f \circ u) + (1_B \circ v) = (f \circ u) + (1_B \circ 0) = f \circ u.$$

Such a factorization u is unique since f is a monomorphism. □

Lemma 2.6.9 *An additive exact category $\mathscr{C}$ is finitely cocomplete.*

Proof By 2.8.1, volume 1, and 2.6.4, this volume, it suffices to prove the existence of coequalizers, which is equivalent to the existence of cokernels (see 1.2.8). Given a morphism $f\colon A \longrightarrow B$, we factor it as $f = i \circ p$ with i a monomorphism and p a regular epimorphism. Since p is an epimorphism, the cokernel of i, which exists by 2.6.8, is also the cokernel of f. □

Lemma 2.6.10 *In an additive exact category, every epimorphism is a cokernel.*

Proof Let $f\colon A \longrightarrow B$ be an epimorphism. Since the category is regular, we can factor f as $f = i \circ p$ where i is a monomorphism and p is a regular epimorphism (see 2.1.3). But the monomorphism i is a kernel by 2.6.8; so it is a strong monomorphism by 4.3.6, volume 1; but i is also an epimorphism, since f is. Finally i is an isomorphism (see 4.3.6, volume 1) and f is a regular epimorphism. Thus $f = \mathsf{Coker}\,(u, v)$ for some pair $u, v\colon P \rightrightarrows A$ and therefore $f = \mathsf{Coker}\,(u - v)$; see 1.2.8. □

Theorem 2.6.11 *The following conditions are equivalent:*

(1) $\mathscr{C}$ is an abelian category;

(2) $\mathscr{C}$ is an additive exact category;

(3) $\mathscr{C}$ is a non-empty, preadditive exact category.

Proof By lemmas 2.6.2 to 2.6.10 and examples 2.4.1 and 2.5.5.d. □

Counterexample 2.6.12
Let us recall that an abelian group A (written additively) is torsion-free when the property

$$na = a + \ldots + a = 0 \quad \Rightarrow \quad a = 0$$

holds for each element $a \in A$ and each non-zero natural number $n \in \mathbb{N}^*$. Let us consider the category $\mathscr{C}$ of torsion-free abelian groups and group homomorphisms between them. We shall prove that $\mathscr{C}$ is regular, but not exact.

First of all observe that the product of two torsion-free groups is again torsion-free as well as every subgroup of this product. This proves already that the category of torsion-free groups has pullbacks which are computed as in the category of abelian groups.

Next observe that given a homomorphism $f\colon A \longrightarrow B$ between torsion-free groups, the coequalizer of its kernel pair in the category of abelian groups is given by the epimorphic part p of the mono–epi-factorization $f = i \circ p$ of f (see 2.1.4). Since every subgroup of the torsion-free group B is still torsion-free, this coequalizer is also the coequalizer in $\mathscr{C}$.

The previous observation shows also that if f is a strong epimorphism in $\mathscr{C}$, i is both a monomorphism and a strong epimorphism in $\mathscr{C}$, thus an isomorphism (see 4.3.6, volume 1). Therefore f is a strong epimorphism in $\mathscr{C}$ precisely when it is a strong epimorphism in the category of abelian groups.

All these observations, joined to the fact that the category of abelian groups is regular (see 2.4.1), let us conclude that $\mathscr{C}$ is regular as well.

Since $\mathscr{C}$ is additive, the exactness of $\mathscr{C}$ would imply its abelianness (see 2.6.11). In particular one would have a short exact sequence

$$0 \longrightarrow 2\mathbb{Z} \xrightarrow{\ i\ } \mathbb{Z} \xrightarrow{\ q\ } Q \longrightarrow 0$$

where i is the canonical inclusion of the even integers in $\mathbb{Z}$ and q is the cokernel of i in $\mathscr{C}$. But from $q(2) = 0$ we deduce $q(1) + q(1) = 0$, thus $q(1) = 0$ since Q is torsion-free. This implies that $q = 0$, with therefore the identity on $\mathbb{Z}$ as kernel (see 1.1.8). This contradicts the fact that i should be the kernel of q (see 1.8.5).

2.7 An embedding theorem

The aim of this section is to prove that every small regular category $\mathscr{C}$ admits a full exact embedding in a Grothendieck topos: as a consequence, to prove an exactness property in a regular category, it suffices to prove it in every Grothendieck topos. And most often, in a topos, an exactness property is proved just as in the category of sets, using the internal logic of the topos (see chapter 6, volume 2).

In this chapter, we assume some familiarity with at least the definition of a Grothendieck topos. Moreover we assume freely the axiom of universes (see 1.1.4, volume 1) so that every regular category can be viewed as a small category with respect to some sufficiently big universe of sets.

Proposition 2.7.1 *Let $\mathscr{C}$ be a small regular category. For each regular epimorphism $f\colon D \twoheadrightarrow C$ in $\mathscr{C}$, consider the subobject $R_f \subseteq \mathscr{C}(-, C)$ of*

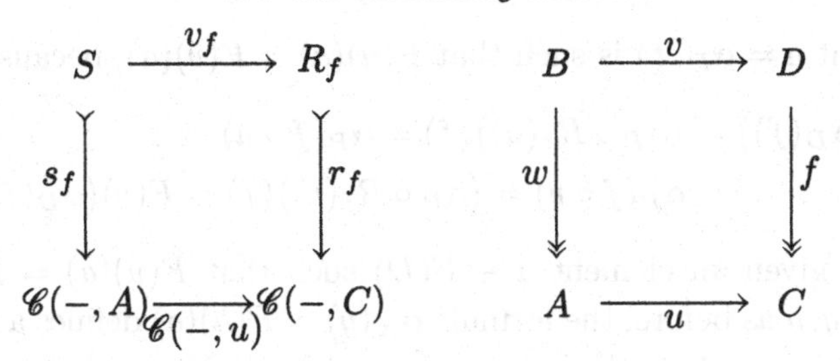

Diagram 2.10

those morphims of $\mathscr{C}$ which factor through f:

$$R_f(X) = \{g \in \mathscr{C}(X,C) \mid \exists h \in \mathscr{C}(X,D) \quad g = f \circ h\}$$

Those subobjects R_f constitute a localizing system in the sense of 3.2.1, volume 3 (i.e. are pullback stable). We shall refer to it as the "localizing system of regular epimorphisms".

Proof Consider a regular epimorphism $f\colon D \twoheadrightarrow C$, an arbitrary morphism $u\colon A \longrightarrow C$ and the pullbacks of diagram 2.10, respectively in $\mathsf{Fun}(\mathscr{C}^*, \mathsf{Set})$ and $\mathscr{C}$. By 2.1.3, w is a regular epimorphism. Moreover given $g \in \mathscr{C}(X,A)$, g factors through w iff $u \circ g$ factors through f, just by the definition of a pullback. With the notation of the statement, this means

$$g \in R_w \Leftrightarrow u \circ g \in R_f \Leftrightarrow g \in \mathscr{C}(-,u)^{-1}(R_f),$$

i.e. $R_w = \mathscr{C}(-,u)^{-1}(R_f)$. □

Lemma 2.7.2 *Let $\mathscr{C}$ be a small regular category. A contravariant functor $F\colon \mathscr{C}^* \longrightarrow \mathsf{Set}$ is a sheaf for the localizing system of regular epimorphisms iff it maps an exact sequence in $\mathscr{C}$ onto an equalizer diagram in Set.*

Proof We use the notation of 2.7.1. Fixing a regular epimorphism $f\colon D \twoheadrightarrow C$ in $\mathscr{C}$, we write u, v for its kernel pair, yielding an exact sequence in $\mathscr{C}$:

$$P \underset{v}{\overset{u}{\rightrightarrows}} D \overset{f}{\twoheadrightarrow} C.$$

Given a presheaf (i.e. a contravariant functor) $F\colon \mathscr{C}^* \longrightarrow \mathsf{Set}$, a natural transformation $\alpha\colon R_f \Rightarrow F$ is completely determined by the single element $\alpha_D(f) \in F(D)$. Indeed, given $X \in \mathscr{C}$ and $g \in R_f(X)$, $g = f \circ h$ for some $h \in \mathscr{C}(X,D)$ and therefore, by naturality of α,

$$\alpha_X(g) = \alpha_X(f \circ h) = \big(\alpha_X \circ R_f(h)\big)(f) = \big(F(h) \circ \alpha_D\big)(f).$$

This element $a = \alpha_D(f)$ is such that $F(u)(a) = F(v)(a)$, because

$$\begin{aligned} F(u)\big(\alpha_D(f)\big) &= \big(\alpha_P \circ R_f(u)\big)(f) = \alpha_P(f \circ u) \\ &= \alpha_P(f \circ v) = \big(\alpha_P \circ R_f(v)\big)(f) = F(v)\big(\alpha_D(f)\big). \end{aligned}$$

Conversely, given an element $a \in F(D)$ such that $F(u)(a) = F(v)(a)$ and arrows g, h as before, the formula $\alpha_X(g) = F(h)(a)$ defines a natural transformation $\alpha\colon R_f \Rightarrow F$, just because, given $x\colon Y \longrightarrow X$,

$$\begin{aligned} \big(F(x) \circ \alpha_X\big)(g) &= \big(F(x) \circ F(h)\big)(a) = F(h \circ x)(a) = \alpha_Y(f \circ h \circ x) \\ &= \big(\alpha_Y \circ \mathscr{C}(x, C)\big)(f \circ h) = \big(\alpha_Y \circ \mathscr{C}(x, C)\big)(g). \end{aligned}$$

In other words, the natural transformations $\alpha\colon R_f \Rightarrow F$ are exactly determined by the elements $a \in F(D)$ such that $F(u)(a) = F(v)(a)$.

By the Yoneda lemma, the natural transformations $\beta\colon \mathscr{C}(-, C) \Rightarrow F$ correspond bijectively with the elements $b \in F(C)$, the correspondence being given by (see 1.3.3, volume 1)

$$b = \beta_C(1_C), \quad \beta_X(g) = F(g)(b).$$

In particular given $h \in \mathscr{C}(X, D)$ and $g = f \circ h \in R_f(X)$,

$$\beta_X(g) = F(f \circ h)(b) = F(h)\big(F(f)(b)\big).$$

Obviously one has

$$F(u)\big(F(f)(b)\big) = F(f \circ u)(b) = F(f \circ v)(b) = \big(F(v)\big)\big(F(f)(b)\big)$$

so that $a = F(f)(b) \in F(D)$ is the element which determines the restriction α of β to R_f.

In conclusion, every natural transformation $\alpha\colon R_f \Rightarrow F$ extends uniquely as $\beta\colon \mathscr{C}(-, C) \Rightarrow F$ iff the correspondence

$$F(C) \longrightarrow F(D), \quad b \mapsto F(f)(b)$$

induces a bijection between $F(C)$ and the elements $a \in F(D)$ such that $F(u)(a) = F(v)(a)$. This is equivalent to saying that the following diagram is an equalizer in the category of sets:

$$F(C) \xrightarrow{F(f)} F(D) \underset{F(v)}{\overset{F(u)}{\rightrightarrows}} F(P). \qquad \square$$

Theorem 2.7.3 *Let $\mathscr{C}$ be a small regular category. The Yoneda embedding*

$$Y\colon \mathscr{C} \longrightarrow \mathsf{Sh}(\mathscr{C}, \mathcal{R})$$

mapping $C \in \mathscr{C}$ to $Y(C) = \mathscr{C}(-, C)$ is a full exact embedding of $\mathscr{C}$ in the topos of sheaves over $\mathscr{C}$ for the localizing system $\mathcal{R}$ of regular epimorphisms.

Proof Let us first mention that every topos is indeed a regular category (see 3.4.14, volume 3).

Given an exact sequence

$$P \underset{v}{\overset{u}{\rightrightarrows}} D \xrightarrow{f} C$$

in $\mathscr{C}$, $f = \mathsf{Coker}\,(u, v)$ so that $\mathscr{C}(f, C) = \mathsf{Ker}\,\big(\mathscr{C}(u, C), \mathscr{C}(v, C)\big)$ by 2.9.5, volume 1. Thus each $\mathscr{C}(-, X)$ is indeed a sheaf so that the Yoneda embedding, which is full and faithful, takes values in the topos of sheaves.

It remains to prove that this Yoneda embedding is exact. The Yoneda embedding of $\mathscr{C}$ in its category of presheaves preserves finite limits (see 2.15.5, volume 1) and the topos of sheaves $\mathsf{Sh}(\mathscr{C}, \mathcal{R})$ is stable under finite limits in the topos of presheaves (see 3.4.3, volume 3). Therefore the Yoneda embedding of the statement preserves finite limits.

By 2.5.7, volume 1, it remains to prove that Y preserves regular epimorphisms. If $f\colon D \twoheadrightarrow C$ is a regular epimorphism in the category $\mathscr{C}$ and $\alpha, \beta\colon \mathscr{C}(-, C) \rightrightarrows F$ are two morphisms in $\mathsf{Sh}(\mathscr{C}, \mathcal{R})$ such that $\alpha \circ \mathscr{C}(-, f) = \beta \circ \mathscr{C}(-, f)$, α, β correspond by the Yoneda lemma (see 1.3.3, volume 1) to elements $a, b \in F(C)$ such that $F(f)(a) = F(f)(b)$. Since F is a sheaf, $F(f)$ is a monomorphism and therefore $a = b$, thus $\alpha = \beta$. So $\mathscr{C}(-, f)$ is an epimorphism, thus a regular epimorphism (see 3.4.13, volume 3). □

Metatheorem 2.7.4 *To prove that a property involving just finite limits and exact sequences holds in every regular category, it suffices to prove it holds in every Grothendieck topos.*

Proof The Yoneda embedding of 2.7.3 is full and faithful, thus it reflects isomorphisms. Since Y preserves finite limits and exact sequences, it reflects them as well (see 2.9.7, volume 1). □

2.8 The calculus of relations

In this section, we make constant use of the metatheorem 2.7.4 and develop all our proofs in a Grothendieck topos via its internal logic (see chapter 6, volume 3). It is an easy but lengthy exercise to write direct proofs involving just the consideration of diagrams in the original regular

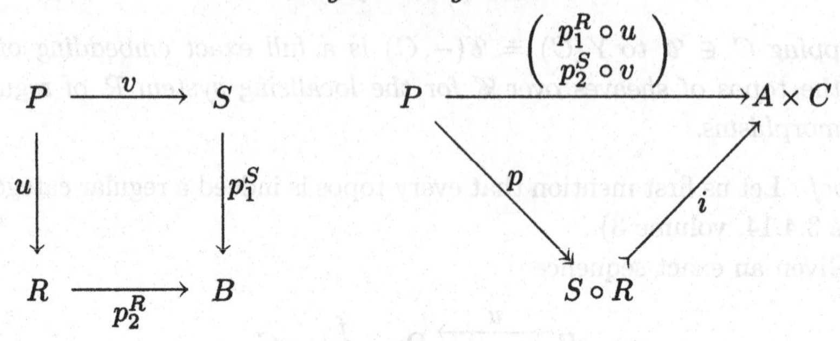

Diagram 2.11

category. Let us mention once more that every topos is a regular category (see 3.4.14, volume 3).

In this section, we shall use the notation $R: A \dashrightarrow B$ to denote a relation from A to B (see 2.8.1). A quick look at theorem 2.8.4 shows that the analogy with the theory of distributors (see 7.8, volume 1) justifies this analogy in the notation.

Finally let us recall that by a subobject, we mean an equivalence class of monomorphisms, for the isomorphism relation (see 4.1.1, volume 1).

Definition 2.8.1 *Let $\mathscr{C}$ be a finitely complete category. By a relation $R: A \dashrightarrow B$, we mean a subobject $R \subseteq A \times B$. We write*

$$p_1^R\colon R \subseteq A \times B \xrightarrow{p_A} A,$$

$$p_2^R\colon R \subseteq A \times B \xrightarrow{p_B} B$$

for the corresponding projections.

In a topos, p_1^R and p_2^R can thus be described by the formulas $p_1^R(a,b) = a$, $p_2^R(a,b) = b$ where $a \in A$, $b \in B$.

Definition 2.8.2 *Let $\mathscr{C}$ be a finitely complete category. Given a relation $R: A \dashrightarrow B$ in $\mathscr{C}$, we define the opposite relation $R^0: B \dashrightarrow A$ as the following composite:*

$$R \subseteq B \times A \cong A \times B,$$

where the isomorphism is the canonical one. In other words, $p_1^{R^0} = p_2^R$ and $p_2^{R^0} = p_1^R$.

In a topos, R^0 can thus be described as

$$R^0 = \{(b,a) \in B \times A \,|\, (a,b) \in R\}.$$

Definition 2.8.3 *Let $\mathscr{C}$ be a finitely complete and regular category. Given two relations $R: A \dashrightarrow B$ and $S: B \dashrightarrow C$ in $\mathscr{C}$, their composite $S \circ R: A \dashrightarrow C$ is defined as in diagram 2.11, where the left-hand square* *is a pullback and the right-hand triangle is an image factorization (see 2.1.4).*

In a topos, one thus has the following descriptions:

$$P = \{(a,b,c) \in A \times B \times C \mid (a,b) \in R \;\; (b,c) \in S\},$$
$$S \circ R = \{(a,c) \in A \times C \mid \exists b \in B \;\; (a,b) \in R \;\; (b,c) \in S\}.$$

Observe that every category $\mathscr{C}$ can be viewed as a "discrete" 2-category with just identity 2-cells (see 7.1.4.d, volume 1). With that in mind, we get the following theorem.

Theorem 2.8.4 *Let $\mathscr{C}$ be a regular, well-powered and finitely complete category. We get a 2-category $\mathsf{Rel}(\mathscr{C})$ by choosing as:*

- *objects = those of $\mathscr{C}$;*
- *arrows = the relations of $\mathscr{C}$, with the composition defined in 2.8.3;*
- *2-cells = the inclusions of relations, viewed as subobjects in $\mathscr{C}$.*

Moreover there is an injective 2-functor $\rho\colon \mathscr{C} \longrightarrow \mathsf{Rel}(\mathscr{C})$ with the property that an arrow $R\colon A \multimap B$ in $\mathsf{Rel}(\mathscr{C})$ has a right adjoint if and only if it has the form $\rho(f)$ for some $f\colon A \longrightarrow B$ in $\mathscr{C}$; this adjoint turns out to be R^0.

Proof By our metatheorem 2.7.4, it suffices to write the proof in a topos, using its internal logic.

Given an object $A \in \mathscr{C}$, write $\Delta_A \subseteq A \times A$ for its diagonal relation. If $R\colon A \multimap B$ is an arbitrary relation,

$$\begin{aligned} R \circ \Delta_A &= \{(a,b) \in A \times B \mid \exists a' \in A \;\; (a,a') \in \Delta_A \;\; (a',b) \in R\} \\ &= \{(a,b) \in A \times B \mid (a,b) \in R\} = R, \end{aligned}$$

since $(a,a') \in \Delta_A$ means $a = a'$; in the same way $\Delta_B \circ R = R$. To prove the associativity of the composition law for relations, consider relations

$$A \overset{R}{\multimap} B \overset{S}{\multimap} C \overset{T}{\multimap} D.$$

We have

$$\begin{aligned} &T \circ (S \circ R) \\ &= \{(a,d) \in A \times D \mid \exists c \in C \;\; (a,c) \in S \circ R \;\; (c,d) \in T\} \\ &= \{(a,d) \in A \times D \mid \exists c \in C \;\; \exists b \in B \;\; (a,b) \in R \;\; (b,c) \in S \;\; (c,d) \in T\} \\ &= \{(a,d) \in A \times D \mid \exists b \in B \;\; (a,b) \in R \;\; (b,d) \in T \circ S\} \\ &= (T \circ S) \circ R. \end{aligned}$$

Thus the objects of $\mathscr{C}$, together with the relations, already constitute a category $\mathsf{Rel}(\mathscr{C})$, with each $\mathsf{Rel}(\mathscr{C})(A,B)$ indeed a set since $\mathscr{C}$ is well-powered.

But in fact each $\mathsf{Rel}(\mathscr{C})(A,B)$ is a poset, thus a category, for the partial ordering given by the inclusion of subobjects. Defining the Godement product of 2-cells reduces to checking that given

$$R \subseteq S \subseteq A \times B, \;\; T \subseteq U \subseteq B \times C,$$

then $T \circ R \subseteq U \circ S$. Indeed

$$\begin{aligned}(a,c) \in T \circ R &\Rightarrow \exists b \in B \;\; (a,b) \in R \;\; (b,c) \in T\\ &\Rightarrow \exists b \in B \;\; (a,b) \in S \;\; (b,c) \in U\\ &\Rightarrow (a,c) \in U \circ S.\end{aligned}$$

The other conditions for having a 2-category are just obvious since they require the equality of some 2-cells; and when a 2-cell exists between two arrows, it is necessarily unique since each $\mathsf{Rel}(\mathscr{C})(A,B)$ is a poset. Thus $\mathsf{Rel}(\mathscr{C})(A,B)$ is already a category.

The embedding ρ is defined by $\rho(A) = A$ for every object A. Given an actual morphism $f\colon A \longrightarrow B$ of $\mathscr{C}$, its "graph" defined by

$$\begin{pmatrix} 1_A \\ f \end{pmatrix} : A \longrightarrow A \times B$$

is a monomorphism, since its composite with the first projection is just 1_A (see 1.7.2, volume 1). So $\binom{1_A}{f}$ is a relation from A to B, which we choose as $\rho(f)$. This relation $\rho(f)$ can thus be described by the formula

$$\rho(f) = \{(a,b) \in A \times B \,|\, b = f(a)\}.$$

In particular

$$\rho(1_A) = \{(a,a') \in A \times A \,|\, a' = 1_A(a)\} = \Delta_A$$

and if $g\colon B \longrightarrow C$ is another morphism of $\mathscr{C}$

$$\begin{aligned}\rho(g) \circ \rho(f) &= \{(a,c) \in A \times C \,|\, \exists b \in B \;\; (a,b) \in \rho(f) \;\; (b,c) \in \rho(g)\}\\ &= \{(a,c) \in A \times C \,|\, \exists b \in B \;\; b = f(a) \;\; g(b) = c\}\\ &= \{(a,c) \in A \times C \,|\, \exists b \in B \;\; b = f(a) \;\; g(f(a)) = c\}\\ &= \{(a,c) \in A \times C \,|\, (g \circ f)(a) = c\}\\ &= \rho(g \circ f).\end{aligned}$$

Thus ρ is indeed a functor.

By construction ρ is injective on the objects. Now if $f, g\colon A \rightrightarrows B$ are morphisms of $\mathscr{C}$ such that $\rho(f) = \rho(g)$, then $\binom{1_A}{f} = \binom{1_A}{g}$ and composing with the second projection yields $f = g$. Thus the functor ρ is injective.

Next consider an arrow $f\colon A \longrightarrow B$. The relation $\rho(f)^0\colon B \longrightarrow A$ is right adjoint to the relation $\rho(f)\colon A \longrightarrow B$ in the 2-category $\mathsf{Rel}(\mathscr{C})$ of relations (see 7.1.2, volume 1) precisely when

$$\rho(f) \circ \rho(f)^0 \subseteq \Delta_B, \quad \Delta_A \subseteq \rho(f)^0 \circ \rho(f).$$

This is indeed the case since

$$\begin{aligned}
\rho(f) \circ \rho(f)^0 &= \{(b,b') \in B \times B \mid \exists a \in A \quad (b,a) \in \rho(f)^0 \quad (a,b') \in \rho(f)\} \\
&= \{(b,b') \in B \times B \mid \exists a \in A \quad b = f(a) \quad b' = f(a)\} \\
&\subseteq \{(b,b') \in B \times B \mid b = b'\} \\
&= \Delta_B, \\
\rho(f)^0 \circ \rho(f) &= \{(a,a') \in A \times A \mid \exists b \in B \quad (a,b) \in \rho(f) \quad (b,a') \in \rho(f)^0\} \\
&= \{(a,a') \in A \times A \mid \exists b \in B \quad b = f(a) \quad b = f(a')\} \\
&= \{(a,a') \in A \times A \mid f(a) = f(a')\} \\
&\supseteq \Delta_A.
\end{aligned}$$

Conversely consider two relations $R\colon A \rightarrow\!\!\!\!\circ\!\!\!\!\rightarrow B$ and $S\colon B \rightarrow\!\!\!\!\circ\!\!\!\!\rightarrow A$ such that

$$R \circ S \subseteq \Delta_B, \quad \Delta_A \subseteq S \circ R.$$

We must prove that $R = \rho(f)$, for some $f\colon A \longrightarrow B$ in $\mathscr{C}$. For this it suffices to prove that

$$\forall a \in A \quad \exists!\, b \in B \quad (a,b) \in R$$

(see 5.10.9, volume 3). And indeed, given $a \in A$,

$$\begin{aligned}
a \in A &\Rightarrow (a,a) \in \Delta_A \\
&\Rightarrow (a,a) \in S \circ R \\
&\Rightarrow \exists b \in B \quad (a,b) \in R \quad (b,a) \in S \\
&\Rightarrow \exists b \in B \quad (a,b) \in R,
\end{aligned}$$

$$\begin{aligned}
&(a,b) \in R \quad (a,b') \in R \\
&\quad\Rightarrow (a,b) \in R \quad (a,b') \in R \quad (a,a) \in \Delta_A \\
&\quad\Rightarrow (a,b) \in R \quad (a,b') \in R \quad (a,a) \in S \circ R \\
&\quad\Rightarrow (a,b) \in R \quad (a,b') \in R \quad \exists b'' \in B \quad (a,b'') \in R \quad (b'',a) \in S \\
&\quad\Rightarrow \exists b'' \in B \quad (b'',a) \in S \quad (a,b) \in R \quad (b'',a) \in S \quad (a,b') \in R
\end{aligned}$$

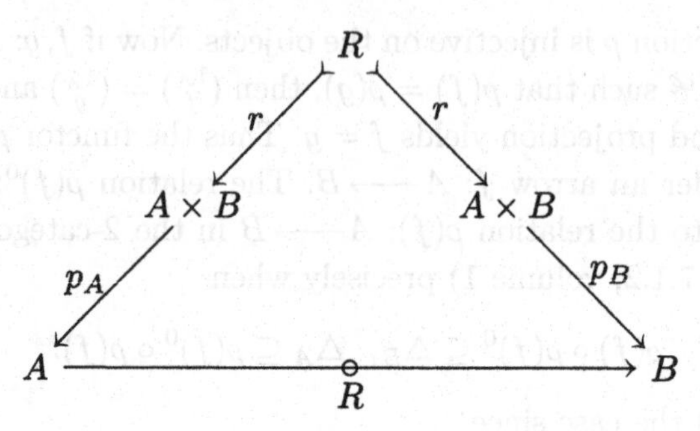

Diagram 2.12

$$\begin{aligned}
&\Rightarrow \exists b'' \in B \quad (b'', b) \in R \circ S \quad (b'', b') \in R \circ S \\
&\Rightarrow \exists b'' \in B \quad (b'', b) \in \Delta_B \quad (b'', b') \in \Delta_B \\
&\Rightarrow \exists b'' \in B \quad b'' = b \quad b'' = b' \\
&\Rightarrow b = b'.
\end{aligned}$$
□

In fact, in 2.8.4, the arrows of the type $\rho(f)$ and their adjoints suffice to generate $\mathsf{Rel}(\mathscr{C})$.

Proposition 2.8.5 *Let $\mathscr{C}$ be a regular, well-powered and finitely complete category. Given a relation $R: A \multimap B$, consider the situation of diagram 2.12 in the category $\mathscr{C}$. In the category $\mathsf{Rel}(\mathscr{C})$ of relations, R can be written as the composite $R = \rho(p_B \circ r) \circ \rho(p_A \circ r)^0$.*

Proof By our metatheorem 2.7.4, it suffices to write down the proof in a topos, using its internal logic.

$$\begin{aligned}
&\rho(p_B \circ r) \circ \rho(p_A \circ r)^0 \\
&\quad = \big\{(a, b) \in A \times B \big| \exists (a', b') \in R \;\; p_A(a', b') = a \;\; p_B(a', b') = b\big\} \\
&\quad = \big\{(a, b) \in A \times B \big| \exists (a', b') \in R \;\; a' = a \;\; b' = b\big\} \\
&\quad = \big\{(a, b) \in A \times B \big| (a, b) \in R\big\} \\
&\quad = R.
\end{aligned}$$
□

Another interesting property of the inclusion in 2.8.4 is a characterization of monomorphisms and epimorphisms.

Proposition 2.8.6 *Let $\mathscr{C}$ be a regular, well-powered and finitely complete category. Given a morphism $f: A \longrightarrow B$ in $\mathscr{C}$:*

(1) f is a monomorphism iff $\rho(f)^0 \circ \rho(f) = \Delta_A$;

(2) f is a regular epimorphism iff $\rho(f) \circ \rho(f)^0 = \Delta_B$.

Proof By our metatheorem 2.7.4, it suffices to write down the proof in a topos, using its internal logic. We recall that in a topos, every epimorphism is regular (see 3.4.13, volume 3).

We know already that given $f\colon A \longrightarrow B$,

$$\begin{aligned}\rho(f)^0 \circ \rho(f) &= \{(a,a') \in A \times A \mid f(a) = f(a')\},\\ \rho(f) \circ \rho(f)^0 &= \{(b,b') \in B \times B \mid \exists a \in A \;\; b = f(a) \;\; b' = f(a)\}\\ &= \{(b,b) \in B \times B \mid \exists a \in A \;\; b = f(a)\}\end{aligned}$$

(see proof of 2.8.4). It follows immediately from these formulas that

$$\begin{aligned}\rho(f)^0 \circ \rho(f) = \Delta_A &\text{ iff } \forall a, a' \in A \;\; f(a) = f(a') \Leftrightarrow a = a'\\ &\text{ iff } f \text{ is a monomorphism}\\ \rho(f) \circ \rho(f)^0 = \Delta_B &\text{ iff } \forall b \in B \;\; \exists a \in A \;\; b = f(a)\\ &\text{ iff } f \text{ is an epimorphism}\end{aligned}$$

(see 5.10.2, volume 3). □

Let us conclude this section with the so-called "modularity laws" for relations. They are identities satisfied in the category $\mathsf{Rel}(\mathscr{C})$ and particularly useful when computing in this category.

Proposition 2.8.7 *Let $\mathscr{C}$ be a regular, well-powered, finitely complete category. Consider three relations*

$$R\colon A \longrightarrow\!\!\!\!\!\!\circ\!\!\!\longrightarrow B, \quad S\colon B \longrightarrow\!\!\!\!\!\!\circ\!\!\!\longrightarrow C, \quad T\colon A \longrightarrow\!\!\!\!\!\!\circ\!\!\!\longrightarrow C.$$

The following identities hold:

$$(S \circ R) \cap T \subseteq S \circ \big(R \cap (S^0 \circ T)\big),$$
$$(S \circ R) \cap T \subseteq \big(S \cap (T \circ R^0)\big) \circ R,$$

where $\cap$ denotes the intersection as subobjects in $\mathscr{C}$.

Proof Once more we use our metatheorem 2.7.4 and the internal logic of toposes. We prove the first relation; the second one is analogous.

$$\begin{aligned}&(S \circ R) \cap T\\ &\quad= \{(a,c) \in A \times C \mid (a,c) \in S \circ R \;\; (a,c) \in T\}\\ &\quad= \{(a,c) \in A \times C \mid \exists b \in B \;\; (a,b) \in R \;\; (b,c) \in S \;\; (a,c) \in T\}\\ &\quad= \{(a,c) \in A \times C \mid \exists b \in B \;\; (a,b) \in R\\ &\qquad\qquad (a,c) \in T \;\; (c,b) \in S^0 \;\; (b,c) \in S\}\\ &\quad\subseteq \{(a,c) \in A \times C \mid \exists b \in B \;\; (a,b) \in R\\ &\qquad\qquad (\exists c' \;\; (a,c') \in T \;\; (c',b) \in S^0) \;\; (b,c) \in S\}\\ &\quad= \{(a,c) \in A \times C \mid \exists b \in B \;\; (a,b) \in R \;\; (a,b) \in S^0 \circ T \;\; (b,c) \in S\}\end{aligned}$$

$$= \{(a,c) \in A \times C \mid \exists b \in B \ (a,b) \in R \cap (S^0 \circ T) \ (b,c) \in S\}$$
$$= S \circ (R \cap (S^0 \circ T)). \qquad \square$$

2.9 Exercises

2.9.1 In the category of abelian groups, prove that pulling back along a morphism does not respect coequalizers. [Hint: consider the two canonical injections $s_1, s_2\colon A \rightrightarrows A \oplus A$ of a biproduct and pull their coequalizer back along a zero morphism.]

2.9.2 Let us call "universal" an epimorphism $f\colon A \longrightarrow B$ whose pullback along every morphism $g\colon C \longrightarrow B$ exists and is again an epimorphism. Show that in definition 2.1.1, the last axiom cannot equivalently be replaced by "every regular epimorphism is universal" [Hint: observe that in the category of topological spaces, every epimorphism is universal]. Nevertheless, show that making this replacement, every morphism still factors as a regular epimorphism followed by a monomorphism.

2.9.3 Prove that the category of sets is both exact and coexact.

2.9.4 Prove that the category of compact Hausdorff spaces is both exact and coexact.

2.9.5 Describe, in terms of finite limits, what it means for a relation $r_1, r_2\colon R \rightrightarrows A$ on an object A of a category to be a partial order.

2.9.6 In a category with binary products, prove that the two projections $p_1, p_2\colon A \times A \rightrightarrows A$ constitute an equivalence relation on A, corresponding to the identity subobject $A \times A \Longrightarrow A \times A$.

2.9.7 Prove that every poset is an exact category.

2.9.8 Show that the category of compact Hausdorff 0-dimensional spaces is regular but not exact.

2.9.9 Let $\mathscr{C}$ be a finitely complete regular category. An object $C \in \mathscr{C}$ has global support when the morphism $C \longrightarrow\!\!\!\!\!\rightarrow 1$ is a regular epimorphism. Prove that the full subcategory $\mathscr{G}$ of objects with global support is again regular, but not in general finitely complete. When $\mathscr{C}$ is exact, $\mathscr{G}$ is exact as well.

2.9.10 Let $\mathscr{C}$ be a regular category. Prove that the following conditions are equivalent:

(1) if R, S are two equivalence relations on an object A, then $R \circ S = S \circ R$;

(2) if $R\colon A \longrightarrow\!\!\!\!\!\!\!\circ\;\; A$ is a relation, then $R = R \circ R^0 \circ R$;

(3) every reflexive relation $R \subseteq A \times A$ is symmetric;
(4) every reflexive relation $R \subseteq A \times A$ is an equivalence relation;
(5) if R, S are two equivalence relations on an object A, then $R \circ S$ is also an equivalence relation on A.

An exact category which satisfies these equivalent conditions is called a Mal'cev category.

2.9.11 Let $\mathscr{C}$ be a finitely complete category with strong-epi–mono factorizations. Prove that $\mathscr{C}$ is regular iff the composition of relations is associative. [Hint: given an epimorphism $f: A \twoheadrightarrow C$ and an arbitrary morphism $g: B \longrightarrow C$, consider the composite $\rho(f) \circ \rho(f)^0 \circ \rho(g)$.]

3
Algebraic theories

In this chapter, we investigate a general approach to those structures like groups, rings, modules, lattices, boolean algebras... characterized by the existence of one or several operations which are defined everywhere and satisfy axioms expressed by equalities.

Let us immediately underline the fact that structures like fields and categories do not fall under the scope of this chapter, even if they have an obvious algebraic nature. Indeed, the theory of fields admits an operation (the inverse for the multiplication) which is not defined everywhere and, clearly, the composition law in the case of a category is just defined for some pairs of arrows (see exercises 3.13.1 and 3.13.2), not for all of them.

3.1 The theory of groups revisited

A group can be defined as a set G provided with a binary operation

$$+: G \times G \longrightarrow G$$

satisfying the axioms

$$\forall x, y, z \in G \;\; (x+y)+z = x+(y+z),$$

$$\exists 0 \in G \;\; \forall x \in G \;\; \exists y \in G \;\; x+0 = x = 0+x \;\;,\;\; x+y = 0 = y+x.$$

The existence of the unit or the opposite can in fact be presented as an axiom, not as a property, yielding the equivalent definition that a group is a set G provided with

- a binary operation $+: G \times G \longrightarrow G$,
- a 1-ary operation $-: G \longrightarrow G$,
- a constant $0 \in G$,

satisfying the axioms

$$\begin{aligned}\forall x, y, z \in G \ \ (x+y)+z &= x+(y+z),\\ \forall x \in G \ \ x+0 = x &= 0+x,\\ \forall x \in G \ \ x+(-x) = 0 &= (-x)+x.\end{aligned}$$

One should obšerve that these axioms are now presented in a very elementary form: just equalities between algebraic composites, without any existential quantifier, implication symbol, conjunction, disjunction or negation.

As usual, we shall write G^n for the product of n copies of G, $n \in \mathbb{N}$; in particular G^0 is the singleton, as observed in 2.3.2.a, volume 1. Observe also that giving a constant $0 \in G$ is equivalent to giving a "0-ary operation" $0\colon G^0 \longrightarrow G$, so that finally the theory of groups has three operations,

- a binary operation $+\colon G^2 \longrightarrow G$,
- a 1-ary operation $-\colon G^1 \longrightarrow G$,
- a 0-ary operation $0\colon G^0 \longrightarrow G$,

and the axioms can be expressed by the commutativity of the various pieces of diagram 3.1.

The previous presentation is somewhat misleading since it does not describe the "theory of groups", but the "theory of a group G"; indeed, our description uses explicitly a "generic" set G to describe the group structure. It is easy to overcome this difficulty. The theory of groups can be described by giving a denumerable set of variables $x, y, z, \ldots$ as well as three formal symbols $+, -, 0$ and the equality. The terms of the theory of groups are then defined inductively by the following rules:

- every variable is a term;
- if s, t are terms, $s+t$ is a term;
- if t is a term, $-t$ is a term;
- 0 is a term.

The axioms of the theory of groups are expressed by equalities between terms: if x, y, z are variables,

$$\begin{aligned}x+(y+z) &= (x+y)+z,\\ x+0 = x &= 0+x,\\ x+(-x) = 0 &= (-x)+x.\end{aligned}$$

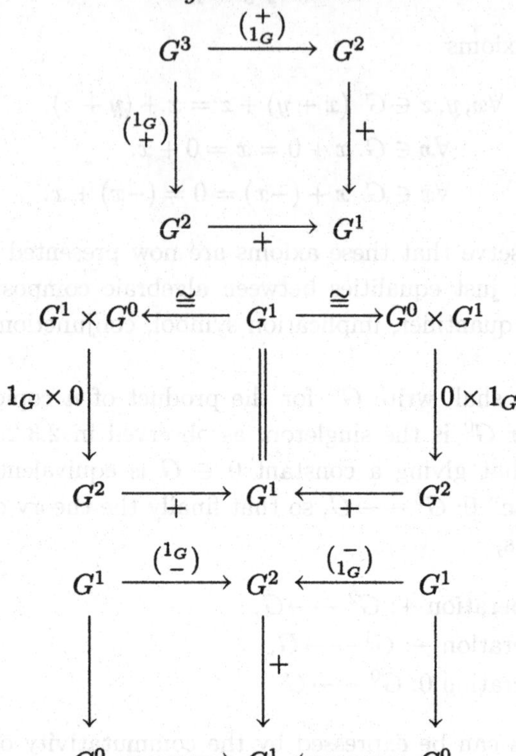

Diagram 3.1

A model of this formal theory of groups is now a set theoretical interpretation. More precisely, we fix a set G and interpret the three symbols $+, -, 0$ by choosing

- a mapping $\dot{+}\colon G \times G \longrightarrow G$,
- a mapping $\dot{-}\colon G \longrightarrow G$,
- an element $\dot{0} \in G$,

where we have written $a \dot{+} b$ for $\dot{+}(a, b)$. The terms are then interpreted inductively:

- a variable can be interpreted as any element of G;
- if the terms s, t are already interpreted as elements $|s|, |t| \in G$, the term $s + t$ is interpreted as the element $|s| \dot{+} |t|$;
- if the term t is already interpreted as an element $|t| \in G$, the term $-t$ is interpreted as $\dot{-}\,|t|$;
- the term 0 is interpreted as the element $\dot{0} \in G$.

The data $(G, \dot{+}, \dot{-}, \dot{0})$ constitute a model of the theory of groups when, for each possible interpretation of the variables, the three equalities

$$|x| \dot{+} (|y| \dot{+} |z|) = (|x| \dot{+} |y|) \dot{+} |z|,$$
$$|x| \dot{+} \dot{0} = |x| = \dot{0} \dot{+} |x|,$$
$$|x| \dot{+} (\dot{-} |x|) = \dot{0} = (\dot{-} |x|) \dot{+} |x|$$

hold between elements of G, for all variables x, y, z.

3.2 A glance at universal algebra

This section is intended to explain the precise link between classical universal algebra and categorical universal algebra. The reader just interested in the latter can go directly to section 3.3.

What has been done in 3.1 could be repeated for the theory of rings with unit: it would be necessary to add a second constant 1 and a second binary operation $\times$. And clearly one can easily imagine theories where one has ternary or even n-ary operations, for $n \in \mathbb{N}$. This leads to the following general definition.

Definition 3.2.1 *A presentation of an algebraic theory $\mathcal{T}$ is a theory with equality, specified by choosing, besides a denumerable set of variables, a set $\mathcal{O}_n$ of "n-ary operations" for each integer $n \in \mathbb{N}$, together with a set of axioms, subject to the following requirements. The terms of the theory are defined inductively:*

- *each variable is a term;*
- *if $\alpha \in \mathcal{O}_n$ and $t_1, \ldots, t_n$ are terms, then $\alpha(t_1, \ldots, t_n)$ is a term.*

An axiom is an equality between two terms.

Observe that the second condition defining the terms, in the case $n = 0$, becomes exactly

- if $\alpha \in \mathcal{O}_0$, then α is a term.

The 0-ary operations are also called *constants*; thus

- every constant is a term.

Definition 3.2.2 *Let $\mathcal{T}$ be a presentation of an algebraic theory in the sense of 3.2.1. By a model of $\mathcal{T}$ we mean the choice of*

- *a set M,*
- *for all $n \in \mathbb{N}$, for all $\alpha \in \mathcal{O}_n$, a mapping $|\alpha| : M^n \longrightarrow M$,*

in such a way that the axioms of $\mathcal{T}$ are realized by this interpretation. More precisely

- *a variable is interpreted as any element of M,*
- *if $\alpha \in \mathcal{O}_n$ and the terms $t_1, \ldots, t_n$ are already interpreted as elements $|t_1|, \ldots, |t_n| \in M$ then $\alpha(t_1, \ldots, t_n)$ is interpreted as the element*

$$|\alpha|\left(|t_1|, \ldots, |t_n|\right) \in M,$$

and an axiom is satisfied in M when, for every possible interpretation of the variables, both sides of the equality have the same interpretation in M.

Definition 3.2.3 *Let $\mathcal{T}$ be a presentation of an algebraic theory in the sense of 3.2.1. If L, M are models of $\mathcal{T}$, a $\mathcal{T}$-homomorphism $f\colon L \longrightarrow M$ is a mapping $f\colon L \longrightarrow M$ such that for every operation $\alpha \in \mathcal{O}_n$ and every elements $x_1, \ldots, x_n \in L$*

$$f\left(|\alpha|\left(x_1, \ldots, x_n\right)\right) = |\alpha|\left(f(x_1), \ldots, f(x_n)\right).$$

It is a completely obvious observation that

Proposition 3.2.4 *Let $\mathcal{T}$ be a presentation of an algebraic theory in the sense of 3.2.1. The models of $\mathcal{T}$ and their homomorphisms, together with the usual composition of mappings, constitute a category.*

What we have just described is the precise object of universal algebra. Now we would like to give an equivalent and very elegant categorical presentation of these notions. To achieve this, we need a series of elementary observations.

Lemma 3.2.5 *Let $\mathcal{T}$ be a presentation of an algebraic theory. There exists a smallest equivalence relation R on the set of terms such that:*

(1) if the axiom $s = t$ holds, then the pair (s, t) is in R;
(2) if the terms s, t are written using the variables $x_1, \ldots, x_n$, the pair (s, t) is in R and $t_1, \ldots, t_n$ are terms, then the pair (s', t') is in R, where s', t' are obtained from s, t by replacing x_i by t_i, $i = 1, \ldots, n$;
(3) if $\alpha \in \mathcal{O}_n$ and the pairs (s_i, t_i) are in R, $(i = 1, \ldots, n)$, then the pair $\left(\alpha(s_1, \ldots, s_n), \alpha(t_1, \ldots, t_n)\right)$ is in R.

Proof Just construct R inductively from the pairs (s, t) given by the axioms, using conditions (2), (3) and the requirements for an equivalence relation. □

Lemma 3.2.6 *Let $\mathcal{T}$ be a presentation of an algebraic theory with set $\{x_1, \ldots, x_n, \ldots\}$ of variables. Let T_n be the set of terms involving only the variables $x_1, \ldots, x_n$. Let F_n be the quotient of T_n by the (restriction* *of the) equivalence relation R of lemma 3.2.5. The set F_n is naturally provided with the structure of a $\mathcal{T}$-model.*

Proof If $\alpha \in \mathcal{O}_m$ and $t_1, \ldots, t_m \in T_n$, then clearly $\alpha(t_1, \ldots, t_m) \in T_n$. By the last condition defining R, this construction is compatible with the equivalence relation, yielding an interpretation

$$|\alpha| : (F_n)^m \longrightarrow F_n.$$

This interpretation satisfies all the axioms of $\mathcal{T}$ whatever the interpretations chosen for the variables are, just by the first two conditions defining the relation R. □

Lemma 3.2.7 *Let $\mathcal{T}$ be a presentation of an algebraic theory. In the category $\mathsf{Mod}_{\mathcal{T}}$ of $\mathcal{T}$-models, F_n is the n-th copower of F_1.*

Proof We use the notation of 3.2.6. For each variable x_i and index $1 \leq i \leq n$, consider all the equivalence classes of all the terms involving just the variable x_i; this yields a $\mathcal{T}$-model $F_1^{(i)}$ isomorphic to F_1, together with an inclusion

$$s_i \colon F_1^{(i)} \longrightarrow F_n$$

which is obviously a $\mathcal{T}$-homomorphism.

Now consider a $\mathcal{T}$-model M together with a family of $\mathcal{T}$-homomorphisms $f_i \colon F_1^{(i)} \longrightarrow M$, $i = 1, \ldots, n$. Interpret the variable x_i, $i = 1, \ldots, n$, in M as $f_i([x_i])$, where $[x_i]$ is the equivalence class of the term x_i in F_n. Define a mapping $T_n \longrightarrow M$ by mapping the term t to its interpretation in M corresponding to the previous interpretation of the variables. Since M is a $\mathcal{T}$-model, this construction is compatible with the axioms of $\mathcal{T}$ and since each f_i is a $\mathcal{T}$-homomorphism, this compatibility extends to the whole relation R, yielding finally a $\mathcal{T}$-homomorphism $f \colon F_n \longrightarrow M$. By construction, $f \circ s_i = f_i$. On the other hand this relation yields $f([x_i]) = f_i([x_i])$, from which we get the uniqueness of f. □

Lemma 3.2.8 *Let $\mathcal{T}$ be a presentation of an algebraic theory. The model F_n is the free model on n generators.*

Proof We consider the "underlying set functor"

$$U \colon \mathsf{Mod}_{\mathcal{T}} \longrightarrow \mathsf{Set}$$

mapping a model M to the underlying set M. The statement means that F_n, together with the mapping

$$\{1, \ldots, n\} \longrightarrow F_n, \quad i \mapsto [x_i],$$

is the reflection of $\{1, \ldots, n\}$ along U.

Giving a mapping $\{1, \ldots, n\} \longrightarrow U(M)$ in the category of sets is just giving n elements $a_1, \ldots, a_n \in M$. Those elements are then chosen as interpretations of $x_1, \ldots, x_n$ in M. This yields a mapping $T_n \longrightarrow M$ mapping a term t to its interpretation in M corresponding to the previous interpretation of the variables. Since M is a model, this construction is compatible with the equivalence relation R of 3.2.5, yielding the required morphism $F_n \longrightarrow M$. □

Here now is the key to a categorical approach to universal algebra.

Proposition 3.2.9 *Let $\mathcal{T}$ be a presentation of an algebraic theory. With the notations of 3.2.7, write $\mathscr{F}$ for the full subcategory of $\mathsf{Mod}_{\mathcal{T}}$ generated by the free models F_n on finitely many generators. The dual category $\mathscr{F}^*$ has finite products and $\mathsf{Mod}_{\mathcal{T}}$ is equivalent to the category of finite product-preserving functors from $\mathscr{F}^*$ to the category of sets, and natural transformations between them.*

Proof For clarity, we work with the category $\mathscr{F}$ which has finite coproducts (see 3.2.7) and consider the contravariant functors $\mathscr{F} \longrightarrow \mathsf{Set}$ transforming finite coproducts into finite products. By 3.2.8, a morphism $F_n \longrightarrow F_m$ in $\mathscr{F}$ is just the choice of n elements of F_m.

Given a contravariant functor $G: \mathscr{F} \longrightarrow \mathsf{Set}$ transforming finite coproducts into finite products, let us construct a $\mathcal{T}$-model by putting $M = G(F_1)$. Every n-ary operation α yields an element $[\alpha(x_1, \ldots, x_n)]$ in F_n, thus a $\mathcal{T}$-homomorphism $\overline{\alpha}$: $F_1 \longrightarrow F_n$. This induces a mapping in Set $G(\overline{\alpha})$: $G(F_n) \longrightarrow G(F_1)$. By assumption on G, $G(F_n) \cong \big(G(F_1)\big)^n$ so that $G(\overline{\alpha})$ is in fact a mapping $|\alpha|: M^n \longrightarrow M$, yielding the interpretation of α in M.

Now suppose the variables $x_1, \ldots, x_n$ are interpreted in the model M as elements $a_1, \ldots, a_n$. The interpretation of the term $\alpha(x_1, \ldots, x_n)$ in M is just $|\alpha|\,(a_1, \ldots, a_n)$. By induction, if t is a term with variables $x_1, \ldots, x_n$, then $[t] \in F_n$ corresponds to a morphism $\overline{t}$: $F_1 \longrightarrow F_n$ and the interpretation of $[t]$ is $G(\overline{t})(a_1, \ldots, a_n) \in M$. In particular if s, t are terms with variables among $\{x_1, \ldots, x_n\}$ and the axiom $s = t$ holds, the relation $[s] = [t]$ holds in F_n by definition of the relation R (see 3.2.5); therefore $\overline{s} = \overline{t}$ and finally $G(\overline{s})(a_1, \ldots, a_n) = G(\overline{t})(a_1, \ldots, a_n)$, proving that s and t have the same interpretation in M. Finally, M is indeed a $\mathcal{T}$-model.

Conversely, let us start with a $\mathcal{T}$-model M. We define $G: \mathscr{F} \longrightarrow \mathsf{Set}$ on the objects by $G(F_n) = M^n$. A morphism f: $F_n \longrightarrow F_m$ corresponds to the choice of n elements $[t_1], \ldots, [t_n] \in F_m$ where the terms $t_1, \ldots, t_n$

have variables $x_1, \ldots, x_m$. In the category of sets, we must construct a mapping $G(f)\colon M^m \longrightarrow M^n$. Given an m-tuple $(a_1, \ldots, a_m) \in M^m$, we interpret each variable x_i $(i = 1, \ldots, m)$ as a_i and map the m-tuple $(a_1, \ldots, a_m)$ to the n-tuple $(|\,t_1\,|, \ldots, |\,t_n\,|)$ in M^n corresponding to that interpretation of the variables. Clearly G is a functor. Moreover, with the notation of 3.2.7, $s_i\colon F_1 \longrightarrow F_m$ corresponds to the element $[x_i] \in F_m$; the corresponding mapping $G(s_i)\colon M^m \longrightarrow M$ maps therefore the m-tuple $(a_1, \ldots, a_m)$ to the interpretation of the term x_i, i.e. to a_i. Thus $G(s_i)$ is just the i-th projection and G transforms finite coproducts into finite products.

The argument is now easily extended to the morphisms. Indeed given two functors $G, H\colon \mathscr{F} \longrightarrow \mathsf{Set}$ transforming finite coproducts into finite products, a natural transformation $\mu\colon G \Rightarrow H$ is a family of mappings $\mu_n\colon G(F_n) \longrightarrow H(F_n)$ satisfying in particular the equality

$$H(s_i) \circ \mu_n = \mu_1 \circ G(s_i) \quad (i = 1, \ldots, n).$$

But since $H(s_i)$ and $G(s_i)$ are just the i-th projections, this means that μ_n is just the n-th power of μ_1. For every n-ary operation α, we can then consider the corresponding morphism $\overline{\alpha}\colon F_1 \longrightarrow F_n$ and the relation

$$H(\overline{\alpha}) \circ (\mu_1)^n = H(\overline{\alpha}) \circ \mu_n = \mu_1 \circ G(\overline{\alpha})$$

proves that $\mu_1\colon G(F_1) \longrightarrow H(F_1)$ is a $\mathcal{T}$-homomorphism.

Conversely, given the $\mathcal{T}$-homomorphism $\mu_1\colon G(F_1) \longrightarrow H(F_1)$, we define $\mu_n\colon G(F_n) \longrightarrow H(F_n)$ as the n-th power of μ_1, yielding not only the relations

$$H(s_i) \circ \mu_n = H(s_i) \circ (\mu_1)^n = \mu_1 \circ G(s_i)$$

for every variable x_i, but also

$$H(\overline{\alpha}) \circ \mu_n = H(\overline{\alpha}) \circ (\mu_1)^n = \mu_1 \circ G(\overline{\alpha})$$

for every n-ary operation α, since μ_1 is a $\mathcal{T}$-homomorphism. By induction, this extends to

$$H(\overline{t}) \circ \mu_n = \mu_1 \circ G(\overline{t})$$

for every term $t \in T_n$. This suffices to prove the naturality of μ since every $f\colon F_n \longrightarrow F_m$ is completely determined by its composites with the canonical morphisms $s_i\colon F_1 \longrightarrow F_n$ of the coproduct. □

To conclude this section, let us observe that given a presentation of an algebraic theory $\mathcal{T}$, we can define a new presentation $\mathcal{T}'$ of an algebraic theory by choosing as n-ary operations all the terms of T_n (notation of

3.2.6) and as axioms all the equations $s = t$ where the pair (s, t) belongs to the relation R described in 3.2.5. Obviously, every $\mathcal{T}'$-model is a $\mathcal{T}$-model. Conversely, if M is a $\mathcal{T}$-model and $t \in \mathcal{T}_n$ is a $\mathcal{T}$-term, one gets a corresponding interpretation of the $\mathcal{T}'$-operation t

$$|t|: M^n \longrightarrow M$$

in the usual way: choosing $(a_1, \ldots, a_n) \in M^n$ we interpret x_i as a_i and map $(a_1, \ldots, a_n)$ to the corresponding interpretation of t in the $\mathcal{T}$-model M. Since M is a $\mathcal{T}$-model, one has $|s| = |t|$ for every pair $(s, t) \in R$, proving that $\mathcal{T}$ and $\mathcal{T}'$ have the same models. In some sense, $\mathcal{T}'$ is a "saturation of $\mathcal{T}$" for all the possible operations and axioms.

But one can get an even more canonical presentation $\mathcal{T}''$ of an algebraic theory, again having the same models as $\mathcal{T}$ and $\mathcal{T}'$, by choosing now the elements of F_n as n-ary operations. For every pair $(s, t) \in R$, the axiom $s = t$ of $\mathcal{T}'$ induces a corresponding axiom $[s] = [t]$ for $\mathcal{T}''\ldots$, but this axiom does not say anything relevant since it has the form $u = u$, for some $u \in F_n$. In fact, for every term $t \in \mathcal{T}_n$ constructed inductively from the variables $x_1, \ldots, x_n$ and the operations of $\mathcal{T}$, one can consider both the $\mathcal{T}''$-term $[t] \in F_n$ and the $\mathcal{T}''$-term $\langle t \rangle \in F_n$ obtained by replacing, in the inductive construction of t, each m-ary $\mathcal{T}$-operation α by the corresponding $\mathcal{T}''$-term $\big[\alpha(x_1, \ldots, x_n)\big]$; one must choose all the equations $[t] = \langle t \rangle$ as axioms for $\mathcal{T}''$. We leave the details to the reader who will observe that this last remark is essentially the content of 3.3.4.

3.3 A categorical approach to universal algebra

With 3.2.9 in mind, we define:

Definition 3.3.1 *By an algebraic theory $\mathcal{T}$ we mean a category $\mathcal{T}$ with a denumerable set $\{T^0, T^1, \ldots, T^n, \ldots\}$ of distinct objects, each object T^n being the n-th power of the object T^1. A model of $\mathcal{T}$ is a functor $F: \mathcal{T} \longrightarrow \mathsf{Set}$ which preserves finite products. A homomorphism of $\mathcal{T}$-models is a natural transformation.*

We shall write $\mathsf{Mod}_{\mathcal{T}}$ for the category of $\mathcal{T}$-models.

Lemma 3.3.2 *Let $\mathcal{T}$ be an algebraic theory. If $\alpha: F \Rightarrow G$ is a morphism in $\mathsf{Mod}_{\mathcal{T}}$, the square in diagram 3.2 commutes, where the isomorphisms are the canonical ones.*

$$\begin{array}{ccc} F(T^n) & \xrightarrow{\alpha_{T^n}} & G(T^n) \\ \cong\downarrow & & \downarrow\cong \\ F(T^1)^n & \xrightarrow[(\alpha_{T^1})^n]{} & G(T^1)^n \end{array}$$

Diagram 3.2

Proof For every projection $p_i\colon T^n \longrightarrow T^1$ of the n-th power $(T^1)^n$, one has

$$G(p_i) \circ \alpha_{T^n} = \alpha_{T^1} \circ F(p_i) = G(p_i) \circ (\alpha_{T^1})^n$$

since the morphisms $G(p_i)$ and $F(p_i)$ are just the i-th projections of the powers $\big(G(T^1)\big)^n$, $\big(F(T^1)\big)^n$. As a consequence $\alpha_{T^n} \cong (\alpha_{T^1})^n$. □

Proposition 3.3.3 *Let $\mathcal{T}$ be an algebraic theory. Consider the functor*

$$U\colon \mathsf{Mod}_{\mathcal{T}} \longrightarrow \mathsf{Set}$$

of evaluation at T^1. Then:

(1) U is representable by $\mathcal{T}(T^1, -)$;
(2) U is faithful;
(3) U reflects isomorphisms;
(4) each finite set with n elements ($n \in \mathbb{N}$) admits $\mathcal{T}(T^n, -)$ as a reflection along U;
(5) $\mathcal{T}(T^1, -)$ is a strong generator for $\mathsf{Mod}_{\mathcal{T}}$.

Proof Each representable functor preserves finite products by 2.9.4, volume 1; therefore each functor $\mathcal{T}(T^n, -)$ is a $\mathcal{T}$-model. By the Yoneda lemma, given a $\mathcal{T}$-model F,

$$\mathsf{Nat}\big(\mathcal{T}(T^1, -), F\big) \cong F(T^1) \cong U(F),$$

proving that U is represented by $\mathcal{T}(T^1, -)$. In the same way

$$\begin{aligned} \mathsf{Nat}\big(\mathcal{T}(T^n, -), F\big) &\cong F(T^n) \\ &\cong \big(F(T^1)\big)^n \\ &\cong \mathsf{Set}\big(\{1, \ldots, n\}, F(T^1)\big) \\ &\cong \mathsf{Set}\big(\{1, \ldots, n\}, U(F)\big), \end{aligned}$$

proving that $\mathcal{T}(T^n, -)$ is the reflection of the set $\{1, \ldots, n\}$ along U; see 3.1, volume 1.

Now consider a morphism of $\mathcal{T}$-models $\alpha: F \Rightarrow G$ such that α_{T^1} is bijective, i.e. an isomorphism in Set (see 1.9.6, volume 1). By 3.3.2, $\alpha_{T^n} \cong (\alpha_{T^1})^n$, thus each α_{T^n} is bijective as well. Therefore when $U(\alpha) = \alpha_{T^1}$ is an isomorphism, so is the natural transformation α and U reflects isomorphisms. In particular $\mathcal{T}(T^1, -)$ is a strong generator (see 4.5.13, volume 1).

Observe also that given two morphisms of $\mathcal{T}$-models $\alpha, \beta: F \Rightarrow G$, the relations $\alpha_{T^n} = (\alpha_{T^1})^n$, $\beta_{T^n} = (\beta_{T^1})^n$ show at the same time that $U\alpha = U\beta$ implies $\alpha = \beta$, proving the faithfulness of U. □

Given an algebraic theory $\mathcal{T}$, we easily get a presentation $\mathcal{T}_1$ of this algebraic theory by choosing the elements of $\mathcal{T}(T^n, T^1)$ as n-ary operations for $\mathcal{T}_1$. Next, given a $\mathcal{T}_1$-term t with variables among $x_1, \ldots, x_n$, one defines inductively the n-ary operation $t \in \mathcal{T}(T^n, T^1)$ associated with t:

- the i-th projection $T^n \longrightarrow T^1$ is associated with x_i;
- if the operations $\beta_1, \ldots, \beta_m$ are associated with terms $s_1, \ldots, s_m$ and α is an m-ary operation, the composite $\alpha \circ (\beta_1, \ldots, \beta_m)$ is associated with the term $\alpha(s_1, \ldots, s_m)$.

One chooses then $t = \tau(x_1, \ldots, x_n)$ as an axiom for $\mathcal{T}_1$.

A warning is necessary here. Clearly, if $(T^n; p_1, \ldots, p_n)$ is the n-th power of T^1 and $\alpha: T^n \longrightarrow T^n$ is an isomorphism, $(T^n, p_1 \circ \alpha, \ldots, p_n \circ \alpha)$ is also an n-th power of T^1, so that the previous description refers to one specified choice of projections. In particular, when $n = 1$, every isomorphism $\alpha: T^1 \longrightarrow T^1$ is such that (T^1, α) is the "1-th" power of T^1. But among all possible choices, there is now a canonical one, namely $\alpha = 1_{T^1}$, the identity arrow on T^1 in the category $\mathcal{T}$. It is convenient to assume that in the description of $\mathcal{T}_1$, this canonical choice has been made. In other words, we require the axiom

$$1_{T^1}(x) = x$$

for the presentation $\mathcal{T}_1$ associated with the algebraic theory $\mathcal{T}$.

Proposition 3.3.4 *Let $\mathcal{T}$ be an algebraic theory and consider the corresponding presentation $\mathcal{T}_1$ of this algebraic theory. The categories of models for $\mathcal{T}$ and $\mathcal{T}_1$ are equivalent and, via this equivalence, the functor*

$$U: \mathsf{Mod}_{\mathcal{T}} \longrightarrow \mathsf{Set}$$

of 3.3.3 maps a $\mathcal{T}_1$-model to its underlying set.

Proof If $F: \mathcal{T} \longrightarrow \mathsf{Set}$ is a $\mathcal{T}$-model, the set $F(T^1)$ provided with the operations

$$F(\alpha): F(T^1)^n \cong F(T^n) \longrightarrow F(T^1)$$

for each $\alpha \in \mathcal{T}(T^n, T^1)$ is obviously a $\mathcal{T}_1$-model; indeed, the satisfaction of the axioms is just the functoriality of F. Moreover it follows immediately from 3.3.2 that giving a morphism $\varphi: F \Rightarrow G$ between $\mathcal{T}$-models is equivalent to giving a morphism $\varphi_{T^1}: F(T^1) \longrightarrow G(T^1)$ between the corresponding $\mathcal{T}_1$-models.

It remains to prove that every $\mathcal{T}_1$-model M is isomorphic to the model $F(T^1)$ arising from a $\mathcal{T}$-model F. Just define F on the objects by $F(T^n) = M^n$. Next if $\alpha: T^n \longrightarrow T^1$ is a morphism in $\mathcal{T}$, α is also a n-ary operation of $\mathcal{T}_1$ and one defines

$$F(\alpha): F(T^n) = M^n \longrightarrow M = F(T^1)$$

to be the realization of the $\mathcal{T}_1$-operation α in the model M. More generally, given $\alpha: T^n \longrightarrow T^m$ one defines

$$F(\alpha): F(T^n) \cong M^n \longrightarrow M^n \cong F(T^m)$$

to be the m-tuple $\big(F(p_1 \circ \alpha), \ldots, F(p_m \circ \alpha)\big)$. The validity of the $\mathcal{T}_1$-axiom in M indicates that the functor F preserves *a priori* the composites of the form $T^n \longrightarrow T^m \longrightarrow T^1$, but this implies immediately the preservation of all composites, just by definition of $F(\alpha)$ for $\alpha: T^n \longrightarrow T^m$. The preservation of the identity on T^1 follows from the axiom $1_{T^1}(x) = x$ of $\mathcal{T}_1$; applying 3.3.2, this implies the preservation of the identity on each T^n. □

Examples 3.3.5

3.3.5.a The considerations of 3.1 have described a presentation of the theory of groups, from which we derive a corresponding algebraic theory, applying 3.2.9; analogous observations hold for the theories of monoids, abelian groups, rings with or without a unit, commutative or not, and so on.

3.3.5.b The most elementary presentation of an algebraic theory is that without any operation and any axiom; a model of it is therefore just a set. Applying 3.2.8 and 3.2.9, we conclude that the corresponding algebraic theory is the dual of the category of finite sets.

3.3.5.c A poset structure on a set X is given not by operations on X, but by a relation on X. In fact, the theory of posets is not algebraic (see exercise 3.13.3). Nevertheless if one requires the poset $(X, \leq)$ to be a $\wedge$-semi-lattice, its structure can now equivalently be defined by the binary "meet" operation

$$\wedge \colon X \times X \longrightarrow X$$

which satisfies the axioms

$$x \wedge (y \wedge z) = (x \wedge y) \wedge z,$$
$$x \wedge y = y \wedge x,$$
$$x \wedge x = x.$$

Indeed, the poset structure is determined by the meet operation via the relation

$$x \leq y \quad \text{iff} \quad x \wedge y = x.$$

Conversely, given a set X together with a meet operation satisfying the three properties we have indicated, the relation $x \leq y$ iff $x \wedge y = x$ is a poset structure on X for which $x \wedge y$ is the infimum of the pair (x, y). Indeed, the relation is reflexive since $x \wedge x = x$; it is transitive because $x \wedge y = x$ and $y \wedge z = y$ imply

$$x \wedge z = (x \wedge y) \wedge z = x \wedge (y \wedge z) = x \wedge y = x;$$

it is antisymmetric since $x \wedge y = x$ and $y \wedge x = y$ imply

$$x = x \wedge y = y \wedge x = y.$$

To observe that $x \wedge y$ is the infimum of the pair (x, y) for the poset structure, notice that

$$x \wedge (x \wedge y) = (x \wedge x) \wedge y = x \wedge y$$

implies $x \wedge y \leq x$ and in the same way $x \wedge y \leq y$, by symmetry of $\wedge$; moreover if $z \leq x$ and $z \leq y$,

$$z = z \wedge z = (z \wedge x) \wedge (z \wedge y) = z \wedge (x \wedge y),$$

proving that $z \leq x \wedge y$. This proves that the theory of $\wedge$-semi-lattices is algebraic. A top element can easily be introduced as a constant 1 (i.e. a 0-ary operation) together with the axiom

$$x \wedge 1 = x.$$

In the same way a bottom element is required by introducing a constant 0 together with the axiom

$$0 \wedge x = 0.$$

3.3.5.d A lattice is a set provided with the structure of both a $\wedge$-semi-lattice and a $\vee$-semi-lattice, those two structures arising from the same poset structure on X. The theory of lattices is algebraic, admitting a presentation given by two binary operations $\wedge, \vee$ which are both associative, commutative and idempotent (see example 3.3.5.c); these operations are connected by the axioms

$$x \wedge (x \vee y) = x,$$
$$(x \wedge y) \vee y = y.$$

We must check that these two additional axioms suffice to prove that the two following poset structures coincide:

$$x \leq y \text{ iff } x \wedge y = x,$$
$$x \preceq y \text{ iff } x \vee y = y.$$

Indeed one has

$$x \leq y \quad \Rightarrow \quad y = (x \wedge y) \vee y = x \vee y \quad \Rightarrow \quad x \preceq y,$$
$$x \preceq y \quad \Rightarrow \quad x = x \wedge (x \vee y) = x \wedge y \quad \Rightarrow \quad x \leq y.$$

3.3.5.e From the two previous examples, one easily defines the theory of distributive lattices, by adding the axioms

$$x \wedge (y \vee z) = (x \wedge y) \vee (x \wedge z),$$
$$x \vee (y \wedge z) = (x \vee y) \wedge (x \vee z),$$

or the theory of boolean algebras, by adding a 1-ary operation $(-)^*$ satisfying the axioms

$$x \vee x^* = 1, \;\; x \wedge x^* = 0.$$

All these theories are therefore algebraic.

3.3.5.f Let us now consider the theory of complete $\bigvee$-lattices, i.e. of those posets $(X, \leq)$ where every subset has a supremum; the morphisms are the mappings preserving those suprema, thus also the poset structure. This is typically a non-finitary theory: the operation of "taking the supremum" applies not to a finite set of elements, but to an arbitrary set of elements. And indeed the theory of complete $\bigvee$-lattices is not algebraic in the sense of 3.3.1 (see exercise 3.13.4).

3.3.5.g If R is a ring, an R-module is an abelian group M together with a scalar multiplication

$$R \times M \longrightarrow M.$$

Such a scalar multiplication does not have the required form of an operation $M^n \longrightarrow M$. Nevertheless the theory of R-modules is algebraic. It is obtained by adding to the theory of abelian groups one 1-ary operation $\overline{r}$ for every element $r \in R$. The interpretation of $\overline{r}$ in the case of an R-module M is just the scalar multiplication by r. The additional axioms in the case of left R-modules are thus

$$\overline{r}(x+y) = \overline{r}(x) + \overline{r}(y),$$
$$\overline{r}\big(\overline{s}(x)\big) = \big(\overline{rs}\big)(x),$$
$$\big(\overline{r+s}\big)(x) = \overline{r}(x) + \overline{s}(x),$$

and when R has a unit

$$\overline{1}(x) = x.$$

3.3.5.h If G is a group (written multiplicatively), a G-set is a set X provided with an action of G,

$$G \times X \longrightarrow X, \quad (g,x) \mapsto gx,$$

satisfying the two axioms

$$(gg')x = g(g'x),$$
$$1x = x.$$

For a fixed group G, the theory of G-sets is algebraic and admits a presentation given by a 1-ary operation $\overline{g}$ for every element $g \in G$, together with the axioms

$$\overline{gg'}(x) = \overline{g}\big(\overline{g'}(x)\big),$$
$$\overline{1}(x) = x.$$

Observe that the empty set is always a model of this theory.

3.3.5.i Consider the presentation of an algebraic theory having just one single constant and no axioms at all. The models are the pointed sets, i.e. the pairs (X, x) where X is a set and $x \in X$ is an element of X; a morphism $f\colon (X,x) \longrightarrow (Y,y)$ is just a mapping $f\colon X \longrightarrow Y$ such that $f(x) = y$.

3.3.5.j Consider the category $\mathcal{T}$ whose objects are the spaces $\mathbb{R}^n$, $(n \in \mathbb{N})$, and whose morphisms are the $\mathscr{C}^\infty$ functions between them; this is obviously an algebraic theory. The models of this theory are called the $\mathscr{C}^\infty$ algebras. Given a $\mathscr{C}^\infty$ differentiable manifold V, the set $\mathscr{C}^\infty(V, \mathbb{R})$ of $\mathscr{C}^\infty$ real-valued functions on V can be canonically provided with the structure of a $\mathscr{C}^\infty$ algebra: given $\mathscr{C}^\infty$ mappings $\alpha\colon \mathbb{R}^n \longrightarrow \mathbb{R}$ and $f_1, \ldots, f_n\colon V \longrightarrow \mathbb{R}$, one defines the composite $\alpha(f_1, \ldots, f_n)$ just as

$$V \xrightarrow{\begin{pmatrix} f_1 \\ \vdots \\ f_n \end{pmatrix}} \mathbb{R}^n \xrightarrow{\ \alpha\ } \mathbb{R}.$$

3.4 Limits and colimits in algebraic categories

Proposition 3.4.1 *Let $\mathcal{T}$ be an algebraic theory. The category $\mathsf{Mod}_\mathcal{T}$ is complete and limits are computed pointwise. In particular, the forgetful functor $U\colon \mathsf{Mod}_\mathcal{T} \longrightarrow \mathsf{Set}$ preserves and reflects limits.*

Proof By 2.15.1, volume 1, limits in $\mathsf{Fun}(\mathcal{T}, \mathsf{Set})$ are computed pointwise; by 2.12.1, volume 1, the limit of a diagram constituted of finite product-preserving functors is again a functor-preserving finite products. This proves that $\mathsf{Mod}_\mathcal{T}$ is complete and that limits in $\mathsf{Mod}_\mathcal{T}$ are computed pointwise. In particular evaluating at T^1 preserves limits, showing that U preserves limits (see 3.3.3). Since U reflects isomorphisms, U reflects limits as well (see 3.3.3, this volume, and 2.9.7, volume 1). □

Proposition 3.4.2 *Let $\mathcal{T}$ be an algebraic theory. The category $\mathsf{Mod}_\mathcal{T}$ has filtered colimits and these are computed pointwise. In particular, the forgetful functor $U\colon \mathsf{Mod}_\mathcal{T} \longrightarrow \mathsf{Set}$ preserves and reflects filtered colimits.*

Proof By 2.15.1, volume 1, colimits in $\mathsf{Fun}(\mathcal{T}, \mathsf{Set})$ are computed pointwise; by 2.13.4, volume 1, the colimit of a filtered diagram constituted of finite product-preserving functors is again a functor preserving finite products. This proves that $\mathsf{Mod}_\mathcal{T}$ admits filtered colimits which are computed pointwise. In particular evaluating at T^1 preserves filtered colimits, showing that U preserves filtered colimits (see 3.3.3). Since U reflects isomorphisms, U reflects filtered colimits as well (see 3.3.3, this volume, and 2.9.7, volume 1). □

Corollary 3.4.3 *Let $\mathcal{T}$ be an algebraic theory. In $\mathsf{Mod}_\mathcal{T}$, finite limits commute with filtered colimits.*

Proof The property holds in Set (see 2.13.4, volume 1) and both finite limits and filtered colimits are computed pointwise in $\mathsf{Mod}_\mathcal{T}$ see 3.4.1, 3.4.2. □

Corollary 3.4.4 *Let $\mathcal{T}$ be an algebraic theory. The forgetful functor $U\colon \mathsf{Mod}_{\mathcal{T}} \longrightarrow \mathsf{Set}$ preserves and reflects monomorphisms.*

Proof Now U preserves and reflects isomorphisms (see 3.3.3) and kernel pairs (see 3.4.1), thus also monomorphisms (see 2.5.6, volume 1). □

Theorem 3.4.5 *Let $\mathcal{T}$ be an algebraic theory. The category $\mathsf{Mod}_{\mathcal{T}}$ is reflective in the category $\mathsf{Fun}(\mathcal{T}, \mathsf{Set})$ and therefore it is both complete and cocomplete.*

Proof $\mathsf{Fun}(\mathcal{T}, \mathsf{Set})$ is both complete and cocomplete (see 2.15.4, volume 1), so that the last assertion follows from 3.5.3,4, volume 1. It remains to prove that $\mathsf{Mod}_{\mathcal{T}}$ is reflective in $\mathsf{Fun}(\mathcal{T}, \mathsf{Set})$. We know already that $\mathsf{Mod}_{\mathcal{T}}$ is complete and the inclusion in $\mathsf{Fun}(\mathcal{T}, \mathsf{Set})$ preserves limits (see 3.4.1); by the adjoint functor theorem (see 3.3.3, volume 1), it remains to check the solution set condition.

Consider two functors $F, G\colon \mathcal{T} \rightrightarrows \mathsf{Set}$ and a natural transformation $\varphi\colon F \Rightarrow G$; suppose G preserves finite products. Consider the set X of all elements of $G(T^1)$ of the form $\varphi_{T^1}(x)$ for some $x \in F(T^1)$; the cardinality of X is bounded by that of $F(T^1)$. Consider now the set Y of all elements of $G(T^1)$ of the form $\alpha(x_1, \ldots, x_n)$, for some integer n, n-ary operation α and elements $x_i \in X$; the cardinality of Y is bounded by that of $\coprod_{n\in\mathbb{N}} \mathcal{T}(T^n, T^1) \times X^n$, thus by the cardinality of $\coprod_{n\in\mathbb{N}} \mathcal{T}(T^n, T^1) \times F(T^1)^n$. By construction, Y is stable for all the $\mathcal{T}$-operations, thus putting

$$H(T^n) = Y^n \subseteq G(T^1)^n \cong G(T^n)$$

defines a subfunctor $H \subseteq G$ which, by construction, preserves finite products. Observe that φ factors through H since given $x \in F(T^n)$, one has

$$G(p_i)\big(\varphi_{T^n}(x)\big) = \varphi_{T^1}\big(F(p_i)(x)\big) \in Y$$

and thus $\varphi_{T^n}(x) \in Y^n$. Therefore one gets a solution set for F by choosing those finite product-preserving functors $H\colon \mathcal{T} \longrightarrow \mathsf{Set}$ such that $H(T^1)$ is a subset of some fixed set Y with cardinality less than the cardinality of $\coprod_{n\in\mathbb{N}} \mathcal{T}(T^n, T^1) \times F(T^1)^n$. Once $H(T^1)$ is fixed, there is indeed (up to a canonical choice of products) just a set of possibilities for constructing a product-preserving functor H since, for every morphism $\beta\colon T^n \longrightarrow T^m$, there is just a set of possible candidates for $H(\beta)$, i.e. the set of mappings $H(T^1)^n \longrightarrow H(T^1)^m$. □

3.5 The exactness properties of algebraic categories

The main purpose of this section is to prove that algebraic categories are regular and even exact in the sense of chapter 2. Using proposition 3.3.4, we find it more convenient to work with the presentation $\mathcal{T}_1$ associated with the algebraic category $\mathcal{T}$.

Lemma 3.5.1 *Let $\mathcal{T}$ be an algebraic theory and $\mathcal{T}_1$ the corresponding presentation of this algebraic theory. In $\mathsf{Mod}_{\mathcal{T}}$, an equivalence relation on an object M is a sub-$\mathcal{T}_1$-model $R \subseteq M \times M$ which is an equivalence relation in the category of sets. Such an R is generally called a "congruence" on M.*

Proof A sub-$\mathcal{T}_1$-model $R \subseteq M \times M$ which is an equivalence relation in Set is obviously an equivalence relation in $\mathsf{Mod}_{\mathcal{T}_1}$ (see 2.5.2). And all equivalence relations in $\mathsf{Mod}_{\mathcal{T}_1}$ are of this type. Indeed from 2.5.2, given an equivalence relation $R \rightrightarrows F$ in $\mathsf{Mod}_{\mathcal{T}}$, $\mathsf{Mod}_{\mathcal{T}}(\mathcal{T}(T_1, -), R)$ is an equivalence relation on $\mathsf{Mod}_{\mathcal{T}}(\mathcal{T}(T_1, -), F)$, but by 3.3.3 this means exactly that the $\mathcal{T}_1$-model $U(R)$ is an equivalence relation on the $\mathcal{T}_1$-model $U(F)$. □

Proposition 3.5.2 *Let $\mathcal{T}$ be an algebraic theory. In $\mathsf{Mod}_{\mathcal{T}}$, equivalence relations are effective and the functor $U\colon \mathsf{Mod}_{\mathcal{T}} \longrightarrow \mathsf{Set}$ preserves and reflects the coequalizers of equivalence relations.*

Proof We work with the corresponding presentation $\mathcal{T}_1$ of the algebraic theory (see 3.3.4). Given an equivalence relation $R \subseteq M \times M$ in $\mathsf{Mod}_{\mathcal{T}_1}$ we consider the corresponding quotient $q\colon M \longrightarrow M/R$ in the category of sets. We shall prove that q underlies a $\mathcal{T}_1$-homomorphism.

Given an operation $\alpha \in \mathcal{T}(T^n, T^1)$ and elements $x_1, \ldots, x_n \in M$, we realize α in M/R by

$$\alpha([x_1], \ldots, [x_n]) = [\alpha(x_1, \ldots, x_n)].$$

This definition makes sense because given $y_1, \ldots, y_n \in M$ such that $[x_i] = [y_i]$, $i = 1, \ldots, n$, one has $(x_i, y_i) \in R$ for each index i and thus

$$(\alpha(x_1, \ldots, x_n), \alpha(y_1, \ldots, y_n)) \in R;$$

in other words,

$$[\alpha(x_1, \ldots, x_n)] = [\alpha(y_1, \ldots, y_n)].$$

It is obvious that M/R provided with these operations is a $\mathcal{T}_1$-model, since M is, and q becomes in this way a $\mathcal{T}_1$-homomorphism.

Let us prove that the $\mathcal{T}_1$-homomorphism q is the coequalizer of the two projections $p_1, p_2 \colon R \rightrightarrows M$ in $\mathsf{Mod}_{\mathcal{T}_1}$. If $g \colon M \longrightarrow L$ is a $\mathcal{T}_1$-homomorphism coequalizing p_1, p_2, we get in the category of sets a unique mapping $h \colon M/R \longrightarrow L$ satisfying $h \circ q = g$. We must prove that h is a $\mathcal{T}_1$-homomorphism. Given $\alpha \in \mathcal{T}(T^n, T^1)$ and $x_1, \ldots, x_n \in M$,

$$\begin{aligned} h\Big(\alpha([x_1], \ldots, [x_n])\Big) &= h\big[\alpha(x_1, \ldots, x_n)\big] \\ &= h \circ q(\alpha(x_1, \ldots, x_n)) \\ &= g(\alpha(x_1, \ldots, x_n)) \\ &= \alpha(g(x_1), \ldots, g(x_n)) \\ &= \alpha((h \circ q)(x_1), \ldots, (h \circ q)(x_n)) \\ &= \alpha(h[x_1], \ldots, h[x_n]), \end{aligned}$$

since g is a $\mathcal{T}_1$-homomorphism.

We have already proved that the coequalizer of $p_1, p_2 \colon R \rightrightarrows M$ is computed in the same way in $\mathsf{Mod}_{\mathcal{T}_1}$ and in Set. Since kernel pairs are also computed in the same way in $\mathsf{Mod}_{\mathcal{T}_1}$ and in Set (see 3.4.1) and equivalence relations are effective in Set (see 2.5.5.a), equivalence relations are effective in $\mathsf{Mod}_{\mathcal{T}_1}$.

We have already observed that the functor $U \colon \mathsf{Mod}_{\mathcal{T}} \longrightarrow \mathsf{Set}$ preserves coequalizers of equivalence relations. Since moreover U reflects isomorphisms (see 3.3.3), it reflects coequalizers of equivalence relations (see 2.9.7, volume 1). □

Corollary 3.5.3 *Let $\mathcal{T}$ be an algebraic theory. The forgetful functor $U \colon \mathsf{Mod}_{\mathcal{T}} \longrightarrow \mathsf{Set}$ preserves and reflects regular epimorphisms.*

Proof Since $\mathsf{Mod}_{\mathcal{T}}$ has kernel pairs (see 3.4.1), every regular epimorphism is the coequalizer of its kernel pair (see 2.5.7, volume 1), i.e. the coequalizer of an equivalence relation (see 2.5.5.e). One derives the conclusion by 3.5.2. □

Theorem 3.5.4 *Let $\mathcal{T}$ be an algebraic theory. The category $\mathsf{Mod}_{\mathcal{T}}$ is regular and exact in the sense of 2.1.1 and 2.6.1.*

Proof $\mathsf{Mod}_{\mathcal{T}}$ is complete and cocomplete (see 3.4.1 and 3.4.5) and $U \colon \mathsf{Mod}_{\mathcal{T}} \longrightarrow \mathsf{Set}$ preserves and reflects pullbacks and regular epimorphisms (see 3.4.1 and 3.5.3). Therefore regular epimorphisms are stable in $\mathsf{Mod}_{\mathcal{T}}$ under pullbacks, since they are in Set (see 2.4.2). Thus $\mathsf{Mod}_{\mathcal{T}}$ is regular and, by 3.5.2, exact. □

Corollary 3.5.5 *Let $\mathcal{T}$ be an algebraic theory. The forgetful functor $U\colon \mathsf{Mod}_{\mathcal{T}} \longrightarrow \mathsf{Set}$ is exact in the sense of 2.3.5.* □

3.6 The algebraic lattices of subobjects

We recall a classical definition

Definition 3.6.1 *Let L be a complete lattice. An element $k \in L$ is compact when $k \leq \bigvee_{i \in I} x_i$ implies the existence of a finite subset $J \subseteq I$ such that $k \leq \bigvee_{i \in J} x_i$. An algebraic lattice is a complete lattice in which every element is a join of compact elements.*

Proposition 3.6.2 *In an algebraic lattice, finite meets distribute over filtered joins.*

Proof Let us consider a filtered family $(x_i)_{i \in I}$, i.e. a non-empty family such that

$$\forall i, j \in I \quad \exists k \in I \quad x_i \leq x_k, \quad x_j \leq x_k.$$

Given an arbitrary element x of the lattice, we must prove the equality

$$x \wedge \left(\bigvee_{i \in I} x_i \right) = \bigvee_{i \in I} (x \wedge x_i).$$

From the relations $x \wedge \left(\bigvee_{i \in I} x_i \right) \geq x \wedge x_i$ we deduce immediately that

$$x \wedge \left(\bigvee_{i \in I} x_i \right) \geq \bigvee_{i \in I} (x \wedge x_i).$$

It remains to prove the converse inequality. We can write

$$x \wedge \left(\bigvee_{i \in i} x_i \right) = \bigvee_{j \in J} k_j$$

for a family of compact elements $(k_j)_{j \in J}$. From $k_j \leq \bigvee_{i \in I} x_i$ we deduce that $k_j \leq \bigvee_{i \in I_j} x_i$ for some finite subset $I_j \subseteq I$. By filteredness of the family $(x_i)_{i \in I}$ we choose $i_j \in I$ such that $x_i \leq x_{i_j}$ for every $i \in I_j$. This implies $k_j \leq x_{i_j}$ and since $k_j \leq x$, we deduce $k_j \leq x \wedge x_{i_j}$. □

Proposition 3.6.3 *Let $\mathcal{T}$ be an algebraic theory. The category $\mathsf{Mod}_{\mathcal{T}}$ is well-powered.*

Proof $\mathsf{Mod}_{\mathcal{T}}$ is complete (see 3.4.1) and possesses a strong generator (see 3.3.3); one derives the conclusion by 4.5.15, volume 1. □

Lemma 3.6.4 *Let $\mathcal{T}$ be an algebraic theory and $\mathcal{T}_1$ its corresponding presentation. Given a family $(M_i \subseteq M)_{i \in I}$ of submodels of a $\mathcal{T}_1$-model M, the union $\bigcup_{i \in I} M_i$ in $\mathsf{Mod}_{\mathcal{T}_1}$ exists and is given by the set of all elements $\alpha(x_1, \ldots, x_n)$ of M, with α an n-ary operation and $x_1, \ldots, x_n$ elements of the set theoretical union of the subsets $M_i \subseteq M$.*

Proof Every submodel $M' \subseteq M$ which contains the M_i's must contain the elements described. On the other hand, those elements obviously constitute a submodel of M.

Proposition 3.6.5 *Let $\mathcal{T}$ be an algebraic theory. The submodels of every $\mathcal{T}$-model constitute an algebraic lattice.*

Proof We work with the corresponding presentation $\mathcal{T}_1$ of the algebraic theory. It is obvious that a set theoretical intersection of submodels is again a submodel; on the other hand lemma 3.6.4 gives an explicit description of unions.

Given a subset $X \subseteq M$ of a $\mathcal{T}_1$-model M, all the elements of the form $\alpha(x_1, \ldots, x_n)$ for $n \in \mathbb{N}$, $x_i \in X$ and α an n-ary operation obviously constitute a submodel $\overline{X}$ of M, called the submodel generated by X. When X has finitely many elements $x_1, \ldots, x_n$, this submodel $\overline{X}$ is a compact element in the lattice of submodels. Indeed if the M_i's are submodels and $\overline{X} \subseteq \bigcup_{i \in I} M_i$, each element $x_k \in X$ has the form $\alpha(y_1, \ldots, y_l)$ for some elements $y_1, \ldots, y_l$ in the set theoretical union of the M_i's (see 3.6.4). Considering the finitely many elements $x_k \in X$, we get altogether finitely many elements y_j in the set theoretical union of the M_i's and obviously those y_j's belong to the set theoretical union of finitely many M_i's, say $M_1, \ldots, M_l$. But the union $M_1 \cup \ldots \cup M_l$ in $\mathsf{Mod}_{\mathcal{T}_1}$ now contains the elements $x_1, \ldots, x_n$ and therefore the submodel $\overline{X}$ they generate.

Every submodel $M' \subseteq M$ is obviously the union in $\mathsf{Mod}_{\mathcal{T}}$ of all the submodels of M' generated by a single element $x \in M'$, which concludes the proof. □

Proposition 3.6.6 *Let $\mathcal{T}$ be an algebraic theory. The forgetful functor $U\colon \mathsf{Mod}_{\mathcal{T}} \longrightarrow \mathsf{Set}$ preserves and reflects arbitrary intersections and filtered unions of subobjects. In particular, in $\mathsf{Mod}_{\mathcal{T}}$, finite intersections distribute over filtered unions.*

Proof Given a filtered family of subobjects $M_i \subseteq M$ in $\mathsf{Mod}_{\mathcal{T}}$, we consider the corresponding filtered colimit $L = \operatorname{colim} M_i$ and the induced factorization $f\colon L \longrightarrow M$. In the category of sets, $U(L)$ is the filtered colimit of the $U(M_i)$'s (see 3.4.2), thus it is their union and $U(f)\colon U(L) \longrightarrow U(M)$ is the canonical inclusion. Therefore f is a monomorphism (see 3.4.4) and L is the union of the M_i's. The rest follows from 3.6.2, 3.4.1 and 3.4.2. □

3.7 Algebraic functors

Definition 3.7.1 *Let $\mathcal{R}$ and $\mathcal{T}$ be algebraic theories, with objects respectively written $R^0, R^1, \ldots, R^n, \ldots$ and $T^0, T^1, \ldots, T^n, \ldots$. A morphism of algebraic theories is a functor $F\colon \mathcal{R} \longrightarrow \mathcal{T}$ which preserves finite products and maps R^n to T^n ($n \in \mathbb{N}$).*

Observe that the requirement $F(R^n) = T^n$ does not follow from the preservation of finite products. Indeed T^0 is the terminal object of $\mathcal{T}$ so that the constant functor on T^0 preserves finite products.

Proposition 3.7.2 *The theory of sets is an initial object in the category of algebraic theories and their morphisms.*

Proof We have seen in 3.3.5.b that the theory $\mathcal{S}$ of sets is (up to an equivalence) the dual of the category of finite sets. In the category of finite sets the only morphisms $\{*\} \longrightarrow \{1, \ldots, n\}$ are the canonical injections of the coproduct $\coprod_{i=1}^{n} \{*\}$. Therefore in $\mathcal{S}$ the only n-ary operations $S^n \longrightarrow S^1$ are the projections, from which follows the existence of a unique morphism $\mathcal{S} \longrightarrow \mathcal{T}$ for every algebraic theory $\mathcal{T}$. □

Corollary 3.7.3 *Let $\mathcal{T}$ be an algebraic theory and $\mathcal{S}$ the theory of sets. The forgetful functor $U\colon \mathsf{Mod}_{\mathcal{T}} \longrightarrow \mathsf{Set}$ is the functor of composition with the unique morphism of algebraic theories $\sigma\colon \mathcal{S} \longrightarrow \mathcal{T}$.*

Proof Given a $\mathcal{T}$-model $G\colon \mathcal{T} \longrightarrow \mathsf{Set}$, the composite $G \circ \sigma$ preserves finite products since G and σ do. The set X corresponding to $G \circ \sigma$ is $G \circ \sigma(S^1) = G(T^1)$, i.e. the set UG; see 3.3.3. □

Definition 3.7.4 *Let $F\colon \mathcal{R} \longrightarrow \mathcal{T}$ be a morphism of algebraic theories. The functor of composition with F*

$$\mathsf{Mod}_{\mathcal{T}} \longrightarrow \mathsf{Mod}_{\mathcal{R}}, \quad G \mapsto G \circ F,$$

is called an algebraic functor.

Observe that this definition makes sense: indeed $G \circ F$ preserves finite products since F and G do.

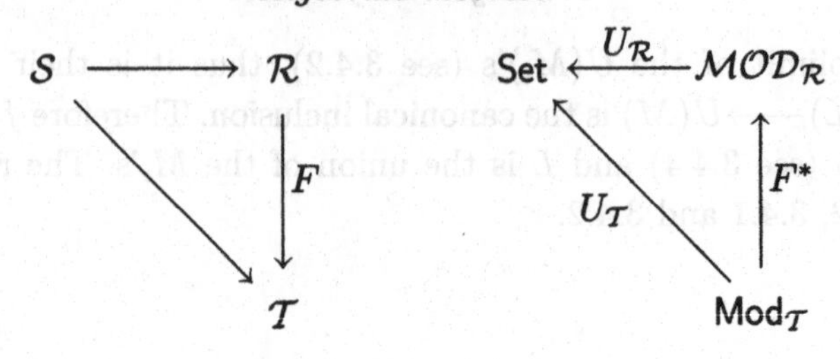

Diagram 3.3

Proposition 3.7.5 *An algebraic functor is faithful and reflects isomorphisms; it preserves and reflects regular epimorphisms, coequalizers of equivalence relations, small limits and filtered colimits.*

Proof Consider a morphism of theories $F\colon \mathcal{R} \longrightarrow \mathcal{T}$. This yields the commutative triangles of diagram 3.3 where $\mathcal{S}$ is the theory of sets, $U_{\mathcal{R}}$, $U_{\mathcal{T}}$ are the forgetful functors and F^* is the algebraic functor of composition with F.

The functor F^* is faithful since $U_{\mathcal{T}}$ is faithful; it reflects isomorphisms, regular epimorphisms, coequalizers of equivalence relations, small limits and filtered colimits since these are preserved by $U_{\mathcal{R}}$ and reflected by $U_{\mathcal{T}}$. The functor F^* preserves also regular epimorphisms, coequalizers of equivalence relations, small limits and filtered colimits since these are preserved by $U_{\mathcal{T}}$ and reflected by $U_{\mathcal{R}}$; see 3.3, 3.4, 3.5. □

In order to prove that every algebraic functor has an adjoint, we need to observe a property of the category of sets (or, more generally, of a cartesian closed category; see chapter 6).

Lemma 3.7.6 *Consider two functors $F\colon \mathcal{A} \longrightarrow \mathsf{Set}$ and $G\colon \mathcal{B} \longrightarrow \mathsf{Set}$, with $\mathcal{A}, \mathcal{B}$ small categories. The colimit of the functor*

$$\mathcal{A} \times \mathcal{B} \xrightarrow{F \times G} \mathsf{Set} \times \mathsf{Set} \xrightarrow{\quad \times \quad} \mathsf{Set}$$

is just $(\operatorname{colim} F) \times (\operatorname{colim} G)$.

Proof We have seen in 3.1.6.f, volume 1, that the functor $A \times -$ admits $(-)^A$ as a right adjoint; therefore the functor $A \times -$ preserves colimits (see 3.2.2, volume 1). So

$$\begin{aligned}(\operatorname*{colim}_A F(A)) \times (\operatorname*{colim}_B G(B)) &\cong \operatorname*{colim}_A \big(F(A) \times \operatorname*{colim}_B G(B)\big) \\ &\cong \operatorname*{colim}_A \operatorname*{colim}_B \big(F(A) \times G(B)\big) \cong \operatorname*{colim}_{(A,B)} \big(F(A) \times G(B)\big)\end{aligned}$$

by associativity of colimits (see 2.12, volume 1). □

Theorem 3.7.7 *Every algebraic functor has a left adjoint.*

Proof Given a morphism $F\colon \mathcal{R} \longrightarrow \mathcal{T}$ of algebraic theories and a model $G\colon \mathcal{R} \longrightarrow \mathsf{Set}$ of $\mathcal{R}$, the reflection of G along the functor of composition with F,

$$F^*\colon \mathsf{Fun}(\mathcal{T}, \mathsf{Set}) \longrightarrow \mathsf{Fun}(\mathcal{R}, \mathsf{Set}),$$

is the left Kan extension K of G along F (see 3.7.2, volume 1). To conclude the proof, it suffices to verify that K preserves finite products.

We write respectively $R^0, R^1, \ldots, R^n, \ldots$ and $T^0, T^1, \ldots, T^n, \ldots$ for the objects of $\mathcal{R}$ and $\mathcal{T}$. Using the construction of 3.7.2, volume 1, we know that $K(T^n)$ is obtained as the colimit of the composite

$$\mathcal{E}_n \xrightarrow{\phi_n} \mathcal{R} \xrightarrow{G} \mathsf{Set}$$

where the objects of $\mathcal{E}_n$ are the pairs (R^m, t) with $t\colon F(R^m) \longrightarrow T^n$; a morphism of $\mathcal{E}_n$ $r\colon (R^m, t) \longrightarrow (R^l, s)$ is a morphism $r\colon R^m \longrightarrow R^l$ in $\mathcal{R}$ such that $s \circ F(r) = t$. Since $F(R^m) = T^m$, $\mathcal{E}_n$ can equivalently be described as the category having for objects the pairs (m, t) with $t\colon T^m \longrightarrow T^n$; a morphism $r\colon (m, t) \longrightarrow (l, s)$ of $\mathcal{E}_n$ is a morphism in $\mathcal{R}$ $r\colon R^m \longrightarrow R^n$ such that $s \circ F(r) = t$. Essentially, we must prove that

$$K(T^n) \cong \operatorname{colim}(G \circ \phi_n) \cong (\operatorname{colim} G \circ \phi_1)^n \cong K(T^1)^n.$$

By using 3.7.6 and the preservation of finite products by G, we have

$$(\operatorname{colim} G \circ \phi_1)^n \cong \operatorname{colim}(G \circ \phi_1)^n \cong \operatorname{colim} G \circ (\phi_1)^n$$

where $(\phi_1)^n$ stands for the composite

$$\mathcal{E}_1 \times \ldots \times \mathcal{E}_1 \xrightarrow{\phi_1 \times \ldots \times \phi_1} \mathcal{R} \times \ldots \times \mathcal{R} \xrightarrow{\quad\times\quad} \mathcal{R}$$

mapping $(m_i, t_i)_{i=1,\ldots,n}$ to $R^{m_1} \times \cdots \times R^{m_n}$. On the other hand we can consider the functor

$$\psi_n\colon \mathcal{E}_1 \times \ldots \times \mathcal{E}_1 \longrightarrow \mathcal{E}_n$$

mapping $(m_i, t_i)_{i=1,\ldots,n}$ to $(\sum_{i=1}^n m_i, \prod_{i=1}^n t_i)$, which is such that the relation $\phi_n \circ \psi_n = (\phi_1)^n$ holds. To conclude the proof, it suffices to observe that the functor ψ_n is cofinal (see 2.11.1, volume 1).

Let us consider the following situation:

$$\mathcal{E}_1 \times \ldots \times \mathcal{E}_1 \xrightarrow{\psi_n} \mathcal{E}_n \xrightarrow{H} \mathcal{X}$$

where H is an arbitrary functor. Writing Δ_X for the constant functor on the object $X \in \mathcal{X}$, a cocone $\alpha\colon H \Rightarrow \Delta_X$ on H induces a cocone

$\alpha * \psi_n\colon H \circ \psi_n \Rightarrow \Delta_X$ on $H \circ \psi_n$. Conversely a cocone $\beta\colon H \circ \psi_n \Rightarrow \Delta_X$ on $H \circ \psi_n$ has the form $\alpha * \psi_n$ for a unique cocone $\alpha\colon H \Rightarrow \Delta_X$. Indeed, given $(m,t) \in \mathcal{E}_n$ the diagonal $\delta\colon R^m \longrightarrow (R^m)^n$ yields a morphism of $\mathcal{E}_n$

$$\delta\colon (m,t) \longrightarrow \left(m \times n, \prod\nolimits_{i=1}^{n} p_i \circ t\right) = \psi_n\big((m, p_i \circ t)_{i=1,\dots,n}\big)$$

so that it is necessary to define

$$\alpha_{(m,t)} = \beta_{(m,p_i \circ t)_{i=1,\dots,n}} \circ H(\delta).$$

The naturality of β implies immediately that of α. Thus the cocones on H are in bijection with the cocones on $H \circ \psi_n$, from which it follows easily that ψ_n is a cofinal functor, as in 2.11.2, volume 1. □

Notice that the second condition for cofinality, as described in 2.11.2, volume 1, has no reason to be satisfied in the previous situation. It was used in 2.11.2, volume 1 to prove the equivalence between the various possible ways of extending the original cone. In the situation of 3.7.7, the diagonal morphisms offer a canonical way to realize the extension.

Corollary 3.7.8 *Let $\mathcal{T}$ be an algebraic theory. The forgetful functor $U\colon \mathsf{Mod}_{\mathcal{T}} \longrightarrow \mathsf{Set}$ has a left adjoint.*

Proof By 3.7.7 and 3.7.3. □

Observe that 3.7.8 could already have been obtained from 3.3.3 and 3.4.5: the singleton admits $\mathcal{T}(T^1, -)$ as a reflection along U from which $\coprod_{x \in X} \mathcal{T}(T^1, -)$ is the reflection of the set X along U; see 3.8.3, volume 1.

3.8 Freely generated models

Definition 3.8.1 *Let $\mathcal{T}$ be an algebraic theory, $U\colon \mathsf{Mod}_{\mathcal{T}} \longrightarrow \mathsf{Set}$ the corresponding forgetful functor and $F\colon \mathsf{Set} \longrightarrow \mathsf{Mod}_{\mathcal{T}}$ its left adjoint.*

(1) By a free $\mathcal{T}$-model we mean, up to isomorphism, a model of the form $F(X)$ for some set X;

(2) by a finitely generated $\mathcal{T}$-model, we mean a model M which is a quotient of a free model $F(n)$ on a finite set n;

(3) by a finitely presentable $\mathcal{T}$-model, we mean a model M which can be obtained via a coequalizer diagram

$$F(m) \rightrightarrows F(n) \longrightarrow\!\!\!\!\!\!\rightarrow M$$

where m, n are finite sets.

Proposition 3.8.2 *Let $\mathcal{T}$ be an algebraic theory. The free model functor $F\colon \mathsf{Set} \longrightarrow \mathsf{Mod}_{\mathcal{T}}$ preserves monomorphisms.*

Proof Let $i\colon X \rightarrowtail Y$ be an injection in Set. If X is not empty, i admits a retraction $r\colon Y \longrightarrow X$ and from $r \circ i = 1_X$, we get $F(r) \circ F(i) = 1_{F(X)}$ proving that $F(i)$ is a monomorphism.

If $X = \emptyset$ is the empty set, i.e. the initial object of Set, $F(\emptyset)$ is the initial object of $\mathsf{Mod}_{\mathcal{T}}$; see 3.2.2, volume 1. If $UF(\emptyset)$ is empty, $UF(\emptyset) \longrightarrow UF(Y)$ is injective and thus $F(\emptyset) \longrightarrow F(Y)$ is a monomorphism; see 3.4.4. If $UF(\emptyset)$ is not empty, every mapping $Y \longrightarrow UF(\emptyset)$ induces by adjunction a $\mathcal{T}$-homomorphism $t\colon F(Y) \twoheadrightarrow F(\emptyset)$; see 3.1.1, volume 1. The relation $t \circ F(i) = 1_{F(\emptyset)}$ holds since $F(\emptyset)$ is an initial object, thus $F(i)$ is a monomorphism. □

Lemma 3.8.3 *Let $\mathcal{T}$ be an algebraic theory. With the previous notations, every free model $F(X)$ can be written as a filtered colimit $F(X) = \operatorname{colim}_{n \subseteq X} F(n)$ where n runs through the finite subsets of X.*

Proof Obviously $X = \operatorname{colim}_{n \subseteq X} n$ holds in Set and F preserves colimits (see 3.2.2, volume 1). □

Lemma 3.8.4 *Let $\mathcal{T}$ be an algebraic theory. With the previous notations, the finitely generated free models are exactly, up to isomorphism, the models $F(n)$ for n a finite set.*

Proof Suppose $F(X)$ is a quotient of some $F(n)$, for n a finite set, and consider the corresponding regular epimorphism $q\colon F(n) \twoheadrightarrow F(X)$. We can write $F(X) = \operatorname{colim}_{m \subseteq X} F(m)$ where m runs through the finite subsets of X. By construction of filtered colimits in $\mathsf{Mod}_{\mathcal{T}}$ (see 3.4.2), for every index i the element $U(q)\big(\eta_n(i)\big)$ arises from an element in some $UF(m_i)$, for $m_i \subseteq X$, m_i finite. Since n is finite, all those elements are already in some $UF(m)$, for $\mu\colon m \hookrightarrow X$ a finite subset of X, namely the union of the m_i's. This yields a mapping $u\colon n \longrightarrow UF(m)$ and thus a corresponding factorization $v\colon F(n) \longrightarrow F(m)$, since $\big(F(n), \eta_n\big)$ is the reflection of n along U. Diagram 3.4 indicates already that $U(v) \circ \eta_n = u$. Therefore

$$UF(\mu) \circ U(v) \circ \eta_n = UF(\mu) \circ u = U(q) \circ \eta_n$$

and $q = F(\mu) \circ v$ by the uniqueness condition in the definition of a reflection; see 3.1.1, volume 1.

Since q is a regular epimorphism, $F(\mu)$ is a regular epimorphism as well (see 3.5.4 and 2.1.6). But since μ is a monomorphism, so is $F(\mu)$ (see 3.8.2). Finally, $F(\mu)$ is an isomorphism (see 3.5.4). □

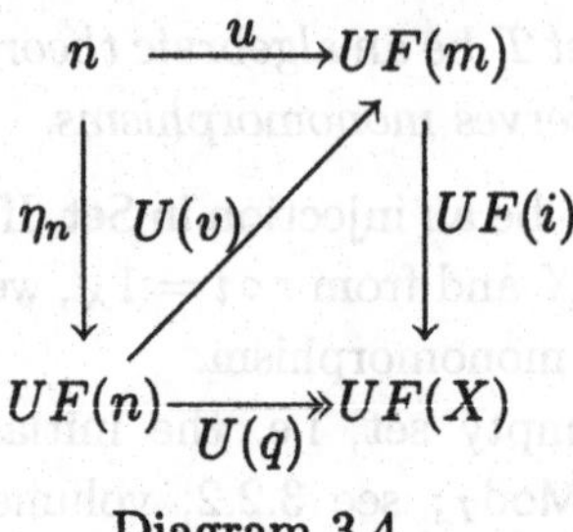

Diagram 3.4

Proposition 3.8.5 *An algebraic theory $\mathcal{T}$ is equivalent to the dual of the full subcategory of* $\mathsf{Mod}_{\mathcal{T}}$ *whose objects are the finitely generated free models.*

Proof With the previous notation, we know that the free model $F(n)$ on n generators is just $\mathcal{T}(T^n, -)$; see 3.3.3. By the Yoneda lemma (see 1.3.3, volume 1)

$$\mathsf{Mod}_{\mathcal{T}}(F(n), F(m)) \cong \mathsf{Nat}(\mathcal{T}(T^n, -), \mathcal{T}(T^m, -)) \cong \mathcal{T}(T^m, T^n). \quad \square$$

Proposition 3.8.6 *The free models of an algebraic theory $\mathcal{T}$ are projective.*

Proof Consider a free model $F(X)$, a strong epimorphism $p\colon M \twoheadrightarrow N$ in $\mathsf{Mod}_{\mathcal{T}}$ and a morphism $f\colon F(X) \longrightarrow N$. The set theoretical mapping $U(p)\colon U(M) \twoheadrightarrow U(N)$ is a surjection (see 3.5.3, 3.5.4 and 2.1.4); thus, by the axiom of choice, it has a section $s\colon U(N) \rightarrowtail U(M)$. By adjunction, f corresponds to a morphism $g\colon X \longrightarrow U(N)$ and the composite $s \circ g\colon X \longrightarrow U(M)$ to a morphism $h\colon F(X) \longrightarrow M$. Let us prove that $p \circ h = f$. By adjunction, this is equivalent to $U(p) \circ s \circ g = g$, which is the case since s is a section of $U(p)$. $\square$

We keep the notation $U\colon \mathsf{Mod}_{\mathcal{T}} \longrightarrow \mathsf{Set}$, $F\colon \mathsf{Set} \longrightarrow \mathsf{Mod}_{\mathcal{T}}$ for the forgetful functor and the free model functor. We write $\eta\colon 1_{\mathsf{Set}} \Rightarrow UF$ and $\varepsilon\colon FU \Rightarrow 1_{\mathsf{Mod}_{\mathcal{T}}}$ for the natural transformations of the adjunction (see 3.1.5, volume 1). We recall they make commutative the triangles of diagram 3.5.

Lemma 3.8.7 *With the previous notation, ε_M is a regular epimorphism for every $\mathcal{T}$-model M.*

Proof By the first triangle, $U(\varepsilon_M)$ is surjective and thus ε_M is a regular epimorphism (see 3.5.3). $\square$

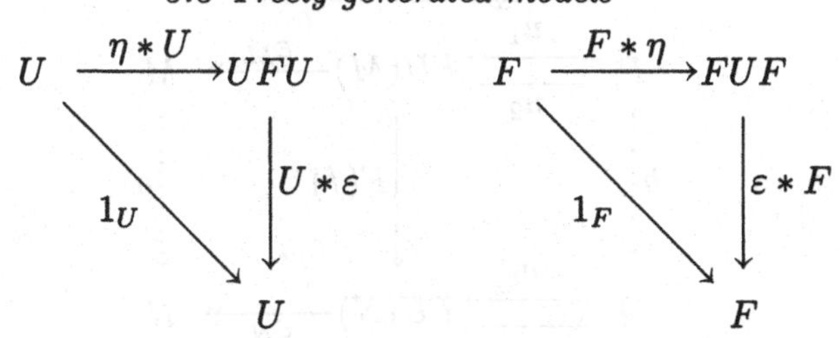

Diagram 3.5

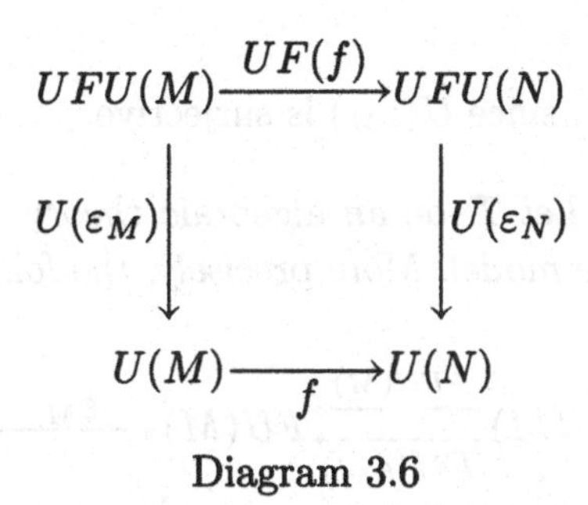

Diagram 3.6

Proposition 3.8.8 *Let $\mathcal{T}$ be an algebraic theory and M, N two $\mathcal{T}$-models. A mapping $f: U(M) \longrightarrow U(N)$ has the form $U(g)$ for a (necessarily unique) $\mathcal{T}$-homomorphism $g: M \longrightarrow N$ iff the square in diagram 3.6 commutes.*

Proof If $f = U(g)$, the commutativity holds by naturality of ε. Conversely, suppose the commutativity of the given diagram. In $\mathsf{Mod}_{\mathcal{T}}$ consider diagram 3.7, where the horizontal lines are kernel pairs. Applying U, we get

$$\begin{aligned} U(\varepsilon_N) \circ UF(f) \circ U(u_1) &= f \circ U(\varepsilon_M) \circ U(u_1) \\ &= f \circ U(\varepsilon_M) \circ U(u_2) \\ &= U(\varepsilon_N) \circ UF(f) \circ U(u_2). \end{aligned}$$

This yields $\varepsilon_N \circ F(f) \circ u_1 = \varepsilon_N \circ F(f) \circ u_2$ by faithfulness of U (see 3.3.3) and therefore a unique factorization h such that $F(f) \circ u_i = v_i \circ h$, $i = 1, 2$.

Now $\varepsilon_N, \varepsilon_M$ are regular epimorphisms (see 3.8.7), so are the coequalizers of their kernel pairs; see 2.5.7, volume 1. The relation $\varepsilon_N \circ F(f) \circ u_1 = \varepsilon_N \circ F(f) \circ u_2$ thus implies also the existence of a unique $g: M \longrightarrow N$ such that $g \circ \varepsilon_M = \varepsilon_N \circ F(f)$. Applying U we get

$$U(g) \circ U(\varepsilon_M) = U(\varepsilon_N) \circ UF(f) = f \circ U(\varepsilon_M)$$

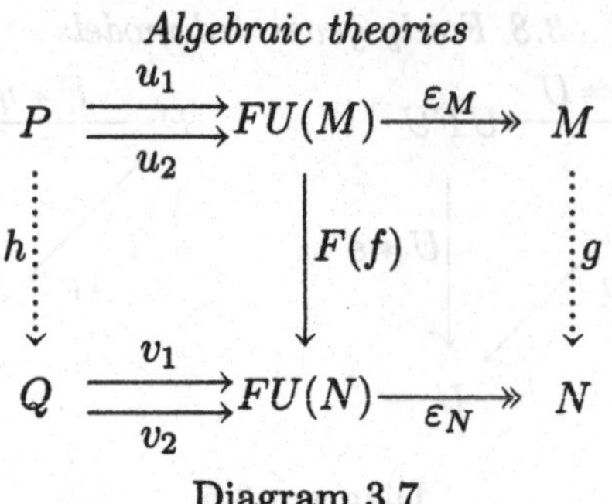

Diagram 3.7

from which $U(g) = f$, since $U(\varepsilon_M)$ is surjective. □

Proposition 3.8.9 *Let $\mathcal{T}$ be an algebraic theory. Every $\mathcal{T}$-model M is a quotient of a free model. More precisely, the following diagram is a coequalizer:*

$$FUFU(M) \underset{FU(\varepsilon_M)}{\overset{\varepsilon_{FU(M)}}{\rightrightarrows}} FU(M) \xrightarrow{\varepsilon_M} M.$$

Proof We know already that ε_M is a regular epimorphism (see 3.8.7), thus M is a quotient of the free model $FU(M)$ and ε_M is the coequalizer of its kernel pair (see 2.5.7, volume 1).

To get the stated canonical coequalizer let us observe the equality $\varepsilon_M \circ \varepsilon_{FU(M)} = \varepsilon_M \circ FU(\varepsilon_M)$, which follows just by naturality of ε. Now given $h: FU(M) \longrightarrow N$ in $\mathsf{Mod}_\mathcal{T}$ such that $h \circ \varepsilon_{FU(M)} = h \circ FU(\varepsilon_M)$, we must construct $g: M \longrightarrow N$ such that $g \circ \varepsilon_M = h$; since ε_M is an epimorphism, g will necessarily be unique. By adjunction, h corresponds to a mapping $f: U(M) \longrightarrow U(N)$ such that the relations $\varepsilon_N \circ F(f) = h$ and $f = U(h) \circ \eta_{U(M)}$ hold. By naturality of η, the triangular identities of the adjunction and the assumption on h,

$$\begin{aligned} f \circ U(\varepsilon_M) &= U(h) \circ \eta_{U(M)} \circ U(\varepsilon_M) = U(h) \circ UFU(\varepsilon_M) \circ \eta_{UFU(M)} \\ &= U(h) \circ U\big(\varepsilon_{FU(M)}\big) \circ \eta_{UFU(M)} = U(h) = U(\varepsilon_N) \circ UF(f). \end{aligned}$$

Therefore the mapping f underlies a $\mathcal{T}$-homomorphism $g: M \longrightarrow N$; see 3.8.8. From $U(g) = f$ we get $U(g) \circ U(\varepsilon_M) = U(h)$, thus $g \circ \varepsilon_M = h$ by faithfulness of U; see 3.3.3. □

Proposition 3.8.10 *Let $\mathcal{T}$ be an algebraic theory. The finitely generated free models constitute a family of dense generators of $\mathsf{Mod}_\mathcal{T}$ (see 3.5.4, volume 1).*

Proof $\mathsf{Mod}_\mathcal{T}$ is reflective in $\mathsf{Fun}(\mathcal{T}, \mathsf{Set})$; see 3.4.5. In $\mathsf{Fun}(\mathcal{T}, \mathsf{Set})$, every $\mathcal{T}$-model $F \in \mathsf{Mod}_\mathcal{T}$ is the colimit of the canonical diagram of all

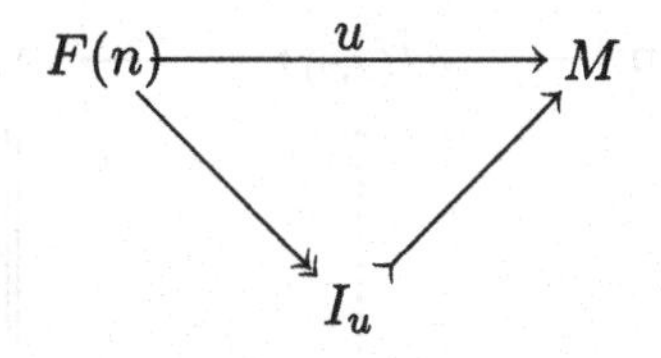

Diagram 3.8

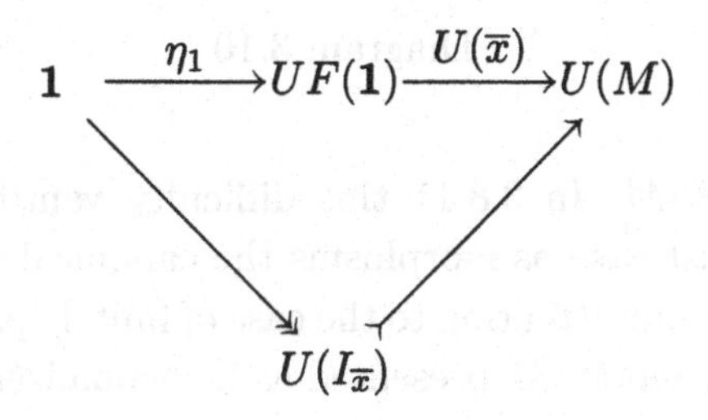

Diagram 3.9

representable functors $\mathcal{T}(T^n, -)$ over F (see 2.15.6, volume 1). Since the representable functors are already in $\mathsf{Mod}_{\mathcal{T}}$, this is a colimit in $\mathsf{Mod}_{\mathcal{T}}$ and one gets the conclusion by 3.3.3(4). □

Proposition 3.8.11 *Let $\mathcal{T}$ be an algebraic theory. Every $\mathcal{T}$-model is the filtered union of its finitely generated $\mathcal{T}$-submodels.*

Proof We use the notation of 3.8.10. For each object $(F(n), u)$ of $\mathscr{F}/M$, we consider the image of u (see 3.5.4 and 2.1.4) as in diagram 3.8. Each I_u is, by construction, a finitely generated submodel of M. The smallest submodel of M containing all the I_u's is M itself since, given an element $x \in U(M)$, we can consider diagram 3.9. In fact $x = U(\overline{x}) \circ \eta_1(*)$, thus $x \in U(I_{\overline{x}})$ and therefore the set theoretical union of the $U(I_u)$ contains already all the elements of $U(M)$.

It remains to prove that the union is filtered. Given finite sets n, m and morphisms $u\colon F(n) \longrightarrow M$ $v\colon F(m) \longrightarrow M$, let us consider the corresponding factorization

$$(u,v)\colon F(n+m) = F(n) \amalg F(m) \longrightarrow M.$$

Observing diagram 3.10, we deduce that $I_u \subseteq I_{(u,v)}$; see 4.4.5, volume 1. In the same way $I_v \subseteq I_{(u,v)}$, which proves the filteredness. □

It should be observed that even if the construction of the previous proof is based on that of 3.8.10, the category $\mathscr{F}/M$ in 3.8.10 is not filtered. Indeed, two parallel morphisms in $\mathscr{F}/M$ have no reason to be

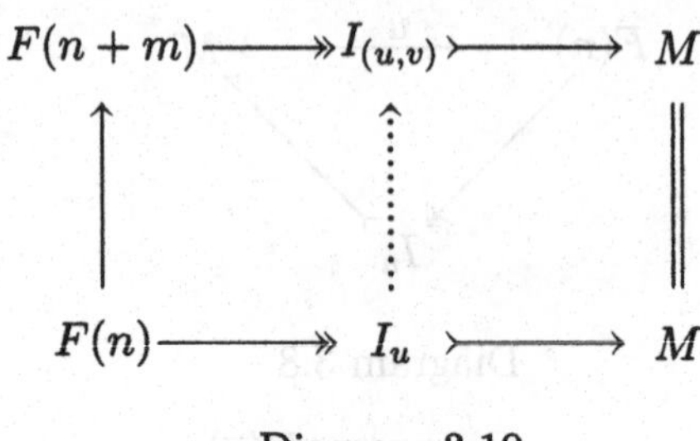

Diagram 3.10

coequalized in $\mathcal{F}/M$. In 3.8.11 this difficulty vanishes since between subobjects, we just take as morphisms the canonical inclusions.

Now let us turn our attention to the case of finitely presentable models. If we have such a model M presented as a coequalizer

$$F(m) \underset{v}{\overset{u}{\rightrightarrows}} F(n) \xrightarrow{q} M$$

with m, n finite, u and v correspond by adjunction to mappings in Set $\overline{u}, \overline{v}\colon m \rightrightarrows UF(n)$, thus to m-tuples $(u_1, \ldots, u_m)$, $(v_1, \ldots, v_m)$ of elements of $UF(n)$. It is immediate that M is the quotient of $F(n)$ by the smallest congruence containing the pairs (u_i, v_i), for $i = 1, \ldots, n$. Thus M is the model obtained from:

- finitely many generators: the elements of n;
- finitely many relations $u_i = v_i$ between terms constructed from those generators (see 3.2.6).

This justifies the terminology "finitely presentable".

Proposition 3.8.12 *Let T be an algebraic theory. The finitely presentable models constitute a dense family of generators of Mod_T, stable under finite colimits. Moreover every T-model is a filtered colimit of finitely presentable ones.*

Proof We write $\mathcal{P}$ for the full subcategory of finitely presentable T-models. Given a model M, we consider as in 4.5.4, volume 1, the category $\mathcal{P}/M$ and the forgetful functor $\phi\colon \mathcal{P}/M \longrightarrow \mathsf{Mod}_T$. We consider the canonical cocone

$$s_{(P,u)}\colon \phi(P,u) \longrightarrow M, \quad s_{(P,u)} = u,$$

where $P \in \mathcal{P}$ and $u\colon P \longrightarrow M$. We must first prove the universality of this cocone.

Given another cocone on ϕ,

$$t_{(P,u)}\colon \phi(P,u) \longrightarrow N$$

consideration of the morphisms $t_{(F(n),u)}$ yields, by 3.8.10, a unique factorization $g\colon M \longrightarrow N$ such that $g \circ s_{(F(n),u)} = t_{(F(n),u)}$ for every finite set n and morphism $u\colon F(n) \longrightarrow M$. Now given $P \in \mathscr{P}$ and $u\colon P \longrightarrow M$, we consider a regular epimorphism $p\colon F(m) \longrightarrow\!\!\!\!\!\rightarrow P$ with m finite. This yields a morphism $p\colon \big(F(m), u \circ p\big) \longrightarrow (P,u)$ in $\mathscr{P}/M$ and the relations

$$t_{(P,u)} \circ p = t_{(F(m),u\circ p)} = g \circ s_{(F(m),u\circ p)} = g \circ s_{(P,u)} \circ p$$

imply $t_{(P,u)} = g \circ s_{(P,u)}$ since p is an epimorphism.

We have already proved that the finitely presentable objects are a dense family of generators. It remains to prove that $\mathscr{P}/M$ is filtered. This will certainly be achieved if we prove that $\mathscr{P}$ is stable in $\mathsf{Mod}_{\mathcal{T}}$ for finite colimits.

Now $\mathscr{P}$ contains the initial object $F(\emptyset)$ and, if P, Q are finitely presentable via

$$F(m) \underset{v}{\overset{u}{\rightrightarrows}} F(n) \xrightarrow{\;p\;}\!\!\!\!\rightarrow P, \quad F(k) \underset{s}{\overset{r}{\rightrightarrows}} F(l) \xrightarrow{\;q\;}\!\!\!\!\rightarrow Q,$$

it follows from 2.12.1, volume 1, that

$$F(m) \amalg F(k) \underset{v \amalg s}{\overset{u \amalg r}{\rightrightarrows}} F(n) \amalg F(l) \xrightarrow{\;p \amalg q\;} P \amalg Q$$

is still a coequalizer. This proves that $P \amalg Q$ is finitely presentable, since $F(m) \amalg F(k) = F(m+k)$ and $F(n) \amalg F(l) = F(n+l)$.

Finally $\mathscr{P}$ is stable under coequalizers. Consider diagram 3.11, where the columns are coequalizers in $\mathsf{Mod}_{\mathcal{T}}$ as well as the bottom row. The sets k, l, m, n are finite, so that P, Q are finitely presentable. We must prove that R is finitely presentable as well. By projectivity of $F(n)$ with respect to regular epimorphisms (see 3.8.6), we choose morphisms x, y such that $q \circ x = u \circ p$, $q \circ y = v \circ p$. We consider then the object $F(n) \amalg F(l) \cong F(n+l)$ and the corresponding factorizations $x \amalg c$, $y \amalg d$. We shall prove that $r \circ q = \mathsf{Coker}\,(x \amalg c, y \amalg d)$, showing that R is finitely presentable. First $r \circ q$ coequalizes x, y, since r coequalizes u, v and $r \circ q$ coequalizes c, d since q coequalizes c, d. Next, if a morphism $z\colon F(k) \longrightarrow Z$ coequalizes $x \amalg c$, $y \amalg d$, z factors as $m \circ q$ since $z \circ c = z \circ d$. Now

$$m \circ u \circ p = m \circ q \circ x = z \circ x = z \circ y = m \circ q \circ y = m \circ v \circ p$$

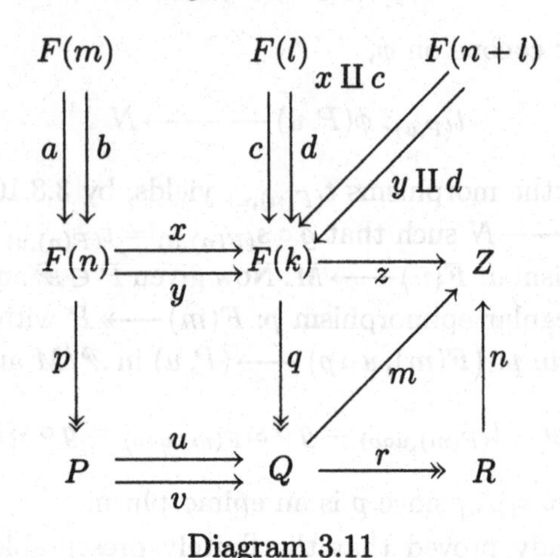

Diagram 3.11

from which $m \circ u = m \circ v$, since p is an epimorphism. Therefore m factors as $n \circ r$. □

We pursue this section with charaterizations of the finitely generated and finitely presentable models.

Proposition 3.8.13 *Let $\mathcal{T}$ be an algebraic theory. A $\mathcal{T}$-model M is finitely generated precisely when the representable functor*

$$\mathsf{Mod}_{\mathcal{T}}(M, -)\colon \mathsf{Mod}_{\mathcal{T}} \longrightarrow \mathsf{Set}$$

preserves filtered unions.

Proof Suppose M is finitely generated and choose a regular epimorphism $q\colon F(n) \twoheadrightarrow M$, with n finite. Given a filtered family of submodels $N_i \subseteq N$, their filtered union is just the corresponding filtered colimit and it is computed as in the category of sets (see 3.6.6). We must prove the isomorphism

$$\bigcup_{i\in I} \mathsf{Mod}_{\mathcal{T}}(M, N_i) \cong \mathsf{Mod}_{\mathcal{T}}\Big(M, \bigcup_{i\in I} N_i\Big).$$

Clearly the left-hand side is contained in the right. Conversely given $f\colon M \longrightarrow \bigcup_{i\in I} N_i$, the composite $f \circ q\colon F(n) \longrightarrow \bigcup_{i\in I} N_i$ corresponds by adjunction to a mapping $n \longrightarrow U\left(\bigcup_{i\in I} N_i\right)$, thus to a family $x_1, \ldots, x_n$ in $U\left(\bigcup_{i\in I} N_i\right)$ of elements. Since U preserves filtered unions, each x_k is in some $U(N_{i_k})$ and by filteredness, there exists an i_0 such that $x_1, \ldots, x_n$ are in $U(N_{i_0})$. This means that $f \circ q$ factors through N_{i_0}. But since $U(q)$ is surjective, this implies that $U(f)$ takes all its values in $U(N_{i_0})$, thus f factors through N_{i_0}. Therefore f is in the left-hand side of the required isomorphism.

Now suppose the previous isomorphism holds for every filtered union. We apply it to the filtered union of the finitely generated submodels M_i of M, which is M itself (see 3.8.11):

$$\bigcup_{i\in I}\mathsf{Mod}_{\mathcal{T}}(M,M_i)\cong\mathsf{Mod}_{\mathcal{T}}(M,M).$$

The identity on M corresponds, in the left-hand side, to some morphism $f: M \longrightarrow M_i$ for some finitely presentable submodel M_i. In other words, the composite

$$M\xrightarrow{\ f\ }M_i\subseteq M$$

is the identity on M. In particular the inclusion $M_i \subseteq M$ is a strong epimorphism, thus an isomorphism (see 4.3.6, volume 1). Thus M is isomorphic to the finitely generated model M_i. □

Proposition 3.8.14 *Let $\mathcal{T}$ be an algebraic theory. A $\mathcal{T}$-model M is finitely presentable precisely when the representable functor*

$$\mathsf{Mod}_{\mathcal{T}}(M,-)\colon \mathsf{Mod}_{\mathcal{T}}\longrightarrow\mathsf{Set}$$

preserves filtered colimits.

Proof Let us consider a filtered colimit $N = \operatorname{colim} N_i$ in $\mathsf{Mod}_{\mathcal{T}}$. The canonical morphisms $s_i: N_i \longrightarrow N$ induce a cocone $\mathsf{Mod}_{\mathcal{T}}(M, s_i)$ thus a canonical factorization

$$\varphi\colon \operatorname{colim}\mathsf{Mod}_{\mathcal{T}}(M,N_i)\longrightarrow\mathsf{Mod}_{\mathcal{T}}(M,\operatorname{colim}N_i).$$

Let us assume that M is finitely presented by the coequalizer

$$F(m)\underset{v}{\overset{u}{\rightrightarrows}}F(n)\xrightarrow{\ q\ }\!\!\!\rightarrow M$$

and let us prove that φ is a bijection.

Given a morphism $f: M \longrightarrow \operatorname{colim} N_i$, we get a composite morphism $f \circ q: F(n) \longrightarrow \operatorname{colim} N_i$, thus by adjunction a corresponding mapping $n \longrightarrow U(\operatorname{colim} N_i)$, thus finally n-elements $x_1, \ldots, x_n \in U(\operatorname{colim} N_i)$. Since U preserves filtered colimits (see 3.5.2), those elements x_j can be written as the equivalence classes of elements $\overline{x_j} \in U(N_{i_0})$ (see 2.13, volume 1; by filteredness, there is no restriction in supposing all the $\overline{x_j}$ in the same $U(N_{i_0})$. This yields a new mapping $n \longrightarrow U(N_{i_0})$ and thus a corresponding morphism $g: F(n) \longrightarrow N_{i_0}$ such that $s_{i_0} \circ g = f \circ q$.

The morphisms $g \circ u: F(m) \longrightarrow N_{i_0}$, $g \circ v: F(m) \longrightarrow N_{i_0}$ correspond in the same way to families $y_1, \ldots, y_m$ and $z_1, \ldots, z_m$ of elements of $U(N_{i_0})$. The equalities

$$s_{i_0}\circ g\circ u=f\circ q\circ u=f\circ q\circ v=s_{i_0}\circ g\circ v$$

imply the corresponding equalities $[y_k] = [z_k]$ between equivalence classes in the colimit. Since there are just finitely many identities of this kind, by filteredness, they are already realized at some further level N_{i_1} of the colimit (see 2.13.4, volume 1). Thus there is a morphism $s: N_{i_0} \longrightarrow N_{i_1}$ in the diagram defining the colimit such that $s(y_k) = s(z_k)$ for every k. By adjunction, this implies now $s \circ g \circ u = s \circ g \circ v$.

Considering the relation $s \circ g \circ u = s \circ g \circ v$, we now get a unique factorization $h: M \longrightarrow N_{i_1}$ through the coequalizer q of u, v; thus $h \circ q = s \circ g$. This immediately implies

$$s_{i_1} \circ h \circ q = s_{i_1} \circ s \circ g = s_{i_0} \circ g = f \circ q,$$

from which $s_{i_1} \circ h = f$ since q is an epimorphism. This means precisely that $h \in \mathsf{Mod}_{\mathcal{T}}(M, N_{i_1})$ and $f = \varphi([h])$, proving that φ is surjective.

The injectivity of φ is analogous and even easier. Given two morphisms $h \in \mathsf{Mod}_{\mathcal{T}}(M, N_{I_1})$ and $k \in \mathsf{Mod}_{\mathcal{T}}(M, N_{i_2})$ such that $\varphi([h]) = \varphi([k])$, we can choose a further level N_{i_3} and assume h and k are defined at this level N_{i_3}. The two morphisms $h \circ q, k \circ q$: $F(n) \rightrightarrows N_{i_3}$ correspond to families of elements $a_1, \ldots, a_n$ and $b_1, \ldots, b_n$ in $U(N_{i_3})$. The relations $[a_j] = x_j = [b_j]$ hold in the colimit since $\varphi([h]) = \varphi([k])$. Therefore, by filteredness, all the relations $[a_j] = [b_j]$ are already realized at some further level N_{i_4}, for some morphism s': $N_{i_3} \longrightarrow N_{i_4}$ in the diagram. These relations, by adjunction, mean precisely $s' \circ h \circ q = s' \circ k \circ q$, thus $s' \circ h = s' \circ k$ since q is an epimorphism. But this means exactly $[h] = [k]$, proving the injectivity of φ.

We must prove now the converse implication. We suppose that φ is an isomorphism for every filtered colimit. We apply this assumption to the case of the filtered colimit of 3.8.12 defining M itself:

$$\varphi\colon \operatorname{colim} \mathsf{Mod}_{\mathcal{T}}\big(M, \phi(P,u)\big) \cong \mathsf{Mod}_{\mathcal{T}}(M, M).$$

The identity on M corresponds to the equivalence class of some morphism $f: M \longrightarrow P$, on the left-hand side; in other words, the composite

$$M \xrightarrow{\ f\ } P = \phi(P,u) \xrightarrow{\ s_{(P,u)}\ } M$$

is the identity on M. Thus, M is a retract of P and therefore the following diagram is a coequalizer:

$$P \underset{f \circ s_{(P,u)}}{\overset{1_P}{\rightrightarrows}} P \xrightarrow{\ s_{(P,u)}\ } M$$

(see 6.5.4). Since finitely presentable objects are stable under finite colimits (see 3.8.12), M is finitely presentable. □

Let us conclude this section with some useful observations related with the consideration of finitely generated free modules on a ring.

Proposition 3.8.15 *Let R be a ring with unit and $\mathcal{R}$ the theory of right R-modules. The free R-module on n generators ($n \in \mathbb{N}$) is just the power R^n with the pointwise operations.*

Proof If M is a right R-module, one immediately gets bijections

$$\mathsf{Mod}_R(R, M) \cong M \cong \mathsf{Set}(\{*\}, M)$$

by mapping a linear mapping $f\colon R \longrightarrow M$ to $f(1) \in M$ and an element $m \in M$ to the linear mapping

$$g\colon R \longrightarrow M, \quad r \mapsto mr.$$

Thus R is the free right R-module on a single generator. Since $n = 1 \amalg \ldots \amalg 1$ in Set and the free module functor preserves coproducts (see 3.2.2, volume 1), the free R-module on n generators is the n-th copower of R, which is R^n since the category of modules is abelian (see 1.4.6.a and 1.6.5). □

Proposition 3.8.16 *Let R be a ring with unit and $\mathcal{R}$ the theory of right R-modules. Then the dual category $\mathcal{R}^*$ is the theory of left R-modules.*

Proof Clearly $\mathcal{R}$ is the dual of the full subcategory of right R-modules generated by the objects R^n; see 3.8.5 and 3.8.15. Since R^n is both an n-th power and an n-th copower, both $\mathcal{R}$ and its dual $\mathcal{R}^*$ are algebraic theories. Since by 3.8.15 R^n, provided with the pointwise operations (on the left), is the free left R-module on n-generators, we must prove the existence of an isomorphism

$$R\text{-}\mathsf{Mod}(R^n, R^m) \cong \mathsf{Mod}_R(R^m, R^n),$$

where R-Mod indicates the category of left modules, and Mod_R the category of right modules. But since the modules involved are free, we can write a left R-linear mapping $R^n \longrightarrow R^m$ as

$$(x_1, \ldots, x_n) \mapsto (x_1, \ldots, x_n)(a_{ij})_{ij}$$

where $(a_{ij})_{ij}$ is an $n \times m$ matrix. In the same way a right R-linear mapping $R^m \longrightarrow R^n$ can be written as

$$\begin{pmatrix} x_1 \\ \vdots \\ x_m \end{pmatrix} \mapsto (a_{ij})_{ij} \begin{pmatrix} x_1 \\ \vdots \\ x_m \end{pmatrix}$$

where $(a_{ij})_{ij}$ is again a $n \times m$ matrix. This obviously yields the required isomorphism and clearly it acts contravariantly with respect to the composition. □

3.9 Characterization of algebraic categories

We want now to characterize those categories which can be presented as categories of models for some algebraic theory.

Theorem 3.9.1 *Let $\mathscr{C}$ be a category and $U: \mathscr{C} \longrightarrow \mathsf{Set}$ a functor. The following conditions are equivalent:*

(1) $\mathscr{C}$ is equivalent to the category of models of some algebraic theory $\mathcal{T}$, with U the corresponding forgetful functor;

(2) the following conditions are satisfied:
- *(a) $\mathscr{C}$ has coequalizers and kernel pairs;*
- *(b) U has a left adjoint F;*
- *(c) U reflects isomorphisms;*
- *(d) U preserves regular epimorphisms;*
- *(e) UF preserves filtered colimits.*

Under these conditions, $\mathcal{T}^*$ is equivalent to the full subcategory of $\mathscr{C}$ generated by the objects $F(n)$, for n running through the finite sets. A category $\mathscr{C}$ as in the statement is called an "algebraic category".

Proof By 3.4.5, 3.4.1, 3.7.8, 3.3.3 and 3.5.3 conditions (2)(a), (b), (c), (d) are necessary. By 3.2.2, volume 1, and 3.4.2, this volume, condition (2)(e) is necessary as well. The very last assertion follows from 3.8.5.

Conversely, if conditions (2)(a) to (2)(e) are satisfied, U preserves kernel pairs (see 3.2.8, volume 1). Since U reflects isomorphisms, it also reflects kernel pairs (see 2.9.7, volume 1). This implies that U reflects regular epimorphism. Indeed consider $f: A \longrightarrow B$ in $\mathscr{C}$ such that $U(f)$ is a regular epimorphism in Set. Write (u, v) for the kernel pair of f. Then $\big(U(u), U(v)\big)$ is the kernel pair of $U(f)$ and since $U(f)$ is a regular epimorphism, it is the coequalizer of $\big(U(u), U(v)\big)$; see 2.5.7, volume 1. If $q: B \longrightarrow Q$ is the coequalizer of (u, v) and $w: Q \longrightarrow B$ the unique morphism such that $w \circ q = f$, then (u, v) is the kernel pair of q; see 2.5.8, volume 1. Therefore $U(q)$ is another regular epimorphism with kernel pair $\big(U(u), U(v)\big)$, thus $U(q)$ is another coequalizer of $\big(U(u), U(v)\big)$; see 2.5.7, volume 1. Finally $U(w)$, and therefore w, is an isomorphism, proving that f, isomorphic to q, is a regular epimorphism.

Writing $0, 1, 2, \ldots, n, \ldots$ for a specific choice of finite sets, we define $\mathcal{T}$ to be the category with formally distinct objects $T^0, T^1, T^2, \ldots, T^n, \ldots$

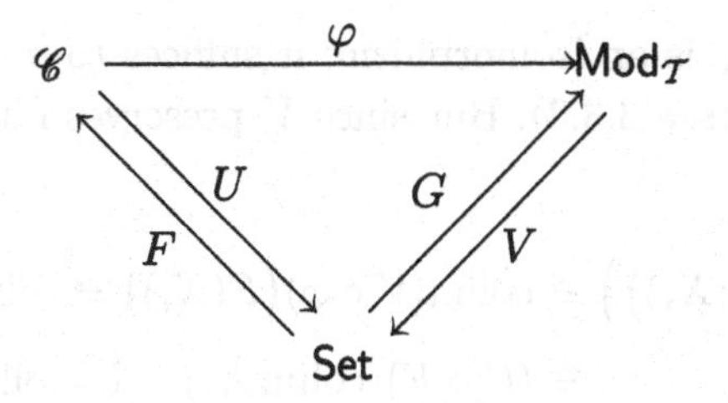

Diagram 3.12

and $\mathcal{T}(T^n, T^m) \cong \mathscr{C}(F(m), F(n))$. The composition is determined by the fact that $\mathcal{T}$ is equivalent to the dual of the full subcategory of $\mathscr{C}$ generated by the objects $F(0), F(1), F(2), \ldots, F(n), \ldots$. By 3.2.2, volume 1, $F(n)$ is the n-th copower of $F(1)$, so that $\mathcal{T}$ is indeed an algebraic theory. For every object $C \in \mathscr{C}$, the contravariant representable functor $\mathscr{C}(-, C)$ transforms finite coproducts into finite products (see 2.9.5, volume 1), thus induces a $\mathcal{T}$-model

$$\mathscr{C}(-, C) \colon \mathcal{T} \longrightarrow \mathsf{Set}, \ \ T^n \mapsto \mathscr{C}\big(F(n), C\big).$$

Every arrow $f \in \mathcal{T}(T^n, T^m)$, i.e. every arrow $f\colon F(m) \longrightarrow F(n)$ in $\mathscr{C}$, induces a corresponding mapping

$$\mathscr{C}\big(F(n), C\big) \longrightarrow \mathscr{C}\big(F(m), C\big), \ \ g \mapsto g \circ f,$$

so that finally we have constructed a functor

$$\varphi\colon \mathscr{C} \longrightarrow \mathsf{Mod}_\mathcal{T}, \ \ C \mapsto \mathscr{C}(-, C), \ \ f \mapsto \mathscr{C}(-, f),$$

and it remains to prove it is an equivalence of categories. To avoid confusion, we write $V\colon \mathsf{Mod}_\mathcal{T} \longrightarrow \mathsf{Set}$ for the canonical forgetful functor and $G\colon \mathsf{Set} \longrightarrow \mathsf{Mod}_\mathcal{T}$ for its left adjoint, as in diagram 3.12.

First of all, let us observe that (see 3.3.3)

$$(V \circ \varphi)(C) \cong \mathscr{C}\big(F(1), C\big) \cong \mathsf{Set}\big(1, U(C)\big) \cong U(C),$$

from which $V \circ \varphi \cong U$. Observe also that for every finite set n (see 3.3.3)

$$\varphi\big(F(n)\big) \cong \mathscr{C}\big(-, F(n)\big) \cong \mathcal{T}(T^n, -) \cong G(n).$$

More generally, every set X can be written as a filtered colimit $X = \operatorname{colim} X_i$ where the X_i's run through the finite subsets of X. Since F preserves colimits (see 3.2.2, volume 1), $F(X) \cong \operatorname{colim} F(X_i)$. We consider the canonical factorization (see 2.12, volume 1)

$$\alpha_X\colon \operatorname{colim} \varphi\big(F(X_i)\big) \longrightarrow \varphi\big(\operatorname{colim} F(X_i)\big).$$

To prove that α_X is an isomorphism, it suffices to prove that $V(\alpha_X)$ is an isomorphism (see 3.3.3). But since V preserves filtered colimits (see 3.4.2)

$$\begin{aligned} V\Big(\operatorname{colim} \varphi\big(F(X_i)\big)\Big) &\cong \operatorname{colim}(V \circ \varphi)\big(F(X_i)\big) \cong \operatorname{colim}(U \circ F)(X_i) \\ &\cong (U \circ F)(\operatorname{colim} X_i) \cong U\big(\operatorname{colim} F(X_i)\big) \\ &\cong V\Big(\varphi\big(\operatorname{colim} F(X_i)\big)\Big) \end{aligned}$$

from which the fact that $V(\alpha_X)$ and thus α_X is an isomorphism. In other words, $\varphi \circ F \cong G$.

Now let us write $\eta\colon 1_{\mathsf{Set}} \Rightarrow UF$, $\varepsilon\colon FU \Rightarrow 1_{\mathscr{C}}$ for the natural transformations associated with the adjunction $F \dashv U$; see 3.1.3, volume 1. From the relation $(U * \varepsilon) \circ (\eta * U) = 1_U$ we deduce that $U(\varepsilon_C)$ is a surjection for every $C \in \mathscr{C}$, thus $\varepsilon_C\colon FU(C) \twoheadrightarrow C$ is a regular epimorphism; let us write $\varepsilon_C = \mathsf{Coker}(u_C, v_C)$. In particular considering the natural bijections (see 3.1.3, volume 1)

$$\mathscr{C}\big(FU(C), D\big) \cong \mathsf{Set}\big(U(C), U(D)\big)$$

and morphisms $f, g\colon C \rightrightarrows D$, the two composites $f \circ \varepsilon_C, g \circ \varepsilon_C \in \mathscr{C}\big(FU(C), D\big)$ correspond to $U(f), U(g) \in \mathsf{Set}\big(U(C), U(D)\big)$ so that

$$U(f) = U(g) \quad \Rightarrow \quad f \circ \varepsilon_C = g \circ \varepsilon_C \quad \Rightarrow \quad f = g$$

since ε_C is an epimorphism. In other words, U is faithful and, since $U = V \circ \varphi$, φ is faithful as well.

Let us also write $\sigma\colon 1_{\mathsf{Set}} \Rightarrow VG$ and $\tau\colon GV \Rightarrow 1_{\mathsf{Mod}_{\mathcal{T}}}$ for the natural trasformations associated with the adjunction $G \dashv V$ (see 3.1.3, volume 1). The same argument as above (or as in 3.8.7) indicates that $\tau_M\colon GV(M) \twoheadrightarrow M$ is a regular epimorphism for each $\mathcal{T}$-model M.

In order to prove that φ is full, consider first a set X and an object $D \in \mathscr{C}$. By the various adjunctions and what is already proved we have

$$\begin{aligned} \mathscr{C}\big(F(X), D\big) &\cong \mathsf{Set}\big(X, U(D)\big) \cong \mathsf{Set}\big(X, (V \circ \varphi)(D)\big) \\ &\cong \mathsf{Mod}_{\mathcal{T}}\big(G(X), \varphi(D)\big) \\ &\cong \mathsf{Mod}_{\mathcal{T}}\big((\varphi \circ F)(X), \varphi(D)\big). \end{aligned}$$

The definitions of the individual isomorphisms indicate that the composite bijection is indeed that mapping a morphism $f\colon F(X) \longrightarrow D$ to $\varphi(f)$. It remains now to replace $F(X)$ by an arbitrary object $C \in \mathscr{C}$.

So let us choose $C, D \in \mathscr{C}$ and a morphism $\alpha \colon \varphi(C) \Rightarrow \varphi(D)$ in $\mathsf{Mod}_{\mathcal{T}}$. We know already that the composite

$$\varphi(FU(C)) \xrightarrow{\varphi(\varepsilon_C)} \varphi(C) \xrightarrow{\alpha} \varphi(D)$$

has the form $\varphi(f)$ for a unique $f \colon FU(C) \longrightarrow D$. We observe that, still writing $\varepsilon_C = \mathsf{Coker}\,(u_C, v_C)$,

$$\begin{aligned}\varphi(f) \circ \varphi(u_C) &= \alpha \circ \varphi(\varepsilon_C) \circ \varphi(u_C) = \alpha \circ \varphi(\varepsilon_C) \circ \varphi(v_C) \\ &= \varphi(f) \circ \varphi(v_C),\end{aligned}$$

from which $f \circ u_C = f \circ v_C$, since φ is faithful. This implies the existence of a unique factorization $g \colon C \longrightarrow D$ through $\varepsilon_C = \mathsf{Coker}\,(u_C, v_C)$, thus yielding $g \circ \varepsilon_C = f$. In particular

$$\begin{aligned}(V \circ \varphi)(g) \circ U(\varepsilon_C) &= U(g) \circ U(\varepsilon_C) = U(f) \\ &= (V \circ \varphi)(f) = V(\alpha) \circ (V \circ \varphi)(\varepsilon_C) = V(\alpha) \circ U(\varepsilon_C),\end{aligned}$$

from which $(V \circ \varphi)(g) = V(\alpha)$ since $U(\varepsilon_C)$ is surjective and $\varphi(g) = \alpha$ since V is faithful. This concludes the proof that φ is full and faithful.

Next we shall prove that φ has a left adjoint ψ. For every $\mathcal{T}$-model M, we consider the kernel pair of τ_M,

$$R(M) \overset{r_M}{\underset{s_M}{\rightrightarrows}} GV(M) \xrightarrow{\tau_M} \!\!\!\!\rightarrow M.$$

Since τ_M is a regular epimorphism, $\tau_M = \mathsf{Coker}\,(r_M, s_M)$ by 2.5.7, volume 1.

Let us consider the following composites:

$$\varphi FVR(M) \cong GVR(M) \xrightarrow{\tau_{R(M)}} R(M) \xrightarrow{r_M} GV(M) \cong \varphi FV(M)$$
$$\varphi FVR(M) \cong GVR(M) \xrightarrow{\tau_{R(M)}} R(M) \xrightarrow{s_M} GV(M) \cong \varphi FV(M).$$

Since φ is full and faithful, we get uniquely determined morphisms a_M, b_M such that

$$\varphi(a_M) = r_M \circ \tau_{RM}, \quad \varphi(b_M) = s_M \circ \tau_{RM}.$$

We define $q_M = \mathsf{Coker}\,(a_M, b_M)$,

$$FVR(M) \overset{a_M}{\underset{b_M}{\rightrightarrows}} FV(M) \xrightarrow{q_M} \!\!\!\!\rightarrow \psi(M).$$

We consider now the kernel pair of q_M,

$$S(M) \overset{m_M}{\underset{n_M}{\rightrightarrows}} FV(M) \xrightarrow{q_M} \!\!\!\!\rightarrow \psi(M).$$

$$\begin{array}{ccccccc}
GVR(M) & \xrightarrow{\tau_{R(M)}} & R(M) & \underset{s_M}{\overset{r_M}{\rightrightarrows}} & GV(M) & \xrightarrow{\tau_M} & M \\
 & & \vdots\,\gamma_M & & \downarrow\cong & & \vdots\,\delta_M \\
 & & \varphi S(M) & \underset{\varphi(n_M)}{\overset{\varphi(m_M)}{\rightrightarrows}} & \varphi FV(M) & \xrightarrow[\varphi(q_M)]{} & \varphi\psi(M)
\end{array}$$

Diagram 3.13

Since q_M is a coequalizer, $q_M = \mathsf{Coker}\,(m_M, n_M)$; see 2.5.7, volume 1. Since U preserves kernel pairs and regular epimorphisms while V reflects them, the image of this diagram under φ produces the kernel pair $(\varphi(m_M), \varphi(n_M))$ of the regular epimorphism $\varphi(q_M)$; as a consequence, $\varphi(q_M)$ is still the coequalizer of $(\varphi(m_M), \varphi(n_M))$.

In $\mathsf{Mod}_{\mathcal{T}}$ we can now consider diagram 3.13, where the horizontal lines are both kernel pairs and coequalizers. The relations

$$\begin{aligned}
\varphi(q_M) \circ r_M \circ \tau_{R(M)} &= \varphi(q_M) \circ \varphi(a_M) = \varphi(q_M) \circ \varphi(b_M) \\
&= \varphi(q_M) \circ s_M \circ \tau_{R(M)}
\end{aligned}$$

imply $\varphi(q_M) \circ r_M = \varphi(q_M) \circ s_M$, since τ_{RM} is an epimorphism. This yields a unique factorization $\gamma_M \colon RM \longrightarrow \varphi SM$ such that

$$\varphi(m_M) \circ \gamma_M = r_M, \quad \varphi(n_M) \circ \gamma_M = s_M.$$

Since the upper line is a coequalizer, we get a unique factorization morphism $\delta_M \colon M \longrightarrow (\varphi \circ \psi)(M)$ such that $\delta_M \circ \tau_M = \varphi(q_M)$.

Let us prove that $(\psi(M), \delta_M)$ is the reflection of M along φ. Given $C \in \mathscr{C}$ and $\alpha \colon M \Rightarrow \varphi(C)$, the composite

$$\varphi FV(M) \cong GV(M) \xrightarrow{\tau_M} M \xrightarrow{\alpha} \varphi(C)$$

has the form $\varphi(x)$ for a unique $x \colon FV(M) \longrightarrow C$, since φ is full and faithful. One has immediately

$$\begin{aligned}
\varphi(x) \circ \varphi(a_M) &= \alpha \circ \tau_M \circ r_M \circ \tau_{R(M)} = \alpha \circ \tau_M \circ s_M \circ \tau_{R(M)} \\
&= \varphi(x) \circ \varphi(b_M),
\end{aligned}$$

from which $x \circ a_M = x \circ b_M$, since the functor φ is faithful. And as $q_M = \mathsf{Coker}\,(a_M, b_M)$, we get a unique $y \colon \psi(M) \longrightarrow C$ such that $y \circ q_M = x$. One has $\varphi(y) \circ \delta_M = \alpha$ since

$$\varphi(y) \circ \delta_M \circ \tau_M = \varphi(y) \circ \varphi(q_M) = \varphi(x) = \alpha \circ \tau_M$$

and τ_M is an epimorphism. We know y must be unique since δ_M is an epimorphism and φ is faithful.

Thus ψ extends in a functor $\psi\colon \mathsf{Mod}_{\mathcal{T}} \longrightarrow \mathscr{C}$ which is left adjoint to the full and faithful functor φ; see 3.1.3, volume 1. In particular $\psi \circ \varphi$ is isomorphic to the identity on $\mathscr{C}$ (see 3.4.1, volume 1) and it remains to prove that the natural transformation $\delta\colon 1_{\mathsf{Mod}_{\mathcal{T}}} \Rightarrow \varphi \circ \psi$ is itself an isomorphism.

Since the pair (r_M, s_M) is a kernel pair, it is jointly monomorphic and therefore γ_M is a monomorphism. More explicitly, if $\gamma_M \circ h = \gamma_M \circ k$, then

$$r_M \circ h = \varphi(m_M) \circ \gamma_M \circ h = \varphi(m_M) \circ \gamma_M \circ k = r_M \circ k,$$
$$s_M \circ h = \varphi(n_M) \circ \gamma_M \circ h = \varphi(n_M) \circ \gamma_M \circ k = s_M \circ k,$$

from which $h = k$, since (r_M, s_M) is a pullback. Thus γ_M is indeed a monomorphism.

Since ψ is left adjoint to φ, consideration of $\gamma_M\colon R(M) \longrightarrow \varphi S(M)$ yields the existence of a unique morphism $h_M\colon \psi R(M) \longrightarrow S(M)$ such that $\varphi(h_M) \circ \delta_{R(M)} = \gamma_M$. As γ_M is a monomorphism, $\delta_{R(M)}$ is a monomorphism as well. But since $\varphi(q_{R(M)})$ is a regular, thus strong epimorphism, and $\varphi(q_{R(M)}) = \delta_{R(M)} \circ \tau_{R(M)}$, $\delta_{R(M)}$ is also a strong epimorphism and therefore an isomorphism (see 4.3.6, volume 1).

Consider now the diagram

$$\varphi\psi R(M) \underset{\varphi(n_M \circ h_M)}{\overset{\varphi(m_M \circ h_M)}{\rightrightarrows}} GV(M) \xrightarrow{\tau_M} M.$$

Composing with the isomorphism δ_{RM} we get

$$\varphi(m_M \circ h_M) \circ \delta_{RM} = \varphi(m_M) \circ \gamma_M = r_M,$$
$$\varphi(n_M \circ h_M) \circ \delta_{RM} = \varphi(n_M) \circ \gamma_M = s_M,$$

from which the above diagram is both a kernel pair and a coequalizer.

Next let us observe that

$$a_M = m_M \circ h_M \circ q_{RM}, \quad b_M = n_M \circ h_M \circ q_{RM}.$$

We prove the first equality

$$\begin{aligned}\varphi(m_M \circ h_M \circ q_{RM}) &= \varphi(m_M) \circ \varphi(h_M) \circ \delta_{RM} \circ \tau_{RM} \\ &= \varphi(m_M) \circ \gamma_M \circ \tau_{RM} = r_M \circ \tau_{RM} \\ &= \varphi(a_M),\end{aligned}$$

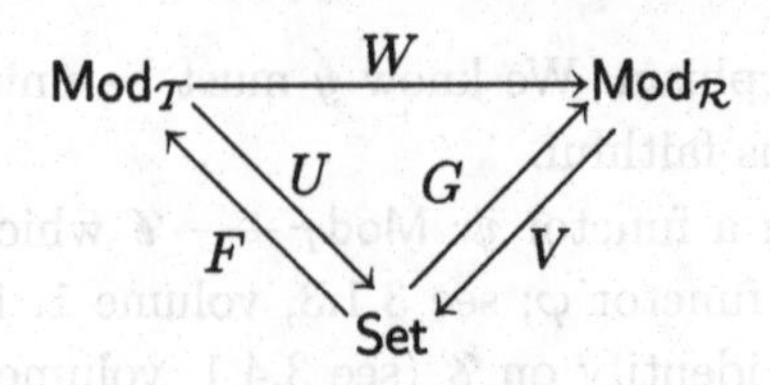

Diagram 3.14

from which we get the required result, since φ is faithful. Therefore

$$q_M = \mathsf{Coker}\,(m_M \circ h_M \circ q_{RM}, n_M \circ h_M \circ q_{RM})$$

and since q_{RM} is an epimorphism

$$q_M = \mathsf{Coker}\,(m_M \circ h_M, n_M \circ h_M).$$

But applying φ to the diagram

$$\psi R(M) \underset{n_M \circ h_M}{\overset{m_M \circ h_M}{\rightrightarrows}} FV(M) \xrightarrow{q_M} \psi(M)$$

yields on the left-hand side the kernel pair of τ_M, as already observed. Since kernel pairs are preserved by V and reflected by U, the composite pair $(m_M \circ h_M, n_M \circ h_M)$ is a kernel pair in $\mathscr{C}$, thus it is the kernel pair of its coequalizer q_M (see 2.5.8, volume 1).

Now φ preserves kernel pairs, since it has a left adjoint (see 3.2.2, volume 1). Thus the previous diagram has for image the kernel pair of $\varphi(q_M)$, with $\varphi(q_M)$ a regular epimorphism since U preserves regular epimorphisms and V reflects them. Thus φq_M is also the coequalizer of the pair $(\varphi(m_M \circ h_M), \varphi(n_M \circ h_M))$, i.e. it is isomorphic to τ_M. In other words, δ_M is an isomorphism. □

Proposition 3.9.2 *Let $\mathcal{T}, \mathcal{R}$ be algebraic theories, with corresponding forgetful functors U, V and free algebra functors F, G, as in diagram 3.14. A functor $W\colon \mathsf{Mod}_{\mathcal{T}} \longrightarrow \mathsf{Mod}_{\mathcal{R}}$ is the algebraic functor induced by a morphism of theories $H\colon \mathcal{R} \longrightarrow \mathcal{T}$ if and only if $V \circ W \cong U$.*

Proof Let us use the notation

$$\eta\colon 1_{\mathsf{Set}} \Rightarrow U \circ F,\ \ \varepsilon\colon F \circ U \Rightarrow 1_{\mathsf{Mod}_{\mathcal{T}}},$$
$$\sigma\colon 1_{\mathsf{Set}} \Rightarrow V \circ G,\ \ \tau\colon G \circ V \Rightarrow 1_{\mathsf{Mod}_{\mathcal{R}}}$$

to denote the natural transformations associated with the two adjunctions $F \dashv U$, $G \dashv V$; see 3.1.3, volume 1.

For every finite set n, the mapping $\eta_n: n \longrightarrow UF(n) = VWF(n)$ corresponds by adjunction with a morphism $\rho_n: G(n) \longrightarrow WF(n)$. We shall prove that $\big(F(n), \rho_n\big)$ is the reflection of $G(n)$ along W. Indeed, given $M \in \mathsf{Mod}_{\mathcal{T}}$ and $f: G(n) \longrightarrow W(M)$, we get by adjunction a mapping $g: n \longrightarrow VW(M) = U(M)$ and thus a morphism $h: F(n) \longrightarrow M$. The relation $W(h) \circ \rho_n = f$ is equivalent to $VW(h) \circ \eta_n = g$, thus to $U(h) \circ \eta_n = g$ which holds by definition of h. Those relations prove at the same time the uniqueness of h.

So each $G(n)$ admits $\big(F(n), \rho_n\big)$ as reflection along W. As in 3.1.3, volume 1, this extends to a functor between the full subcategories generated by those objects: given an arrow $f: G(n) \longrightarrow G(m)$, it is mapped to the unique arrow $h: F(n) \longrightarrow F(m)$ such that $\rho_m \circ f = W(h) \circ \rho_n$. Considering 3.8.5, $\mathcal{T}$ is the dual of the full subcategory of $\mathsf{Mod}_{\mathcal{T}}$ generated by the objects $F(n)$ for n finite; in the same way $\mathcal{R}$ is the dual of the full subcategory of $\mathsf{Mod}_{\mathcal{R}}$ generated by the $G(n)$'s. In particular, the previous construction defines a functor $H: \mathcal{R} \longrightarrow \mathcal{T}$ which is a morphism of theories, since $H\big(G(n)\big) \cong F(n)$.

From the equivalence φ of 3.9.1, an object $M \in \mathsf{Mod}_{\mathcal{T}}$ can be identified with the functor

$$\mathsf{Mod}_{\mathcal{T}}(-, M): \mathcal{T} \longrightarrow \mathsf{Set}, \quad F(n) \mapsto \mathsf{Mod}_{\mathcal{T}}\big(F(n), M\big).$$

Its composite with H produces the functor

$$\mathcal{R} \longrightarrow \mathsf{Set}, \quad G(n) \mapsto \mathsf{Mod}_{\mathcal{T}}\big(F(n), M\big).$$

Considering the isomorphisms

$$\mathsf{Mod}_{\mathcal{T}}\big(F(n), M\big) \cong \mathsf{Mod}_{\mathcal{T}}\big(HG(n), M\big) \cong \mathsf{Mod}_{\mathcal{R}}\big(G(n), W(M)\big)$$

we deduce that this last functor corresponds to $W(M) \in \mathsf{Mod}_{\mathcal{R}}$ again via the equivalence φ of 3.9.1. Thus composing with H indeed corresponds to applying W. □

Corollary 3.9.3 *Let $f: R \longrightarrow T$ be a ring homomorphism. The functor "restricting the scalars along f",*

$$\mathsf{Mod}_{\mathcal{T}} \longrightarrow \mathsf{Mod}_{\mathcal{R}},$$

which maps a right T-module M to M provided with the scalar multiplication

$$M \times R \xrightarrow{1_M \times f} M \times T \longrightarrow M$$

is algebraic. □

$$\begin{array}{ccc} T^{n\times m} & \xrightarrow{\beta^n} & T^n \\ {\scriptstyle \alpha^m}\downarrow & & \downarrow{\scriptstyle \alpha} \\ T^m & \xrightarrow[\beta]{} & T^1 \end{array}$$

Diagram 3.15

3.10 Commutative theories

A group, written additively, is commutative when the axiom $x+y=y+x$ holds for each pair x,y of elements. More generally an n-ary operation $\alpha\colon A^n \longrightarrow A$ on a set A is commutative when, for every n-tuple $(a_1,\dots,a_n)\in A^n$, the value of $\alpha(a_1,\dots,a_n)$ is unaffected by a permutation of $a_1,\dots,a_n$. One could be interested in an algebraic theory T in which every morphism $\alpha\colon T^n \longrightarrow T^1$ gives rise, in the models, to a commutative operation in the previous sense. But, except in the degenerate cases, this never happens since the projections $p_i\colon T^n \longrightarrow T^1$ already fail to satisfy this commutativity requirement.

So, by a commutative theory, we do not mean a theory where every operation is commutative, but one where every operation commutes with every operation, in particular with itself. Let us give a precise definition.

Definition 3.10.1 *Let T be an algebraic theory. We call T a commutative theory when, given morphisms $\alpha\colon T^n \longrightarrow T^1$ and $\beta\colon T^m \longrightarrow T^1$, the square in diagram 3.15 commutes.*

To write the previous definition in terms of variables, as in 3.2.3, it suffices to consider a matrix $(x_{ij})_{1\le i\le n, 1\le j\le m}$ of elements and the axiom

$$\begin{aligned}\beta\big(\alpha(x_{11},\dots,x_{n1}),\dots,\alpha(x_{1m},\dots,x_{nm})\big)\\ =\alpha\big(\beta(x_{11},\dots,x_{1m}),\dots,\beta(x_{n1},\dots,x_{nm})\big).\end{aligned}$$

For example, if α,β are two binary operations written $+$ and $\times$, the previous axiom becomes

$$(x_{11}+x_{21})\times(x_{12}+x_{22})=(x_{11}\times x_{12})+(x_{21}\times x_{22}),$$

which already indicates that the theory of commutative rings is definitely not a commutative theory. Now if α and β are both the binary operation $+$, the axiom becomes

$$(x_{11}+x_{21})+(x_{12}+x_{22})=(x_{11}+x_{12})+(x_{21}+x_{22});$$

when the operation + is associative and admits a 0 element, this reduces to the usual commutativity law

$$x_{21} + x_{12} = x_{12} + x_{21}.$$

The aim of this section is to give several characterizations of commutative theories.

First of all, we generalize a well known definition.

Definition 3.10.2 *Let $\mathcal{T}$ be a presentation of an algebraic theory and A, B, C three $\mathcal{T}$-models. A set theoretic mapping $f\colon A \times B \longrightarrow C$ is a $\mathcal{T}$-bihomomorphism when*

$$\forall a \in A \quad B \longrightarrow C, \ b \mapsto f(a,b),$$

$$\forall b \in B \quad A \longrightarrow C, \ a \mapsto f(a,b),$$

are $\mathcal{T}$-homomorphisms.

Theorem 3.10.3 *Let $\mathcal{T}$ be an algebraic theory and $\mathcal{T}_1$ its corresponding canonical presentation. The following conditions are equivalent:*

(1) $\mathcal{T}$ is a commutative theory;

(2) for every operation $\alpha\colon T^n \longrightarrow T^1$ and every $\mathcal{T}_1$-model A, the mapping

$$A^n \longrightarrow A, \ (a_1, \ldots, a_n) \mapsto \alpha(a_1, \ldots, a_n),$$

is a $\mathcal{T}_1$-homomorphism;

(3) for every pair A, B of $\mathcal{T}_1$-models, the set $\mathsf{Mod}_{\mathcal{T}_1}(A,B)$ of $\mathcal{T}_1$-homomorphisms from A to B is provided with the structure of a $\mathcal{T}_1$-model when we define

$$\alpha(f_1, \ldots, f_n)(a) = \alpha\big(f_1(a), \ldots, f_n(a)\big)$$

for each operation $\alpha\colon T^n \longrightarrow T^1$, $\mathcal{T}_1$-homomorphisms $f_i\colon A \longrightarrow B$ and element $a \in A$.

Under these conditions, the construction defined in (3) is part of a symmetric monoidal closed structure on $\mathsf{Mod}_{\mathcal{T}}$ (see 6.1.3). The corresponding unit object is the free model on one generator. Moreover, given two objects $A, B \in \mathsf{Mod}_{\mathcal{T}_1}$, the tensor product $A \otimes B$ gives rise to natural bijections

$$\mathcal{T}_1 - \mathsf{Bihom}(A \times B, C) \cong \mathsf{Mod}_{\mathcal{T}_1}(A \otimes B, C)$$

where $A, B, C \in \mathsf{Mod}_{\mathcal{T}_1}$ and Bihom *indicates the set of bimorphisms.*

Proof If β is an m-ary operation, the second condition means precisely the relation

$$\beta\big(\alpha(a_{11}, \ldots, a_{n1}), \ldots, \alpha(a_{1m}, \ldots, a_{nm})\big)$$
$$= \alpha\big(\beta(a_{11}, \ldots, a_{1m}), \ldots, \beta(a_{n1}, \ldots, a_{nm})\big)$$

for every matrix $(a_{ij})_{i,j}$ of elements in every $\mathcal{T}_1$-model A. Thus it holds when the theory $\mathcal{T}$ is commutative. Conversely applying this relation to the $\mathcal{T}$-model $\mathcal{T}(T^{n\times m}, -)$ and the projections $p_{ij}: T^{n\times m} \longrightarrow T^1$ as elements yields precisely the relation $\beta \circ \alpha^m = \alpha \circ \beta^n$.

To prove (1) $\Rightarrow$ (3), let $\alpha: T^n \longrightarrow T^1$ and $\beta: T^m \longrightarrow T^1$ be operations. Considering homomorphisms $f_1, \ldots, f_n: A \longrightarrow B$ and elements $a, a_1, \ldots, a_m \in A$, we define

$$\alpha(f_1, \ldots, f_n)(a) = \alpha\big(f_1(a), \ldots, f_n(a)\big)$$

and prove that $\alpha(f_1, \ldots, f_n)$ is a $\mathcal{T}_1$-homomorphism. Indeed, since the f_i's are $\mathcal{T}$-homomorphisms and (1) holds

$$\begin{aligned}
&\alpha(f_1, \ldots, f_n)\big(\beta(a_1, \ldots, a_m)\big)\\
&\quad = \alpha\big(f_1(\beta(a_1, \ldots, a_m)), \ldots, f_n(\beta(a_1, \ldots, a_m))\big)\\
&\quad = \alpha\Big(\beta\big(f_1(a_1), \ldots, f_1(a_m)\big), \ldots, \beta\big(f_n(a_1), \ldots, f_n(a_m)\big)\Big)\\
&\quad = \beta\Big(\alpha\big(f_1(a_1), \ldots, f_n(a_1)\big), \ldots, \alpha\big(f_1(a_m), \ldots, f_n(a_m)\big)\Big)\\
&\quad = \beta\big(\alpha(f_1, \ldots, f_n)(a_1), \ldots, \alpha(f_1, \ldots, f_n)(a_m)\big).
\end{aligned}$$

The pointwise definition of the operations implies immediately that all the $\mathcal{T}_1$-axioms are satisfied in $\mathsf{Mod}_{\mathcal{T}_1}(A, B)$, since they are in B; thus the set $\mathsf{Mod}_{\mathcal{T}_1}(A, B)$ has been provided with the structure of a $\mathcal{T}_1$-model. Conversely, let us prove (3)$\Rightarrow$(2). For any operation $\alpha: T^m \longrightarrow T^1$ and every $\mathcal{T}_1$-model A, consider the projections $p_1, \ldots, p_n: A^n \longrightarrow A$. By assumption the mapping

$$\alpha(p_1, \ldots, p_n): A^n \longrightarrow A, \ \ (a_1, \ldots, a_n) \mapsto \alpha(a_1, \ldots, a_n)$$

is a $\mathcal{T}_1$-homomorphism, which is precisely the content of (2).

Given a commutative theory $\mathcal{T}$, the previous construction extends immediately to a bifunctor

$$(\mathsf{Mod}_{\mathcal{T}_1})^* \times \mathsf{Mod}_{\mathcal{T}_1} \longrightarrow \mathsf{Mod}_{\mathcal{T}_1}, \ \ (A, B) \mapsto \mathsf{Mod}_{\mathcal{T}_1}(A, B).$$

Indeed, given a morphism $f: B \longrightarrow C$, one defines

$$\mathsf{Mod}_{\mathcal{T}_1}(A, f): \mathsf{Mod}_{\mathcal{T}_1}(A, B) \longrightarrow \mathsf{Mod}_{\mathcal{T}_1}(A, C), \ \ g \longrightarrow f \circ g.$$

This is a $\mathcal{T}_1$-homomorphism as given an operation $\alpha: T^n \longrightarrow T^1$, $\mathcal{T}_1$-

homomorphisms $g_1, \dots, g_n \colon A \longrightarrow B$ and an element $a \in A$,

$$\begin{aligned} f \circ \big(\alpha(g_1, \dots, g_n)\big)(a) &= f\Big(\alpha\big(g_1(a), \dots, g_n(a)\big)\Big) \\ &= \alpha\big(f \circ g_1(a), \dots, f \circ g_n(a)\big) \\ &= \alpha(f \circ g_1, \dots, f \circ g_n)(a), \end{aligned}$$

since f is a $\mathcal{T}_1$-homomorphism. Next given $h \colon D \longrightarrow A$, a $\mathcal{T}_1$-homomorphism, one constructs in the same way $\mathsf{Mod}_{\mathcal{T}_1}(h, B)$ and it is obvious we have defined a bifunctor.

Observe now that the free $\mathcal{T}_1$-algebra $F(1)$ on one generator (see 3.8.1) is a "unit" for this bifunctor. Indeed we have bijections, for every $\mathcal{T}_1$-model A,

$$\mathsf{Mod}_{\mathcal{T}_1}\big(F(1), A\big) \cong \mathsf{Set}\big(1, U(A)\big) \cong U(A).$$

This composite bijection maps a $\mathcal{T}_1$-homomorphism $f \colon F(1) \longrightarrow A$ to the element $f(*) \in A$, where $*$ stands for the generator of $F(1)$. The structure of $\mathsf{Mod}_{\mathcal{T}_1}\big(F(1), A\big)$ indicates that this is a $\mathcal{T}_1$-homomorphism, thus a $\mathcal{T}_1$-isomorphism since U reflects isomorphisms (see 3.3.3). Let us also recall that U is precisely the functor represented by that "unit" $F(1)$; see 3.3.3.

To construct the tensor product $A \otimes B$ of two $\mathcal{T}_1$-models A, B, consider the free $\mathcal{T}_1$-model $F\big(U(A \times B)\big)$ on the set $U(A \times B) \cong U(A) \times U(B)$; see 3.8.1 and 3.4.1. We consider also the canonical morphism of the adjunction

$$\eta_{U(A) \times U(B)} \colon U(A) \times U(B) \longrightarrow UF\big(U(A) \times U(B)\big), \quad (a, b) \mapsto \langle a, b \rangle,$$

and for simplicity write $\langle a, b \rangle = \eta_{U(A) \times U(B)}(a, b)$. For every operation $\alpha \colon T^n \longrightarrow T^1$ and elements $a_1, \dots, a_n \in A$, $b \in B$ we consider the elements

$$\big\langle \alpha(a_1, \dots, a_n), b \big\rangle, \ \ \alpha\big(\langle a_1, b \rangle, \dots, \langle a_n, b \rangle\big),$$

where the operation α is thus evaluated in A and $F\big(U(A) \times U(B)\big)$. In the same way, for elements $a \in A$ and $b_1, \dots, b_n \in B$ we consider the elements

$$\big\langle a, \alpha(b_1, \dots, b_n) \big\rangle, \ \ \alpha\big(\langle a, b_1 \rangle, \dots, \langle a, b_n \rangle\big).$$

We consider the smallest congruence on $F\big(U(A) \times U(B)\big)$ generated by all those pairs

$$\Big(\big\langle \alpha(a_1, \dots, a_n), b \big\rangle, \alpha\big(\langle a_1, b \rangle, \dots, \langle a_n, b \rangle\big)\Big),$$

$$\Big(\big\langle a, \alpha(b_1, \dots, b_n) \big\rangle, \alpha\big(\langle a, b_1 \rangle, \dots, \langle a, b_n \rangle\big)\Big),$$

i.e. the intersection of all sub-$\mathcal{T}_1$-models of $F(U(A) \times U(B))$ which are equivalence relations and contain all those pairs (see 3.5.1). Write R for this congruence. By definition $A \otimes B$ is the coequalizer in $\mathsf{Mod}_{\mathcal{T}_1}$ of the two projections $R \rightrightarrows F(U(A) \times U(B))$; write $a \otimes b$ to denote the equivalence class of $\langle a, b\rangle$ in this coequalizer (see 3.5.5). The mapping

$$A \times B \longrightarrow A \otimes B, \quad (a, b) \mapsto a \otimes b$$

is certainly a $\mathcal{T}_1$-bihomomorphism, by definition of the congruence R. Next if $f\colon A \times B \longrightarrow C$ is a $\mathcal{T}_1$-bihomomorphism, the set theoretic mapping

$$U(f)\colon U(A) \times U(B) \cong U(A \times B) \longrightarrow U(C)$$

correspond, by adjunction, with a $\mathcal{T}_1$-homomorphism

$$h\colon F(U(A) \times U(B)) \longrightarrow C$$

such that, in particular, $h(\langle a, b\rangle) = f(a, b)$. Since f is a bihomomorphism, h identifies all the pairs of elements which generate R; since the kernel pair of h is a congruence (see 2.5.6.g), the minimality of R implies it is contained in that kernel pair. In particular h coequalizes the two projections $R \rightrightarrows F(U(A) \times U(B))$ and thus factors uniquely through the coequalizer $A \otimes B$ via a $\mathcal{T}_1$-homomorphism g. We have thus defined a mapping

$$\mathcal{T}_1\text{-}\mathsf{Bihom}(A \times B, C) \longrightarrow \mathsf{Mod}_{\mathcal{T}_1}(A \otimes B, C), \quad f \mapsto g,$$

where $\mathcal{T}_1$-Bihom stands for the set of $\mathcal{T}_1$-bihomomorphisms and the mapping g is given by $g(a \otimes b) = f(a, b)$ for every pair $(a, b) \in A \times B$. This last equality proves the injectivity of the mapping. The surjectivity is obvious since, given a $\mathcal{T}_1$-homomorphism $g\colon A \otimes B \longrightarrow C$, the relation $f(a, b) = g(a \otimes b)$ defines a $\mathcal{T}_1$-bihomomorphism $f\colon A \times B \longrightarrow C$ which is mapped to g by the previous construction.

Let us now observe the existence of a bijection

$$\mathcal{T}_1\text{-}\mathsf{Bihom}(A \times B, C) \cong \mathsf{Mod}_{\mathcal{T}_1}(A, \mathsf{Mod}_{\mathcal{T}_1}(B, C)).$$

A $\mathcal{T}_1$-bihomomorphism $f\colon A \times B \longrightarrow C$ is mapped to

$$\varphi\colon A \longrightarrow \mathsf{Mod}_{\mathcal{T}_1}(B, C)$$

defined by

$$\varphi(a)\colon B \longrightarrow C, \quad b \mapsto f(a, b).$$

Each $\varphi(a)$ is a $\mathcal{T}_1$-homomorphism because f is a $\mathcal{T}_1$-bihomomorphism. Moreover, given an operation $\alpha\colon T^n \longrightarrow T$ and elements $a_1, \ldots, a_n \in A$ and $b \in B$,

$$\begin{aligned}\varphi\big(\alpha(a_1,\ldots,a_n)\big)(b) &= f\big(\alpha(a_1,\ldots,a_n), b\big) = \alpha\big(f(a_1,b),\ldots,f(a_n,b)\big)\\ &= \alpha\big(\varphi(a_1)(b),\ldots,\varphi(a_n)(b)\big)\\ &= \alpha\big(\varphi(a_1),\ldots,\varphi(a_n)\big)(b)\end{aligned}$$

again since f is a $\mathcal{T}_1$-bihomomorphism; this shows that φ itself is a $\mathcal{T}_1$-homomorphism. This correspondence is of course bijective; its inverse maps a $\mathcal{T}_1$-homomorphism $h\colon A \longrightarrow \mathsf{Mod}_{\mathcal{T}_1}(B,C)$ to

$$\psi(h)\colon A \times B \longrightarrow C,\quad (a,b) \longrightarrow h(a)(b),$$

which is indeed a $\mathcal{T}_1$-homomorphism in the first variable, since h is, and a $\mathcal{T}_1$-homomorphism in the second, since each $h(a)$ is.

Putting together the previous results, we have got bijections

$$\mathsf{Mod}_{\mathcal{T}_1}(A \otimes B, C) \cong \mathsf{Mod}_{\mathcal{T}_1}\big(A, \mathsf{Mod}_{\mathcal{T}_1}(B,C)\big).$$

It is simple to check their naturality as well as the compatibility conditions proving that $\mathsf{Mod}_{\mathcal{T}}$ has thus been provided with the structure of a symmetric monoidal closed category. □

The reader will have observed that the construction of the tensor product and the proof of its factorizing property for $\mathcal{T}_1$-bihomomorphisms does not require the commutativity of the theory. But the tensor product obtained in this way is often not interesting when the theory is not "commutative enough". For example in the case of two groups A, B one must have

$$\begin{aligned}(a+a')\otimes(b+b') &= \big((a+a')\otimes b\big) + \big((a+a')\otimes b'\big)\\ &= (a\otimes b) + (a'\otimes b) + (a\otimes b') + (a'\otimes b'),\\ (a+a')\otimes(b+b') &= \big(a\otimes(b+b')\big) + \big(a'\otimes(b+b')\big)\\ &= (a\otimes b) + (a\otimes b') + (a'\otimes b) + (a'\otimes b'),\end{aligned}$$

which indicates a high degree of commutativity in $A \otimes B$.

Examples 3.10.4

3.10.4.a The theory of sets is certainly commutative since its only operations are projections, which obviously satisfy condition (2) of 3.10.3.

3.10.4.b The theory of abelian groups is commutative since, given two abelian groups A, B, one gets a structure of abelian group on $\mathsf{Ab}(A, B)$

by defining

$$(f+g)(a) = f(a) + g(a),$$
$$(-f)(a) = -(f(a)),$$
$$0(a) = 0.$$

Observe that, for example, the relations

$$\begin{aligned}(f+g)(a+a') &= f(a+a') + f(a+a') \\ &= f(a) + f(a') + g(a) + g(a') \\ &= f(a) + g(a) + f(a') + g(a') \\ &= (f+g)(a) + (f+g)(a')\end{aligned}$$

indeed require the commutativity of B.

3.10.4.c The theory of modules over a commutative ring R is a commutative theory. If $f\colon A \longrightarrow B$ is a linear mapping and $r \in R$, the fact that

$$rf\colon A \longrightarrow B,\ a \mapsto r(f(a))$$

is R-linear indeed requires the commutativity of R; given $s \in R$

$$(rf)(sa) = r(f(sa)) = r(sf(a)) = s(rf(a)) = s(rf)(a).$$

3.10.4.d A commutative theory has at most one constant, i.e. one 0-ary operation. Indeed given 0-ary operations $\alpha, \beta\colon T^0 \longrightarrow T^1$, the relation $\alpha \circ \beta^0 = \beta \circ \alpha^0$ for commutativity reduces to $\alpha = \beta$. In particular, the theory of pointed sets (see 3.3.5.i) is commutative.

Counterexample 3.10.5

Let $\mathcal{T}$ be an algebraic theory. The fact that $\mathsf{Mod}_{\mathcal{T}}$ accepts some structure of a symmetric monoidal closed category does not imply *a priori* the commutativity of the theory $\mathcal{T}$: condition 3.10.3.(3) refers indeed to a specific such structure. And in fact this implication is not valid at all. For a counterexample, consider a non-commutative group G and the corresponding theory of G-sets (see 3.3.5.h); since G is not commutative, the 1-ary operations do not commute with each other and the theory is not commutative. Nevertheless we shall see in 5.2.6, volume 3, that the category of G-sets is cartesian closed, thus symmetric monoidal closed in a very canonical way (see 6.1.5).

3.11 Tensor product of theories

In this section, we investigate a rather useful way to construct new theories from given ones. One first trivial way to realize this is to proceed by limits or colimits; this is indeed possible since we have:

Proposition 3.11.1 *The category of algebraic theories and their morphisms is complete and cocomplete.*

Proof Let us write Th for the category of algebraic theories. For every pair (n, m) of natural numbers, we get a functor

$$M_{n,m}: \mathsf{Th} \longrightarrow \mathsf{Set}, \quad \mathcal{T} \mapsto \mathcal{T}(T^n, T^m).$$

Given any functor $\tau: \mathscr{I} \longrightarrow \mathsf{Th}$ with $\mathscr{I}$ a small category, we define a theory $\mathcal{T}$ by

$$\mathcal{T}(T^n, T^m) = \lim(M_{n,m} \circ \tau).$$

In more explicit terms

$$\mathcal{T}(T^n, T^m) = \lim_I \tau(I)(T^n, T^m).$$

Since the morphisms of Th are functors, they respect composition and identities, and the category structure on each $\tau(I)$ induces a category structure on $\mathcal{T}$. We shall prove that $\mathcal{T}$ is the limit of τ. Since the morphisms of Th respect the projections of the products, the family of i-th projections $T^n \longrightarrow T^1$ in each $\tau(I)$ produces a corresponding projection in $\mathcal{T}$, proving finally that $\mathcal{T}$ is an algebraic theory. Observe that the terminal object of Th is thus the degenerate theory $\mathcal{T}$ where $\mathcal{T}(T^n, T^m)$ is always a singleton; the unique arrow $T^n \longrightarrow T^m$ is therefore an isomorphism and the unique $\mathcal{T}$-model is the singleton.

The construction of colimits in Th is easy as well, but more technical. By 3.7.2, the theory of sets is an initial object in Th. Now if $F, G: \mathcal{T}_1 \rightrightarrows \mathcal{T}_2$ are two morphisms of Th, we get a presentation of a new algebraic theory by choosing all the operations and axioms of $\mathcal{T}_2$, together with the axiom

$$\big(F(\alpha)\big)(x_1, \ldots, x_n) = \big(G(\alpha)\big)(x_1, \ldots, x_n)$$

for every operation $\alpha: T^n \longrightarrow T^1$ of $\mathcal{T}_1$; we write $\mathcal{T}$ for the corresponding algebraic theory. By 3.2.6 every operation $\beta: T^m \longrightarrow T$ of $\mathcal{T}_2$ induces an m-ary operation of $\mathcal{T}$ (see 3.2.9) and this correspondence is easily seen to define an algebraic functor $H: \mathcal{T}_2 \longrightarrow \mathcal{T}$. Just by the construction of $\mathcal{T}$ as in 3.2.6, H is the coequalizer of F, G in Th. To construct the coproduct $\mathcal{T} = \coprod_{i \in I} \mathcal{T}_i (I \neq \emptyset)$ in Th, one takes as operations the disjoint union of all the operations of all the individual theories and one keeps all the axioms of the individual theories. Let us observe that given two indices i, j, a projection $p_k\colon \mathcal{T}^n \longrightarrow \mathcal{T}^1$ in $\mathcal{T}_i$ and the corresponding projection $p'_k\colon \mathcal{T}^n \longrightarrow \mathcal{T}^1$ in $\mathcal{T}_j$, the two axioms

$$p_k(x_1, \dots, x_n) = x_k, \;\; p'_k(x_1, \dots, x_n) = x_k$$

imply automatically the axiom (see 3.2.5)

$$p_k(x_1, \dots, x_n) = p'_k(x_1, \dots, x_n)$$

in $\mathcal{T}$. This indicates that the obvious embeddings $\mathcal{T}_i \longrightarrow \mathcal{T}$ are morphisms of theories. Just by construction, $\mathcal{T}$ is the coproduct of the $\mathcal{T}_i$'s. □

In fact the most useful construction on algebraic theories is the "tensor product" of two theories, sometimes combined with a coequalizer to construct the tensor product of two theories over a third one. To understand the importance of these constructions let us start with an apparently different question.

Definition 3.11.2 *Let $\mathscr{C}$ be a category with finite products. If $\mathcal{T}$ is an algebraic theory, a model of $\mathcal{T}$ in $\mathscr{C}$ is a functor $F\colon \mathcal{T} \longrightarrow \mathscr{C}$ preserving finite products. A morphism of $\mathcal{T}$-models in $\mathscr{C}$ is just a natural transformation.*

We shall not develop universal algebra in a category $\mathscr{C}$ with finite products; let us nevertheless indicate that generalizing to this context the results of the previous sections requires rather strong assumptions on $\mathscr{C}$. Our interest is mainly in the following theorem.

Theorem 3.11.3 *Let $\mathcal{T}$ and $\mathcal{R}$ be algebraic theories. There exists an algebraic theory, written $\mathcal{T} \otimes \mathcal{R}$, such that the following three categories are equivalent:*
(1) the category of $\mathcal{T}$-models in $\mathsf{Mod}_{\mathcal{R}}$;
(2) the category $\mathsf{Mod}_{\mathcal{T}\otimes\mathcal{R}}$ of $\mathcal{T} \otimes \mathcal{R}$-models in Set;
(3) the category of $\mathcal{R}$-models in $\mathsf{Mod}_{\mathcal{T}}$.
Moreover there are canonical morphisms of theories

$$\mathcal{T} \longrightarrow \mathcal{T} \otimes \mathcal{R}, \;\; \mathcal{R} \longrightarrow \mathcal{T} \otimes \mathcal{R}.$$

Proof Let us first define $\mathcal{T} \otimes \mathcal{R}$ as the theory constructed by performing the coproduct of $\mathcal{T}$ and $\mathcal{R}$ and imposing the requirement that every $\mathcal{T}$-operation commutes with every $\mathcal{R}$-operation. More precisely, a presentation of $\mathcal{T} \otimes \mathcal{R}$ is obtained in the following way. One takes as operations all the operations of $\mathcal{T}$ and of $\mathcal{R}$; one takes as axioms all the axioms of

$\mathcal{T}$ and of $\mathcal{R}$; moreover, for every $\mathcal{T}$-operation $\alpha\colon T^n \longrightarrow T^1$ and every $\mathcal{R}$-operation $\beta\colon R^m \longrightarrow R^1$, one requires the axiom $\alpha \circ \beta^n = \beta \circ \alpha^m$ in $\mathcal{T} \otimes \mathcal{R}$ (see 3.10.1), i.e. the axiom

$$\begin{aligned}\alpha\big(\beta(x_{11},\ldots,x_{1m}),\ldots,\beta(x_{n1},\ldots,x_{nm})\big) \\ = \beta\big(\alpha(x_{11},\ldots,x_{n1}),\ldots,\alpha(x_{1m},\ldots,x_{nm})\big)\end{aligned}$$

for every $n \times m$ matrix $(x_{ij})_{ij}$ of variables. We recall that if $p_i'\colon T^n \longrightarrow T^1$ and $p_i''\colon R^n \longrightarrow R^1$ are the i-th projections in $\mathcal{T}$ and $\mathcal{R}$, the axioms

$$p_i'(x_1,\ldots,x_n) = x_i,\ \ p_i''(x_1,\ldots,x_n) = x_i$$

in $\mathcal{T}$ and $\mathcal{R}$ imply immediately the axiom (see 3.2.5)

$$p_i'(x_1,\ldots,x_n) = p_i''(x_1,\ldots,x_n)$$

in $\mathcal{T} \otimes \mathcal{R}$ and both p_i' and p''_i are identified in $\mathcal{T} \otimes \mathcal{R}$ with the i-th projection $p_i\colon S^n \longrightarrow S^1$. In particular the obvious mappings

$$s_{\mathcal{T}}\colon \mathcal{T} \longrightarrow \mathcal{T} \otimes \mathcal{R},\ \ s_{\mathcal{R}}\colon \mathcal{R} \longrightarrow \mathcal{T} \otimes \mathcal{R}$$

are functors, because $\mathcal{T} \otimes \mathcal{R}$ satisfies all the axioms of $\mathcal{T}$ and $\mathcal{R}$; and morphisms of theories, because they respect projections. Since the previous construction is perfectly symmetric in $\mathcal{T}$ and $\mathcal{R}$, it suffices now to prove the equivalence of (1) and (2).

Let us work with the presentations $\mathcal{T}_1$, $\mathcal{R}_1$ of the theories as in 3.3.4. The considerations of 3.3.4 can be repeated here without any change to conclude that a $\mathcal{T}_1$-model in $\mathsf{Mod}_{\mathcal{R}_1}$ is an $\mathcal{R}_1$-model M together with an $\mathcal{R}_1$-homomorphism $\alpha_M\colon M^n \longrightarrow M$ for each $\mathcal{T}_1$-operation $\alpha\colon T^n \longrightarrow T$; these $\mathcal{R}_1$-homomorphisms α_M are required to satisfy all the equalities corresponding to the $\mathcal{T}_1$-axioms. Forgetting the $\mathcal{R}_1$-structure of M, the mappings α_M provide the set M with the structure of a $\mathcal{T}_1$-model. Thus all $\mathcal{T}_1$- and all $\mathcal{R}_1$-operations are realized in the set M and all $\mathcal{T}_1$- and all $\mathcal{R}_1$-axioms are satisfied. To have a $(\mathcal{T} \otimes \mathcal{R})_1$-model, it remains to check the commutativity condition corresponding to the choice of operations $\alpha\colon T^n \longrightarrow T^1$ in $\mathcal{T}$ and $\beta\colon R^m \longrightarrow R^1$ in $\mathcal{R}$. Given a matrix $(x_{ij})_{ij}$ of elements of M ($i \leq n$, $j \leq m$), the relation

$$\begin{aligned}\beta\big((x_{11},\ldots,x_{n1}),\ldots,(x_{1m},\ldots,x_{nm})\big) \\ = \big(\beta(x_{11},\ldots,x_{1m}),\ldots,\beta(x_{n1},\ldots,x_{nm})\big)\end{aligned}$$

holds in the product $M^n \cong M \times \ldots \times M$ (see 3.4.1), so that, since α_M

is an $\mathcal{R}_1$-homomorphism,

$$\begin{aligned}\alpha_M\big(\beta(x_{11},\ldots,x_{1m}),\ldots,\beta(x_{n1},\ldots,x_{nm})\big)\\ &= \alpha_M\Big(\beta\big((x_{11},\ldots,x_{n1}),\ldots,(x_{1m},\ldots,x_{nm})\big)\Big)\\ &= \beta\big(\alpha_M(x_{11},\ldots,x_{n1}),\ldots,\alpha_M(x_{1m},\ldots,x_{nm})\big),\end{aligned}$$

which is the required commutation property.

Conversely, assume the set M is provided with the structure of a $(\mathcal{T}\otimes\mathcal{R})$-model; it is in particular an $\mathcal{R}$-model, and the realization of a $\mathcal{T}$-operation $\alpha\colon T^n \longrightarrow T^1$ is a mapping $\alpha_M\colon M^n \longrightarrow M$ with the property

$$\begin{aligned}\alpha_M\big(\beta(x_{11},\ldots,x_{1m}),\ldots,\beta(x_{n1},\ldots,x_{nm})\big)\\ &= \beta\big(\alpha_M(x_{11},\ldots,x_{n1}),\ldots,\alpha_M(x_{1m},\ldots,x_{nm})\big)\end{aligned}$$

for every $\mathcal{R}$-operation $\beta\colon R^m \longrightarrow R^1$ and every matrix $(x_{ij})_{ij}$ of elements of M. Because of the pointwise action of the operation β on the product $M^n \cong M \times \ldots \times M$, this relation becomes

$$\begin{aligned}\alpha_M\Big(\beta\big((x_{11},\ldots,x_{n1}),\ldots,(x_{1m},\ldots,x_{nm})\big)\Big)\\ &= \beta\big(\alpha_M(x_{11},\ldots,x_{n1}),\ldots,\alpha_M(x_{1m},\ldots,x_{nm})\big).\end{aligned}$$

But saying this for each β is just saying that α_M is an $\mathcal{R}_1$-homomorphism and thus M is a $\mathcal{T}_1$-model in $\mathsf{Mod}_{\mathcal{R}_1}$.

The case of morphisms is easy since given two $(\mathcal{T}\otimes\mathcal{R})_1$-models M and N, a $(\mathcal{T}\otimes\mathcal{R})_1$-homomorphism is a mapping which commutes with all the operations of $(\mathcal{T}\otimes\mathcal{R})_1$, thus with all the operations of $\mathcal{T}_1$ and all operations of $\mathcal{R}_1$. This is the same as giving an $\mathcal{R}_1$-homomorphism which commutes with all the operations of $\mathcal{T}$. □

Proposition 3.11.4 *The tensor product of theories as defined in 3.11.3 is associative, commutative and admits the theory of sets as a unit.*

Proof We have already observed that the construction of $\mathcal{T}\otimes\mathcal{R}$ in 3.11.3 is completely symmetric in $\mathcal{T}$ and $\mathcal{R}$, from which follows the commutativity of the tensor product.

Let us now compute the tensor product of a theory $\mathcal{T}$ with the theory of sets. The only operations of the theory of sets are the projections, thus $(\mathcal{T}\otimes\mathcal{S})_1$ has the same operations as $\mathcal{T}$. If $\alpha\colon T^n \longrightarrow T^1$ is a $\mathcal{T}$-operation and $p_k\colon T^m \longrightarrow T^1$ is a projection, the commutativity requirement becomes

$$\alpha(x_{1k},\ldots,x_{nk}) = \alpha(x_{1k},\ldots,x_{nk})$$

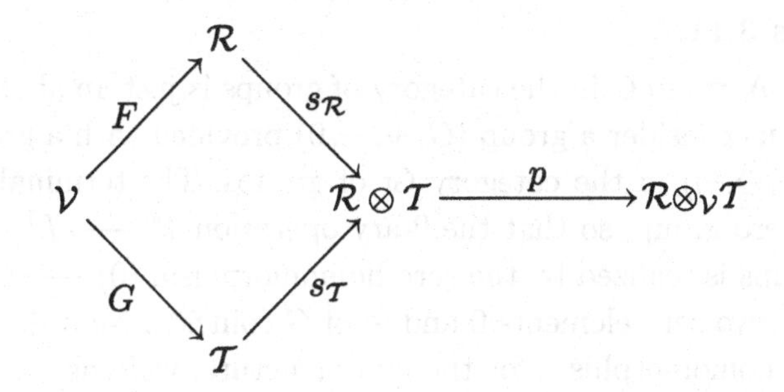

Diagram 3.16

which is a tautology. Thus $\mathcal{T}\otimes\mathcal{S}$ is just $\mathcal{T}$.

For the associativity condition, consider three theories $\mathcal{T},\mathcal{R},\mathcal{V}$. Both theories $\mathcal{T}\otimes(\mathcal{R}\otimes\mathcal{V})$ and $(\mathcal{T}\otimes\mathcal{R})\otimes\mathcal{V}$ admit a presentation given by all the operations of $\mathcal{T},\mathcal{R}$ and $\mathcal{V}$ together with the axioms of these theories and the requirement that every operation of a theory commutes with every operation of one of the other two theories. This proves the associativity. □

A useful generalization of the notion of tensor product given in 3.11.3 is expressed in the following definition.

Definition 3.11.5 *Let $\mathcal{T},\mathcal{R},\mathcal{V}$ be algebraic theories and $F\colon \mathcal{V}\longrightarrow\mathcal{R}$, $G\colon \mathcal{V}\longrightarrow\mathcal{T}$ morphisms of theories. By the tensor product $\mathcal{R}\otimes_{\mathcal{V}}\mathcal{T}$ we mean the theory obtained as the coequalizer in diagram 3.16. In other words, it is the theory obtained from $\mathcal{R}\otimes\mathcal{T}$ by adding the axiom*

$$\big(F(\alpha)\big)(x_1,\ldots,x_n) = \big(G(\alpha)\big)(x_1,\ldots,x_n)$$

for every operation $\alpha\colon V^n\longrightarrow V^1$ of $\mathcal{V}$.

In order to study some examples, it is useful to note the following lemma.

Lemma 3.11.6 *Let G be a set provided with two binary additions $+,\oplus$ with the same zero element 0. Suppose the two additions commute, i.e.*

$$(a+b)\oplus(c+d) = (a\oplus c)+(b\oplus d)$$

for elements $a,b,c,d\in G$. In these conditions the two additions coincide, are associative and commutative.

Proof Choosing $b=0=c$ yields $a\oplus d = a+d$. Putting $b=0$ implies the associativity and putting $a=d=0$ gives the commutativity. □

Examples 3.11.7

3.11.7.a A group G in the category of groups is just an abelian group. Indeed let us consider a group $(G, +, -, 0)$ provided with a group structure $(G, \oplus, \ominus, \odot)$ in the category Gr of groups. The terminal object of Gr is the zero group, so that the 0-ary operation $T^0 \longrightarrow T^1$ of the theory of groups is realized by the zero homomorphism $(0) \longrightarrow G$; in other words, the two zero elements 0 and $\odot$ of G coincide. Next $\oplus: G^2 \longrightarrow G$ is a group homomorphism for the first structure, yielding

$$\begin{aligned}(a+b) \oplus (c+d) &= \oplus(a+b, c+d) = \oplus\big((a,c)+(b,d)\big)\\ &= \oplus(a,c) + \oplus(b,d) = (a \oplus c) + (b \oplus d).\end{aligned}$$

Applying lemma 3.11.6, we conclude that both group structures coincide and are commutative.

3.11.7.b Let R, T be two rings with units. A left R-module in the category of right T-modules is thus a set M provided with the structure of both a left R-module and a right T-module, the operations of the left R-module being T-linear. As in 3.11.7.(a), lemma 3.11.6 implies that both additions coincide. It remains to express the statement that left multiplying by $r \in R$ is right T-linear, which means $r(mt) = (rm)t$ for elements $m \in M$, $t \in T$. We have just described the notion of a R-T-bimodule.

Now write $\mathcal{R}$ for the theory of left R-modules and $\mathcal{T}$ for the theory of right T-modules. We shall observe that $\mathcal{R} \otimes \mathcal{T}$ is just the theory of $(R \otimes T)$-modules. Recall that the tensor product $R \otimes T$ (as abelian groups) is indeed a ring, with multiplication given by

$$\Big(\sum_i r_i \otimes t_i\Big) \times \Big(\sum_j r'_j \otimes t'_j\Big) = \sum_{i,j} (r_i \times r'_j) \otimes (t_i \times t'_j).$$

We must prove the coincidence between the notions of $R \otimes T$-module and R-T-bimodule. If M is an $(R \otimes T)$-module, we provide it with both an R- and a T-multiplication via the formulae

$$rm = (r \otimes 1)m,$$
$$mt = (1 \otimes t)m,$$

where $r \in R, m \in M, t \in T$. It is obvious that

$$r(mt) = (r \otimes 1)(1 \otimes t)m = (r \otimes t)m = (1 \otimes t)(r \otimes 1)m = (rm)t.$$

Conversely if M is an R-T-bimodule it suffices to define

$$\Big(\sum_i r_i \otimes t_i\Big) m = \sum_i r_i m t_i$$

to get the structure of a $(R \otimes T)$-module. The remaining verifications are straightforward.

3.11.7.c Analogously to the previous example, consider now two homomorphisms $f\colon V \longrightarrow R$, $g\colon V \longrightarrow T$ of commutative rings with units. Write $\mathcal{R}$, $\mathcal{T}$, $\mathcal{V}$ for the corresponding theories of modules and

$$F\colon \mathcal{V} \longrightarrow \mathcal{R}, \quad G\colon \mathcal{V} \longrightarrow \mathcal{T}$$

for the corresponding morphisms of theories (see 3.12). One deduces immediately from 3.11.7.b that $\mathcal{R} \otimes_{\mathcal{V}} \mathcal{T}$ is the theory of $(R \otimes_V T)$-modules.

3.12 A glance at Morita theory

The Morita problem for algebraic theories is simple to state: find conditions on two theories $\mathcal{R}, \mathcal{T}$, so that the corresponding categories $\mathsf{Mod}_{\mathcal{R}}$ and $\mathsf{Mod}_{\mathcal{T}}$ of models are equivalent. The difficulty of the problem lies in the fact that a such non-obvious equivalence can never been realized via an algebraic functor. Indeed we have

Proposition 3.12.1 *An algebraic functor $W\colon \mathsf{Mod}_{\mathcal{T}} \longrightarrow \mathsf{Mod}_{\mathcal{R}}$ between two algebraic categories is an equivalence of categories if and only if the corresponding morphism of theories is an isomorphism.*

Proof Write $H\colon \mathcal{R} \longrightarrow \mathcal{T}$ for the corresponding morphism of theories. If H is an isomorphism, W is an equivalence. Conversely consider the situation and the notation of 3.9.2, where it is proved that H maps the free $\mathcal{R}$-model $G(n)$ on n generators to the free $\mathcal{T}$-model $F(n)$ on n generators, which is the reflection of $G(n)$ along W. But if W is an equivalence, it has a left adjoint which is an equivalence and the restriction H of this left adjoint to the corresponding full subcategories is also an equivalence. But since by definition a morphism H of theories is bijective on the objects, it is an isomorphism as long as it is an equivalence (see 3.4.3, volume 1). □

The classical Morita theorem is concerned with the case where both theories $\mathcal{T}, \mathcal{R}$ are theories of modules on a ring. We assume some familiarity with classical module theory.

Theorem 3.12.2 *Let R, T be two rings with unit. The following conditions are equivalent:*

(1) the categories Mod_R and Mod_T of right module are equivalent;

(2) there exist an R-T-bimodule P and a T-R-bimodule Q such that the following isomorphisms of bimodules hold:

$$P\otimes_T Q \cong R, \quad Q\otimes_R P \cong T.$$

Two rings of this kind are called "Morita equivalent".

Proof Suppose we are given an equivalence $\varphi\colon \mathsf{Mod}_R \longrightarrow \mathsf{Mod}_T$. The functor φ has an adjoint (see 3.4.3, volume 1), thus it preserves the zero morphisms and all finite limits and colimits (see 3.2.2, volume 1). In particular φ preserves the addition of morphisms (see 9.6.2). The structure of left R-module of R given by the multiplication

$$R \times R \longrightarrow R, \quad (r, r') \mapsto rr'$$

can equivalently be given by the group homomorphism

$$R \longrightarrow \mathsf{Mod}_R(R, R), \quad r \mapsto (r' \mapsto rr').$$

Since the functor φ is an additive equivalence of categories, the abelian group $\mathsf{Mod}_R(R, R)$ is isomorphic to $\mathsf{Mod}_T\big(\varphi(R), \varphi(R)\big)$ so that we get a group homomorphism

$$R \longrightarrow \mathsf{Mod}_T\big(\varphi(R), \varphi(R)\big)$$

thus a structure of left R-module on $\varphi(R)$ such that left multiplying by $r \in R$ is right T-linear. In other words, we have provided $\varphi(R)$ with the structure of an R-T-bimodule (see 3.11.7.b). We denote this bimodule by P.

The functor

$$-\otimes_R P\colon \mathsf{Mod}_R \longrightarrow \mathsf{Mod}_T$$

admits a right adjoint, namely

$$\mathsf{Mod}_T(P, -)\colon \mathsf{Mod}_T \longrightarrow \mathsf{Mod}_R,$$

thus $-\otimes_R P$ preserves colimits. But the equivalence of categories φ admits also a right adjoint (see 3.4.3, volume 1), thus preserves colimits (see 3.2.2, volume 1). Those two functors coincide on the free R-module on one generator $F(1) \cong R$ (see 3.8.4) just because $R\otimes_R P \cong P \cong \varphi(R)$. By preservation of colimits they coincide on each $F(X) = \coprod_{x\in X} F(\{x\})$, thus finally on each $M \in \mathsf{Mod}_R$; see 3.8.9. Observing that a right linear mapping $f\colon R \longrightarrow R$ is just left multiplying by $r = f(1)$, one concludes immediately that $\varphi(f) \cong f\otimes_R P$, from which by the previous colimit argument φ and $-\otimes_R P$ coincide on the morphisms as well.

In a completely analogous way one proves that "the" inverse equivalence $\psi\colon \mathsf{Mod}_T \longrightarrow \mathsf{Mod}_R$ has the form $-\otimes_T Q$ for some T-R-bimodule Q. The relations $\psi \circ \varphi \cong \mathrm{id}$, $\varphi \circ \psi \cong \mathrm{id}$ applied respectively to R and T yield the required isomorphisms $P \otimes_T Q \cong R$, $Q \otimes_R P \cong T$. The converse implication is obvious. □

Corollary 3.12.3 *Let R be a commutative ring with a unit. Then R is Morita equivalent to the ring $R^{n\times n}$ of $n \times n$ matrices.*

Proof Now R^n is an R-$R^{n\times n}$-module: R acts componentwise on R^n while $R^{n\times n}$ acts by right matrix multiplication, viewing an element of R^n as a line matrix. In the same way R^n is an $R^{n\times n}$-R-module: R acts componentwise on R^n while $R^{n\times n}$ acts by left matrix multiplication, viewing an element of R^n as a column matrix. And it is well known that

$$R^n \otimes_R R^n \cong R^{n\times n}, \quad R^n \otimes_{R^{n\times n}} R^n \cong R.$$

By 3.12.2, R and $R^{n\times n}$ are thus Morita equivalent. □

To avoid the impression that it is quite common to have Morita equivalent rings, let us study the case of commutative rings.

Lemma 3.12.4 *Two rings with units which are Morita equivalent have isomorphic centres.*

Proof We use the notation of 3.12.2. The conditions on P, Q imply immediately that

$$Q \otimes_R -\colon R\text{-}\mathsf{Mod} \longrightarrow T\text{-}\mathsf{Mod}, \quad P \otimes_R -\colon T\text{-}\mathsf{Mod} \longrightarrow R\text{-}\mathsf{Mod}$$

are also equivalences between the corresponding categories of left modules. In particular

$$T\text{-}\mathsf{Mod}(T, T) \cong R\text{-}\mathsf{Mod}(P, P)$$

while in 3.12.2 we had found the isomorphism

$$\mathsf{Mod}_R(R, R) \cong \mathsf{Mod}_T(P, P).$$

In this last case, left multiplying $s \in R$ by the element $r \in R$ is applying $r(\cdot)\colon R \longrightarrow R$, while left multiplying $p \in P$ by $r \in R$ is by definition applying $\varphi\big(r(\cdot)\big)$. Since φ is an equivalence of categories, the last isomorphism thus restricts to an isomorphism

$$R\text{-}\mathsf{Mod}_R(R, R) \cong R\text{-}\mathsf{Mod}_T(P, P)$$

between the subgroups of left R-linear mappings. An analogous argument in the first case yields

$$T\text{-}\mathsf{Mod}_{\mathcal{T}}(T,T) \cong R\text{-}\mathsf{Mod}_{\mathcal{T}}(P,P).$$

A right R-linear mapping $f\colon R \longrightarrow R$ is left multiplying by $f(1)$, while f is left linear when it coincides with right multiplying by $f(1)$. Thus f is both left and right R-linear when $rf(1) = f(1)r$ for every $r \in R$, hence when $f(1)$ belongs to the centre $Z(R)$ of R. The previous two isomorphisms thus imply that when R, T are Morita equivalent, their centres are isomorphic. □

Corollary 3.12.5 *Two commutative rings with units are Morita equivalent if and only if they are isomorphic.* □

Counterexample 3.12.6
If R is a ring with a unit and $\mathcal{R}$ is the corresponding theory of right modules, R is also the free module on one generator (see 3.8.14). The monoid $\mathcal{R}(R^1, R^1)$ of 1-ary operations is thus the dual of the monoid $\mathsf{Mod}_{\mathcal{R}}(R, R)$ (see 3.8.5). But the right linear mappings $R \longrightarrow R$ are just the mappings

$$\overline{r}\colon R \longrightarrow R, \;\; x \mapsto rx$$

for every $r \in R$. Given another element $s \in R$

$$\overline{r}\big(\overline{s}(x)\big) = \overline{r}(sx) = rsx.$$

Hence $\mathcal{R}(R^1, R^1)$ is just the ring R with the opposite multiplication as composition of the arrows.

Now consider a commutative ring with unit R and the ring $R^{n\times n}$ of $n \times n$ matrices ($n \geq 2$). Write $\mathcal{R}$ for the theory of R-modules and $\mathcal{T}$ for the theory of $R^{n\times n}$-modules. The monoid $\mathcal{R}(R^1, R^1)$ is commutative since R is, but the monoid $\mathcal{T}(T^1, T^1)$ is not commutative because $R^{n\times n}$ is not. Thus the theory $\mathcal{R}$ is certainly not isomorphic to the theory $\mathcal{T}$. Nevertheless the corresponding categories of models are equivalent, as attested by corollary 3.12.3.

3.13 Exercises

3.13.1 Prove that the category of fields is not algebraic. [Hint: there is no terminal object.]

3.13.2 The category of small categories is not algebraic. [Hint: it is not regular, see 2.4.6.]

3.13.3 The category of posets is not algebraic. [Hint: it is not regular; consider the quotient of $\{a \leq b, c \leq d\}$ identifying b, c and pull it back along the inclusion of $\{[a] \leq [d]\}$.]

3.13.4 The category of complete lattices is not algebraic. [Hint: filtered unions do not commute with finite intersections; consider the colimit of the intervals $\{0, \ldots, n\}$ of $\mathbb{N}$, which is $\mathbb{N} \cup \{\infty\}$, and pull it back over the singleton $\{\infty\}$.]

3.13.5 Let $\mathcal{T}$ be an algebraic theory; write $U\colon \mathsf{Mod}_{\mathcal{T}} \longrightarrow \mathsf{Set}$ for the corresponding forgetful functor, $F\colon \mathsf{Set} \longrightarrow \mathsf{Mod}_{\mathcal{T}}$ for the free model functor and $\eta\colon 1_{\mathsf{Set}} \Rightarrow UF$ for the canonical natural transformation of the adjunction. Prove that

(1) F is faithful iff each η_X is injective,

(2) some η_X is not injective iff the axiom $x = y$ holds in $\mathcal{T}$ for two distinct variables x, y.

3.13.6 Prove that given an algebraic theory $\mathcal{T}$, the free model on one generator is a regular generator for $\mathsf{Mod}_{\mathcal{T}}$.

3.13.7 Given an algebraic theory $\mathcal{T}$, prove that the forgetful functor $U\colon \mathsf{Mod}_{\mathcal{T}} \longrightarrow \mathsf{Set}$ does not preserve epimorphisms. [Hint: consider the theory of commutative rings with a unit; see 1.8.5.f, volume 1.]

3.13.8 If R is a commutative ring with unit, prove that the theory of R-modules is a category which admits both a covariant and a contravariant isomorphism with its dual.

3.13.9 In an algebraic category, prove that a free model P does not in general satisfy the condition: *given an epimorphism $p\colon X \twoheadrightarrow Y$ and a morphism $f\colon P \longrightarrow Y$, f factors through p.* Compare with 4.6.1, volume 1 and 3.8.6, this volume. [Hint: see 3.13.7.]

3.13.10 Prove that $\mathbb{Z}$ is not a dense generator in the category of abelian groups (compare with 3.8.10).

3.13.11 A functor $G\colon \mathscr{X} \longrightarrow \mathsf{Set}$ is dominated by a family $(X_i)_{i \in I}$ of objects of $\mathscr{X}$ when, for every $Y \in \mathscr{X}$ and every $y \in G(Y)$, there exist $i \in I$, $x \in G(X_i)$ and $f\colon X_i \longrightarrow Y$ such that $y = G(f)(x)$.

(1) Prove that a functor $G\colon \mathsf{Set} \longrightarrow \mathsf{Set}$ which preserves filtered colimits is dominated by the finite sets.

(2) Prove that the previous implication is not an equivalence. [Hint: consider the identity on the finite sets and extend it by the value $\{*\}$ on the infinite sets.]
(3) Replace condition (2)(e) of 3.9.1 by "UF is dominated by the finite sets" and prove the theorem still holds.

3.13.12 Let $\mathbf{M}$ be the ring of infinite real matrices with the property that each line has only a finite number of non-zero elements; addition is pointwise and multiplication is the matrix product. Prove that if n, m are positive integers, the free $\mathbf{M}$-module on n generators is isomorphic to the free $\mathbf{M}$-module on m generators. [Hint: given two matrices A, B, construct another whose lines are successively the first line of A, the first line of B, the second line of A, the second line of B, ...]

3.13.13 Prove that tensoring with T over R produces the left adjoint to the algebraic functor of 3.9.3.

3.13.14 Let $\mathcal{T}$ be an algebraic theory and consider the forgetful functor $U\colon \mathsf{Mod}_{\mathcal{T}} \longrightarrow \mathsf{Set}$. Prove that $\mathcal{T}$ is isomorphic to the category whose objects are the functors $U^n (n \in \mathbb{N})$ and whose morphisms are the natural transformations between those functors.

3.13.15 We consider two universes $\mathcal{U} \in \mathcal{V}$ and write Set for the category of $\mathcal{U}$-sets. We consider the category $\mathscr{X}$ whose objects are the pairs $(\mathscr{C}, U)$ where

(1) $\mathscr{C}$ is a category with $\mathscr{C} \in \mathcal{V}$ and $\mathscr{C}(X,Y) \in \mathcal{U}$ for $X, Y \in \mathscr{C}$,
(2) $U\colon \mathscr{C} \longrightarrow \mathsf{Set}$ is a functor such that the natural endo-transformations $U \Rightarrow U$ constitute a $\mathcal{U}$-set.

The morphisms $W\colon (\mathscr{C}, U) \longrightarrow (\mathscr{D}, V)$ are just the functors $W\colon \mathscr{C} \longrightarrow \mathscr{D}$ such that $V \circ W = U$. If Th is the category of algebraic theories in the universe $\mathcal{U}$, we get a contravariant functor $\mathsf{Th} \longrightarrow \mathscr{X}$ mapping a theory to the corresponding category of models provided with its forgetful functor and a morphism of theories on the corresponding algebraic functor. Prove that this functor has an adjoint mapping the pair $(\mathscr{C}, U)$ to the full subcategory of $\mathsf{Fun}(\mathscr{C}, \mathsf{Set})$ generated by the functors $U^n (n \in \mathbb{N})$. (This is the so-called "structure semantics" adjunction.) By 3.13.4, the functor $\mathsf{Th} \longrightarrow \mathscr{X}$ is full and faithful.

3.13.16 Let R, T be two rings with units and corresponding theories $\mathcal{R}, \mathcal{T}$ of right modules. Consider an R-T-bimodule M viewed as a finite product preserving functor $F\colon \mathcal{R}^0 \longrightarrow \mathsf{Mod}_{\mathcal{T}}$; see 3.8.15 and 3.11.7. Prove that the functor $-\otimes_R M\colon \mathsf{Mod}_{\mathcal{R}} \longrightarrow \mathsf{Mod}_{\mathcal{T}}$ is the left Kan extension of F along the Yoneda embedding $\mathcal{R}^0 \hookrightarrow \mathsf{Mod}_{\mathcal{R}}$.

3.13.17 Consider an algebraic theory $\mathcal{T}$ and a full subcategory $\mathscr{C}$ of $\mathsf{Mod}_{\mathcal{T}}$. Prove that the following conditions are equivalent:

(1) $\mathscr{C}$ is an algebraic category and the inclusion $\mathscr{C} \subseteq \mathsf{Mod}_{\mathcal{T}}$ is an algebraic functor;

(2) $\mathscr{C}$ is closed in $\mathsf{Mod}_{\mathcal{T}}$ under arbitrary products and the inclusion creates monomorphisms and strong epimorphisms (i.e. if $f\colon A \longrightarrow B$ is a monomorphism in $\mathsf{Mod}_{\mathcal{T}}$ and $B \in \mathscr{C}$, then $A \in \mathscr{C}$; if $f\colon A \longrightarrow B$ is a regular epimorphism in $\mathsf{Mod}_{\mathcal{T}}$ and $A \in \mathscr{C}$, then $B \in \mathscr{C}$).

[Hint: given a $\mathcal{T}$-model M, consider all the regular epimorphisms of the form $M \twoheadrightarrow C$ with $C \in \mathscr{C}$; compute the corresponding product $\prod C$ indexed by all those morphisms and let $\tau(M)$ be the image of the induced factorization $M \longrightarrow \prod C$; prove that τ yields the left adjoint to the inclusion; apply 3.13.11.]

3.13.18 Given a commutative ring R with a unit, consider a retract $r, s\colon S \rightleftarrows R^n$, $r \circ s = 1_S$, of some finite power of R in Mod_R. When S is a generator of Mod_R, prove that R is Morita equivalent to the ring $\mathsf{Mod}_R(S, S)$ of S-linear endomorphisms of S. This generalizes 3.12.3.

4
Monads

The classical definition of a monoid is to give a set M together with a binary operation

$$M \times M \longrightarrow M, \quad (x, y) \mapsto xy$$

which is associative and admits a unit element written 1. As a consequence, with every finite sequence $(x_1, \ldots, x_n)$ of elements of M is associated a composite element $x_1 \ldots x_n$ defined inductively in the usual way:

- with the empty sequence is associated the element 1;
- with the sequence $(x_1, \ldots, x_n)$ is associated the element xx_n, where x is the element associated with the sub-sequence $(x_1, \ldots, x_{n-1})$.

In other words, a monoid can also be seen as a set where "every finite sequence has been given a composite". Let us make this a precise definition.

Given a set M, we write $T(M)$ for the set of finite sequences of elements of M. Given a mapping $f\colon M \longrightarrow N$, $T(f)\colon T(M) \longrightarrow T(N)$ is the mapping sending the sequence $(x_1, \ldots, x_n)$ to the sequence of corresponding images $\big(f(x_1), \ldots, f(x_n)\big)$.

We want to provide the set M with the structure of a monoid via a mapping

$$\xi\colon T(M) \longrightarrow M$$

which associates with every finite sequence $(x_1, \ldots, x_n)$ of elements of M a new element $\xi(x_1, \ldots, x_n) \in M$ which we call the "composite of the sequence". We must investigate the axioms requiring ξ to be exactly the composition law induced by a monoid structure on M.

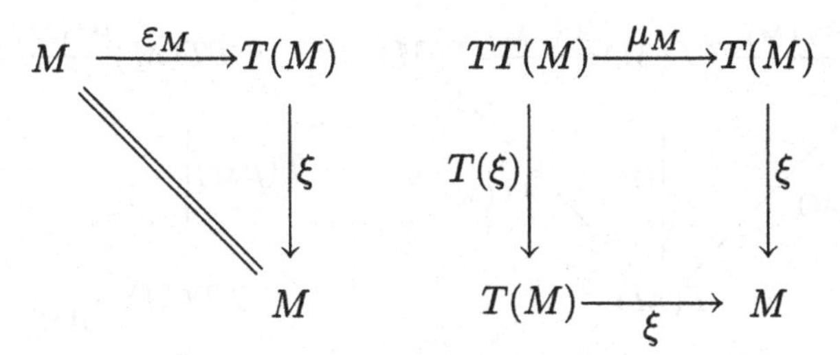

Diagram 4.1

First of all the "normalization condition" $\xi(x) = x$ must certainly be satisfied. We shall express it by considering the mapping

$$\varepsilon_M\colon M \longrightarrow T(M),\quad x \mapsto (x)$$

and requiring as an axiom the commutativity of the triangle part in diagram 4.1.

Next a general associativity condition of the type

$$\begin{aligned}&\xi\big(\xi(a_1^1,\ldots,a_{n_1}^1),\ldots,\xi(a_1^m,\ldots,a_{n_m}^m)\big)\\&\qquad = \xi\big(a_1^1,\ldots,a_{n_1}^1,a_1^2,\ldots,a_{n_{m-1}}^{m-1},a_1^m,\ldots,a_{n_m}^m\big)\end{aligned}$$

must be satisfied as well. Observe that choosing a finite sequence of finite sequences of elements of M is the same as choosing an element of $TT(M)$.

The previous axiom thus involves the concatenation mapping

$$\begin{aligned}&\mu_M\colon TT(M) \longrightarrow T(M),\\&\big((a_1^1,\ldots,a_{n_1}^1),\ldots,(a_1^m,\ldots,a_{n_m}^m)\big)\\&\qquad \mapsto \big(a_1^1,\ldots,a_{n_1}^1,a_1^2,\ldots,a_{n_{m-1}}^{m-1},a_1^m,\ldots,a_{n_m}^m\big),\end{aligned}$$

which constructs a "composite sequence" from a sequence of sequences. The general associativity axiom can be expressed by the commutativity of the square in diagram 4.1.

It is now easy to verify that the previous definition is equivalent to the classical definition of a monoid. It suffices to define a multiplication on M via

$$M \times M \longrightarrow M,\quad (x,y) \mapsto \xi(x,y).$$

The associativity rule is just

$$\begin{aligned}(xy)z &= \xi\big(\xi(x,y),z\big) = \xi\big(\xi(x,y),\xi(z)\big) = \xi(x,y,z),\\ x(yz) &= \xi\big(x,\xi(y,z)\big) = \xi\big(\xi(x),\xi(y,z)\big) = \xi(x,y,z).\end{aligned}$$

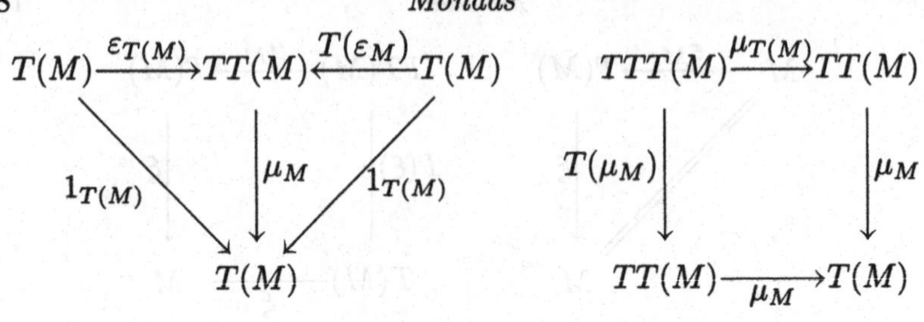

Diagram 4.2

Now put $1 = \xi(\ )$, the composite of the empty sequence.

$$\begin{aligned} 1x &= \xi(\ )\xi(x) = \xi\big(\xi(\ ), \xi(x)\big) = \xi(\ \ x) \\ &= \xi(x) = x \end{aligned}$$

and in the same way $x = x1$.

Thus M has been provided with the structure of a monoid and one proves inductively that ξ is just the composition for this monoid structure:

- $\xi(\ )$ is indeed the element 1;
- if $\xi(x_1, \ldots, x_{n-1}) = x_1 \ldots x_{n-1}$, then

$$\begin{aligned} \xi(x_1, \ldots, x_n) &= \xi\big(\xi(x_1, \ldots, x_{n-1}), \xi(x_n)\big) \\ &= \xi(x_1 \ldots x_{n-1}, x_n) = (x_1 \ldots x_{n-1})x_n \\ &= x_1 \ldots x_{n-1}x_n. \end{aligned}$$

Before leaving this example, one should observe that the two mappings ε_M, μ_M which have been used to describe the two axioms themselves satisfy interesting relations, namely the commutativity of all pieces of diagram 4.2. The commutativity of the triangles means that the concatenation of both sequences

$$\big((x_1, \ldots, x_n)\big), \ \big((x_1), \ldots, (x_n)\big)$$

yields the sequence $(x_1, \ldots, x_n)$; the commutativity of the square expresses the associativity of the concatenation process.

4.1 Monads and their algebras

With the previous example in mind, we define:

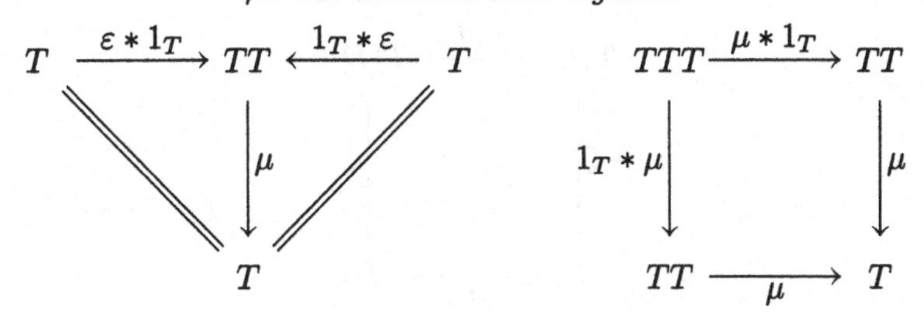

Diagram 4.3

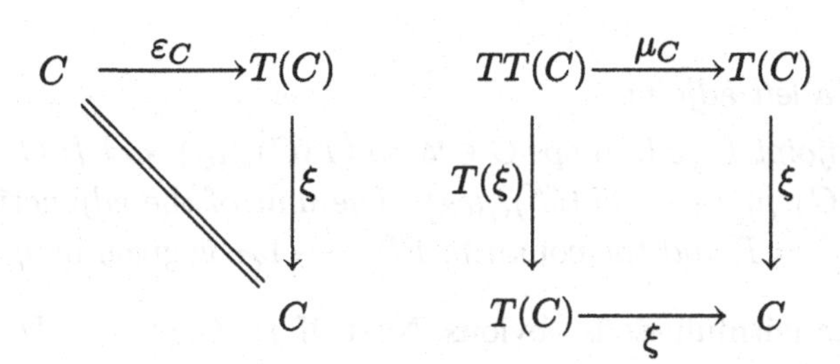

Diagram 4.4

Definition 4.1.1 *A monad on a category $\mathscr{C}$ is a triple (T, ε, μ) where $T: \mathscr{C} \longrightarrow \mathscr{C}$ is a functor and $\varepsilon: 1_{\mathscr{C}} \Rightarrow T$, $\mu: TT \Rightarrow T$ are natural transformations satisfying the commutativity conditions*

$$\mu \circ (\varepsilon * 1_T) = 1_T = \mu \circ (1_T * \varepsilon), \quad \mu \circ (\mu * 1_T) = \mu \circ (1_T * \mu)$$

(see diagram 4.3).

Definition 4.1.2 *Let $\mathbb{T}(T, \varepsilon, \mu)$ be a monad on a category $\mathscr{C}$. By an algebra on this monad is meant a pair (C, ξ) where $C \in \mathscr{C}$, $\xi: T(C) \longrightarrow C$ and $\xi \circ \varepsilon_C = 1_C$, $\xi \circ T(\xi) = \xi \circ \mu_C$; see diagram 4.4. If (D, ζ) is another $\mathbb{T}$-algebra, a morphism $f: (C, \xi) \longrightarrow (D, \zeta)$ of $\mathbb{T}$-algebras is a morphism $f: C \longrightarrow D$ of $\mathscr{C}$ such that $f \circ \xi = \zeta \circ T(f)$; see diagram 4.5.*

Proposition 4.1.3 *Let $\mathbb{T} = (T, \varepsilon, \mu)$ be a monad on a category $\mathscr{C}$. The $\mathbb{T}$-algebras and their morphisms constitute a category, written $\mathscr{C}^{\mathbb{T}}$.* □

The category $\mathscr{C}^{\mathbb{T}}$ is also called the "Eilenberg–Moore" category of the monad.

Proposition 4.1.4 *Let $\mathbb{T} = (T, \varepsilon, \mu)$ be a monad on a category $\mathscr{C}$. Consider the forgetful functor*

$$U: \mathscr{C}^{\mathbb{T}} \longrightarrow \mathscr{C}, \quad (C, \xi) \mapsto C, \quad f \mapsto f$$

(1) U is faithful;
(2) U reflects isomorphisms;

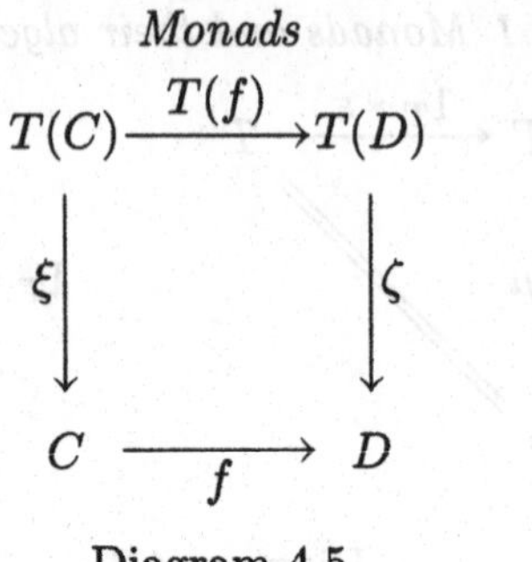

Diagram 4.5

(3) U has a left adjoint.
The left adjoint F of U maps $C \in \mathscr{C}$ to $\big(T(C), \mu_C\big)$ and $f\colon C \longrightarrow C'$ to $T(f) : \big(T(C), \mu_C\big) \longrightarrow \big(T(C'), \mu_{C'}\big)$. The unit of the adjunction is just $\varepsilon\colon 1_{\mathscr{C}} \Rightarrow UF = T$ and the counit $\eta\colon FU \longrightarrow 1_{\mathscr{C}^{\mathbb{T}}}$ is given by $\eta_{(C,\xi)} = \xi$.

Proof The faithfulness is obvious. Next, if $f\colon (C, \xi) \longrightarrow (D, \zeta)$ is such that f is an isomorphism in $\mathscr{C}$,

$$f \circ \xi \circ T(f^{-1}) = \zeta \circ T(f) \circ T(f^{-1}) = \zeta = f \circ f^{-1} \circ \zeta,$$

from which $\xi \circ T(f^{-1}) = f^{-1} \circ \zeta$ and $f^{-1}\colon (D, \zeta) \longrightarrow (C, \xi)$ is a morphism of $\mathbb{T}$-algebras, inverse to f.

Now if $C \in \mathscr{C}$, let us first observe that the axioms for being a monad imply immediately that $\big(T(C), \mu_C\big)$ is a $\mathbb{T}$-algebra. We shall prove that $\big(T(C), \mu_C\big)$, together with the morphism $\varepsilon_C\colon C \longrightarrow T(C)$, constitutes the reflection of C along U; see 3.1.1, volume 1. Given a $\mathbb{T}$-algebra (D, ζ) and a morphism $f\colon C \longrightarrow D$ in $\mathscr{C}$, we must prove the existence of a unique $g\colon \big(T(C), \mu_C\big) \longrightarrow (D, \zeta)$ in $\mathscr{C}^{\mathbb{T}}$ such that $g \circ \varepsilon_C = f$. For the uniqueness, it suffices to observe that, with the conditions imposed on g,

$$g = g \circ \mu_C \circ T(\varepsilon_C) = \zeta \circ T(g) \circ T(\varepsilon_C) = \zeta \circ T(f).$$

For the existence it suffices to check that $g = \zeta \circ T(f)$ satisfies the required conditions. It is indeed a morphism of $\mathbb{T}$-algebras, since

$$g \circ \mu_C = \zeta \circ T(f) \circ \mu_C = \zeta \circ \mu_D \circ TT(f) = \zeta \circ T(\zeta) \circ TT(f) = \zeta \circ T(g),$$

and it is such that $g \circ \varepsilon_C = f$, because

$$g \circ \varepsilon_C = \zeta \circ T(f) \circ \varepsilon_C = \zeta \circ \varepsilon_D \circ f = f.$$

Observe that given a morphism $h\colon C \longrightarrow C'$ in $\mathscr{C}$, putting $(D, \zeta) = \big(T(C'), \mu_{C'}\big)$ and $f = \varepsilon_{C'} \circ h$ in the previous construction yields the relation $g = \mu_{C'} \circ T(\varepsilon_{C'}) \circ T(h) = T(h)$; thus by 3.1.3, volume 1,

$$T(h): \big(T(C), \mu_C\big) \longrightarrow \big(T(C'), \mu_{C'}\big)$$

is the value of the left adjoint F of U on $h: C \longrightarrow C'$. By construction, $U \circ F = T$.

By 3.1.5, volume 1, the unit of the adjunction is just ε. If η is the counit of the adjunction, $\eta_{(C,\xi)}$ is obtained by putting $f = 1_C$ in the previous construction (see 3.1.5, volume 1), yielding $\eta_{(C,\xi)} = \xi$. □

Definition 4.1.5 *Let $\mathbb{T} = (T, \varepsilon, \mu)$ be a monad on a category $\mathscr{C}$. Consider the forgetful functor $U = \mathscr{C}^{\mathbb{T}} \longrightarrow \mathscr{C}$ and its left adjoint $F: \mathscr{C} \longrightarrow \mathscr{C}^{\mathbb{T}}$, mapping C to $\big(T(C), \mu_C\big)$. A $\mathbb{T}$-algebra is free when it is isomorphic to one of the form $F(C) = \big(T(C), \mu_C\big)$, for some object $C \in \mathscr{C}$.*

Proposition 4.1.6 *Let $\mathbb{T} = (T, \varepsilon, \mu)$ be a monad on a category $\mathscr{C}$. The full subcategory of $\mathscr{C}^{\mathbb{T}}$ generated by the free $\mathbb{T}$-algebras is equivalent to the following category $\mathscr{C}_{\mathbb{T}}$:*

- *the objects of $\mathscr{C}_{\mathbb{T}}$ are those of $\mathscr{C}$;*
- *a morphism $f: C \longrightarrow D$ in $\mathscr{C}_{\mathbb{T}}$ is a morphism $f: C \longrightarrow T(D)$ in $\mathscr{C}$;*
- *the composite of two morphisms $f: A \longrightarrow B$, $g: B \longrightarrow C$ in $\mathscr{C}_{\mathbb{T}}$ is given in $\mathscr{C}$ by the composite*

$$A \xrightarrow{\ f\ } T(B) \xrightarrow{\ T(g)\ } TT(C) \xrightarrow{\ \mu_C\ } T(C);$$

- *the identity on an object C of $\mathscr{C}_{\mathbb{T}}$ is just $\varepsilon_C: C \longrightarrow T(C)$ in $\mathscr{C}$.*

Proof Choosing another morphism $h: C \longrightarrow D$ in $\mathscr{C}_{\mathbb{T}}$, the associativity of the composition is attested by the following relations in $\mathscr{C}$:

$$\begin{aligned}\mu_D \circ T(h) \circ \big(\mu_C \circ T(g) \circ f\big) &= \mu_D \circ \mu_{T(D)} \circ TT(h) \circ T(g) \circ f \\ &= \mu_D \circ T\mu_D \circ TT(h) \circ T(g) \circ f \\ &= \mu_D \circ T\big(\mu_D \circ T(h) \circ g\big) \circ f;\end{aligned}$$

while the identity axioms, for a morphism $f: C \longrightarrow D$ in $\mathscr{C}_{\mathbb{T}}$, mean in $\mathscr{C}$,

$$\mu_D \circ T(f) \circ \varepsilon_C = \mu_D \circ \varepsilon_{T(D)} \circ f = f, \;\; \mu_D \circ T(\varepsilon_D) \circ f = f.$$

Writing $\mathscr{F}_{\mathbb{T}}$ for the full subcategory of $\mathscr{C}^{\mathbb{T}}$ generated by the free $\mathbb{T}$-algebras, we get a functor $\varphi: \mathscr{C}_{\mathbb{T}} \longrightarrow \mathscr{F}_{\mathbb{T}}$ just by putting

- $\varphi(C) = \big(T(C), \mu_C\big)$ for $C \in \mathscr{C}_{\mathbb{T}}$,
- $\varphi(f) = \mu_D \circ T(f)$ for $f: C \longrightarrow T(D)$ in $\mathscr{C}$, i.e. $f: C \longrightarrow D$ in $\mathscr{C}_{\mathbb{T}}$.

Observe that $\varphi(f)\colon (T(C),\mu_C) \longrightarrow (T(D),\mu_D)$ is indeed a morphism in $\mathscr{F}_{\mathbb{T}}$ since

$$\begin{aligned}\mu_D \circ T\varphi(f) &= \mu_D \circ T(\mu_D) \circ TT(f) = \mu_D \circ \mu_{T(D)} \circ TT(f) \\ &= \mu_D \circ T(f) \circ \mu_C = \varphi(f) \circ \mu_C.\end{aligned}$$

Now φ is a functor: indeed it preserves identities since

$$\varphi(\varepsilon_C) = \mu_C \circ T(\varepsilon_C) = 1_{T(C)}$$

and also composition since, with the notation of the statement,

$$\begin{aligned}\varphi(g) \circ \varphi(f) &= \mu_C \circ T(g) \circ \mu_B \circ T(f) = \mu_C \circ \mu_{T(C)} \circ TT(g) \circ T(f) \\ &= \mu_C \circ T(\mu_C) \circ TT(g) \circ T(f) = \mu_C \circ T(\mu_C \circ T(g) \circ f) = \varphi(g \circ f).\end{aligned}$$

By the choice of $\mathscr{F}_{\mathbb{T}}$, every object of $\mathscr{F}_{\mathbb{T}}$ is isomorphic to an object of the form φC; thus it remains to prove that φ is full and faithful (see 3.4.3, volume 1). This is essentially 4.1.4(3). If $f, g\colon C \rightrightarrows T(D)$ are such that $\mu_D \circ T(f) = \mu_D \circ T(g)$,

$$f = \mu_D \circ \varepsilon_{T(D)} \circ f = \mu_D \circ T(f) \circ \varepsilon_C = \mu_D \circ T(g) \circ \varepsilon_C = \mu_D \circ \varepsilon_{T(D)} \circ g = g,$$

so that φ is faithful. The functor φ is also full since, given

$$h\colon (T(C),\mu_C) \longrightarrow (T(D),\mu_D)$$

in $\mathscr{F}_{\mathbb{T}}$, the composite $h \circ \varepsilon_C\colon C \longrightarrow T(D)$ is such that

$$\mu_D \circ T(h \circ \varepsilon_C) = \mu_D \circ T(h) \circ T(\varepsilon_C) = h \circ \mu_C \circ T(\varepsilon_C) = h. \qquad \square$$

The category $\mathscr{C}_{\mathbb{T}}$ is also called the "Kleisli" category of the monad $\mathbb{T}$. Using only proposition 4.1.6, one has a full and faithful functor

$$\mathscr{C}_{\mathbb{T}} \longrightarrow \mathscr{C}^{\mathbb{T}}, \quad C \mapsto (T(C), \mu_C),$$

identifying, up to equivalence, $\mathscr{C}_{\mathbb{T}}$ with the full subcategory of $\mathscr{C}^{\mathbb{T}}$ generated by the free $\mathbb{T}$-algebras.

Corollary 4.1.7 *Let $\mathbb{T} = (T, \varepsilon, \mu)$ be a monad on a category $\mathscr{C}$. The functor*

$$\mathscr{C}_{\mathbb{T}} \longrightarrow \mathscr{C}, \quad C \mapsto T(C), \quad (f\colon C \to D) \mapsto \mu_D \circ T(f)$$

is faithful, reflects isomorphisms and has a left adjoint given by

$$\mathscr{C} \longrightarrow \mathscr{C}_{\mathbb{T}}, \quad C \mapsto C, \quad (f\colon C \to D) \mapsto \varepsilon_D \circ f.$$

Proof Via the equivalence of 4.1.6, this is just the situation of proposition 4.1.4 restricted to the full subcategory of free $\mathbb{T}$-algebras. $\square$

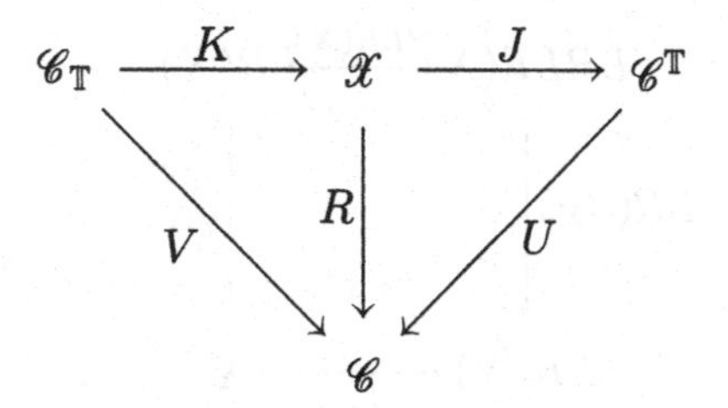

Diagram 4.6

4.2 Monads and adjunctions

Given a monad (T, ε, μ) on a category $\mathscr{C}$, we have produced two adjoint pairs: the Eilenberg–Moore adjunction and the Kleisli adjunction. Let us fix notation. We have already written

$$U: \mathscr{C}^{\mathbb{T}} \longrightarrow \mathscr{C}, \quad F: \mathscr{C} \longrightarrow \mathscr{C}^{\mathbb{T}}, \quad F \dashv U$$

for the first adjoint pair; one of the canonical natural transformations of the adjunction (see 3.1.5, volume 1) is just $\varepsilon: 1_{\mathscr{C}} \Rightarrow U \circ F = T$ (see 4.1.4); we shall write $\eta: F \circ U \Rightarrow 1_{\mathscr{C}^{\mathbb{T}}}$ for the other one. Let us also write

$$V: \mathscr{C}_{\mathbb{T}} \longrightarrow \mathscr{C}, \quad G: \mathscr{C} \longrightarrow \mathscr{C}_{\mathbb{T}}, \quad G \dashv V$$

for the Kleisli adjoint pair (see 4.1.7); the first canonical natural transformation of this adjunction is again $\varepsilon: 1_{\mathscr{C}} \Rightarrow V \circ G = T$, while we write $\gamma: G \circ V \Rightarrow 1_{\mathscr{C}_{\mathbb{T}}}$ for the other one.

The object of this section is to prove that every adjunction induces a monad and to compare the original adjunction with the Eilenberg–Moore and Kleisli adjunctions of the monad.

Proposition 4.2.1 *Let $R: \mathscr{X} \longrightarrow \mathscr{C}$ and $L: \mathscr{C} \longrightarrow \mathscr{X}$ constitute an adjoint pair, with L left adjoint to R; write $\alpha: 1_{\mathscr{C}} \Rightarrow R \circ L$, $\beta: L \circ R \Rightarrow 1_{\mathscr{X}}$ for the canonical natural transformations of this adjunction. Under those conditions, putting*

$$T = R \circ L, \quad \varepsilon = \alpha, \quad \mu = 1_R * \beta * 1_L$$

yields a monad (T, ε, μ) on $\mathscr{C}$. Moreover there exist functors $J: \mathscr{C}_{\mathbb{T}} \longrightarrow \mathscr{X}$ and $K: \mathscr{X} \longrightarrow \mathscr{C}^{\mathbb{T}}$ such that the isomorphisms $R \circ K \cong V$, $U \circ J \cong R$ hold (see diagram 4.6). Moreover $J \circ K$ is isomorphic to the canonical inclusion of $\mathscr{C}_{\mathbb{T}}$ in $\mathscr{C}^{\mathbb{T}}$ (see 4.1.6), J is full and K is full and faithful.

Proof The statement defines $T: \mathscr{C} \longrightarrow \mathscr{C}$, $\varepsilon: 1_{\mathscr{C}} \Rightarrow T$ and $\mu: T \circ T \Rightarrow T$. Let us check the three conditions for a monad (see 4.1.1), which we deduce from the triangular identities of the adjunction (see 3.1.5, volume 1)

$$\begin{array}{ccc} LRLR(X) & \xrightarrow{\beta_{LR(X)}} & LR(X) \\ {\scriptstyle LR(\beta_X)}\downarrow & & \downarrow{\scriptstyle \beta_X} \\ LR(X) & \xrightarrow[\beta_X]{} & X \end{array}$$

Diagram 4.7

and the naturality of β; this naturality implies the commutativity of the square in diagram 4.7 for every object $X \in \mathscr{X}$. Next

$$\begin{aligned}
\mu \circ (\varepsilon * 1_T) &= (1_R * \beta * 1_L) \circ (\alpha * 1_R * 1_L) \\
&= \big((1_R * \beta) \circ (\alpha * 1_R)\big) * 1_L = 1_R * 1_L \\
&= 1_T, \\
\mu \circ (1_T * \varepsilon) &= (1_R * \beta * 1_L) \circ (1_R * 1_L * \alpha) \\
&= 1_R * \big((\beta * 1_L) \circ (1_L * \alpha)\big) = 1_R * 1_L \\
&= 1_T, \\
\mu \circ (\mu * 1_T) &= (1_R * \beta * 1_L) \circ (1_R * \beta * 1_L * 1_R * 1_L) \\
&= 1_R * \big(\beta \circ (\beta * 1_L * 1_R)\big) * 1_L = 1_R * \big(\beta \circ (1_L * 1_R * \beta)\big) * 1_L \\
&= (1_R * \beta * 1_L) \circ (1_R * 1_L * 1_R * \beta * 1_L) \\
&= \mu \circ (1_T * \mu).
\end{aligned}$$

We are now able to define K and J.

$$K\colon \mathscr{C}_{\mathbb{T}} \longrightarrow \mathscr{X}, \quad K(C) = L(C), \quad K(f) = \beta_{LD} \circ L(f),$$

where $f\colon C \longrightarrow D$ in $\mathscr{C}_{\mathbb{T}}$ is thus an arrow $f\colon C \longrightarrow RL(D)$ in $\mathscr{C}$; $K(f)$ is thus the composite

$$L(C) \xrightarrow{\;Lf\;} LRL(D) \xrightarrow{\;\beta_{L(D)}\;} L(D).$$

Observe that K is a functor since, given $g\colon D \longrightarrow E$ in $\mathscr{C}_{\mathbb{T}}$ (see 4.1.6),

$$\begin{aligned}
K(g) \circ K(f) &= \beta_{L(E)} \circ L(g) \circ \beta_{LD} \circ L(f) \\
&= \beta_{L(E)} \circ \beta_{LRL(E)} \circ LRL(g) \circ L(f) \\
&= \beta_{L(E)} \circ LR(\beta_{L(E)}) \circ LRL(g) \circ L(f) \\
&= \beta_{L(E)} \circ L(\mu_E \circ T(g) \circ f) = \beta_{L(E)} \circ L(g \circ f) \\
&= K(g \circ f),
\end{aligned}$$

$$K(1_D) = \beta_{L(D)} \circ L(1_D) = \beta_{L(D)} \circ L(\varepsilon_D) = \beta_{L(D)} \circ L(\alpha_D)$$
$$= 1_{L(D)}.$$

Moreover, given an arrow $f: C \longrightarrow D$ in $\mathscr{C}_{\mathbb{T}}$,

$$RK(C) = RL(C) = T(C) = V(C),$$
$$RK(f) = R(\beta_{L(D)} \circ L(f)) = R(\beta_{L(D)}) \circ RL(f) = \mu_D \circ T(f) = V(f),$$

which proves the relation $RK \cong V$.

Next let us construct J;

$$J: \mathscr{X} \longrightarrow \mathscr{C}^{\mathbb{T}}, \quad X \mapsto \big(R(X), R(\beta_X)\big), \quad x \mapsto R(x),$$

where $x: X \longrightarrow Y$ is a morphism of $\mathscr{X}$. Observe first that

$$R\beta_X: TR(X) =\!=\!= RLR(X) \longrightarrow R(X)$$

provides $R(X)$ with the structure of a $\mathbb{T}$-algebra.

$$R(\beta_X) \circ \varepsilon_{R(X)} = R(\beta_X) \circ \alpha_{R(X)}$$
$$= 1_{R(X)},$$
$$R(\beta_X) \circ \mu_{R(X)} = R(\beta_X) \circ R(\beta_{LR(X)}) = R(\beta_X \circ \beta_{LR(X)})$$
$$= R(\beta_X \circ LR(\beta_X))$$
$$= R(\beta_X) \circ T(R(\beta_X)).$$

On the other hand $R(x)$: $\big(R(X), R(\beta_X)\big) \longrightarrow \big(R(Y), R(\beta_Y)\big)$ is a morphism of $\mathbb{T}$-algebras since

$$R(\beta_Y) \circ TR(x) = R(\beta_Y) \circ RLR(x)$$
$$= R(\beta_Y \circ LR(x)) = R(x \circ \beta_X)$$
$$= R(x) \circ R(\beta_X).$$

J is obviously a functor, since R is, and by definition $U \circ J \cong R$ and $J \circ K \cong \varphi$, where $\varphi: \mathscr{C}_{\mathbb{T}} \longrightarrow \mathscr{C}^{\mathbb{T}}$ is the functor defined in 4.1.6.

Since $J \circ K$ is full and faithful, K is faithful and J is full. But given a morphism $h: L(C) \longrightarrow L(D)$ in $\mathscr{X}$, h corresponds by adjunction with a morphism $f: C \longrightarrow RL(D) = T(D)$ in $\mathscr{C}$, i.e. a morphism $f: C \longrightarrow D$ in $\mathscr{C}_{\mathbb{T}}$ such that $K(f) = h$. So K is full as well. □

To complete the statement of proposition 4.2.1, let us consider the cases where the original adjunction is already the Eilenberg–Moore or the Kleisli adjunction of a monad.

Proposition 4.2.2 *Let $\mathbb{T} = (T, \varepsilon, \mu)$ be a monad on a category $\mathscr{C}$. Via the construction of 4.2.1, both the Eilenberg–Moore and the Kleisli adjunctions of $\mathbb{T}$ generate the monad $\mathbb{T}$.*

Proof We keep the notation described at the beginning of this section. By construction, we have $UF = T$ with $\varepsilon\colon 1_{\mathscr{C}} \Rightarrow UF$ as the unit of the adjunction. It remains to prove that $\mu = U * \eta * F$, where η is the counit of the adjunction. But given a $\mathbb{T}$-algebra (X, ξ), $\eta_{(X,\xi)} = \xi$ by 4.1.4, thus for $C \in \mathscr{C}$, $U(\eta_{F(C)}) = U(\eta_{(T(C),\mu_C)}) = \mu_C$. The case of the Kleisli adjunction follows immediately from the Eilenberg–Moore case and proposition 4.1.6. □

Combining propositions 4.1.1 and 4.2.2 we find that every adjunction generates a monad and, among all the adjunctions which generate the same monad, the Kleisli adjunction and the Eilenberg–Moore adjunction are somehow "extremal", in the precise sense explained in 4.2.1.

Now let us emphasize a special case of interest.

Proposition 4.2.3 *Let $\mathbb{T} = (T, \varepsilon, \mu)$ be a monad on a category $\mathscr{C}$. With the notation of the beginning of this section, the following conditions are equivalent:*

(1) the forgetful functor $U\colon \mathscr{C}^{\mathbb{T}} \longrightarrow \mathscr{C}$ is full and faithful;

(2) the counit $\eta\colon F \circ U \Rightarrow 1_{\mathscr{C}^{\mathbb{T}}}$ of the adjunction $F \dashv U$ is an isomorphism;

(3) the multiplication $\mu\colon T \circ T \Rightarrow T$ of the monad is an isomorphism;

(4) for every $\mathbb{T}$-algebra (X, ξ), $\xi\colon T(X) \longrightarrow X$ is an isomorphism in $\mathscr{C}$.

Such a monad is called an idempotent monad.

Proof (1) and (2) are equivalent by 3.4.1, volume 1. (2) implies (3) since we have observed in proving 4.2.2 that $\mu = U * \eta * F$. And (4) implies (3) since $(T(X), \mu_X)$ is a $\mathbb{T}$-algebra for every $X \in \mathscr{C}$; see 4.1.5.

Going back to 3.1.5, volume 1, we know that given a $\mathbb{T}$-algebra (X, ξ), the morphism $\eta_{(X,\xi)}\colon FU(X,\xi) \longrightarrow (X, \xi)$ is the unique one with the property $U(\eta_{(X,\xi)}) \circ \varepsilon_{U(X,\xi)} = 1_{U(X,\xi)}$; so $\eta_{(X,\xi)}\colon T(X) \longrightarrow X$ is the unique morphism yielding $\xi \circ T(\eta_{(X,\xi)}) = \eta_{(X,\xi)} \circ \mu_X$ and $\eta_{(X,\xi)} \circ \varepsilon_X = 1_X$. But ξ is such a morphism, thus $\eta_{(X,\xi)} = \xi$. This proves the equivalence of (2) and (4).

It remains to prove, say, that (3) implies (4). From $\mu_X \circ \varepsilon_{T(X)} = 1_{T(X)} = \mu_X \circ T(\varepsilon_X)$ we deduce that $\varepsilon_{T(X)} = T(\varepsilon_X)$ is an isomorphism. From $\xi \circ \varepsilon_X = 1_X$ we deduce $T(\xi) \circ T(\varepsilon_X) = 1_{T(X)}$, thus $T(\xi) = \mu_X$ is an isomorphism as well. From $T(\xi) \circ \varepsilon_{T(X)} = \varepsilon_X \circ \xi$ we get that $\varepsilon_X \circ \xi$ is an isomorphism, thus in particular ξ is a monomorphism. But since

$\xi \circ \varepsilon_X = 1_X$, this implies that ξ is an isomorphism with inverse ε_X; see 1.7.5, volume 1. □

Corollary 4.2.4 *Let $\mathscr{C}$ be a category. There is a coincidence, up to equivalences of categories, between*

(1) the reflective subcategories of $\mathscr{C}$ (see 3.5.2, volume 1),

(2) the categories of $\mathbb{T}$-algebras for the idempotent monads $\mathbb{T}$ on $\mathscr{C}$.

Proof Consider a reflective subcategory $i\colon \mathscr{D} \hookrightarrow \mathscr{C}$ with corresponding reflection $r\colon \mathscr{C} \longrightarrow \mathscr{D}$; write, as in 4.2.1, $\alpha\colon 1_{\mathscr{C}} \Rightarrow i \circ r$ and $\beta\colon r \circ i \Rightarrow 1_{\mathscr{D}}$ for the canonical natural transformations of the adjunction. By 3.4.1, volume 1, β is an isomorphism. If $\mathbb{T} = (T, \varepsilon, \mu)$ is the corresponding monad as in 4.2.1, then μ is an isomorphism as well since $\mu = 1_i \circ \beta \circ 1_r$. Thus the monad $\mathbb{T}$ is idempotent. Moreover the comparison functor $J\colon \mathscr{D} \longrightarrow \mathscr{C}^{\mathbb{T}}$ described in 4.2.1 is now an equivalence of categories. Indeed for every $\mathbb{T}$-algebra (C, ξ), the isomorphism $\xi\colon ir(C) \cong T(C) \longrightarrow C$ indicates that C is in the replete subcategory $\mathscr{D}$; on the other hand giving a morphism $x\colon X \longrightarrow Y$ in $\mathscr{D}$ is equivalent to giving a morphism $x\colon (X, \beta_X) \longrightarrow (Y, \beta_Y)$ in $\mathscr{C}^{\mathbb{T}}$, just by naturality of β. Thus every reflective subcategory of $\mathscr{C}$ is equivalent to the Eilenberg–Moore category of the idempotent monad it generates.

On the other hand, applying 4.2.3 and 3.4.1, volume 1, we know already that an idempotent monad $\mathbb{T}$ on $\mathscr{C}$ admits as category of $\mathbb{T}$-algebras, up to equivalence, a reflective subcategory of $\mathscr{C}$. □

4.3 Limits and colimits in categories of algebras

Proposition 4.3.1 *Let $\mathbb{T} = (T, \varepsilon, \mu)$ be a monad on a category $\mathscr{C}$. Let $G\colon \mathscr{D} \longrightarrow \mathscr{C}^{\mathbb{T}}$ be a functor such that $UG\colon \mathscr{D} \longrightarrow \mathscr{C}$ has a limit; then G has a limit which is preserved by $U\colon \mathscr{C}^{\mathbb{T}} \longrightarrow \mathscr{C}$. In particular if $\mathscr{C}$ admits some type of limits, $\mathscr{C}^{\mathbb{T}}$ admits the same type of limits and they are preserved by U.*

Proof Let $G\colon \mathscr{D} \longrightarrow \mathscr{C}^{\mathbb{T}}$ be a functor such that $U \circ G\colon \mathscr{D} \longrightarrow \mathscr{C}$ has a limit; we must prove that $G\colon \mathscr{D} \longrightarrow \mathscr{C}^{\mathbb{T}}$ has a limit, which will of course be preserved by U since U has a left adjoint (see 4.1.4 and 3.2.2, volume 1).

Write $\big(p_D\colon L \longrightarrow UG(D)\big)_{D \in \mathscr{D}}$ for the limit of $UG\colon \mathscr{D} \longrightarrow \mathscr{C}$, where the forgetful functor $U\colon \mathscr{C}^{\mathbb{T}} \longrightarrow \mathscr{C}$ has been defined in 4.1.4. For each

$D \in \mathscr{D}$, let us write $G(D) = (UG(D), \xi_D)$. The family of morphisms

$$T(L) \xrightarrow{T(p_D)} TUG(D) \xrightarrow{\quad \xi_D \quad} UG(D)$$

is a cone on UG since, given $d\colon D \longrightarrow D'$ in $\mathscr{D}$,

$$UG(d) \circ \xi_D \circ T(p_D) = \xi_{D'} \circ TUG(d) \circ T(p_D) = \xi_{D'} \circ T(p_{D'}).$$

Therefore we get a unique factorization $\xi\colon T(L) \longrightarrow L$ such that $p_D \circ \xi = \xi_D \circ T(p_D)$ for each $D \in \mathscr{D}$.

It is now routine to check that $(p_D\colon (L,\xi) \longrightarrow (UG(D), \xi_D))$ is the limit of G. First,

$$p_D \circ \xi \circ \varepsilon_L = \xi_D \circ T(p_D) \circ \varepsilon_L = \xi_D \circ \varepsilon_{UG(D)} \circ p_D = p_D,$$

from which $\xi \circ \varepsilon_L = 1_L$, by definition of a limit. In the same way

$$\begin{aligned} p_D \circ \xi \circ \mu_L &= \xi_D \circ T(p_D) \circ \mu_L = \xi_D \circ \mu_{UG(D)} \circ TT(p_D) \\ &= \xi_D \circ T(\xi_D) \circ TT(p_D) = \xi_D \circ T(p_D) \circ T(\xi) \\ &= p_D \circ \xi \circ T(\xi), \end{aligned}$$

from which we obtain $\xi \circ \mu_L = \xi \circ T(\xi)$. Thus (L,ξ) is already a $\mathbb{T}$-algebra and the relation $\xi_D \circ T(p_D) = p_D \circ \xi$ means precisely that the arrow $p_D\colon (L,\xi) \longrightarrow (UG(D), \xi_D)$ is a morphism of $\mathbb{T}$-algebras. Since $U\colon \mathscr{C}^{\mathbb{T}} \longrightarrow \mathscr{C}$ is faithful (see 4.1.4), these morphisms constitute a cone on the functor G.

Now given another cone $(q_D\colon (M,\zeta) \longrightarrow (UG(D), \xi_D))_{D \in \mathscr{D}}$ on G, we get in $\mathscr{C}$ a unique factorization $m\colon M \longrightarrow L$ such that for each $D \in \mathscr{D}$, $p_D \circ m = q_D$. It remains to prove that $m\colon (M,\zeta) \longrightarrow (L,\xi)$ is a morphism in $\mathscr{C}^{\mathbb{T}}$. This follows immediately from the relations

$$p_D \circ m \circ \zeta = q_D \circ \zeta = \xi_D \circ T(q_D) \circ \xi_D \circ T(p_D) \circ T(m) = p_D \circ \xi \circ T(m). \ \square$$

Proposition 4.3.2 *Let $\mathbb{T} = (T, \varepsilon, \mu)$ be a monad on a category $\mathscr{C}$. Let $G\colon \mathscr{D} \longrightarrow \mathscr{C}^{\mathbb{T}}$ be a functor such that $UG\colon \mathscr{D} \longrightarrow \mathscr{C}$ has a colimit preserved by T and by $T \circ T$; then G has a colimit which is preserved by $U\colon \mathscr{C}^{\mathbb{T}} \longrightarrow \mathscr{C}$. In particular if $\mathscr{C}$ has some type of colimits preserved by T, then $\mathscr{C}^{\mathbb{T}}$ has the same type of colimits and these are preserved by $U\colon \mathscr{C}^{\mathbb{T}} \longrightarrow \mathscr{C}$.*

Proof Let $G\colon \mathscr{D} \longrightarrow \mathscr{C}^{\mathbb{T}}$ be a functor such that $U \circ G\colon \mathscr{D} \longrightarrow \mathscr{C}$ has a colimit preserved by T and $T \circ T$; we must prove that $G\colon \mathscr{D} \longrightarrow \mathscr{C}^{\mathbb{T}}$ has a colimit preserved by U.

Write $(s_D\colon UG(D) \longrightarrow L)_{D \in \mathscr{D}}$ for the colimit of $UG\colon \mathscr{D} \longrightarrow \mathscr{C}$, where $U\colon \mathscr{C}^{\mathbb{T}} \longrightarrow \mathscr{C}$ is the forgetful functor of 4.1.4. To fix the notation, or each $D \in \mathscr{D}$, let us write $G(D) = \big(UG(D), \xi_D\big)$. Observe that the morphisms $\xi_D\colon TUG(D) \longrightarrow UG(D)$ are the components of a natural transformation $TUG \Rightarrow UG$, just because for every $d\colon D \longrightarrow D'$ in $\mathscr{D}$, $G(d)$ is a morphism in $\mathscr{C}^{\mathbb{T}}$. Therefore we get a unique factorization $\xi\colon \operatorname{colim} TUG \longrightarrow \operatorname{colim} UG$, i.e. a unique morphism $\xi\colon T(L) \longrightarrow L$ such that $\xi \circ T(s_D) = s_D \circ \xi_D$ for each $D \in \mathscr{D}$.

It is now routine to check that

$$\Big(s_D\colon \big(UG(D), \xi_D\big) \longrightarrow (L, \xi)\Big)_{D \in \mathscr{D}}$$

is the colimit of G. First of all,

$$\xi \circ \varepsilon_L \circ s_D = \xi \circ T(s_D) \circ \varepsilon_{UG(D)} = s_D \circ \xi_D \circ \varepsilon_{UG(D)} = s_D,$$

from which $\xi \circ \varepsilon_L = 1_L$, by definition of a colimit. Next, since the colimit of $TTUG$ is just $\big(TT(L), TT(s_D)\big)_{D \in \mathscr{D}}$,

$$\begin{aligned}\xi \circ \mu_L \circ TT(s_D) &= \xi \circ T(s_D) \circ \mu_{UG(D)} = s_D \circ \xi_D \circ \mu_{UG(D)}\\ &= s_D \circ \xi_D \circ T(\xi)_D = \xi \circ T(s_D) \circ T(\xi_D)\\ &= \xi \circ T(\xi) \circ TT(s_D)\end{aligned}$$

implies $\xi \circ \mu_L = \xi \circ T(\xi)$, from which (L, ξ) is already a $\mathbb{T}$-algebra. The relation $\xi \circ T(s_D) = s_D \circ \xi_D$ implies that $s_D\colon \big(UG(D), \xi_D\big) \longrightarrow (L, \xi)$ is a morphism of $\mathbb{T}$-algebras.

Now given another cocone $\big(r_D\colon \big(UG(D), \xi_D\big) \longrightarrow (M, \zeta)\big)_{D \in \mathscr{D}}$ on G, we get in $\mathscr{C}$ a unique factorization $m\colon L \longrightarrow M$ such that for each $D \in \mathscr{D}$, $m \circ s_D = r_D$. It remains to prove that $m\colon (L, \xi) \longrightarrow (M, \zeta)$ is a morphism in $\mathscr{C}^{\mathbb{T}}$. Since $\big(T(L), T(s_D)\big)_{D \in \mathscr{D}}$ is the colimit of TG, this follows immediately from the relations

$$\zeta \circ T(m) \circ T(s_D) = \zeta \circ T(r_D) = r_D \circ \xi_D = m \circ s_D \circ \xi_D = m \circ \xi \circ T(s_D). \ \square$$

Lemma 4.3.3 *Let $\mathbb{T} = (T, \varepsilon, \mu)$ be a monad on a category $\mathscr{C}$. For every $\mathbb{T}$-algebra (C, ξ), the following diagram is a coequalizer in $\mathscr{C}^{\mathbb{T}}$:*

$$\big(TT(C), \mu_{T(C)}\big) \underset{T(\xi)}{\overset{\mu_C}{\rightrightarrows}} \big(T(C), \mu_C\big) \xrightarrow{\ \ \xi\ \ } (C, \xi).$$

This coequalizer is transformed by the forgetful functor $U\colon \mathscr{C}^{\mathbb{T}} \longrightarrow \mathscr{C}$ in the absolute coequalizer of diagram 4.8.

Proof Now $\big(TT(C), \mu_{T(C)}\big)$ and $\big(T(C), \mu_C\big)$ are $\mathbb{T}$-algebras (see 4.1.4 and 4.1.5); μ_C is a morphism in $\mathscr{C}^{\mathbb{T}}$ by definition of a monad (see 4.1.1); $T(\xi)$ is a morphism in $\mathscr{C}^{\mathbb{T}}$ by naturality of μ and ξ is a morphism in $\mathscr{C}^{\mathbb{T}}$

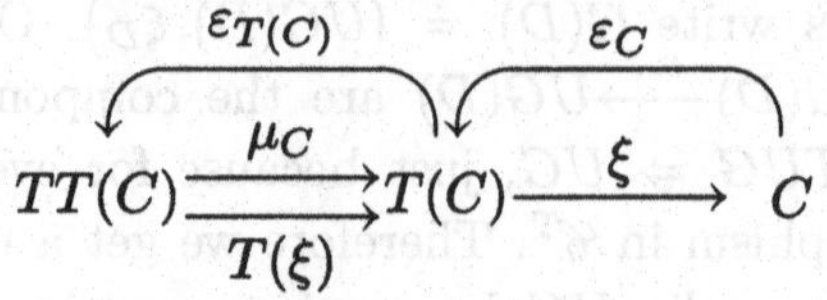

$$\xymatrix{TT(C) \ar@<0.5ex>[r]^{\mu_C} \ar@<-0.5ex>[r]_{T(\xi)} & T(C) \ar[r]^{\xi} \ar@/_1pc/[l]_{\varepsilon_{T(C)}} & C \ar@/_1pc/[l]_{\varepsilon_C}}$$

Diagram 4.8

by definition of a $\mathbb{T}$-algebra (see 4.1.2). Moreover in $\mathscr{C}$, the consideration of the morphisms

$$TT(C) \xleftarrow{\varepsilon_{T(C)}} T(C) \xleftarrow{\varepsilon_C} C$$

yields $\xi \circ \mu_C = \xi \circ T(\xi)$ by definition of a $\mathbb{T}$-algebra; $\mu_C \circ \varepsilon_{T(C)} = 1_{T(C)}$ by definition of a monad (see 4.1.1); $\xi \circ \varepsilon_C = 1_C$ by definition of a $\mathbb{T}$-algebra (see 4.1.2) and $T(\xi) \circ \varepsilon_{T(C)} = \varepsilon_C \circ \xi$ by naturality of ε. Thus in $\mathscr{C}$, $U(\xi)$ is the absolute coequalizer of $(\mu_C, T(\xi))$; see 2.10.2, volume 1, and diagram 4.8.

It remains to prove that we also have a coequalizer in $\mathscr{C}^{\mathbb{T}}$. We already have $\xi \circ \mu_C = \xi \circ T(\xi)$ and if $f\colon (T(C), \mu_C) \longrightarrow (D, \zeta)$ is a morphism in $\mathscr{C}^{\mathbb{T}}$ such that $f \circ \mu_C = f \circ T(\xi)$, the composite

$$C \xrightarrow{\varepsilon_C} T(C) \xrightarrow{f} D$$

is the unique factorization of f through ξ in $\mathscr{C}$; see 2.10.2, volume 1. It suffices to prove that $f \circ \varepsilon_C\colon (C, \xi) \longrightarrow (D, \zeta)$ is a morphism of $\mathbb{T}$-algebras. Indeed

$$\begin{aligned} \zeta \circ T(f) \circ T(\varepsilon_C) &= f \circ \mu_C \circ T(\varepsilon_C) = f \\ &= f \circ \mu_C \circ \varepsilon_{T(C)} = f \circ T(\xi) \circ \varepsilon_{T(C)} \\ &= f \circ \varepsilon_C \circ \xi. \end{aligned}$$

□

It can be useful to observe that in the previous lemma, the morphism $T(\varepsilon_C)\colon T(C) \longrightarrow TT(C)$ is in fact a common section for μ_C and $T(\xi)$,

$$\mu_C \circ T(\varepsilon_C) = 1_{T(C)}, \quad T(\xi) \circ T(\varepsilon_C) = 1_{T(C)},$$

by definition of a monad (see 4.1.1) and a $\mathbb{T}$-algebra (see 4.1.2).

Proposition 4.3.4 *Let $\mathbb{T} = (T, \varepsilon, \mu)$ be a monad on a cocomplete category $\mathscr{C}$. The following conditions are equivalent:*

(1) the category $\mathscr{C}^{\mathbb{T}}$ of $\mathbb{T}$-algebras has coequalizers;

(2) the category $\mathscr{C}^{\mathbb{T}}$ of $\mathbb{T}$-algebras is cocomplete.

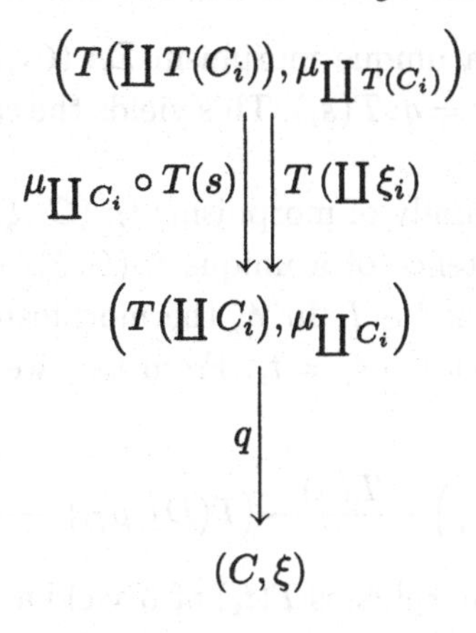

Diagram 4.9

Proof By 2.8.1, volume 1, it suffices to prove that if $\mathscr{C}^{\mathbb{T}}$ has coequalizers, then $\mathscr{C}^{\mathbb{T}}$ has also coproducts. The proof is inspired by the statement of lemma 4.3.3.

Consider a family $(C_i, \xi_i)_{i\in I}$ of objects of $\mathscr{C}^{\mathbb{T}}$ and compute its coproduct $\left(s_i\colon C_i \longrightarrow \coprod_{i\in I} C_i\right)_{i\in I}$ in $\mathscr{C}$. Write $\left(\sigma_i\colon T(C_i) \longrightarrow \coprod_{i\in I} T(C_i)\right)_{i\in I}$ for the corresponding coproduct in $\mathscr{C}$ of the images along T. The morphisms $T(s_i)$ yield a unique $\mathscr{C}$-morphism $s\colon \coprod_{i\in I} T(C_i) \longrightarrow T\left(\coprod_{i\in I} C_i\right)$ such that $s \circ \sigma_i = T(s_i)$ for each $i \in I$.

Let us now define the coproduct $(C, \xi) = \coprod_{i\in I}(C_i, \xi_i)$ in $\mathscr{C}^{\mathbb{T}}$ via the coequalizer $\mathscr{C}^{\mathbb{T}}$ given by diagram 4.9 where, for clarity, we have omitted writing "$i \in I$" all the time. The two first objects are just free $\mathbb{T}$-algebras (see 4.1.4 and 4.1.5) and to give perfect sense to the definition of (C, ξ), it remains to prove that the two left arrows are morphisms in $\mathscr{C}^{\mathbb{T}}$. The arrow $T\left(\coprod \xi_i\right)$ is a morphism in $\mathscr{C}^{\mathbb{T}}$ just by naturality of μ while $\mu_{\amalg C_i} \circ T(s)$ is a morphism in $\mathscr{C}^{\mathbb{T}}$ because

$$
\begin{aligned}
\mu_{\amalg C_i} \circ T\left(\mu_{\amalg C_i}\right) \circ TT(s) &= \mu_{\amalg C_i} \circ \mu_{T(\amalg C_i)} \circ TT(s) \\
&= \mu_{\amalg C_i} \circ T(s) \circ \mu_{\amalg T(C_i)}.
\end{aligned}
$$

Thus (C, ξ) has been correctly defined.

For each index $i \in I$ $T(s_i)\colon \left(T(C_i), \mu_{C_i}\right) \longrightarrow \left(T\left(\coprod C_i\right), \mu_{\amalg C_i}\right)$ is a morphism of $\mathbb{T}$-algebras, by naturality of μ. Because of lemma 4.3.3, the relations

$$
\begin{aligned}
q \circ T(s_i) \circ \mu_{C_i} &= q \circ \mu_{\amalg C_i} \circ TT(s_i) = q \circ \mu_{\amalg C_i} \circ T(s) \circ T(\sigma_i) \\
&= q \circ T\left(\coprod \xi_i\right) \circ T(\sigma_i) = q \circ T(s_i) \circ T(\xi_i)
\end{aligned}
$$

imply the existence of a unique morphism $\Sigma_i: (C_i, \xi_i) \longrightarrow (C, \xi)$ of $\mathbb{T}$-algebras such that $\Sigma_i \circ \xi_i = q \circ T(s_i)$. This yields the canonical morphisms of the coproduct.

Now let us choose a family of morphisms $t_i: (C_i, \xi_i) \longrightarrow (D, \zeta)$ in $\mathscr{C}^{\mathbb{T}}$; we must prove the existence of a unique $t: (C, \xi) \longrightarrow (D, \zeta)$ such that $t \circ \Sigma_i = t_i$ for each index $i \in I$. In $\mathscr{C}$, the morphisms t_i yield a unique $u: \coprod C_i \longrightarrow D$ such that $u \circ s_i = t_i$. From this we get, by 4.1.4, morphisms of $\mathbb{T}$-algebras

$$\left(T\left(\coprod C_i\right), \mu_{\amalg C_i}\right) \xrightarrow{T(u)} (T(D), \mu_D) \xrightarrow{\ \zeta\ } (D, \zeta).$$

On the other hand the morphisms $T(t_i)$ of $\mathscr{C}$ yield a unique factorization $v: \coprod T(C_i) \longrightarrow T(D)$ such that $v \circ \sigma_i = T(t_i)$. And from the relations

$$u \circ \left(\coprod \xi_i\right) \circ \sigma_i = u \circ s_i \circ \xi_i = t_i \circ \xi_i = \zeta \circ T(t_i) = \zeta \circ v \circ \sigma_i$$

one deduces that $u \circ (\coprod \xi_i) = \zeta \circ v$. In an analogous way the relations

$$\zeta \circ T(u) \circ s \circ \sigma_i = \zeta \circ T(u) \circ T(s_i) = \zeta \circ T(t_i) = \zeta \circ v \circ \sigma_i$$

imply $\zeta \circ T(u) \circ s = \zeta \circ v$. We are now ready to prove that $\zeta \circ T(u)$ coequalizes $\mu_{\amalg C_i} \circ T(s)$ and $T(\coprod \xi_i)$:

$$\begin{aligned}\zeta \circ T(u) \circ \mu_{\amalg C_i} \circ T(s) &= \zeta \circ \mu_D \circ TT(u) \circ T(s)\\ &= \zeta \circ T(\zeta) \circ TT(u) \circ T(s) = \zeta \circ T(\zeta) \circ T(v)\\ &= \zeta \circ T(u) \circ T\left(\coprod \xi_i\right).\end{aligned}$$

By definition of (C, ξ), we get $t: (C, \xi) \longrightarrow (D, \zeta)$, a unique morphism of $\mathbb{T}$-algebras such that $t \circ q = \zeta \circ T(u)$. This implies

$$t \circ \Sigma_i \circ \xi_i = t \circ q \circ T(s_i) = \zeta \circ T(u) \circ T(s_i) = \zeta \circ T(t_i) = t_i \circ \xi_i,$$

from which $t \circ \Sigma_i = t_i$ since ξ_i is an epimorphism ($\xi_i \circ \varepsilon_{C_i} = 1_{C_i}$).

It remains to prove the uniqueness of t. Let $r: (C, \xi) \longrightarrow (D, \zeta)$ be such that $r \circ \Sigma_i = t_i$ for every index $i \in I$. Observe first that

$$r \circ q \circ s \circ \sigma_i = r \circ q \circ T(s_i) = r \circ \Sigma_i \circ \xi_i = t_i \circ \xi_i = u \circ s_i \circ \xi_i = u \circ \left(\coprod \xi_i\right) \circ \sigma_i.$$

This implies $r \circ q \circ s = u \circ (\coprod \xi_i)$ and thus

$$\begin{aligned}r \circ q &= r \circ q \circ T\left(\coprod \xi_i\right) \circ T\left(\coprod \varepsilon_{C_i}\right)\\ &= r \circ q \circ \mu_{\amalg C_i} \circ T(s) \circ T\left(\coprod \varepsilon_{C_i}\right)\\ &= r \circ \xi \circ T(q) \circ T(s) \circ T\left(\coprod \varepsilon_{C_i}\right)\end{aligned}$$

$$= \zeta \circ T(r) \circ T(q) \circ T(s) \circ T\left(\coprod \varepsilon_{C_i}\right)$$
$$= \zeta \circ T(u) \circ T\left(\coprod \xi_i\right) \circ T\left(\coprod \varepsilon_{C_i}\right)$$
$$= \zeta \circ T(u),$$

from which $r = t$. □

Theorem 4.3.5 *Let $\mathscr{C}$ be a complete and cocomplete regular category in which every regular epimorphism has a section. Under these conditions, for every monad $\mathbb{T} = (T, \varepsilon, \mu)$ on $\mathscr{C}$:*

(1) the category $\mathscr{C}^{\mathbb{T}}$ of $\mathbb{T}$-algebras is complete, cocomplete and regular;
(2) the forgetful functor $U: \mathscr{C}^{\mathbb{T}} \longrightarrow \mathscr{C}$ is exact;
(3) the forgetful functor $U: \mathscr{C}^{\mathbb{T}} \longrightarrow \mathscr{C}$ preserves and reflects regular epimorphisms;
(4) a pair of morphisms $(u, v): (C, \xi) \rightrightarrows (D, \zeta)$ is a kernel pair in $\mathscr{C}^{\mathbb{T}}$ if and only if (u, v) is a kernel pair in $\mathscr{C}$;
(5) $\mathscr{C}^{\mathbb{T}}$ is exact as long as $\mathscr{C}$ is exact.

In particular, assuming the axiom of choice, all those properties are satisfied when $\mathscr{C}$ is the category Set *of sets.*

Proof Applying 6.5.4, volume 1, a regular epimorphism $p: A \longrightarrow B$ with section $s: B \longrightarrow A$ is the coequalizer of the pair $(s \circ p, 1_A)$. Since $\mathscr{C}(A, A)$ is a set, there is just – up to isomorphism – a set of regular quotients of A, corresponding to the idempotents $e \in \mathscr{C}(A, A)$; see 6.5.4, volume 1. On the other hand, every functor F with domain $\mathscr{C}$ preserves regular epimorphisms, since $F(p)$ has a section $F(s)$ and 6.5.4, volume 1, again leads us to conclude that $F(p)$ is a coequalizer.

Now $\mathscr{C}^{\mathbb{T}}$ is complete by 4.3.1; to show it is cocomplete, we need to prove the existence of coequalizers (see 4.3.4). Given $f, g: (C, \xi) \rightrightarrows (D, \zeta)$ in $\mathscr{C}^{\mathbb{T}}$, we consider all the morphisms $h: (D, \zeta) \longrightarrow (X, \chi)$ in $\mathscr{C}^{\mathbb{T}}$ with the following properties:

(1) $h \circ f = h \circ g$;
(2) $h: D \longrightarrow X$ is a regular epimorphism in $\mathscr{C}$.

We know there is just such a set of morphisms h so that we can compute in $\mathscr{C}^{\mathbb{T}}$ the product of all the $\mathbb{T}$-algebras (X, χ), for all the possible h. The family of morphisms h induces a factorization k through the product and we consider its image factorization $k = i \circ p$ in $\mathscr{C}$ as in diagram 4.10. We shall prove that I is in fact a sub-$\mathbb{T}$-algebra of the product $\prod_h (X, \chi)$ and p is the coequalizer of f, g.

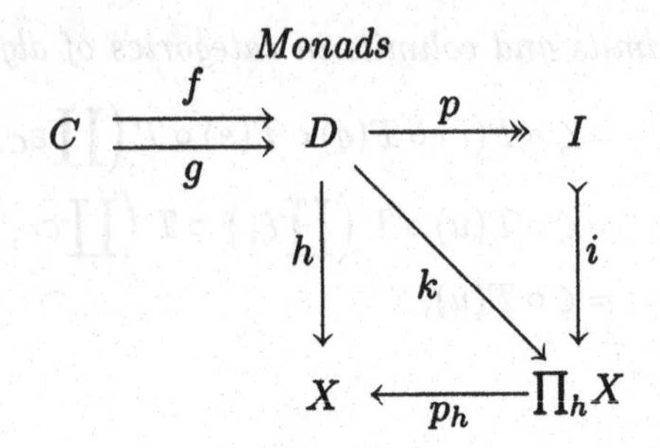

Diagram 4.10

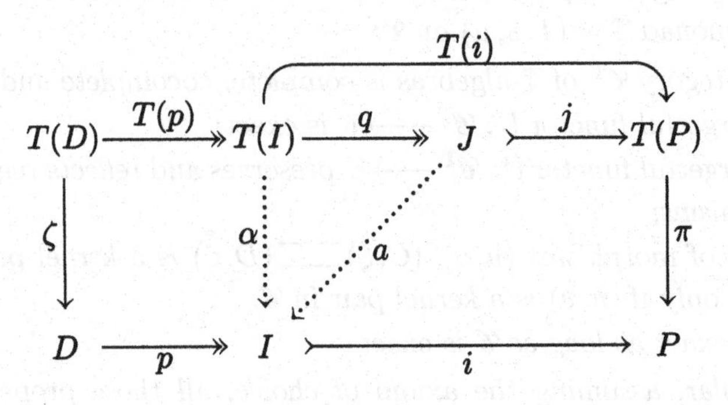

Diagram 4.11

Let us write $(P,\pi) \cong \prod_h (X,\chi)$ for the product in $\mathscr{C}^{\mathbb{T}}$. The factorization k: $(D,\zeta) \longrightarrow (P,\pi)$ is a morphism of $\mathbb{T}$-algebras, thus the outer part of diagram 4.11 in $\mathscr{C}$ is commutative, where we know that $T(p)$ is a regular epimorphism. The arrow $T(i)$ has no reason to be a monomorphism, but nevertheless by 4.4.5, volume 1, we get a factorization a: $J \longrightarrow I$ making the diagram commutative. Putting $\alpha = a \circ q$, we shall prove that (I,α) is a $\mathbb{T}$-algebra; by definition of α, this will make p: $(D,\zeta) \longrightarrow (I,\alpha)$ and i: $(I,\alpha) \longrightarrow (P,\pi)$ morphisms of $\mathbb{T}$-algebras.

Let us recall that $p, T(p)$ and $TT(p)$ are (regular) epimorphisms. By naturality of ε, μ,

$$\begin{aligned} \alpha \circ \varepsilon_I \circ p &= \alpha \circ T(p) \circ \varepsilon_D = p \circ \zeta \circ \varepsilon_D = p, \\ \alpha \circ \mu_I \circ TT(p) &= \alpha \circ T(p) \circ \mu_D = p \circ \zeta \circ \mu_D = p \circ \zeta \circ T(\zeta) \\ &= \alpha \circ T(p) \circ T(\zeta) = \alpha \circ T(\alpha) \circ TT(p), \end{aligned}$$

from which $\alpha \circ \varepsilon_I = 1_I$ and $\alpha \circ \mu_I = \alpha \circ T(\alpha)$.

Thus we have produced p: $(D,\zeta) \longrightarrow (I,\alpha)$ in $\mathscr{C}^{\mathbb{T}}$. From the relations

$$p_h \circ i \circ p \circ f = p_h \circ k \circ f = h \circ f = h \circ g = p_h \circ k \circ g = p_h \circ i \circ p \circ g$$

we deduce $i \circ p \circ f = i \circ p \circ g$ and thus $p \circ f = p \circ g$, since i is a monomorphism. Moreover, if the morphism $h: (D,\zeta) \longrightarrow (X,\chi)$ in $\mathscr{C}^{\mathbb{T}}$ is such that $h \circ f = h \circ g$, by construction we have $p_h \circ i: (I,\alpha) \longrightarrow (X,\chi)$ in $\mathscr{C}^{\mathbb{T}}$ such that $(p_h \circ i) \circ p = h$. This factorization is unique since p is an epimorphism in $\mathscr{C}$. Thus $p: (D,\zeta) \longrightarrow (I,\alpha)$ is the coequalizer of (f,g) in $\mathscr{C}^{\mathbb{T}}$ and $\mathscr{C}^{\mathbb{T}}$ is cocomplete.

The functor $U: \mathscr{C}^{\mathbb{T}} \longrightarrow \mathscr{C}$ preserves limits (see 4.3.1) but also regular epimorphisms, since the construction of the coequalizer $p = \mathsf{Coker}\,(f,g)$ in the first part of the proof indicates in particular that p is a regular epimorphism in $\mathscr{C}$. By 2.5.7, volume 1, U preserves coequalizers of kernel pairs since it preserves kernel pairs and regular epimorphisms. But since U reflects isomorphisms (see 4.1.4), U reflects coequalizers of kernel pairs (see 2.9.7, volume 1) and thus reflects regular epimorphisms.

As $U: \mathscr{C}^{\mathbb{T}} \longrightarrow \mathscr{C}$ preserves and reflects pullbacks and regular epimorphisms, the pullback of a regular epimorphism in $\mathscr{C}^{\mathbb{T}}$ along any morphism is again a regular epimorphism, since this is the case in $\mathscr{C}$. Thus $\mathscr{C}^{\mathbb{T}}$ is regular (see 2.1.1).

Let us now consider an exact sequence

$$C \underset{v}{\overset{u}{\rightrightarrows}} D \xrightarrow{\quad q \quad} Q$$

in $\mathscr{C}$ (see 2.3.1) and a section $s: Q \longrightarrow D$ of the regular epimorphism q. Since $q \circ 1_D = q \circ s \circ q$, we get a unique factorization $t: D \longrightarrow C$ through the kernel pair (u,v) such that $1_D = u \circ t$, $s \circ q = v \circ t$. In other words every exact sequence in $\mathscr{C}$ is an absolute coequalizer (see 2.10.2, volume 1).

Next, if $u, v: (C,\xi) \rightrightarrows (D,\zeta)$ is a pair of morphisms in $\mathscr{C}^{\mathbb{T}}$ such that $u, v: C \rightrightarrows D$ is a kernel pair in $\mathscr{C}$, with coequalizer $q: D \longrightarrow Q$, we have just observed that this coequalizer is preserved by every functor, thus in particular by T and $T \circ T$. Therefore 4.3.2 applies, showing that Q can be provided with the structure (Q,α) of a $\mathbb{T}$-algebra in such a way that $q: (D,\xi) \longrightarrow (Q,\alpha)$ is the cokernel of (u,v) in $\mathscr{C}^{\mathbb{T}}$. Since U reflects kernel pairs (see 4.3.1), (u,v) is the kernel pair of q in $\mathscr{C}^{\mathbb{T}}$.

Finally if $u, v = (R,\rho) \rightrightarrows (C,\xi)$ is an equivalence relation in $\mathscr{C}^{\mathbb{T}}$, then $u, v: R \rightrightarrows C$ is an equivalence relation in $\mathscr{C}$ since U preserves limits (see 2.5.4). If $\mathscr{C}$ is exact, (u,v) is a kernel pair in $\mathscr{C}$ (see 2.5.3) and thus (u,v) is also a kernel pair in $\mathscr{C}^{\mathbb{T}}$. □

Another interesting example of a category $\mathscr{C}$ satisfying the requirements of 4.3.5 is a Grothendieck topos $\mathscr{C}$ satisfying the axiom of choice (see 3.4.3 and 7.5.1, volume 3).

Here is a final example where the cocompleteness of a category of $\mathbb{T}$-algebras can be proved; this example will take its full meaning in the context of locally presentable categories (see 5.5.8, volume 1).

Proposition 4.3.6 *Let $\mathscr{C}$ be a complete and cocomplete category and $\mathbb{T} = (T, \varepsilon, \mu)$ a monad on $\mathscr{C}$. If the functor $T: \mathscr{C} \longrightarrow \mathscr{C}$ preserves κ-filtered colimits, for some regular cardinal κ, the category $\mathscr{C}^{\mathbb{T}}$ of $\mathbb{T}$-algebras is complete and cocomplete.*

Proof The completeness is attested by 4.3.1. To prove the cocompleteness it suffices, by 4.3.4, to prove the existence of coequalizers. If $\mathscr{K}$ is the category $\bullet \rightrightarrows \bullet$ used to describe coequalizers (see 2.6.7.b, volume 1) we must prove that the functor

$$\Delta: \mathscr{C}^{\mathbb{T}} \longrightarrow \mathsf{Fun}(\mathscr{K}, \mathscr{C}^{\mathbb{T}}),$$

mapping (A, ξ) to the constant functor on (A, ξ), has a left adjoint (see 3.2.3, volume 1). But we know already that $\mathscr{C}^{\mathbb{T}}$ is complete and, moreover, that Δ preserves limits since these are computed pointwise in $[\mathscr{K}, \mathscr{C}^{\mathbb{T}}]$; see 2.15.1, volume 1. By the adjoint functor theorem, it remains to prove the solution set condition (see 3.3.3, volume 1). Giving a functor $F: \mathscr{K} \longrightarrow \mathscr{C}^{\mathbb{T}}$ means giving a pair $f, g: (A, \alpha) \rightrightarrows (B, \beta)$ of arrows in $\mathscr{C}^{\mathbb{T}}$. A set of objects $(D_i, \delta_i) \in \mathscr{C}^{\mathbb{T}}$, $i \in I$, is a solution set for F when for each morphism $h: (B, \beta) \longrightarrow (C, \gamma)$ in $\mathscr{C}^{\mathbb{T}}$ such that $h \circ f = h \circ g$, h factors through some (D_i, δ_i) in $\mathscr{C}^{\mathbb{T}}$. (Indeed, giving a cocone on F is equivalent to giving such a h; see 2.6.7.b, volume 1.) We shall construct such a solution set reduced to a single object of $\mathscr{C}^{\mathbb{T}}$. This will be done via a transfinite induction.

For every cardinal λ, we shall define objects P_λ, Q_λ and arrows p_λ, q_λ, u_λ, v_λ as in diagram 4.12. The diagram will be commutative with $q_\lambda \circ f = q_\lambda \circ g$, $p_\lambda \circ T(f) = p_\lambda \circ T(g)$. Given cardinals $\lambda \leq \nu$, there will be transition morphisms $p_{\lambda,\nu}: P_\lambda \longrightarrow P_\nu$, $q_{\lambda,\nu}: Q_\lambda \longrightarrow Q_\nu$ still making diagram 4.12 commutative, with $q_{\lambda,\nu} \circ q_\lambda = q_\nu$, $p_{\lambda,\nu} \circ p_\lambda = p_\nu$ and $q_{\nu\nu} = 1_{Q_\nu}$, $p_{\nu\nu} = 1_{P_\nu}$. We shall prove that $T(Q_\kappa) \cong P_\kappa$ and $\{(Q_\kappa, u_\kappa)\}$ is the required solution set.

For the initial step of the induction, we just define $q_0 = \mathsf{Coker}\,(f, g)$ and $p_0 = \mathsf{Coker}\,(T(f), T(g))$. From the relations

$$q_0 \circ \beta \circ T(f) = q_0 \circ f \circ \alpha = q_0 \circ g \circ \alpha = q_0 \circ \beta \circ T(g)$$

we deduce the existence of a unique u_0 such that $u_0 \circ p_0 = q_0 \circ \beta$. From $T(q_0) \circ T(f) = T(q_0) \circ T(g)$ we deduce the existence of a unique v_0 such that $v_0 \circ p_0 = T(q_0)$.

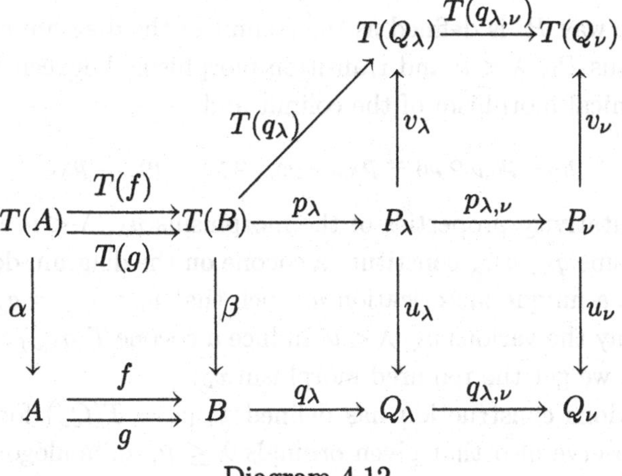

Diagram 4.12

Now suppose the construction is achieved for every ordinal $\lambda \leq \nu$; we shall extend it to the level $\nu + 1$. We define $u_{\nu+1}$ as the coequalizer

$$T(P_\nu) \underset{T(u_\nu)}{\overset{\mu_{Q_\nu} \circ T(v_\nu)}{\rightrightarrows}} T(Q_\nu) \xrightarrow{u_{\nu+1}} Q_{\nu+1}$$

and simply put

$$q_{\nu,\nu+1} = u_{\nu+1} \circ \varepsilon_{Q_\nu}, \quad P_{\nu+1} = T(Q_\nu), \quad p_{\nu,\nu+1} = v_\nu, \quad v_{\nu+1} = T(q_{\nu,\nu+1})$$

and necessarily

$$q_{\nu+1} = q_{\nu,\nu+1} \circ q_\nu \quad p_{\nu+1} = p_{\nu,\nu+1} \circ p_\nu.$$

One has the required commutativities since

$$\begin{aligned} q_{\nu,\nu+1} \circ u_\nu &= u_{\nu+1} \circ \varepsilon_{Q_\nu} \circ u_\nu = u_{\nu+1} \circ T(u_\nu) \circ \varepsilon_{P_\nu} \\ &= u_{\nu+1} \circ \mu_{Q_\nu} \circ T(v_\nu) \circ \varepsilon_{P_\nu} = u_{\nu+1} \circ \mu_{Q_\nu} \circ \varepsilon_{T(Q_\nu)} \circ v_\nu \\ &= u_{\nu+1} \circ v_\nu = u_{\nu+1} \circ p_{\nu,\nu+1} \end{aligned}$$

and

$$v_{\nu+1} \circ p_{\nu,\nu+1} = T(q_{\nu,\nu+1}) \circ v_\nu.$$

Now suppose ν is a limit ordinal and the construction is realized for every ordinal $\lambda < \nu$. We define Q_ν as the colimit of the diagram constituted of the various Q_λ, $\lambda < \nu$, and transition morphisms between them; $q_{\lambda,\nu}$ is the canonical morphism to the colimit and

$$q_\nu = q_{0,\nu} \circ q_0 = q_{\lambda,\nu} \circ q_{0,\lambda} \circ q_0 = q_{\lambda,\nu} \circ q_\lambda.$$

In the same way P_ν is defined as the colimit of the diagram constituted of the various P_λ, $\lambda < \nu$ and transition morphisms between them; $p_{\lambda,\nu}$ is the canonical morphism of the colimit and

$$p_\nu = p_{0,\nu} \circ p_0 = p_{\lambda,\nu} \circ p_{0,\lambda} \circ p_0 = p_{\lambda,\nu} \circ p_\lambda.$$

The commutativity properties of the morphisms u_λ, $\lambda < \nu$, imply that the morphisms $q_{\lambda,\nu} \circ u_\lambda$ constitute a cocone on the diagram defining P_ν; thus we get a unique factorization u_ν such that $u_\nu \circ p_{\lambda,\nu} = q_{\lambda,\nu} \circ u_\lambda$. In the same way the various v_λ, $\lambda < \nu$, induce a cocone $T(q_{\lambda,\nu}) \circ v_\lambda$, $\lambda < \nu$, from which we get the required morphism v_ν.

The previous construction has defined $P_{\nu+1} = T(Q_\nu)$ for every ordinal ν. Observe also that given ordinals $\lambda \leq \nu$, an analogous formula $p_{\lambda+1,\nu+1} = T(q_{\lambda,\nu})$ holds for the transition morphisms. Indeed, by the previous construction and the commutativity of diagram 4.12,

$$\begin{aligned} p_{\lambda+1,\nu+1} &= p_{\nu,\nu+1} \circ p_{\lambda+1,\nu} = v_\nu \circ p_{\lambda+1,\nu} \\ &= T(q_{\lambda+1,\nu}) \circ v_{\lambda+1} = T(q_{\lambda+1,\nu}) \circ T(q_{\lambda,\lambda+1}) \\ &= T(q_{\lambda,\nu}). \end{aligned}$$

Next we note that if ν is a limit ordinal, the successor ordinals $\lambda+1 < \nu$ constitute a cofinal diagram among all the ordinals $\lambda < \nu$ (see 2.11.2, volume 1); therefore computing the colimit on $\lambda < \nu$ is equivalent to computing a colimit on $\lambda + 1 < \nu$. But since $T(Q_\nu) = P_{\nu+1}$ and $T(q_{\lambda,\nu}) = p_{\lambda+1,\nu+1}$ for every ordinal $\lambda \leq \mu$, the previous remark immediately implies the relation

$$P_\kappa \cong \operatorname{colim}_{\lambda<\kappa} T(Q_\lambda),$$

where the regular cardinal κ has been identified with the smallest ordinal of that cardinality. By regularity of κ, the ordinals strictly less than κ constitute a κ-filtered diagram from which, by our assumption on the functor T,

$$P_\kappa \cong \operatorname{colim}_{\lambda<\kappa} T(Q_\lambda) = T\big(\operatorname{colim}_{\lambda<\kappa} Q_\lambda\big) \cong T(Q_\kappa).$$

Since colimits are defined up to isomorphism, we can choose the colimits in such a way that this isomorphism (which is v_κ) is the identity.

The previous isomorphism allows us to consider the pair (Q_κ, u_κ), where $u_\kappa \colon T(Q_\kappa) \longrightarrow Q_\kappa$. We prove it is a $\mathbb{T}$-algebra. Let us consider the coequalizer diagram defining $u_{\nu+1}$. The naturality of μ implies the commutativity of diagram 4.13, for every $\nu < \kappa$, thus the fact that μ_{Q_κ} is the colimit of the various μ_{Q_ν}. Computing the colimit of the

$$\begin{array}{ccc} TT(Q_\nu) & \xrightarrow{\mu_{Q_\nu}} & T(Q_\nu) \\ {\scriptstyle TT(q_{\nu,\kappa})}\downarrow & & \downarrow{\scriptstyle T(q_{\nu,\kappa})} \\ TT(Q_\kappa) & \xrightarrow[\mu_{Q_\kappa}]{} & T(Q_\kappa) \end{array}$$

Diagram 4.13

various coequalizer diagrams and transition morphisms, for the successor ordinals $\nu+1<\kappa$, we get

$$TT(Q_\kappa) \underset{T(u_\kappa)}{\overset{\mu_{Q_\kappa}}{\rightrightarrows}} T(Q_\kappa) \xrightarrow{u_\kappa} Q_\kappa$$

since T preserves the κ-filtered colimits involved and v_κ is the identity. Thus $u_\kappa \circ \mu_{Q_\kappa} = u_\kappa \circ T(u_\kappa)$, which is one of the axioms for being a $\mathbb{T}$-algebra (see 4.1.2). For the other axiom observe that by definition, for every ordinal λ the relation $q_{\lambda,\lambda+1} = u_{\lambda+1} \circ \varepsilon_{Q_\lambda}$ holds. As for μ, the naturality of ε implies that ε_{Q_κ} is the colimit of the various ε_{Q_λ}, $\lambda<\kappa$. On the other hand the colimit of the transition morphisms $q_{\lambda,\lambda+1}$ is just the identity on Q_κ, since computing the colimit on $\lambda<\kappa$ is equivalent to computing it on $\lambda+1<\kappa$. Finally the definition of $q_{\lambda,\lambda+1}$ yields, at the colimit, $1_{Q_\kappa} = u_\kappa \circ \varepsilon_{Q_\kappa}$, which is the required axiom.

Next we prove that $q_\kappa\colon (B,\beta) \longrightarrow (Q_\kappa, u_\kappa)$ is a morphism of $\mathbb{T}$-algebras. By commutativity of diagram 4.12, we know already that $u_\kappa \circ p_\kappa = q_\kappa \circ \beta$ and it remains to prove that $p_\kappa = T(q_\kappa)$. This is an immediate consequence of the relation $v_\lambda \circ p_\lambda = T(q_\lambda)$, since computing the colimit for $\lambda<\kappa$ yields $v_\kappa \circ p_\kappa = T(q_\kappa)$ and the isomorphism v_κ has been chosen to be the identity (see 4.1.2).

Finally let us choose a morphism $h\colon (B,\beta) \longrightarrow (C,\gamma)$ of $\mathbb{T}$-algebras, such that $h \circ f = h \circ g$. We just need to find a morphism of $\mathbb{T}$-algebras $k\colon (Q_\kappa, u_\kappa) \longrightarrow (C,\gamma)$ such that $k \circ q_\kappa = h$. We construct k by transfinite induction. More precisely we shall construct, for every ordinal λ, a morphism $k_\lambda\colon Q_\lambda \longrightarrow C$ in $\mathscr{C}$ such that diagram 4.14 is commutative and $k_\nu = k_\lambda \circ q_{\lambda,\nu}$ for every ordinal $\lambda \le \nu$. We shall prove that $k = k_\kappa$ is the required factorization.

To initialize the process, we choose $k_0\colon Q_0 \longrightarrow C$ as the unique factorization of h through the coequalizer q_0 of (f,g) in $\mathscr{C}$. One has imme-

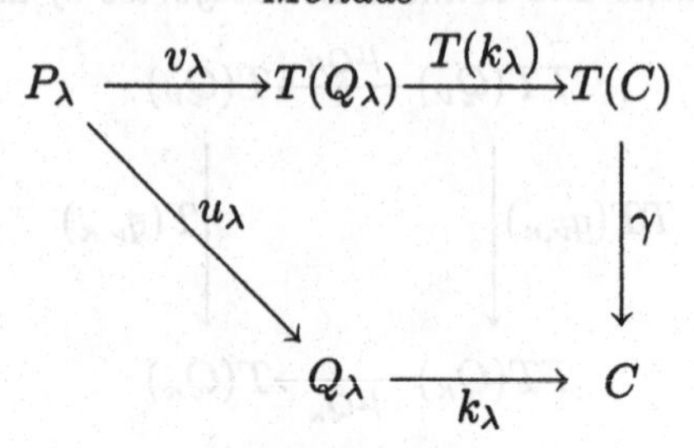

Diagram 4.14

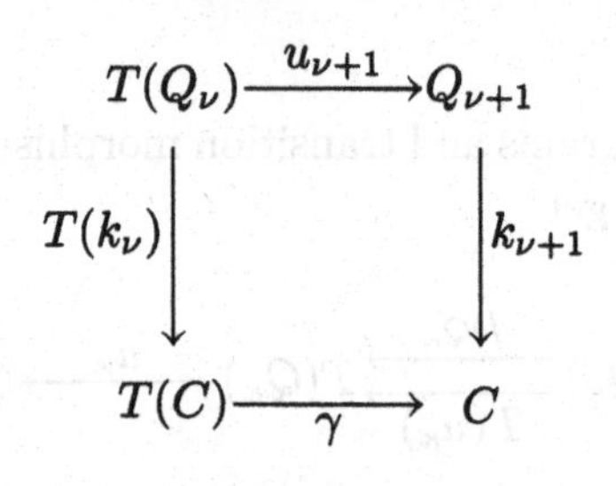

Diagram 4.15

diately, since h is a morphism of $\mathbb{T}$-algebras,

$$\gamma \circ T(k_0) \circ v_0 \circ p_0 = \gamma \circ T(k_0) \circ T(q_0) = \gamma \circ T(h)$$
$$= h \circ \beta = k_0 \circ q_0 \circ \beta = k_0 \circ u_0 \circ p_0,$$

from which $\gamma \circ T(k_0) = k_0 \circ u_0$ since p_0 is an epimorphism (see 2.4.3, volume 1).

Now suppose the construction is achieved for every ordinal $\lambda \leq \nu$; we shall extend it to the level $\nu + 1$. The relations

$$\gamma \circ T(k_\nu) \circ \mu_{Q_\nu} \circ T(v_\nu) = \gamma \circ \mu_C \circ TT(k_\nu) \circ T(v_\nu)$$
$$= \gamma \circ T(\gamma) \circ TT(k_\nu) \circ T(v_\nu)$$
$$= \gamma \circ T(k_\nu) \circ T(u_\nu)$$

imply that the composite $\gamma \circ T(k_\nu)$ factors through the coequalizer $u_{\nu+1}$ of the pair $(\mu_{Q_\nu} \circ T(v_\nu), T(u_\nu))$, yielding a morphism $k_{\nu+1}$ such that diagram 4.15 commutes. The stated pentagonal relation is satisfied since

$$\gamma \circ T(k_{\nu+1}) \circ v_{\nu+1} = \gamma \circ T(k_{\nu+1}) \circ T(q_{\nu,\nu+1})$$
$$= \gamma \circ T(k_{\nu+1}) \circ T(u_{\nu+1}) \circ T(\varepsilon_{Q_\nu})$$
$$= \gamma \circ T(\gamma) \circ TT(k_\nu) \circ T(\varepsilon_{Q_\nu})$$
$$= \gamma \circ T(\gamma) \circ T(\varepsilon_C) \circ T(k_\nu) = \gamma \circ T(k_\nu)$$
$$= k_{\nu+1} \circ u_{\nu+1}.$$

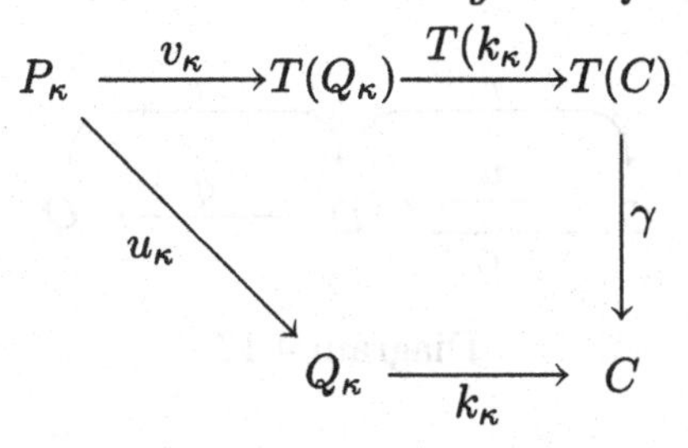

Diagram 4.16

On the other hand the construction is compatible with the transition morphisms since

$$\begin{aligned} k_{\nu+1} \circ q_{\nu,\nu+1} &= k_{\nu+1} \circ u_{\nu+1} \circ \varepsilon_{Q_\nu} \\ &= \gamma \circ T(k_\nu) \circ \varepsilon_{Q_\nu} = \gamma \circ \varepsilon_C \circ k_\nu \\ &= k_\nu. \end{aligned}$$

Next suppose the construction has been realized for every ordinal $\lambda < \nu$, with ν a limit ordinal. Because of the compatibility conditions with the transition morphisms $q_{\lambda,\nu}$, the morphisms $k_\lambda\colon Q_\lambda \longrightarrow C$ define a cocone on the diagram defining Q_ν as a colimit; therefore we get a factorization $k_\nu\colon Q_\nu \longrightarrow C$ such that $k_\nu \circ q_{\lambda,\nu} = k_\lambda$ for each $\lambda < \nu$. This already takes care of the compatibility with the transition morphisms. To prove the pentagonal condition, observe that P_ν itself has been defined as a colimit, so that it suffices to prove the commutativity after composition with each $p_{\lambda,\nu}$, $\lambda < \nu$.

$$\begin{aligned} \gamma \circ T(k_\nu) \circ v_\nu \circ p_{\lambda,\nu} &= \gamma \circ T(k_\nu) \circ T(q_{\lambda,\nu}) \circ v_\lambda = \gamma \circ T(k_\lambda) \circ v_\lambda \\ &= k_\lambda \circ u_\lambda = k_\nu \circ q_{\lambda,\nu} \circ u_\lambda \\ &= k_\nu \circ u_\nu \circ p_{\lambda,\nu} \end{aligned}$$

Considering now the morphism k_κ, we have

$$k_\kappa \circ q_\kappa = k_\kappa \circ q_{0,\kappa} \circ q_0 = k_0 \circ q_0 = h$$

and so just need to prove that $k_\kappa\colon (Q_\kappa, u_\kappa) \longrightarrow (C, \gamma)$ is a morphism of $\mathbb{T}$-algebras. But again since T preserves κ-filtered colimits, computing the colimit for $\lambda < \kappa$ of the pentagonal diagrams, together with the connecting morphisms, yields a commutative pentagon as in diagram 4.16. Since by choice of the colimits, v_κ is just the identity, this proves that k_κ is indeed a morphism in $\mathscr{C}^{\mathbb{T}}$. □

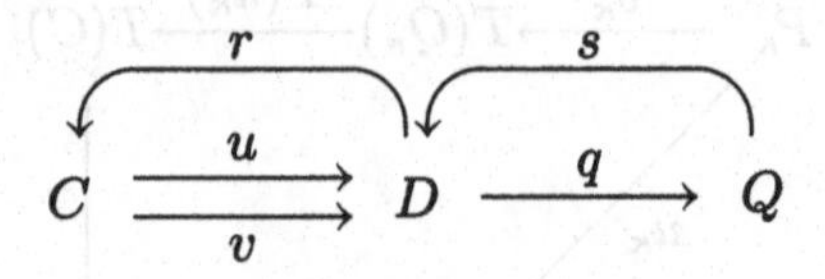

Diagram 4.17

4.4 Characterization of monadic categories

We want now to characterize those categories which can be described as categories of algebras for some monad on a base category.

Definition 4.4.1 *A functor $R\colon \mathscr{X} \longrightarrow \mathscr{C}$ is monadic when there exist a monad $\mathbb{T} \cong (T, \varepsilon, \mu)$ on $\mathscr{C}$ and an equivalence of categories $J\colon \mathscr{X} \longrightarrow \mathscr{C}^{\mathbb{T}}$ such that $U \circ J$ is isomorphic to R, where $U\colon \mathscr{C}^{\mathbb{T}} \longrightarrow \mathscr{C}$ is the forgetful functor of 4.1.4.*

Definition 4.4.2 *By a split coequalizer in a category $\mathscr{C}$ we mean a situation as in diagram 4.17 where $q \circ u = q \circ v$, $q \circ s = 1_Q$, $u \circ r = 1_D$, $v \circ r = s \circ q$.*

The terminology in 4.4.2 is justified by the fact that q is then the coequalizer of u, v and, moreover, this coequalizer is absolute, i.e. preserved by every functor defined on $\mathscr{C}$; see 2.10.2, volume 1.

It is useful to recall here part of the statement of lemma 4.3.3:

Lemma 4.4.3 *Let $\mathbb{T} = (T, \varepsilon, \mu)$ be a monad on a category $\mathscr{C}$. For every $\mathbb{T}$-algebra (X, ξ), diagram 4.8 is a split coequalizer in $\mathscr{C}$.* □

Theorem 4.4.4 *Let $R\colon \mathscr{X} \longrightarrow \mathscr{C}$ be a functor. The following conditions are equivalent:*

(1) R is monadic;

(2) (a) R has a left adjoint L;

(b) R reflects isomorphisms;

(c) if a pair $u, v\colon X \rightrightarrows Y$ in $\mathscr{X}$ is such that $(R(u), R(v))$ has a split coequalizer in $\mathscr{C}$, then (u, v) has a coequalizer in $\mathscr{X}$ which is preserved by R.

Proof To prove (1) ⇒ (2), it suffices to show that given a monad $\mathbb{T} = (T, \varepsilon, \mu)$ on $\mathscr{C}$, the forgetful functor $U\colon \mathscr{C}^{\mathbb{T}} \longrightarrow \mathscr{C}$ satisfies (2)(a), (b), (c). Conditions (2)(a) and (2)(b) are already attested by 4.1.4. So let us choose $u, v\colon (C, \xi) \rightrightarrows (D, \zeta)$ in $\mathscr{C}^{\mathbb{T}}$ and morphisms q, r, s in $\mathscr{C}$ producing a split coequalizer as in 4.4.2. This split coequalizer is preserved by

$$\begin{array}{c} \alpha_{RL(C)} \qquad\qquad \alpha_C \\ RLRL(C) \underset{RL(\xi)}{\overset{R\beta_{L(C)}}{\rightrightarrows}} RL(C) \xrightarrow{\quad \xi \quad} C \end{array}$$

Diagram 4.18

every functor (see 2.10.2, volume 1), thus in particular by T and $T \circ T$. Therefore 4.3.2 applies, proving that the coequalizer of (u, v) exists in $\mathscr{C}^{\mathbb{T}}$ and is preserved by U.

Conversely, we use the notation of section 4.2; it suffices to prove that the comparison functor $J: \mathscr{X} \longrightarrow \mathscr{C}^{\mathbb{T}}$ is an equivalence of categories, where $\mathbb{T}$ is the monad generated by the adjunction $L \dashv R$. Since the composite $J \circ K$ is just the inclusion $\mathscr{C}_{\mathbb{T}} \subseteq \mathscr{C}^{\mathbb{T}}$, we know already that J is full. On the other hand given a $\mathbb{T}$-algebra (C, ξ), we consider the pair $\beta_{L(C)}, L(\xi)$: $LRL(C) \rightrightarrows L(C)$ in $\mathscr{X}$ which, by lemma 4.4.3, satisfies the conditions of (2)(c); (see diagram 4.18), since $T = R \circ L$, $\varepsilon = \alpha$ and $\mu = R * \beta * L$; see 4.2.1. Applying our assumption (2)(c), we get a coequalizer in $\mathscr{X}$,

$$LRL(C) \underset{L(\xi)}{\overset{\beta_{L(C)}}{\rightrightarrows}} L(C) \xrightarrow{\quad q \quad} Q,$$

such that, up to isomorphism, $R(Q) = C$ and $R(q) = \xi$. In particular $R(q)$ has a section α_C, thus $LR(q)$ has a section as well and is therefore an epimorphism (see 1.7.4). By definition of q and naturality of β,

$$q \circ LR(q) = q \circ L(\xi) = q \circ \beta_{LR(Q)} = \beta_Q \circ LR(q),$$

from which $q = \beta_Q$, since $LR(q)$ is an epimorphism. Finally, we have obtained

$$J(Q) = \big(R(Q), R(\beta_Q)\big) \cong \big(C, R(q)\big) = (C, \xi).$$

It remains to prove that J is faithful. Since $U \circ J \cong R$, it suffices to prove that R is faithful. Choose x, y: $X \rightrightarrows Y$ in $\mathscr{X}$ such that $R(x) = R(y)$. By naturality of β,

$$x \circ \beta_X = \beta_Y \circ LR(x) = \beta_Y \circ LR(y) = y \circ \beta_X,$$

so that it suffices to prove that β_X is an epimorphism. For this consider

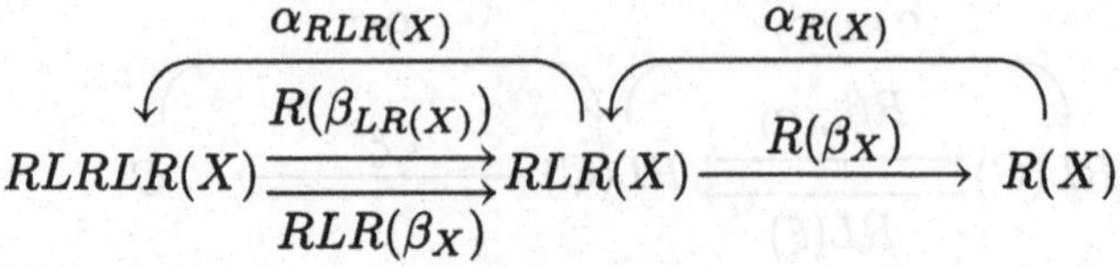

$$\alpha_{RLR(X)} \qquad\qquad \alpha_{R(X)}$$

$$RLRLR(X) \underset{RLR(\beta_X)}{\overset{R(\beta_{LR(X)})}{\rightrightarrows}} RLR(X) \xrightarrow{R(\beta_X)} R(X)$$

Diagram 4.19

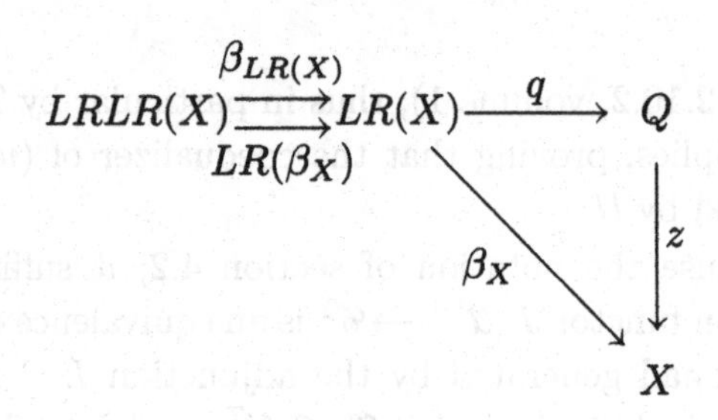

Diagram 4.20

the diagram

$$LRLR(X) \underset{LR(\beta_X)}{\overset{\beta_{LR(X)}}{\rightrightarrows}} LR(X) \xrightarrow{\beta_X} X$$

in $\mathscr{X}$, whose image under R is a split coequalizer in $\mathscr{C}$ (see diagram 4.19), just by the triangular identities of the adjunction $L \dashv R$ (see 3.1.5, volume 1) and the naturality of α. By our assumption (2)(c), we get diagram 4.20 in $\mathscr{X}$ where q is the coequalizer of $(\beta_{LR(X)}, LR(\beta_X))$ and z is the unique factorization yielding $\beta_X = z \circ q$; moreover (2)(c) asserts that $R(q)$ is just $R(\beta_X)$, up to isomorphism. In particular $R(z)$ an isomorphism and thus z is one too, by our assumption (2)(b). Thus β_X is an epimorphism since q is, and the proof is complete. □

It is natural to wonder if the composite of two monadic functors is again monadic. This is not the case in general, as attested by counterexample 4.6.4. Observe that in 4.4.4, conditions (2)(a) and (2)(b) are obviously stable under composition, but condition (2)(c) is not.

In the case of monadic categories over Set, we get a more remarkable result.

Theorem 4.4.5 *Let $\mathscr{C}$ be a category. The following conditions are equivalent:*

(1) there exists a monadic functor $U: \mathscr{C} \longrightarrow$ **Set***;*

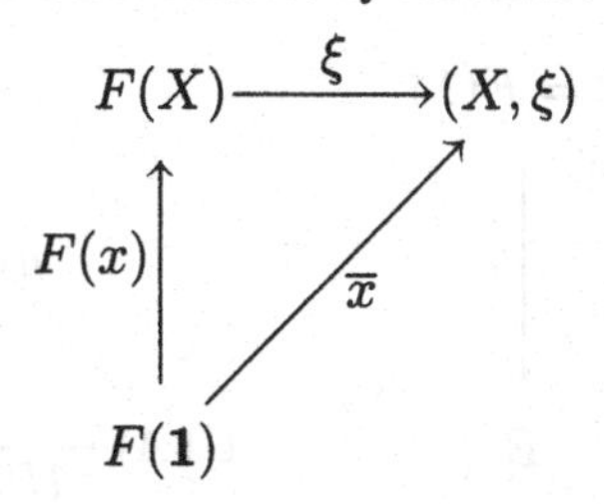

Diagram 4.21

(2) (a) $\mathscr{C}$ *has finite limits;*
(b) $\mathscr{C}$ *is exact;*
(c) $\mathscr{C}$ *has a regular generator* P*;*
(d) P *is projective;*
(e) *the copower* $\coprod_X P$ *exists for every set* X*.*

Proof Given a monadic functor $U\colon \mathscr{C} \longrightarrow \mathsf{Set}$ and its left adjoint $F\colon \mathsf{Set} \longrightarrow \mathscr{C}$ (see 4.4.4), put $P = F(\mathbf{1})$, where $\mathbf{1}$ is the singleton. From the bijections, for every $C \in \mathscr{C}$,

$$U(C) \cong \mathsf{Set}\big(\mathbf{1}, U(C)\big) \cong \mathscr{C}\big(F(\mathbf{1}), C\big),$$

we deduce that U is the functor represented by P. Now $\mathscr{C}$ has finite limits by 4.3.1, $\mathscr{C}$ is exact by 4.3.5 and P is a generator since U is faithful (see 4.1.4, this volume, and 4.5.9, volume 1). The copowers $\coprod_X P$ exist since $\mathscr{C}$ is cocomplete (see 4.3.5).

Now consider the monad $\mathbb{T} = (T, \varepsilon, \mu)$ generating $\mathscr{C}$. Given a $\mathbb{T}$-algebra (X, ξ) and an element $x \in X$, we view x as a mapping $x\colon \mathbf{1} \longrightarrow U(X, \xi)$ and we write $\overline{x}\colon F(\mathbf{1}) \longrightarrow (X, \xi)$ for the corresponding morphism of $\mathbb{T}$-algebras obtained by adjunction. Diagram 4.21 is commutative by 3.1.5, volume 1, since $\xi\colon FU(X,\xi) \longrightarrow (X, \xi)$ is the component at (X, ξ) of the canonical natural transformation $FU \Rightarrow \mathrm{id}_{\mathscr{C}}$ of the adjunction $F \dashv U$ (see 3.1.5, volume 1, and 4.1.4, this volume). But since $X = \coprod_X \mathbf{1}$ and F preserves copowers (see 3.2.2, volume 1), we have just observed, since $X \cong U(X, \xi) \cong \mathscr{C}\big(P, (X, \xi)\big)$, that

$$\xi\colon F(X) \cong \coprod\nolimits_{\mathscr{C}(P,(X,\xi))} P \longrightarrow (X, \xi)$$

is the morphism which, composed with the injection $s_{\overline{x}}$ of the coproduct, reproduces $\overline{x}$. The object P is a regular generator when this morphism is a regular epimorphism (see 4.5.3, volume 1),which is the case by 4.3.3.

It remains to prove that P is projective; by regularity, strong and regular epimorphisms coincide (see 2.1.4). We know that U preserves regular

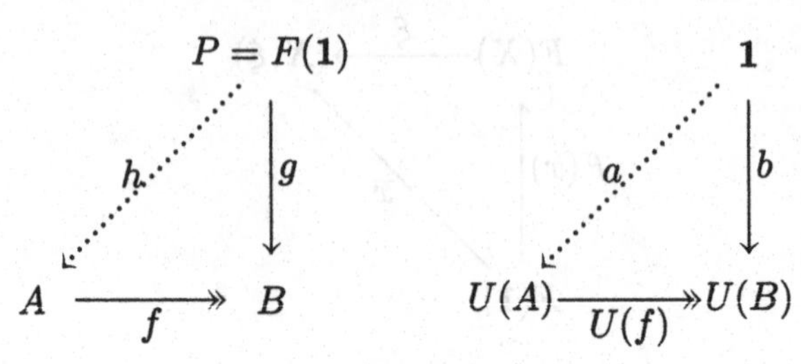

Diagram 4.22

epimorphisms (see 4.3.5). Given a regular epimorphism $f: A \longrightarrow B$ in $\mathscr{C}$ and a morphism $g: P \longrightarrow B$ (see diagram 4.22), g corresponds by adjunction with an element $b \in U(B)$, which has the form $U(f)(a)$ for $a \in U(A)$, since $U(f)$ is surjective. The arrow h is the morphism corresponding with a by adjunction and from $U(f)(a) = b$, we get $f \circ h = g$.

Conversely, suppose the conditions (2)(a), (b), (c), (d), (e) are satisfied. We shall apply 4.4.4 to prove that the representable functor

$$U \equiv \mathscr{C}(P, -): \mathscr{C} \longrightarrow \mathsf{Set}$$

is monadic.

First of all, U has a left adjoint. The reflection of a set X along U is just $(\coprod_X P, \alpha_X)$ where

$$\alpha_X: X \longrightarrow U\left(\coprod\nolimits_X P\right) \cong \mathscr{C}\left(P, \coprod\nolimits_X P\right)$$

maps the element $x \in X$ to the corresponding canonical morphism $s_x: P \longrightarrow \coprod_X P$ of the copower. Indeed, given $C \in \mathscr{C}$ and a mapping $f: X \longrightarrow \mathscr{C}(P, C)$, the family of morphisms $f(x): P \longrightarrow C$, for $x \in X$, yields a unique factorization $g: \coprod_X P \longrightarrow C$ such that $g \circ s_x = f(x)$, i.e. such that $\mathscr{C}(P, g) \circ \alpha_X = f$; see 3.1.1, volume 1. We write F for the left adjoint of U. By construction, $P = F(\mathbf{1})$.

Since P is a generator U is faithful, (see 4.5.9, volume 1). In particular U reflects monomorphisms and epimorphisms (see 1.7.7, volume 1). Let us prove that U reflects regular epimorhisms. Given $f: A \longrightarrow B$ in $\mathscr{C}$ such that $U(f)$ is surjective, consider diagram 4.23 where p_B is the unique morphism such that $p_B \circ s_{\overline{b}} = \overline{b}$, for every $\overline{b} \in \mathscr{C}(P, B)$. By assumption, p_B is a regular epimorphism (see 4.5.3, volume 1). But every $\overline{b} \in \mathscr{C}(F(\mathbf{1}), B) \cong \mathsf{Set}(\mathbf{1}, U(B))$ corresponds then to an element $b \in U(B)$, which can be written as $U(f)(a)$ for some $a \in U(A)$, since $U(f)$ is surjective. The element a, viewed as a mapping $a: \mathbf{1} \longrightarrow U(A)$, corresponds to a morphism $\overline{a}: P = F(\mathbf{1}) \longrightarrow A$. These various mor-

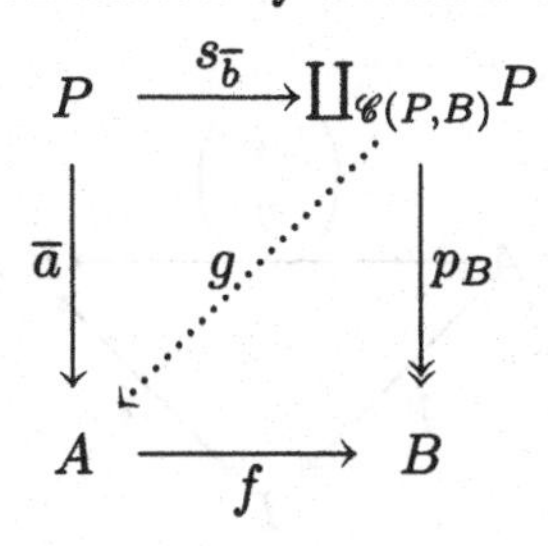

Diagram 4.23

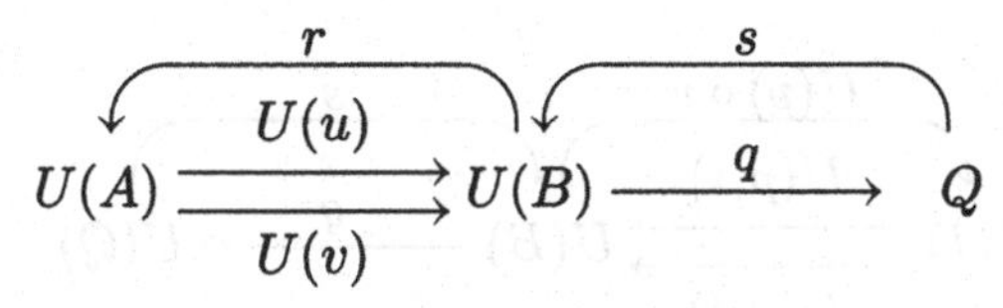

Diagram 4.24

phisms $\overline{a}$ yield a factorization g as in diagram 4.23 such that $g \circ s_{\overline{b}} = \overline{a}$. In particular

$$f \circ g \circ s_{\overline{b}} = f \circ \overline{a} = p_B \circ s_{\overline{b}}$$

since $U(f)(a) = b$, and thus $f \circ g = p_B$. Therefore f is a regular epimorphism (see 2.1.6). So f reflects monomorphisms and regular epimorphisms, hence it reflects isomorphisms (see 2.1.6).

Let us now observe that U is an exact functor (see 2.3.5); it preserves finite limits, since it has a left adjoint F (see 3.2.2, volume 1); we must still prove that U preserves regular (or strong, see 2.1.4) epimorphisms (see 2.3.7). But this is exactly the assumption that P is projective; indeed this means that given $f: A \longrightarrow B$, a regular epimorphism, $\mathscr{C}(P,f): \mathscr{C}(P,A) \longrightarrow \mathscr{C}(P,B)$ is surjective, or in other words $U(f)$ is surjective.

It remains to prove condition (2)(c) of 4.4.4. Consider thus a pair $u, v: A \rightrightarrows B$ in $\mathscr{C}$ such that $\big(U(u), U(v)\big)$ admits a split coequalizer as in diagram 4.24,

$$q \circ U(u) = q \circ U(v), \quad q \circ s = 1_Q, \quad U(u) \circ r = 1_{U(B)}, \quad s \circ q = U(v) \circ r.$$

We must prove that (u, v) has a coequalizer preserved by U. There is no restriction in supposing the pair (u, v) monomorphic, thus in supposing that A is a relation on B see 2.5.1. Indeed consider diagram 4.25, which is an image factorization in $\mathscr{C}$; see 2.1.4. This yields the relation $p_1 \circ i, p_2 \circ i: I \rightrightarrows B$ which has the same coequalizer as u, v, as long as

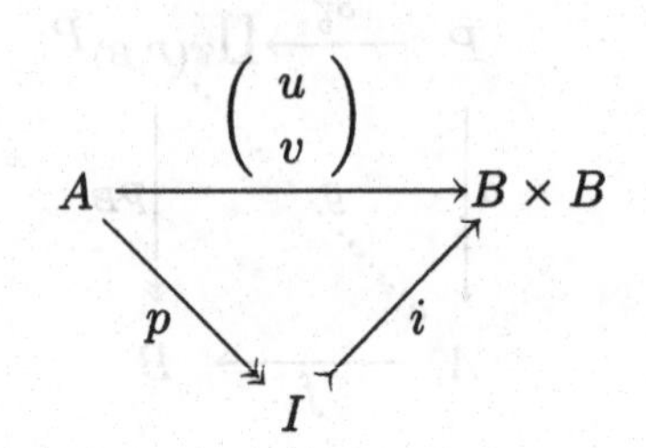

Diagram 4.25

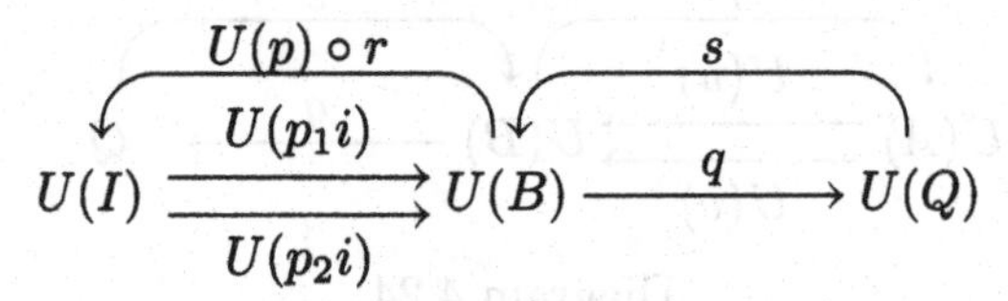

Diagram 4.26

one of those coequalizers exists, just because p is an epimorphism. Now diagram 4.26 obviously remains a split coequalizer. So let us assume that $u, v: A \rightrightarrows B$ is a relation on B.

Since U preserves finite limits and reflects isomorphisms, it also reflects finite limits (see 2.9.7, volume 1). So a relation $r_1, r_2: R \rightrightarrows B$ in $\mathscr{C}$ is an equivalence relation in $\mathscr{C}$ if and only if $U(r_1), U(r_2)$ is an equivalence relation in Set; see 2.5.4. Since U preserves finite limits, it preserves the construction of the dual relation R^0 (see 2.8.2) and since it is exact, it preserves the composite of relations (see 2.8.3 and 2.3.6).

Let us come back to our relation $u, v: A \rightrightarrows B$ such that $U(u), U(v)$ has a split coequalizer. We consider the relation $A \circ A^0$ on B. This is an equivalence relation, since $U(A) \circ U(A^0)$ is one in Set. First of all, if $b \in B$, $r(b) \in U(A)$ and

$$r(b) = (ur(b), vr(b)) = (b, sq(b)).$$

In Set, the equivalence relation generated by $U(A)$ is just the kernel pair of q, thus contains $U(A) \circ U(A^0)$. If we prove that the kernel pair of q is contained in $U(A) \circ U(A^0)$, we will have equality. But if $q(b) = q(b')$ for elements $b, b' \in B$, then

$$(b, sq(b)) \in U(A), \quad (sq(b), b') = (sq(b'), b'), \in U(A^0)$$

from which $(b, b') \in U(A) \circ U(A^0)$. This proves that in Set, $U(A) \circ U(A^0)$ is the equivalence relation generated by $U(A)$, so that in $\mathscr{C}$, $A \circ A^0$ is the smallest equivalence relation containing A.

Since $\mathscr{C}$ is exact, the equivalence relation $A \circ A^0$ on B has a coequalizer p, yielding an exact sequence

$$A \circ A^0 \overset{r_1}{\underset{r_2}{\rightrightarrows}} B \xrightarrow{\quad p \quad} P.$$

This exact sequence is preserved by U, proving $U(p) \cong q$. It remains to observe that p is also the coequalizer of (u, v). But given a morphism $f\colon B \longrightarrow C$ in $\mathscr{C}$ one has, by faithfulness of U and the relations $q = \mathsf{Coker}\,\big(U(u), U(v)\big) = \mathsf{Coker}\,\big(U(r_1), U(r_2)\big)$,

$$\begin{aligned} f \circ u = f \circ v &\Leftrightarrow U(f) \circ U(u) = U(f) \circ U(v) \\ &\Leftrightarrow \exists h\colon Q \longrightarrow U(C) \quad U(f) = h \circ q \\ &\Leftrightarrow U(f) \circ U(r_1) = U(f) \circ U(r_2) \\ &\Leftrightarrow f \circ r_1 = f \circ r_2. \end{aligned}$$

□

Let us conclude this section with an interesting link between monads, comonads (the dual notion of a monad) and adjoint functors.

Proposition 4.4.6 *Let $\mathbb{T} = (T, \varepsilon, \mu)$ be a monad on a category $\mathscr{C}$. If the functor $T\colon \mathscr{C} \longrightarrow \mathscr{C}$ has a right adjoint, then the forgetful functor $U\colon \mathscr{C}^{\mathbb{T}} \longrightarrow \mathscr{C}$ is comonadic for a comonad $\mathbb{S} = (S, \rho, \delta)$ on $\mathscr{C}$, where the functor S is right adjoint to T.*

Proof We shall apply 4.4.4. Observe that the reflection of isomorphisms by U is asserted by 4.1.4 while the dual of condition (2)(c) of 4.4.4 is just a special case of 4.3.1. So it suffices to prove that $U\colon \mathscr{C}^{\mathbb{T}} \longrightarrow \mathscr{C}$ has a right adjoint $G\colon \mathscr{C} \longrightarrow \mathscr{C}^{\mathbb{T}}$ such that $U \circ G$ is right adjoint to T.

Let us write $S\colon \mathscr{C} \longrightarrow \mathscr{C}$ for a right adjoint to T, with $\alpha\colon T \circ S \Rightarrow 1_{\mathscr{C}}$ and $\beta\colon 1_{\mathscr{C}} \Rightarrow S \circ T$ the two canonical natural transformations of the adjunction $T \dashv S$; see 3.1.5, volume 1. Given an object $D \in \mathscr{C}$, we must construct its coreflection along U; see 3.1.1, volume 1. The composite

$$TTS(D) \xrightarrow{\mu_{S(D)}} TS(D) \xrightarrow{\quad \alpha_D \quad} D$$

corresponds by adjunction with a morphism $\sigma\colon TS(D) \longrightarrow S(D)$; we shall prove first that $\big(S(D), \sigma\big)$ is a $\mathbb{T}$-algebra. We refer heavily to 3.1.5, volume 1, without mentioning it all the time; in particular the morphism corresponding with σ by adjunction is $\alpha_D \circ T(\sigma)$, which proves that $\alpha_D \circ T(\sigma) = \alpha_D \circ \mu_{SD}$. The first axiom for a $\mathbb{T}$-algebra is $\sigma \circ \varepsilon_{S(D)} = 1_{S(D)}$; via the adjunction, this is proved by

$$\alpha_D \circ \mu_{S(D)} \circ T\big(\varepsilon_{S(D)}\big) = \alpha_D \circ 1_{TS(D)} = \alpha_D.$$

Next we must verify that $\sigma \circ T(\sigma) = \sigma \circ \mu_{S(D)}$; via the adjunction, this is proved by

$$\begin{aligned}\alpha_D \circ \mu_{S(D)} \circ TT(\sigma) &= \alpha_D \circ T(\sigma) \circ \mu_{TS(D)} = \alpha_D \circ \mu_{S(D)} \circ \mu_{TS(D)} \\ &= \alpha_D \circ \mu_{S(D)} \circ T(\mu_{S(D)}).\end{aligned}$$

Let us prove now that $(S(D), \sigma)$, together with the composite

$$S(D) \xrightarrow{\varepsilon_{S(D)}} TS(D) \xrightarrow{\alpha_D} D,$$

is the coreflection of D along U. Given a $\mathbb{T}$-algebra (C, ξ) and a morphism $f\colon C \longrightarrow D$, we prove that the required factorization is given by the composite

$$C \xrightarrow{\beta_C} ST(C) \xrightarrow{S(\xi)} S(C) \xrightarrow{S(f)} S(D).$$

First of all, this composite is a morphism of $\mathbb{T}$-algebras from (C, ξ) to $(S(D), \sigma)$; this means the equality

$$\sigma \circ TS(f) \circ TS(\xi) \circ T(\beta_C) = S(f) \circ S(\xi) \circ \beta_C \circ \xi,$$

which follows by adjunction from the following relations:

$$\begin{aligned}\alpha_D \circ \mu_{S(D)} &\circ TTS(f) \circ TTS(\xi) \circ TT(\beta_C) \\ &= \alpha_D \circ TS(f) \circ TS(\xi) \circ T(\beta_C) \circ \mu_C \\ &= f \circ \alpha_C \circ TS(\xi) \circ T(\beta_C) \circ \mu_C \\ &= f \circ \xi \circ \alpha_{T(C)} \circ T(\beta_C) \circ \mu_C = f \circ \xi \circ \mu_C \\ &= f \circ \xi \circ T(\xi) = f \circ \xi \circ \alpha_{T(C)} \circ T(\beta_C) \circ T(\xi) \\ &= f \circ \alpha_C \circ TS(\xi) \circ T(\beta_C) \circ T(\xi) \\ &= \alpha_D \circ TS(f) \circ TS(\xi) \circ T(\beta_C) \circ T(\xi).\end{aligned}$$

On the other hand this morphism of $\mathbb{T}$-algebras is a factorization of f through $\alpha_D \circ \varepsilon_{S(D)}$ since

$$\begin{aligned}\alpha_D \circ \varepsilon_{S(D)} \circ S(f) \circ S(\xi) \circ \beta_C &= \alpha_D \circ TS(f) \circ TS(\xi) \circ T(\beta_C) \circ \varepsilon_C \\ &= f \circ \xi \circ \alpha_{T(C)} \circ T(\beta_C) \circ \varepsilon_C \\ &= f \circ \xi \circ \varepsilon_C \\ &= f.\end{aligned}$$

It remains to prove the uniqueness of that factorization. Given another morphism $h\colon (C, \xi) \longrightarrow (S(D), \sigma)$ in $\mathscr{C}^{\mathbb{T}}$ such that $(\alpha_D \circ \varepsilon_{S(D)}) \circ h = f$,

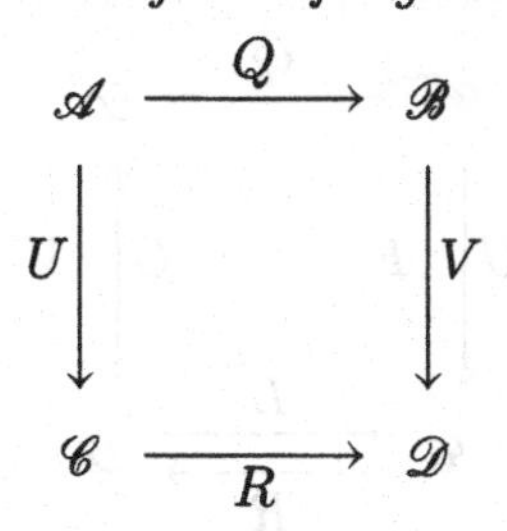

Diagram 4.27

$$\begin{aligned}
S(f)\circ S(\xi)\circ\beta_C &= S(\alpha_D)\circ S\big(\varepsilon_{S(D)}\big)\circ S(h)\circ S(\xi)\circ\beta_C\\
&= S(\alpha_D)\circ S\big(\varepsilon_{S(D)}\big)\circ S(\sigma)\circ ST(h)\circ\beta_C\\
&= S(\alpha_D)\circ ST(\sigma)\circ S\big(\varepsilon_{TS(D)}\big)\circ ST(h)\circ\beta_C\\
&= S(\alpha_D)\circ S\big(\mu_{S(D)}\big)\circ S\big(\varepsilon_{TS(D)}\big)\circ ST(h)\circ\beta_C\\
&= S(\alpha_D)\circ ST(h)\circ\beta_C = S\alpha_D\circ\beta_{S(D)}\circ h\\
&= h.
\end{aligned}$$

Thus U has a right adjoint $G\colon \mathscr{C}\longrightarrow\mathscr{C}^{\mathbb{T}}$ given by $G(D) = \big(S(D),\sigma\big)$ on the objects. But by 3.1.3, volume 1, if $g\colon D\longrightarrow E$ is a morphism in $\mathscr{C}$, $G(g)$ is the unique morphism of $\mathscr{C}^{\mathbb{T}}$ such that the equality

$$\big(\alpha_E\circ\varepsilon_{S(E)}\big)\circ G(g) = g\circ\big(\alpha_D\circ\varepsilon_{S(D)}\big)$$

holds. Observing that

$$\alpha_E\circ\varepsilon_{S(E)}\circ S(g) = \alpha_E\circ TS(g)\circ\varepsilon_{S(D)} = g\circ\alpha_D\circ\varepsilon_{S(D)}$$

we get $G(g) = S(g)$. Therefore $U\circ G = S$, which concludes the proof. □

4.5 The adjoint lifting theorem

The present section is devoted to answering the following question:

> *Consider a commutative square of functors $R\circ U = V\circ Q$ as in diagram 4.27, where U and V are monadic. If R has a left adjoint, does Q have a left adjoint?*

The answer will be yes, provided $\mathscr{A}$ has (enough) coequalizers (see exercise 4.8.5).

First of all, let us fix once for all the situation and the notation of this section. We consider diagram 4.28, made of categories and functors where $\mathbb{T} = (T,\varepsilon,\mu)$ is a monad on $\mathscr{C}$; $\mathbb{S} = (S,\zeta,\eta)$ is a monad on $\mathscr{D}$; U, V

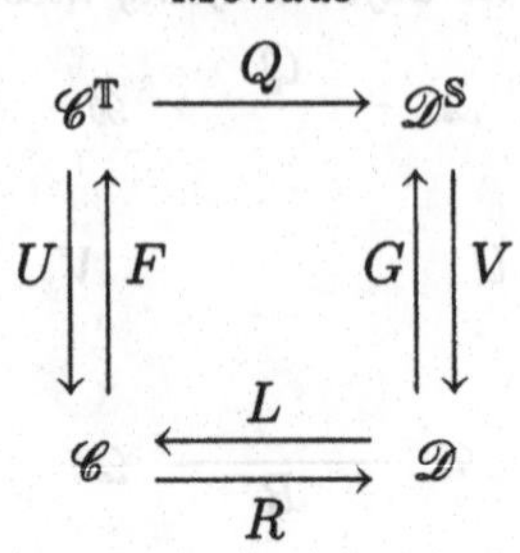

Diagram 4.28

are the corresponding forgetful functors from the categories $\mathscr{C}^{\mathbb{T}}$ and $\mathscr{D}^{\mathbb{S}}$ of algebras; and F, G are the free algebra functors. Choose R and Q such that $R \circ U = V \circ Q$ and L is left adjoint to R, with canonical natural transformations $\alpha\colon 1_{\mathscr{D}} \Rightarrow R \circ L$ and $\beta\colon L \circ R \Rightarrow 1_{\mathscr{C}}$; see 3.1.5, volume 1.

The canonical natural transformations of the adjunction $F \dashv U$ are $\varepsilon\colon 1_{\mathscr{C}} \Rightarrow U \circ F$ and $\tau\colon F \circ U \Rightarrow 1_{\mathscr{C}^{\mathbb{T}}}$, where $\mu = U * \tau * F$ (see 4.2.2); moreover $T = F \circ U$. In the same way the canonical natural transformations of the adjunction $G \dashv V$ are $\zeta\colon 1_{\mathscr{D}} \Rightarrow V \circ G$ and $\sigma\colon G \circ V \Rightarrow 1_{\mathscr{D}^{\mathbb{S}}}$, where $\eta = V * \sigma * G$; moreover $S = G \circ V$.

By 4.1.4 we also know that if (C, ξ) is a $\mathbb{T}$-algebra, then

$$\xi = \tau_{(C,\xi)}\colon F(C) = FU(C,\xi) = (T(C), \mu_C) \longrightarrow (C, \xi).$$

In particular, for every $X \in \mathscr{C}^{\mathbb{T}}$ one has $U(\tau_X) \circ \mu_{U(X)} = U(\tau_X) \circ TU(\tau_X)$, by definition of a $\mathbb{T}$-algebra (see 4.1.2). In the same way if $Y \in \mathscr{D}^{\mathbb{S}}$, one has $V(\sigma_Y) \circ \eta_{V(Y)} = V(\sigma_Y) \circ SV(\sigma_Y)$.

To avoid too heavy a notation, in the rest of this section we shall use notation like $V\sigma G$ to denote what should formally be written $1_V * \sigma * 1_G$; for the same reason, we shall often omit parentheses, thus writing UTX instead of $UT(X)$ or $V\sigma G_C$ instead of $(1_V * \sigma * 1_G)_C$.

Lemma 4.5.1 *In the situation we have just described, there exists a natural transformation $\lambda\colon SR \Rightarrow RT$ which makes diagram 4.29 commutative.*

Proof First of all define a natural transformation $\delta\colon GR \Rightarrow QF$ as the composite

$$GR \xrightarrow{GR\varepsilon} GRT = GRUF = GVQF \xrightarrow{\sigma QF} QF.$$

We define $\lambda\colon SR \Rightarrow RT$ as the composite

$$SR = VGR \xrightarrow{V\delta} VQF = RUF = RT.$$

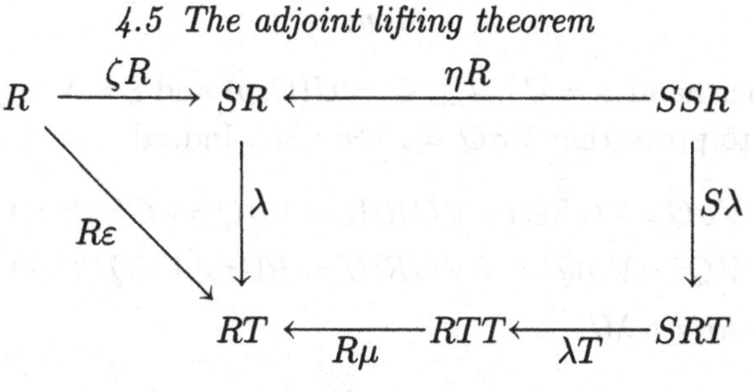

Diagram 4.29

For the left-hand triangle, we have

$$\begin{aligned}\lambda \circ \zeta R &= V\delta \circ \zeta R = V\sigma QF \circ VGR\varepsilon \circ \zeta R\\ &= V\sigma QF \circ \zeta RT \circ R\varepsilon = V\sigma QF \circ \zeta VQF \circ R\varepsilon\\ &= R\varepsilon.\end{aligned}$$

And for the right-hand rectangle, we have

$$\begin{aligned}\lambda \circ \eta R &= V\delta \circ \eta R\\ &= R\mu \circ R\varepsilon T \circ V\delta \circ \eta R = R\mu \circ \lambda T \circ \zeta RT \circ V\delta \circ \eta R\\ &= R\mu \circ \lambda T \circ \zeta RT \circ V\sigma QF \circ VGR\varepsilon \circ \eta R\\ &= R\mu \circ \lambda T \circ \zeta RT \circ V\sigma QF \circ SR\varepsilon \circ \eta R\\ &= R\mu \circ \lambda T \circ \zeta RT \circ V\sigma QF \circ \eta RT \circ SSR\varepsilon\\ &= R\mu \circ \lambda T \circ \zeta RT \circ V\sigma QF \circ \eta VQF \circ SSR\varepsilon\\ &= R\mu \circ \lambda T \circ \zeta RUF \circ V\sigma QF \circ \eta VQF \circ SSR\varepsilon\\ &= R\mu \circ \lambda T \circ \zeta VQF \circ V\sigma QF \circ \eta VQF \circ SSR\varepsilon\\ &= R\mu \circ \lambda T \circ \zeta VQF \circ V\sigma QF \circ SV\sigma QF \circ SSR\varepsilon\\ &= R\mu \circ \lambda T \circ SV\sigma QF \circ SSR\varepsilon\\ &= R\mu \circ \lambda T \circ SV\sigma QF \circ SVGR\varepsilon = R\mu \circ \lambda T \circ SV\delta\\ &= R\mu \circ \lambda T \circ S\lambda.\end{aligned}$$

□

The reader will have observed that lemma 4.5.1 does not depend on the existence of the left adjoint L to R. The same remark applies to the next lemma.

Lemma 4.5.2 *In the situation we have described, if (C, ρ) is a $\mathbb{T}$-algebra, the $\mathbb{S}$-algebra $Q(C, \rho)$ is given by $\big(R(C), R(\rho) \circ \lambda_C\big)$.*

Proof Let us write $Q(C, \rho) = (D, \xi)$. We have

$$D = V(Q), \quad (C, \rho) = RU(C, \rho) = R(C).$$

On the other hand $\rho = U\tau_{(C,\rho)}$, $C = U(C,\rho)$ and $\xi = V\big(\sigma_{Q(C,\rho)}\big)$; thus it remains to prove that $V\sigma Q = RU\tau \circ \lambda U$. Indeed

$$\begin{aligned} V\sigma Q &= V\sigma Q \circ VGRU\tau \circ VGR\varepsilon U = V\sigma Q \circ VGVQ\tau \circ VGR\varepsilon U \\ &= VQ\tau \circ V\sigma QFU \circ VGR\varepsilon U = RU\tau \circ V\sigma QFU \circ VGR\varepsilon U \\ &= RU\tau \circ \lambda U. \end{aligned}$$ □

Lemma 4.5.3 *In the situation we have described, there exist natural transformations $\omega\colon FLS \Rightarrow FL$ and $\varphi\colon G \Rightarrow QFL$ such that $\varphi \circ \sigma G = Q\omega \circ \varphi VG$.*

Proof Keeping the previous notation, we define a natural transformation π as the composite

$$LS \xrightarrow{LS\alpha} LSRL \xrightarrow{L\lambda L} LRTL \xrightarrow{\beta TL} TL$$

and ω is then defined as the composite

$$FLS \xrightarrow{F\pi} FTL = FUFL \xrightarrow{\tau FL} FL.$$

On the other hand φ is the composite

$$G \xrightarrow{G\alpha} GRL \xrightarrow{\delta L} QFL.$$

To prove the stated equality, it suffices to prove that $V\varphi \circ V\sigma G = VQ\omega \circ V\varphi VG$, since V is faithful (see 4.1.4).

$$\begin{aligned} V\varphi \circ V\sigma G &= V\delta L \circ VG\alpha \circ V\sigma G = \lambda L \circ S\alpha \circ \eta \\ &= \lambda L \circ \eta RL \circ SS\alpha = R\mu L \circ \lambda TL \circ S\lambda L \circ SS\alpha \\ &= R\mu L \circ \lambda TL \circ SR\beta TL \circ S\alpha RTL \circ S\lambda L \circ SS\alpha \\ &= R\mu L \circ \lambda TL \circ SR\beta TL \circ SRL\lambda L \circ S\alpha SRL \circ SS\alpha \\ &= R\mu L \circ \lambda TL \circ SR\beta TL \circ SRL\lambda L \circ SRLS\alpha \circ S\alpha S \\ &= R\mu L \circ \lambda TL \circ SR\pi \circ S\alpha S \\ &= RU\tau FL \circ \lambda TL \circ SR\pi \circ S\alpha S \\ &= VQ\tau FL \circ \lambda TL \circ SR\pi \circ S\alpha S \\ &= VQ\tau FL \circ VQF\pi \circ \lambda LVG \circ S\alpha S \\ &= VQ\omega \circ \lambda LVG \circ S\alpha S = VQ\omega \circ V\delta LVG \circ VG\alpha VG \\ &= VG\omega \circ V\varphi VG. \end{aligned}$$ □

The next two lemmas are not necessary for the proof of the adjoint lifting theorem, but they will explain the construction used in that proof.

Lemma 4.5.4 *In the situation described at the beginning of this section, suppose further that Q has a left adjoint K. In these conditions, $\omega \cong K\sigma G$ and $\varphi = \nu G$, where $\nu\colon 1_{\mathscr{D}^{\mathbb{S}}} \Rightarrow Q \circ K$ and $\theta\colon K \circ Q \Rightarrow 1_{\mathscr{C}^{\mathbb{T}}}$ are the canonical natural transformations of the adjunction $K \dashv Q$.*

Proof From $R \circ U = V \circ Q$, we deduce the corresponding relation $F \circ L \cong K \circ G$ between the left adjoints; there is no restriction in assuming – via the axiom of choice – that the left adjoint K has been chosen so that the equality $F \circ L = K \circ G$ holds. It is immediate from 3.2.1, volume 1, that the canonical natural transformations of the composite adjunction $K \circ G \dashv V \circ Q$ are

$$1_{\mathscr{D}} \xrightarrow{\ \zeta\ } VG \xrightarrow{\ V\nu G\ } VQKG, \quad KGVQ \xrightarrow{\ K\sigma Q\ } KQ \xrightarrow{\ \theta\ } 1_{\mathscr{C}^{\mathbb{T}}}$$

while those of the adjunction $F \circ L \dashv R \circ U$ are

$$1_{\mathscr{D}} \xrightarrow{\ \alpha\ } RL \xrightarrow{\ R\varepsilon L\ } RUFL, \quad FLRU \xrightarrow{\ F\beta U\ } FU \xrightarrow{\ \tau\ } 1_{\mathscr{C}^{\mathbb{T}}}.$$

In particular $R\varepsilon L \circ \alpha = V\nu G \circ \zeta$ and $\tau \circ F\beta U = \theta \circ K\sigma Q$. Let us first observe that $V\nu G = \lambda L \circ S\alpha$, i.e. $V\nu G = V\delta \circ VG\alpha$. Via the adjunction $G \dashv V$, we know that $\nu G = \delta \circ G\alpha$ if and only if $V\nu G \circ \zeta = V\delta \circ VG\alpha \circ \zeta$, which is the case since by lemma 4.5.1 and previous relations

$$V\nu G \circ \zeta = R\varepsilon L \circ \alpha = \lambda L \circ \zeta RL \circ \alpha = \lambda L \circ S\alpha \circ \zeta.$$

By 3.1.3, volume 1, $K\sigma G$ is the unique natural transformation such that $QK\sigma G \circ \nu GVG = \nu G \circ \sigma G$. So we must prove that $Q\omega \circ \nu GVG = \nu G \circ \sigma G$, which is equivalent to $VQ\omega \circ V\nu GVG = V\nu G \circ V\sigma G$, since V is faithful. As just observed, this reduces to $VQ\omega \circ \lambda LVG \circ S\alpha VG = \lambda L \circ S\alpha \circ V\sigma G$. And indeed, by lemma 4.5.1,

$$\begin{aligned}
&VQ\omega \circ \lambda LVG \circ S\alpha VG \\
&\quad = RU\omega \circ \lambda LS \circ S\alpha S = RU\tau FL \circ RUF\pi \circ \lambda LS \circ S\alpha S \\
&\quad = R\mu L \circ RUF\pi \circ \lambda LS \circ S\alpha S = R\mu L \circ \lambda TL \circ SR\pi \circ S\alpha S \\
&\quad = R\mu L \circ \lambda TL \circ SR\beta TL \circ SRL\lambda L \circ SRLS\alpha \circ S\alpha S \\
&\quad = R\mu L \circ \lambda TL \circ SR\beta TL \circ SRL\lambda L \circ S\alpha VGRL \circ SS\alpha \\
&\quad = R\mu L \circ \lambda TL \circ SR\beta TL \circ S\alpha RTL \circ S\lambda L \circ SS\alpha \\
&\quad = R\mu L \circ \lambda TL \circ S\lambda L \circ SS\alpha \\
&\quad = \lambda L \circ \eta RL \circ SS\alpha = \lambda L \circ S\alpha \circ \eta \\
&\quad = \lambda L \circ S\alpha \circ V\sigma G.
\end{aligned}$$

Again since V is faithful, it suffices to prove $V\varphi = V\nu G$ to get $\varphi = \nu G$. And indeed

$$V\varphi = V\delta L \circ VG\alpha = \lambda L \circ S\alpha = V\nu G. \qquad \square$$

Lemma 4.5.5 *In the situation described at the beginning of this section, suppose further that Q has a left adjoint K. Under these conditions, given a $\mathbb{S}$-algebra (D,ξ), $K(D,\xi)$ is given in $\mathscr{C}^{\mathbb{T}}$ by a coequalizer diagram of the form*

$$FLS(D) \underset{FL\xi}{\overset{\omega_D}{\rightrightarrows}} FL(D) \longrightarrow K(D,\xi).$$

Proof In $\mathscr{D}^{\mathbb{S}}$, the coequalizer diagram of lemma 4.3.3 can be written, with our notation, as

$$GS(D) \underset{G\xi}{\overset{\sigma G_D}{\rightrightarrows}} G(D) \xrightarrow{\sigma_{(D,\xi)}} (D,\xi).$$

Now K preserves this coequalizer, since it has a right adjoint Q. Moreover $K \circ G = F \circ L$ since $R \circ U = V \circ Q$ and $K\sigma G_D = \omega_D$ by lemma 4.5.4. Therefore we get the stated coequalizer in $\mathscr{C}^{\mathbb{T}}$. $\square$

We are ready to prove the adjoint lifting theorem.

Theorem 4.5.6 *Let $R \circ U = V \circ Q$ be a commutative diagram of functors, where U and V are monadic, as in diagram 4.27. If $\mathscr{A}$ has coequalizers, then Q has a left adjoint as soon as R has a left adjoint.*

Proof We use the notation defined at the beginning of this section and in lemmas 4.5.1 and 4.5.3. Given a $\mathbb{S}$-algebra (D,ξ), we define a $\mathbb{T}$-algebra X by the following coequalizer in $\mathscr{C}^{\mathbb{T}}$:

$$FLS(D) \underset{FL\xi}{\overset{\omega_D}{\rightrightarrows}} FL(D) \xrightarrow{\;x\;} X;$$

compare with 4.5.5. Moreover by lemma 4.5.3, the left-hand part of diagram 4.30 is commutative while the top row is a coequalizer diagram by 4.3.3. This implies the existence of a unique factorization χ such that $Q(x) \circ \varphi_D = \chi \circ \xi$. We shall prove that X, together with $\chi\colon (D,\xi) \longrightarrow QX$, is the reflection of (D,ξ) along Q.

Choose a $\mathbb{T}$-algebra (C,ρ) and a morphism $y\colon (D,\xi) \longrightarrow Q(C,\rho)$ of $\mathbb{T}$-algebras. By 4.5.2, we know already that $Q(C,\rho) = \big(R(C), R(\rho) \circ \lambda_C\big)$, thus in particular we get a morphism $V(y)\colon D \longrightarrow R(C)$ such that the relation $R(\rho) \circ \lambda_C \circ SV(y) = V(y) \circ \xi$ holds. Let us then consider the composite in $\mathscr{C}^{\mathbb{T}}$,

$$FL(D) \xrightarrow{FLV(y)} FLR(C) \xrightarrow{F(\beta_C)} F(C) \xrightarrow{\;\rho\;} (C,\rho),$$

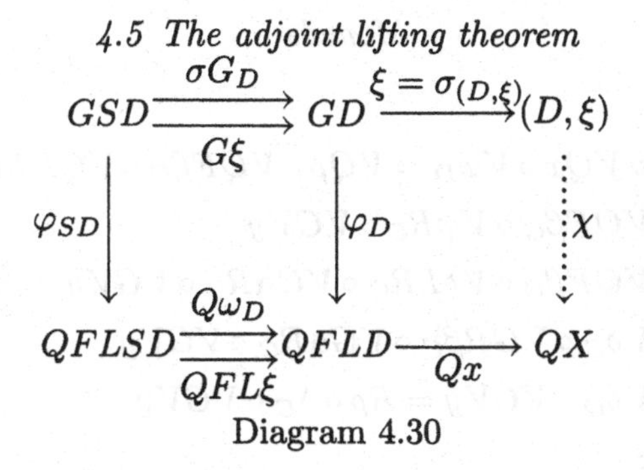

Diagram 4.30

and prove it coequalizes ω_D and $FL\xi$. Since U is faithful, it suffices to prove the equality

$$\rho \circ UF\beta_C \circ UFLVy \circ U\omega_D = \rho \circ UF\beta_C \circ UFLVy \circ UFL\xi.$$

Indeed

$$\begin{aligned}
&\rho \circ UF\beta_C \circ UFLVy \circ U\omega_D \\
&= \rho \circ UF\beta_C \circ U\omega R_C \circ UFLSVy \\
&= \rho \circ UF\beta_C \circ U\tau FLR_C \circ UF\pi R_C \circ UFLSVy \\
&= \rho \circ T\beta_C \circ \mu LR_C \circ UF\pi R_C \circ UFLSVy \\
&= \rho \circ \mu_C \circ TT\beta_C \circ UF\pi R_C \circ UFLSVy \\
&= \rho \circ \mu_C \circ TT\beta_C \circ T\beta TLR_C \circ UFL\lambda LR_C \circ UFLS\alpha R_C \circ UFLSVy \\
&= \rho \circ \mu_C \circ T\beta T_C \circ TLRT\beta_C \circ UFL\lambda LR_C \circ UFLS\alpha R_C \circ UFLSVy \\
&= \rho \circ \mu_C \circ T\beta T_C \circ TL\lambda_C \circ TLSR\beta_C \circ UFLS\alpha R_C \circ UFLSVy \\
&= \rho \circ \mu_C \circ T\beta T_C \circ TL\lambda_C \circ UFLSVy \\
&= \rho \circ T\rho \circ T\beta T_C \circ TL\lambda_C \circ UFLSVy \\
&= \rho \circ T\beta_C \circ TLR\rho \circ TL\lambda_C \circ UFLSVy \\
&= \rho \circ T\beta_C \circ TLVy \circ TL\xi \\
&= \rho \circ UF\beta_C \circ UFLVy \circ UFL\xi.
\end{aligned}$$

By this coequalizing property and the definition of X, we get a unique morphism $z\colon X \longrightarrow (C, \rho)$ in $\mathscr{C}^{\mathbb{T}}$ such that $z \circ x = \rho \circ F\beta_C \circ FLVy$.

We must prove that z is the required factorization, i.e. the unique morphism of $\mathbb{T}$-algebras such that $Qz \circ \chi = y$. Since $\xi\colon GD \longrightarrow (D, \xi)$ is a coequalizer in $\mathscr{D}^{\mathbb{S}}$, this is equivalent to $Qz \circ \chi \circ \xi = y \circ \xi$. And since V is faithful this is further equivalent to $VQz \circ V\chi \circ \xi = Vy \circ \xi$. And in fact

$$
\begin{aligned}
&VQz \circ V\chi \circ \xi \\
&\quad = VQz \circ VQx \circ V\varphi_D = VQ\rho \circ VQF\beta_C \circ VQFLVy \circ V\varphi_D \\
&\quad = R\rho \circ VQF\beta_C \circ V\varphi R_C \circ VGVy \\
&\quad = R\rho \circ VQF\beta_C \circ V\delta LR_C \circ VG\alpha R_C \circ VGVy \\
&\quad = R\rho \circ V\delta_C \circ VGR\beta_C \circ VG\alpha R_C \circ VGVy \\
&\quad = R\rho \circ V\delta_C \circ VGVy = R\rho \circ \lambda_C \circ VGVy \\
&\quad = Vy \circ \xi.
\end{aligned}
$$

It remains to check the uniqueness of the factorization. Given another morphism $w\colon X \longrightarrow (C,\rho)$ of $\mathbb{T}$-algebras such that $Qw \circ \chi = y$, we must prove that $z = w$ or, equivalently, $z \circ x = w \circ x$ since x is a coequalizer. Since ξ has a section ε_D, this is also equivalent to proving $z \circ x \circ FL\xi = w \circ x \circ FL\xi$ or, by faithfulness of U, to $Uz \circ Uy \circ UFL\xi = Uw \circ Ux \circ UFL\xi$. Indeed, writing $X = (UX, \tau_X)$, $w\colon (UX,\tau_X) \longrightarrow (C,\rho)$ is a morphism of $\mathbb{T}$-algebras and one has

$$
\begin{aligned}
&Uz \circ Ux \circ UFL\xi \\
&\quad = \rho \circ UF\beta_C \circ UFLVy \circ UFL\xi \\
&\quad = \rho \circ UF\beta_C \circ UFLVQw \circ UFLV\chi \circ UFL\xi \\
&\quad = \rho \circ UF\beta_C \circ UFLRUw \circ UFLV\chi \circ UFL\xi \\
&\quad = \rho \circ UFUw \circ UF\beta UX \circ UFLV\chi \circ UFL\xi \\
&\quad = \rho \circ UFUw \circ UF\beta UX \circ TLVQx \circ TLV\varphi_D \\
&\quad = \rho \circ UFUw \circ TUx \circ T\beta TL_D \circ TLV\varphi_D \\
&\quad = \rho \circ UFUw \circ TUx \circ T\beta TL_D \circ UFLV\delta L_D \circ UFLVG\alpha_D \\
&\quad = \rho \circ UFUw \circ TUx \circ T\beta TL_D \circ UFL\lambda L_D \circ UFLVG\alpha_D \\
&\quad = \rho \circ TUw \circ TUx \circ UF\pi_D = Uw \circ U\tau X \circ TUx \circ UF\pi_D \\
&\quad = Uw \circ Ux \circ U\tau FL_D \circ UF\pi_D = Uw \circ Ux \circ U\omega_D \\
&\quad = Uw \circ Ux \circ UFL\xi. \qquad \square
\end{aligned}
$$

Corollary 4.5.7 *Let $U = V \circ Q$ be a commutative triangle of functors, where U and V are monadic, as in diagram 4.31. If $\mathscr{A}$ has coequalizers, Q is monadic as well.*

Proof In 4.5.6 put $R = 1_{\mathscr{C}}$, which yields the existence of a left adjoint to Q. It is now easy to get the conclusion by applying 4.4.4. First, Q reflects isomorphisms, because V preserves them and U reflects them. Now if a pair $u, v\colon X \rightrightarrows Y$ is such that $Q(u), Q(v)$ has a split coequalizer $p\colon Q(Y) \longrightarrow P$ in $\mathscr{B}$, then $\big(U(u), U(v)\big) = \big(VQ(u), VQ(v)\big)$ has a

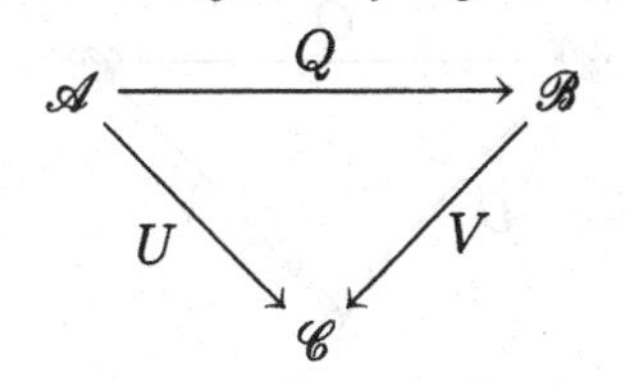

Diagram 4.31

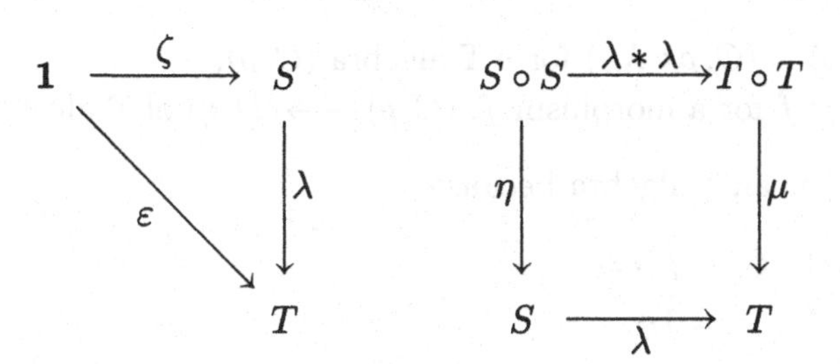

Diagram 4.32

split coequalizer in $\mathscr{C}$ and thus (u, v) has a coequalizer $z\colon Y \longrightarrow Z$ preserved by U. If $w\colon P \longrightarrow Q(Z)$ is the unique factorization such that $w \circ p = Q(z)$, then both $U(z) = VQ(z)$ and $V(p)$ are coequalizers of $\big(U(u), U(v)\big)$ in $\mathscr{C}$; therefore $U(w)$ is an isomorphism and thus w is an isomorphism. Thus the coequalizer z of (u, v) is preserved by Q. □

The previous result, via lemma 4.5.1, is in close relation with the notion of a "morphism of monads".

Definition 4.5.8 *Let $\mathbb{T} = (T, \varepsilon, \mu)$ and $\mathbb{S} = (S, \zeta, \eta)$ be two monads on a same category $\mathscr{C}$. By a morphism of monads $\mathbb{S} \longrightarrow \mathbb{T}$, we mean a natural transformation $\lambda\colon S \Rightarrow T$ such that $\lambda \circ \zeta = \varepsilon$, $\mu \circ (\lambda \circ \lambda) = \lambda \circ \eta$; see diagram 4.32.*

Proposition 4.5.9 *Let $\mathbb{T} = (T, \varepsilon, \mu)$ and $\mathbb{S} = (S, \zeta, \eta)$ be two monads on a same category $\mathscr{C}$. There is a bijection between:*

(1) the morphisms of monads $\lambda\colon \mathbb{S} \longrightarrow \mathbb{T}$;

(2) the functors $Q\colon \mathscr{C}^{\mathbb{T}} \longrightarrow \mathscr{C}^{\mathbb{S}}$ such that $V \circ Q = U$, where U and V are the corresponding forgetful functors (see diagram 4.33).

Proof By lemma 4.5.1 (putting $R = 1_{\mathscr{C}}$), every functor Q such that $VQ = U$ induces a natural transformation $\lambda\colon S \Rightarrow T$ with the required properties. Indeed $(\lambda * \lambda)_C = \lambda_{T(C)} \circ S(\lambda_C) = T(\lambda_C) \circ \lambda_{S(C)}$, by 1.3.4, volume 1.

Conversely given a natural transformation $\lambda\colon S \Rightarrow T$ as in 4.5.8, we define a functor $Q\colon \mathscr{C}^{\mathbb{T}} \longrightarrow \mathscr{C}^{\mathbb{S}}$ by (see lemma 4.5.2)

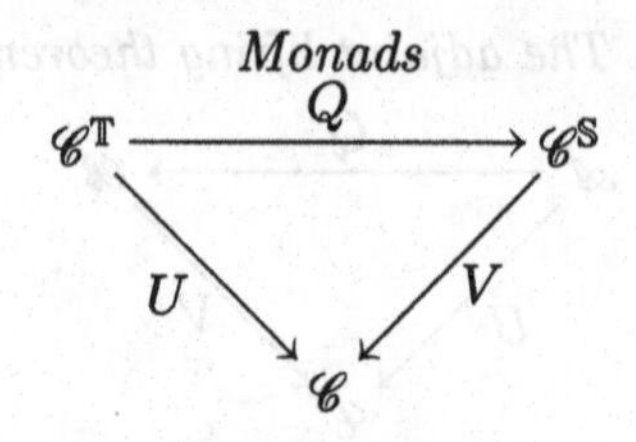

Diagram 4.33

- $Q(C,\rho) = (C, \rho \circ \lambda_C)$ for a $\mathbb{T}$-algebra (C,ρ),
- $Q(f) = f$ for a morphism $f\colon (C,\rho) \longrightarrow (D,\xi)$ of $\mathbb{T}$-algebras.

$(C, \rho \circ \lambda_C)$ is an $\mathbb{S}$-algebra because

$$\begin{aligned}(\rho \circ \lambda_C) \circ \zeta_C &= \rho \circ \varepsilon_C \\ &= 1_C, \\ (\rho \circ \lambda_C) \circ \eta_C &= \rho \circ \mu_C \circ (\lambda * \lambda)_C = \rho \circ \mu_C \circ \lambda_{T(C)} \circ S\lambda_C \\ &= \rho \circ T(\rho) \circ \lambda_{T(C)} \circ S(\lambda_C) = \rho \circ \lambda_C \circ S(\rho) \circ S(\lambda_C) \\ &= (\rho \circ \lambda_C) \circ S(\rho \circ \lambda_C),\end{aligned}$$

and $Q(f)$ is a morphism of $\mathbb{S}$-algebras because

$$f \circ (\rho \circ \lambda_C) = \xi \circ T(f) \circ \lambda_C = (\xi \circ \lambda_D) \circ S(f).$$

On the other hand, the functoriality of Q is obvious as well as the relation $V \circ Q = U$.

It remains to prove that the two constructions are mutually inverse. Let us start with λ and construct Q. With the notation of 4.5.1 we have $F(C) = (T(C), \mu_C)$ and thus $\sigma QF_C = \mu_C \circ \lambda_{TC}$. Therefore the natural transformation $\overline{\lambda}\colon S \Rightarrow T$ constructed from Q as in 4.5.1 is such that

$$\begin{aligned}\overline{\lambda}_C &= V\sigma QF_C \circ VG\varepsilon_C \\ &= \mu_C \circ \lambda_{TC} \circ S\varepsilon_C = \mu_C \circ T\varepsilon_C \circ \lambda_C \\ &= \lambda_C.\end{aligned}$$

Conversely start from Q such that $VQ = U$ and construct λ as in 4.5.1. Construct now $\overline{Q}$ from λ as at the beginning of the proof, thus $\overline{Q}(C,\rho) = (C, \rho \circ \lambda_C)$. By lemma 4.5.2, $\overline{Q}(C,\rho) = Q(C,\rho)$. The case of morphisms is obvious. □

Corollary 4.5.10 *Let $\mathbb{T}$ and $\mathbb{S}$ be two monads on the category of sets. There are bijections between:*

(1) the morphisms of monads $\mathbb{S} \longrightarrow \mathbb{T}$;
(2) the functors $Q\colon \mathsf{Set}^{\mathbb{T}} \longrightarrow \mathsf{Set}^{\mathbb{S}}$ such that $V \circ Q = U$;

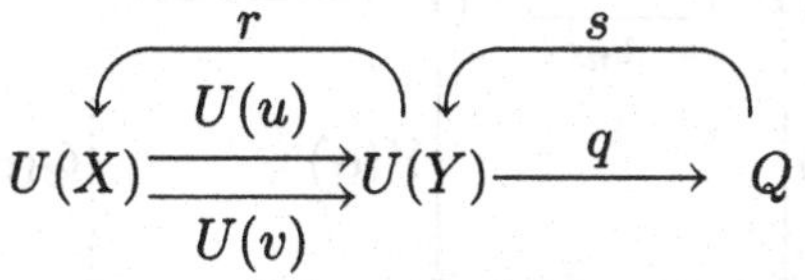

$$\begin{array}{c} U(X) \overset{U(u)}{\underset{U(v)}{\rightrightarrows}} U(Y) \xrightarrow{q} Q \end{array}$$

with $r: U(Y) \to U(X)$ and $s: Q \to U(Y)$.

Diagram 4.34

(3) the monadic functors Q: $\mathsf{Set}^{\mathbb{T}} \longrightarrow \mathsf{Set}^{\mathbb{S}}$ such that $V \circ Q = U$;
where U: $\mathsf{Set}^{\mathbb{T}} \longrightarrow \mathsf{Set}$ and V: $\mathsf{Set}^{\mathbb{S}} \longrightarrow \mathsf{Set}$ are the corresponding forgetful functors.

Proof The result follows from 4.3.5, 4.5.9 and 4.5.6. □

4.6 Monads with rank

Let us first introduce some classical terminology.

Definition 4.6.1 *A monad $\mathbb{T} = (T, \varepsilon, \mu)$ on a category $\mathscr{C}$ has rank α, for some regular cardinal α, when the functor T: $\mathscr{C} \longrightarrow \mathscr{C}$ preserves α-filtered colimits. When $\alpha = \aleph_0$, thus when T preserves filtered colimits, one also says that T has finite rank.*

Proposition 4.6.2 *There is a coincidence between;*
(1) the categories $\mathsf{Mod}_{\mathcal{T}}$ of models of an algebraic theory $\mathcal{T}$, in the sense of 3.3.1;
(2) the categories of $\mathbb{T}$-algebras, where $\mathbb{T}$ is a monad with finite rank on the category on sets.

Proof Let $\mathcal{T}$ be an algebraic theory in the sense of 3.3.1. By 3.9.1, the forgetful functor U: $\mathsf{Mod}_{\mathcal{T}} \longrightarrow \mathsf{Set}$ has a left adjoint and reflects isomorphisms. In order to apply 4.4.4, let us also consider a pair of morphisms u, v: $X \rightrightarrows Y$ in $\mathsf{Mod}_{\mathcal{T}}$ such that $(U(u), U(v))$ admits a split coequalizer as in diagram 4.34:

$$q \circ U(u) = q \circ U(v), \quad U(u) \circ r = 1_{U(Y)}, \quad q \circ s = 1_Q, \quad U(v) \circ r = s \circ q.$$

Raising this diagram to powers $n, m \in \mathbb{N}$ yields other coequalizer diagrams (see 2.10.2, volume 1) so that every arrow $\alpha \in \mathcal{T}(T^n, T^m)$ yields a corresponding factorization $Q(\alpha)$: $Q^n \longrightarrow Q^m$ as in diagram 4.35, making Q a $\mathcal{T}$-model and q: $Y \longrightarrow Q$ a morphism of $\mathcal{T}$-models. By construction $q \circ u = q \circ v$; on the other hand, if w: $Y \longrightarrow Z$ in $\mathsf{Mod}_{\mathcal{T}}$ is such that $w \circ u = w \circ v$, we get in the category of sets a unique map t: $Q \longrightarrow U(Z)$ such that $t \circ q = w$. To prove that t is a $\mathcal{T}$-homomorphism,

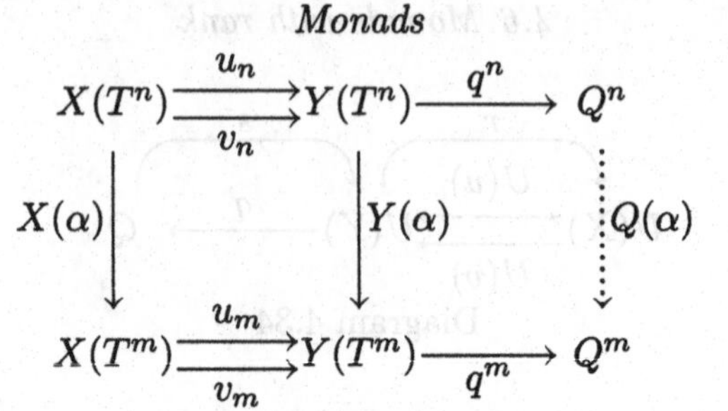

Diagram 4.35

we must check that $Z(\alpha) \circ t^n = t^m \circ Q(\alpha)$ for every $\alpha \in \mathcal{T}(T^n, T^m)$. Since q^n is a surjection, this follows from

$$Z(\alpha) \circ t^n \circ q^n = Z(\alpha) \circ w^n = w^m \circ Y(\alpha) = t^m \circ q^m \circ Y(\alpha) = t^m \circ Q(\alpha) \circ q^n.$$

Thus $q\colon Y \longrightarrow Q$ is the coequalizer of (u, v) in $\mathsf{Mod}_{\mathcal{T}}$ and by 4.4.4, $U\colon \mathsf{Mod}_{\mathcal{T}} \longrightarrow \mathsf{Set}$ is monadic. If F is left adjoint to U, the corresponding monad admits UF as functor part and UF preserves filtered colimits (see 3.9.1).

Conversely, let us suppose that $\mathbb{T}$ is a monad with finite rank over Set. Using 4.1.4, 4.3.2 and 4.3.5, we can conclude the proof by 3.9.1. □

Corollary 4.6.3 *Every algebraic functor, in the sense of 3.7.4, is monadic and the corresponding monad has finite rank.*

Proof By 3.9.2 and 4.5.7, every algebraic functor is monadic. By 3.7.5, the corresponding monad has finite rank, since an algebraic functor preserves filtered colimits while its left adjoint preserves all colimits (see 3.2.2, volume 1). □

Counterexample 4.6.4
The category of abelian groups is monadic over the category of sets (see 4.6.2). On the other hand the category of torsion free abelian groups is a localization of the category of abelian groups (see section 1.12 and theorem 1.13.5), thus is monadic over the category of abelian groups (see 4.2.4). But the composite functor

$$\text{Torsion free abelian groups} \subseteq \text{Abelian groups} \xrightarrow{\;U\;} \mathsf{Set}$$

is not monadic. Indeed the left adjoint to U is the free abelian group functor ... and every free abelian group is certainly torsion free! Thus the composite functor again admits the free abelian group functor as a left adjoint and the corresponding monad therefore admits the category of abelian groups as category of algebras. Therefore the composite functor is not monadic (see 4.4.1 and 4.2.2).

Replacing $\aleph_0$ by an arbitrary regular cardinal α, one can consider the α-algebraic theories $\mathcal{T}$ defined as categories with objects $T^n (n < \alpha)$, where T^n is the n-th power of T^1; a $\mathcal{T}$-model is then a functor $\mathcal{T} \longrightarrow \mathsf{Set}$ which preserves α-products. Proposition 4.6.3 generalizes easily to give a coincidence between the α-algebraic categories and the categories of $\mathbb{T}$-algebras for a monad $\mathbb{T}$ of rank α on the category of sets.

But not every monad on the category of sets has a rank. Here is a classical and easy counterexample

Proposition 4.6.5 *Let $\mathscr{C}$ be the category of $\bigvee$-lattices, i.e. the objects of $\mathscr{C}$ are the complete lattices and the morphisms are the mappings preserving arbitrary suprema. The forgetful functor is monadic over the category of sets but the corresponding monad does not have a rank.*

Proof Consider the covariant power set functor T: $\mathsf{Set} \longrightarrow \mathsf{Set}$ mapping a set X to its power set $\mathcal{P}(X)$ and a mapping f: $X \longrightarrow Y$ to the direct image mapping

$$\mathcal{P}(f)\colon \mathcal{P}(X) \longrightarrow \mathcal{P}(Y), \quad Z \mapsto f(Z).$$

One also defines ε: $1_{\mathsf{Set}} \Rightarrow T$ by

$$\varepsilon_X\colon X \longrightarrow \mathcal{P}(X), \quad x \mapsto \{x\}$$

and μ: $T \circ T \Rightarrow T$ by

$$\mu_X\colon \mathcal{P}(\mathcal{P}(X)) \longrightarrow \mathcal{P}(X), \quad (Z_i)_{i\in I} \mapsto \bigcup_{i\in I} Z_i.$$

It is easy to check that $\mathbb{T} = (T, \varepsilon, \mu)$ is a monad.

Let us now define a comparison functor J: $\mathscr{C} \longrightarrow \mathsf{Set}^{\mathbb{T}}$ just by:

- $J(C) = (C, \bigvee)$, where $\bigvee$: $T(C) \longrightarrow C$ maps a subset $Z \subseteq C$ to its supremum;
- $J(f) = f$ if f: $C \longrightarrow D$ is a morphism of $\mathscr{C}$.

The two relations $\bigvee\{x\} = x$ for $x \in C$ and

$$\bigvee\Big\{\bigvee Z_i \Big| \, Z_i \subseteq C, \; i \in I\Big\} = \bigvee\Big(\bigcup_{i\in I} Z_i\Big)$$

indicate that $J(C)$ is a $\mathbb{T}$-algebra. On the other hand $J(f)$ is a morphism of $\mathbb{T}$-algebras just because f preserves suprema.

More generally, observe that a mapping $g: J(C) \longrightarrow J(D)$ is a morphism of $\mathbb{T}$-algebras precisely when g preserves suprema; thus J is full and faithful.

Finally given a $\mathbb{T}$-algebra (X, ξ), define $x \leq y$ when $\xi(\{x, y\}) = y$. From $\xi(\{x, x\}) = \xi(\{x\}) = (\xi \circ \varepsilon_X)(x) = x$, we get $x = x$. If $x \leq y$ and $y \leq x$, we have $\xi(\{x, y\}) = y$ and $\xi(\{y, x\}) = x$, thus $x = y$. Next if $x \leq y$ and $y \leq z$, the relations $\xi(\{x, y\}) = y$ and $\xi(\{y, z\}) = z$ imply

$$\begin{aligned}\xi\{x, z\} &= \xi\{\xi\{x, x\}, \xi\{y, z\}\} = (\xi \circ T(\xi))\{\{x, x\}, \{y, z\}\} \\ &= \xi \circ \mu_X\{\{x, x\}, \{y, z\}\} = \xi\{x, y, z\} \\ &= (\xi \circ \mu_X)\{\{x, y\}, \{y, z\}\} = (\xi \circ T(\xi))\{\{x, y\}, \{y, z\}\} \\ &= \xi\{\xi\{x, y\}, \xi\{y, z\}\} = \xi\{y, z\} \\ &= z;\end{aligned}$$

thus $x \leq z$. So (X, ξ) is already a poset. Now $\xi: T(X) \longrightarrow X$ is just the supremum operation for this poset structure. Indeed, given $Z \subseteq X$ and $x \in Z$,

$$\begin{aligned}\xi\{x, \xi(Z)\} &= \xi\{\xi\{x\}, \xi(Z)\} = (\xi \circ T(\xi))\{\{x\}, Z\} \\ &= (\xi \circ \mu_X)\{\{x\}, Z\} = \xi(\{x\} \cup Z) \\ &= \xi(Z);\end{aligned}$$

thus $x \leq \xi(Z)$. Next, if $y \in X$ is such that $x \leq y$ for every $x \in Z$,

$$\begin{aligned}\xi\{\xi(Z), y\} &= \xi\{\xi(Z), \xi\{y\}\} = (\xi \circ T(\xi))\{Z, \{y\}\} \\ &= (\xi \circ \mu_X)\{Z, \{y\}\} = \xi(Z \cup \{y\}) \\ &= \xi\left(\bigcup_{x \in Z}\{x, y\}\right) = (\xi \circ \mu_X)\left(\bigcup_{x \in Z}\{x, y\}\right) \\ &= (\xi \circ T(\xi))\left(\bigcup_{x \in Z}\{x, y\}\right) = \xi\{\xi\{x, y\} | x \in Z\} \\ &= \xi\{y | x \in Z\} = \xi\{y\} \\ &= y;\end{aligned}$$

hence $\xi(Z) \leq y$. We have thus checked that $\xi(Z)$ is just the supremum of Z, which indicates that (X, ξ) is just $J(X, \leq)$. This concludes the proof that $\mathscr{C}$ is equivalent to $\mathsf{Set}^{\mathbb{T}}$.

Now let us observe that the functor T: Set $\longrightarrow$ Set does not preserve α-filtered colimits, whatever the regular cardinal α. Just choose a set X whose cardinal is at least α and write X as the union of all its subsets Y with cardinality strictly less than α; this is an α-filtered union, because α is regular, thus an α-filtered colimit in the category of sets (see 4.13.3).

Applying T to that colimit does not yield a α-filtered colimit. Indeed for every inclusion $i\colon Y \longrightarrow X$ in the original diagram and every $Z \in T(Y)$, $Ti(Z)$ has cardinality strictly less than α. In particular $X \in T(X)$ cannot be presented as the equivalence class in the colimit of an element $Ti(Z)$, which proves that the second diagram, which is still α-filtered, does not admit $T(Z)$ as a colimit (see again 4.13.3 and the construction of the α-filtered colimit $\mathrm{colim}_Y T(Y)$ in 6.4, volume 1). □

Clearly it would have been faster to apply 4.4.4 to prove the monadicity of the category of $\bigvee$-lattices; we found it more instructive to give an example of explicit computations with a monad.

Comparing 4.6.5 and 4.6.2, we can say that the theory of $\bigvee$-lattices has operations of arbitrarily large arity: for every cardinal β and every β-family of elements, there must exist a "composite" of that family and the axioms force this composite to be exactly the supremum of the family for a poset structure. This should be compared with examples 3.3.5.c, d, e, f.

With the previous remark in mind, one might now wonder if a theory which can be described via operations of arity β, for every cardinal β, and axioms expressed by equalities, is necessarily induced by a monad on Set. The answer is no (see exercise 4.8.9).

Let us conclude this section with another interesting example of a monadic category over Set (in fact, a monad without rank).

Proposition 4.6.6 *The category of compact Hausdorff spaces is monadic over sets.*

Proof Write Comp for the category of compact Hausdorff spaces and $U\colon \mathsf{Comp} \longrightarrow \mathsf{Set}$ for the underlying set functor. We apply 4.4.4 to prove that U is monadic.

In 3.3.9.c, volume 1, we proved that the inclusion $\mathsf{Comp} \subsetneq \mathsf{Top}$ of Comp in the category of topological spaces has a left adjoint (the Stone–Čech compactification). In 3.1.6.j, volume 1, we proved that the underlying set functor $\mathsf{Top} \longrightarrow \mathsf{Set}$ has a left adjoint (the discrete topology functor). By 3.2.1, volume 1, our functor $U\colon \mathsf{Comp} \longrightarrow \mathsf{Set}$ thus has a left adjoint.

A continuous bijection between compact Hausdorff spaces is automatically a homeomorphism; therefore U reflects isomorphisms.

Finally consider two continuous mappings $u, v\colon X \rightrightarrows Y$ with X, Y compact Hausdorff spaces and suppose they have a split coequalizer in the category of sets, as in diagram 4.36:

$$q \circ s = 1_Q, \quad u \circ r = 1_Y, \quad q \circ u = q \circ v, \quad v \circ r = s \circ q.$$

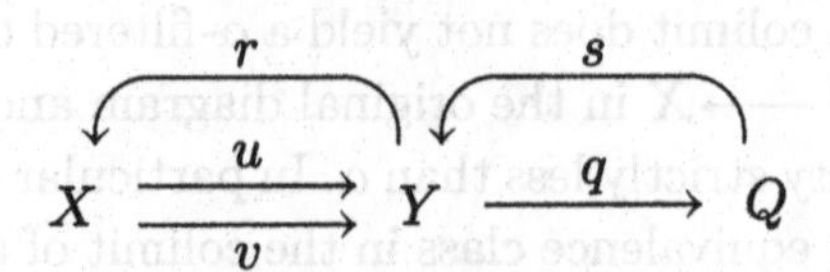

Diagram 4.36

Let us provide Q with the quotient topology, induced by the topology of Y. If we prove that Q is Hausdorff, Q will be compact Hausdorff as continuous image of the compact Hausdorff space Y and q will be the coequalizer of (u, v) in **Comp** by 2.4.6.e, volume 1.

This is equivalent to proving that the kernel pair of q is a closed equivalence relation in $Y \times Y$; see 2.4.6.f, volume 1.

Write R for the relation on Y defined by

$$R = \left\{ (u(x), v(x)) \,\middle|\, x \in X \right\};$$

it is the image of X under the continuous mapping $\begin{pmatrix} u \\ v \end{pmatrix}: X \longrightarrow Y \times Y$, thus is compact, since X is compact, and therefore closed, since Y is Hausdorff.

Now choose $y \in Y$. One has immediately

$$(y, sq(y)) = (ur(y), vr(y)) \in R.$$

Therefore if $y, y' \in Y$ with $q(y) = q(y')$

$$(y, sq(y)) \in R \text{ and } (sq(y'), y') \in R^0 \Rightarrow (y', y) \in R^0 \circ R$$

where R^0 is the inverse relation of R (see 2.8.2) and $R \circ R^0$ is the composite of the relations R, R^0, as defined in 2.8.3. Conversely given a pair $(y', y) \in R^0 \circ R$, there exists $y'' \in Y$ such that $(y, y'') \in R$ and $(y'', y') \in R$, thus $q(y) = q(y'') = q(y')$ since $q \circ u = q \circ v$. This proves that $R^0 \circ R$ is the kernel pair of q.

By 2.8.3, the composite relation $R^0 \circ R$ is given in **Set** by diagram 4.37 where the square is a pullback and the triangle is an image factorization. Since U: **Comp** $\longrightarrow$ **Set** preserves limits (see 3.2.2, volume 1), the pullback can equivalently be computed in **Comp**, but also the product $Y \times Y$. Since a continuous image of a compact space is compact, $R^0 \circ R$ is compact in the Hausdorff space $Y \times Y$, thus is closed. □

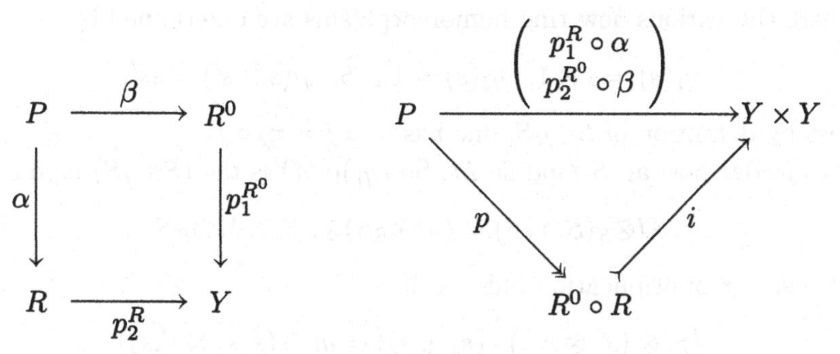

Diagram 4.37

4.7 A glance at descent theory

If $U\colon \mathscr{A} \longrightarrow \mathscr{B}$ is a monadic functor with left adjoint functor $F\colon \mathscr{B} \longrightarrow \mathscr{A}$ (see 4.1.4), descent theory along U is intended to characterize the objects of the form $F(B)$ as objects of $\mathscr{A}$ provided with a convenient structure, and analogously for morphisms. We recall here some basic facts about the classical case of modules over a ring and show how the "descent data" can be elegantly described using the comonad on $\mathscr{A}$ associated with the adjunction $F \dashv U$. We assume some familiarity with classical module theory.

Throughout this section, we fix a homomorphism $f\colon R \longrightarrow S$ of commutative rings with unit. We consider the corresponding algebraic functor (see 3.9.3)

$$f^*\colon \mathsf{Mod}_S \longrightarrow \mathsf{Mod}_R,\ \ M \mapsto M$$

where scalar multiplication on $M \in \mathsf{Mod}_R$ is given by $r \cdot m = f(r) \cdot m$ for $r \in R$, $m \in M$. It is well known that this functor admits as a left adjoint

$$f_!\colon \mathsf{Mod}_R \longrightarrow \mathsf{Mod}_S,\ \ N \mapsto N \otimes_R S$$

where, in the tensor product, S is provided with the R-module structure given by $s \cdot r = s \cdot f(r)$ for $s \in S$, $r \in R$.

Let us also consider the ring $S \otimes_R S$, where the multiplication is induced by

$$(s_1 \otimes s_2)(s_1' \otimes s_2') = s_1 s_1' \otimes s_2 s_2'.$$

This yields the following diagram in the category of commutative rings with unit:

$$R \xrightarrow{\ f\ } S \overset{\eta_1}{\underset{\eta_2}{\rightrightarrows}} S \otimes_R S \xrightarrow{\ \mu\ } S,$$

where the various new ring homomorphisms are determined by

$$\eta_1(s) = s \otimes 1, \ \ \eta_2(s) = 1 \otimes S, \ \ \mu(s \otimes s') = ss'.$$

Just by definition of $S \otimes_R S$, one has $\eta_1 \circ f = \eta_2 \circ f$.

Consider now an S-module M. So $(\eta_1)_!(M)$ is the $(S \otimes_R S)$-module

$$M \otimes_S (S \otimes_R S) \cong (M \otimes_S S) \otimes_R S \cong M \otimes_R S$$

with scalar multiplication induced by

$$\big(m \otimes (s_1' \otimes s_2')\big) \cdot (s_1 \otimes s_2) = m \otimes (s_1' s_1 \otimes s_2' s_2)$$

together with the requirement that

$$\begin{aligned} m \otimes (s \otimes s') &= m \otimes \big((s \otimes 1)(1 \otimes s')\big) = m \otimes \big(\eta_1(s) \cdot (1 \otimes s')\big) \\ &= ms \otimes (1 \otimes s'). \end{aligned}$$

So, viewing $(\eta_1)_!(M)$ as $M \otimes_R S$, scalar multiplication is determined by

$$(m \otimes s)(s_1 \otimes s_2) = ms_1 \otimes ss_2.$$

We shall often write $M \otimes_R S$ instead of $(\eta_1)_!(M)$.

In an analogous way we can consider, for an S-module M, the $(S \otimes_R S)$-module $(\eta_2)_!(M)$ which is just

$$M \otimes_S (S \otimes_R S) \cong (M \otimes_S S) \otimes_R S \cong M \otimes_R S$$

with scalar multiplication induced by

$$\big(m \otimes (s_1' \otimes s_2')\big)(s_1 \otimes s_2) = m \otimes (s_1' s_1 \otimes s_2' s_2)$$

together with the requirement that

$$\begin{aligned} m \otimes (s \otimes s') &= m \otimes \big((1 \otimes s')(s \otimes 1)\big) = m \otimes \big(\eta_2(s') \cdot (s \otimes 1)\big) \\ &= ms' \otimes (s \otimes 1). \end{aligned}$$

So, viewing $(\eta_2)_!(M)$ as $M \otimes_R S$, scalar multiplication is determined by

$$(m \otimes s)(s_1 \otimes s_2) = ms_2 \otimes ss_1.$$

As the tensor product over R is symmetric (up to isomorphism), it is more sensible to write $(\eta_2)_!(M) \cong S \otimes_R M$ together with the multiplication

$$(s \otimes m)(s_1 \otimes s_2) = ss_1 \otimes ms_2.$$

We shall often write $S \otimes_R M$ instead of $(\eta_2)_!(M)$.

Now if N, N' are two R-modules, let us write

$$\sigma_{NN'} \colon N \otimes_R N' \longrightarrow N' \otimes_R N$$

for the symmetry isomorphism determined by $\sigma_{NN'}(n \otimes n') = n' \otimes n$; it is an R-linear mapping. In particular given an S-module M, we get an R-linear isomorphism

$$\sigma_{MS}\colon (\eta_1)_!(M) \longrightarrow (\eta_2)_!(M).$$

Let us insist on the fact that σ_{MS} is only R-linear, even if $(\eta_1)_!(M)$ and $(\eta_2)_!(M)$ are $S \otimes_R S$-modules.

To end the fixing of notation, let us still consider, for an S-module M, the following R-linear mappings:

$$(\eta_1)_!(M) = M \otimes_R S \xrightarrow{\rho_M} M, \quad m \otimes s \mapsto ms,$$
$$(\eta_2)_!(M) = S \otimes_R M \xrightarrow{\lambda_M} M, \quad s \otimes m \mapsto ms.$$

With our notation settled, we come back to the descent problem. We want to investigate when an S-module M has the form $f_!(N)$, for some R-module N. Since $\eta_1 \circ f = \eta_2 \circ f$, we get an isomorphism $(\eta_1)_! f_!(N) \cong (\eta_2)_! f_!(N)$ for every R-module N. This isomorphism of $(S \otimes_R S)$-modules

$$(\eta_1)_! f_!(N) \cong (N \otimes_R S) \otimes_S (S \otimes_R S) \xrightarrow{\varphi_N} (N \otimes_R S) \otimes_S (S \otimes_R S) \cong (\eta_2)_! f_!(N)$$

is just determined by

$$\varphi_N\big((n \otimes s) \otimes (s_1 \otimes s_2)\big) = (n \otimes s) \otimes (s_2 \otimes s_1).$$

Indeed, φ_N is correctly defined since, given an element $s' \in S$,

$$\begin{aligned}
\varphi_N\big((n \otimes s)s' \otimes (s_1 \otimes s_2)\big) &= \varphi_N\big((n \otimes ss') \otimes (s_1 \otimes s_2)\big) \\
&= (n \otimes ss') \otimes (s_2 \otimes s_1) \\
&= \big((n \otimes s)s'\big) \otimes (s_2 \otimes s_1) \\
&= \big((n \otimes s) \otimes (1 \otimes s')(s_2 \otimes s_1)\big) \\
&= (n \otimes s) \otimes (s_2 \otimes s's_1) \\
&= \varphi_N\big((n \otimes s) \otimes (s's_1 \otimes s_2)\big) \\
&= \varphi_N\big((n \otimes s) \otimes (s' \otimes 1)(s_1 \otimes s_2)\big),
\end{aligned}$$

and obviously φ_N is an isomorphism with inverse defined by the same formula. Via the previous isomorphisms

$$(N \otimes_R S) \otimes_S (S \otimes_R S) \cong (N \otimes_R S) \otimes_R S,$$
$$(n \otimes s) \otimes (s_1 \otimes s_2) \mapsto n \otimes ss_1 \otimes s_2,$$
$$(N \otimes_R S) \otimes_S (S \otimes_R S) \cong S \otimes_R (N \otimes_R S),$$
$$(n \otimes s) \otimes (s_1 \otimes s_2) \mapsto s_1 \otimes n \otimes ss_2,$$

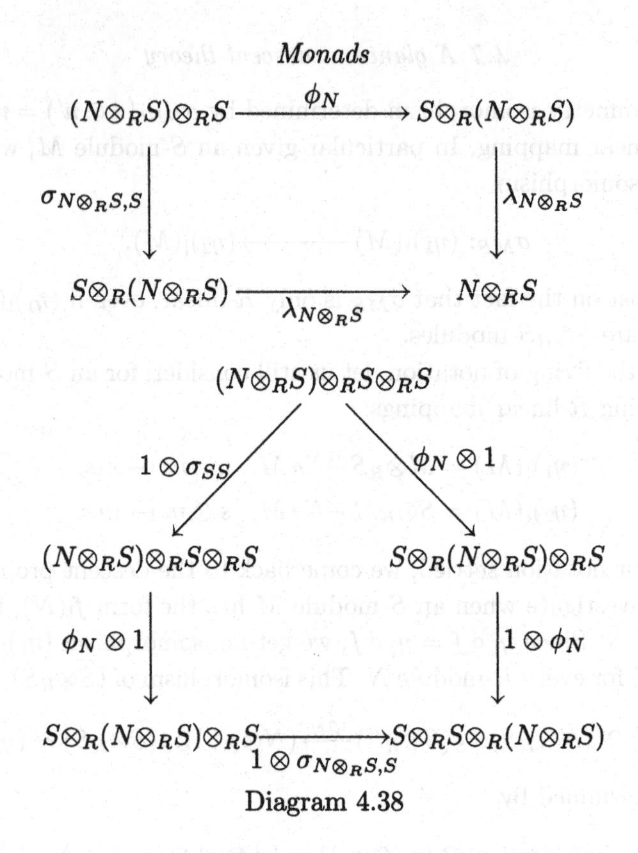

Diagram 4.38

used to calculate $(\eta_1)_!(N\otimes_R S)$ and $(\eta_2)_!(N\otimes_R S)$, our isomorphism φ_N becomes

$$(\eta_1)_! f_!(N) \cong (N\otimes_R S)\otimes_R S \xrightarrow{\phi_N} S\otimes_R(N\otimes_R S) \cong (\eta_2)_! f_!(N)$$

and is determined by

$$\phi_N(n\otimes s_1\otimes s_2) = s_1\otimes n\otimes s_2.$$

It is now an obvious matter to check the commutativity of both parts of diagram 4.38. In the first one, an element $n\otimes s_1\otimes s_2$ is mapped to $n\otimes s_1 s_2$ while in the second one an element $n\otimes s_1\otimes s_2\otimes s_3$ is mapped to $s_1\otimes s_2\otimes n\otimes s_3$. Moreover given an R-linear mapping g: $N\longrightarrow\overline{N}$ and considering the corresponding isomorphism $\phi_{\overline{N}}$ obtained from $\overline{N}$, we get the commutative diagram 4.39, where an element $n\otimes s_1\otimes s_2$ is mapped to $s_1\otimes f(n)\otimes s_2$.

Definition 4.7.1 *Let f: $R\longrightarrow S$ be a homomorphism of commutative rings with units; we use the previous notation. By a descent datum on an S-module M we mean an $(S\otimes_R S)$-linear isomorphism*

$$\phi\colon M\otimes_R S\longrightarrow S\otimes_R M$$

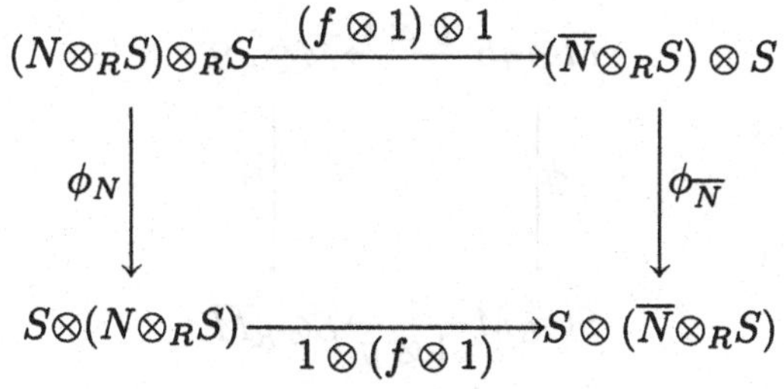

$$\begin{array}{ccc}(N\otimes_R S)\otimes_R S & \xrightarrow{(f\otimes 1)\otimes 1} & (\overline{N}\otimes_R S)\otimes S\\ \downarrow{\scriptstyle \phi_N} & & \downarrow{\scriptstyle \phi_{\overline{N}}}\\ S\otimes(N\otimes_R S) & \xrightarrow[1\otimes(f\otimes 1)]{} & S\otimes(\overline{N}\otimes_R S)\end{array}$$

Diagram 4.39

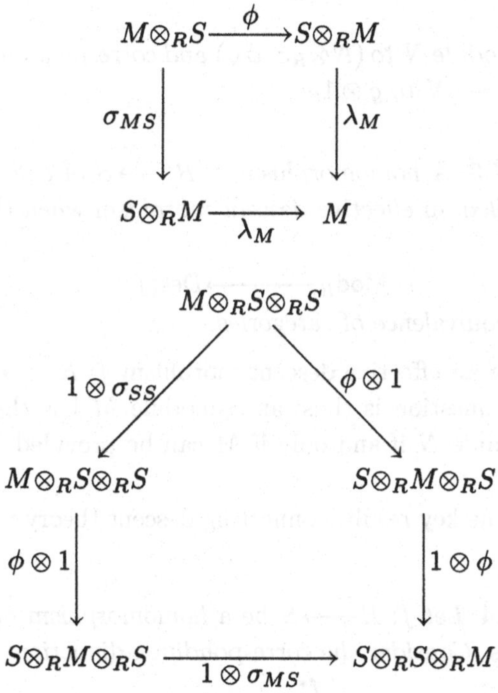

$$\begin{array}{ccc}M\otimes_R S & \xrightarrow{\phi} & S\otimes_R M\\ \downarrow{\scriptstyle \sigma_{MS}} & & \downarrow{\scriptstyle \lambda_M}\\ S\otimes_R M & \xrightarrow[\lambda_M]{} & M\end{array}$$

$$\begin{array}{ccc} & M\otimes_R S\otimes_R S & \\ {\scriptstyle 1\otimes\sigma_{SS}}\swarrow & & \searrow{\scriptstyle \phi\otimes 1}\\ M\otimes_R S\otimes_R S & & S\otimes_R M\otimes_R S\\ \downarrow{\scriptstyle \phi\otimes 1} & & \downarrow{\scriptstyle 1\otimes\phi}\\ S\otimes_R M\otimes_R S & \xrightarrow[1\otimes\sigma_{MS}]{} & S\otimes_R S\otimes_R M\end{array}$$

Diagram 4.40

which makes the two parts of diagram 4.40 commutative. If ϕ is a descent datum on the S-module M and $\overline{\phi}$ is a descent datum on the S-module $\overline{M}$, by a morphism $g\colon (M,\phi)\longrightarrow(\overline{M},\overline{\phi})$ of descent data we mean an S-linear mapping $g\colon M\longrightarrow\overline{M}$ making diagram 4.41 commutative.

The previous considerations immediately imply the following result.

Proposition 4.7.2 *Let $f\colon R\longrightarrow S$ be a homomorphism of commutative rings with units. Write $\mathsf{Des}(f)$ for the category of descent data described in 4.7.1. There exists a "comparison" functor*

$$\kappa\colon \mathsf{Mod}_R\longrightarrow\mathsf{Des}(f)$$

$$\begin{array}{ccc} M\otimes_R S & \xrightarrow{g\otimes 1} & \overline{M}\otimes_R S \\ \phi\downarrow & & \downarrow\overline{\phi} \\ S\otimes_R M & \xrightarrow[1\otimes g]{} & S\otimes_R\overline{M} \end{array}$$

Diagram 4.41

mapping a R-module N to $(N\otimes_R S, \phi_N)$ and correspondingly an R-linear mapping $g: N \longrightarrow \overline{N}$ to $g\otimes 1_S$. □

Definition 4.7.3 *A homomorphism $f: R\longrightarrow S$ of commutative rings with unit is called an effective descent morphism when the comparison functor*

$$\mathsf{Mod}_R \longrightarrow \mathsf{Des}(f)$$

of 4.7.2 is an equivalence of categories.

In the case of an effective descent morphism $f: R\longrightarrow S$, the answer to our original question is thus: an S-module M has the form $N\otimes_R S$ for some R-module N if and only if M can be provided with a descent datum.

Here is now the key result, connecting descent theory with the theory of (co)monads.

Theorem 4.7.4 *Let $f: R\longrightarrow S$ be a homomorphism of commutative rings with units. Consider the corresponding adjunction*

$$\mathsf{Mod}_R \underset{f_!}{\overset{f^*}{\rightleftarrows}} \mathsf{Mod}_S, \quad f_! \dashv f^*,$$

where $f^*(M) = M$ and $f_!(N) = N\otimes_S R$. Write $\mathbb{T} = (T, \varepsilon, \mu)$ for the comonad on Mod_S induced by this adjunction (see 4.2.1). The category $\mathsf{Des}(f)$ of descent data for f is equivalent to the category $\mathsf{Mod}_S^{\mathbb{T}}$ of coalgebras for the comonad $\mathbb{T}$; moreover, via this equivalence, the comparison functor of proposition 4.7.3 is just the comparison functor refered to in 4.2.1.

Proof Using the construction of 4.2.1, the comonad $\mathbb{T}$ can be described as follows. The functor T is given by

$$T: \mathsf{Mod}_S \longrightarrow \mathsf{Mod}_S, \quad M \mapsto M\otimes_R S,$$

where the S-module structure on $M \otimes_R S$ is just

$$(m \otimes s)s' = m \otimes ss'.$$

The two natural transformations ε, μ are given by

$$\varepsilon\colon T \Rightarrow \mathrm{id}, \quad \varepsilon_M\colon M \otimes_R S \longrightarrow M, \quad m \otimes s \mapsto ms,$$

$$\mu\colon T \Rightarrow T \circ T, \quad \mu_M\colon M \otimes_R S \longrightarrow M \otimes_R S \otimes_R S, \quad m \otimes s \mapsto m \otimes 1 \otimes s.$$

The axioms for a comonad are obviously satisfied by ε, μ.

A $\mathbb{T}$-coalgebra is a pair (M, ξ) where M is an S-module and the S-linear mapping $\xi\colon M \longrightarrow M \otimes_R S$ satisfies the conditions dual to those of 4.1.2. The S-linearity of ξ means

$$\forall m \in M \quad \forall s \in S \quad \xi(ms) = \xi(m) \cdot (1 \otimes s).$$

Now writing $\xi(m) = \sum_{i=1}^n m_i \otimes s_i$, the two axioms for a $\mathbb{T}$-coalgebra become

$$\forall m \in M \quad \sum_{i=1}^n \xi(m_i) \otimes s_i = \sum_{i=1}^n m_i \otimes 1 \otimes s_i,$$

$$\forall m \in M \quad m = \sum_{i=1}^n m_i s_i.$$

Starting with such a $\mathbb{T}$-coalgebra (M, ξ), let us construct a descent datum $\phi\colon M \otimes_R S \longrightarrow S \otimes_R M$. With the previous notation, ϕ is the composite

$$M \otimes_R S \xrightarrow{\xi \otimes 1} M \otimes_R S \otimes_R S \xrightarrow{1 \otimes \sigma_{SS}} M \otimes_R S \otimes_R S \ldots$$

$$\ldots \xrightarrow{\rho_M \otimes 1} M \otimes_R S \xrightarrow{\sigma_{MS}} S \otimes_R M.$$

The action of this composite on an element $m \otimes s$ is thus given by

$$m \otimes s \mapsto \sum_{i=1}^n m_i \otimes s_i \otimes s \mapsto \sum_{i=1}^n m_i \otimes s \otimes s_i \ldots$$

$$\ldots \mapsto \sum_{i=1}^n m_i s \otimes s_i \mapsto \sum_{i=1}^n s_i \otimes m_i s.$$

Observe that ϕ is $(S \otimes_R S)$-linear since, given α, β in S,

$$\begin{aligned}(\xi \otimes 1)(\alpha m \otimes \beta s) &= \xi(\alpha m) \otimes \beta s = \xi(m) \cdot (1 \otimes \alpha) \otimes \beta s \\ &= \left(\sum_{i=1}^n m_i \otimes s_i\right) \cdot (1 \otimes \alpha) \otimes \beta s \\ &= \sum_{i=1}^n m_i \otimes s_i \alpha \otimes \beta s,\end{aligned}$$

from which one deduces immediately that

$$\begin{aligned}\phi(\alpha m \otimes \beta s) &= \sum_{i=1}^n s_i \alpha \otimes m_i \beta s = \left(\sum_{i=1}^n s_i \otimes m_i s\right) \cdot (\alpha \otimes \beta) \\ &= \phi(m \otimes s) \cdot (\alpha \otimes \beta).\end{aligned}$$

To prove that ϕ is an isomorphism, just observe that it admits for inverse the following composite ψ:

$$S\otimes_R M \xrightarrow{1\otimes\xi} S\otimes_R M\otimes_R S \xrightarrow{\sigma_{SM}\otimes 1} M\otimes_R S\otimes_R S \xrightarrow{\rho_M\otimes 1} M\otimes_R S.$$

Indeed, given $s\in S$ and $m\in M$,

$$\begin{aligned}(1\otimes\xi)\phi(m\otimes s) &= (1\otimes\xi)\Big(\sum_{i=1}^n s_i\otimes m_i s\Big) = \sum_{i=1}^n s_i\otimes\xi(m_i s)\\ &= \sum_{i=1}^n s_i\otimes\big(\xi(m_i)\cdot(1\otimes s)\big)\\ &= \sum_{i=1}^n \big(s_i\otimes\xi(m_i)\big)\cdot(1\otimes 1\otimes s)\\ &= \sum_{i=1}^n (s_i\otimes m_i\otimes 1)\cdot(1\otimes 1\otimes s)\\ &= \sum_{i=1}^n s_i\otimes m_i\otimes s\end{aligned}$$

from which it follows immediately that,

$$(\psi\circ\phi)(m\otimes s) = \sum_{i=1}^n m_i s_i\otimes s = m\otimes s.$$

The relation $(\phi\circ\psi)(s\otimes m) = s\otimes m$ is proved in a completely analogous way.

The two conditions for a descent datum (see 4.7.1) are proved by the same type of computation. The first condition is just obvious since

$$\begin{aligned}(\lambda_M\circ\phi)(m\otimes s) &= \lambda_M\Big(\sum_{i=1}^n s_i\otimes m_i s\Big) = \sum_{i=1}^n s_i m_i s = ms\\ (\lambda_M\circ\sigma_{MS})(m\otimes s) &= \lambda_M(s\otimes m) = sm.\end{aligned}$$

And the pentagonal condition is proved in the following way:

$$\begin{aligned}(1\otimes\phi)(\phi\otimes 1)(m\otimes s\otimes t) &= (1\otimes\phi)(\phi(m\otimes s)\otimes t)\\ &= (1\otimes\phi)\Big(\sum_{i=1}^n s_i\otimes m_i s\otimes t\Big)\\ &= \sum_{i=1}^n s_i\otimes\phi(m_i s\otimes t).\end{aligned}$$

To compute $\phi(m_i s\otimes t)$, we observe that

$$\begin{aligned}\sum_{i=1}^n s_i\otimes(\xi\otimes 1)(m_i s\otimes t) &= \sum_{i=1}^n s_i\otimes\xi(m_i s)\otimes t\\ &= \sum_{i=1}^n s_i\otimes\xi(m_i)\cdot(1\otimes s)\otimes t\\ &= \sum_{i=1}^n (s_i\otimes\xi(m_i)\otimes 1)(1\otimes 1\otimes s\otimes t)\\ &= \sum_{i=1}^n (s_i\otimes m_i\otimes 1\otimes 1)(1\otimes 1\otimes s\otimes t)\\ &= \sum_{i=1}^n s_i\otimes m_i\otimes s\otimes t,\end{aligned}$$

from which it follows that

$$(1\otimes\phi)(\phi\otimes 1)(m\otimes s\otimes t)=\sum_{i=1}^{n} s_i\otimes s\otimes m_i t.$$

On the other hand

$$\begin{aligned}(1\otimes\sigma_{MS})&(\phi\otimes 1)(1\otimes\sigma_{SS})(m\otimes s\otimes t)\\ &=(1\otimes\sigma_{MS})(\phi\otimes 1)(m\otimes t\otimes s)\\ &=(1\otimes\sigma_{MS})\big(\phi(m\otimes t)\otimes s\big)\\ &=(1\otimes\sigma_{MS})\Big(\sum_{i=1}^{n} s_i\otimes m_i t\otimes s\Big)\\ &=\sum_{i=1}^{n} s_i\otimes s\otimes m_i t.\end{aligned}$$

We have already proved that starting with a $\mathbb{T}$-coalgebra (M,ξ), the morphism ϕ we have defined is a descent datum on M. Given another $\mathbb{T}$-coalgebra $(\overline{M},\overline{\xi})$, consider the corresponding descent datum $\overline{\phi}$. Given a morphism of $\mathbb{T}$-coalgebras $g\colon (M,\xi)\longrightarrow(\overline{M},\overline{\xi})$, let us observe that $g\colon (M,\phi)\longrightarrow(\overline{M},\overline{\phi})$ is also a morphism of descent data, which will imply immediately that we have constructed a functor

$$\theta\colon \mathsf{Mod}_S^{\mathbb{T}}\longrightarrow\mathsf{Des}(f).$$

Given $s\in S$ and $m\in M$, we must prove that

$$(1\otimes g)\phi(m\otimes s)=\overline{\phi}(g\otimes 1)(m\otimes s),$$

while our assumptions are

$$g(ms)=g(m)\cdot s,\quad (g\otimes 1)\xi(m)=\overline{\xi}g(m).$$

Let us again use the notation $\xi(m)=\sum_{i=1}^{n} m_i\otimes s_i$ while we write $\overline{\xi}\big(g(m)\big)=\sum_{j=1}^{l}\overline{m}_j\otimes\overline{s}_j$. Our second assumption can then be written

$$\sum_{i=1}^{n} g(m_i)\otimes s_i=\sum_{j=1}^{l}\overline{m}_j\otimes\overline{s}_j.$$

It is now easy to prove that g is a morphism of descent data:

$$\begin{aligned}(1\otimes g)\phi(m\otimes s)&=(1\otimes g)\Big(\sum_{i=1}^{n} s_i\otimes m_i s\Big)=\sum_{i=1}^{n} s_i\otimes g(m_i s)\\ &=\sum_{i=1}^{n} s_i\otimes g(m_i)s=\Big(\sum_{i=1}^{n} s_i\otimes g(m_i)\Big)(1\otimes s)\\ &=\Big(\sum_{j=1}^{l}\overline{s}_j\otimes\overline{m}_j\Big)(1\otimes s)=\sum_{j=1}^{l}\overline{s}_j\otimes\overline{m}_j s\\ &=\overline{\phi}\big(g(m)\otimes s\big)=\overline{\phi}(g\otimes 1)(s).\end{aligned}$$

Conversely, consider now a descent datum $\phi: M\otimes_R S \longrightarrow S\otimes_R M$ over an S-module M. For simplicity, we shall write $\phi(m\otimes 1) = \sum_{\alpha=1}^{\beta} s_\alpha \otimes m_\alpha$. The $(S\otimes_R S)$-linearity of ϕ means that, given $s, t \in S$,

$$\phi(mt \otimes s) = \sum_{\alpha=1}^{\beta} s_\alpha t \otimes m_\alpha s.$$

With this notation, the first condition in 4.7.1 for being a descent datum, applied to an element $m \otimes 1$, becomes

$$m = \sum_{\alpha=1}^{\beta} s_\alpha m_\alpha,$$

while the pentagonal condition gives

$$\begin{aligned}
&(1\otimes\phi)(\phi\otimes 1)(m\otimes 1\otimes 1)\\
&\qquad = (1\otimes\phi)(\phi(m\otimes 1)\otimes 1)\\
&\qquad = (1\otimes\phi)\left(\sum_{\alpha=1}^{\beta} s_\alpha \otimes m_\alpha \otimes 1\right)\\
&\qquad = \sum_{\alpha=1}^{\beta} s_\alpha \otimes \phi(m_\alpha \otimes 1),\\
&(1\otimes\sigma_{MS})(\phi\otimes 1)(1\otimes\sigma_{SS})(m\otimes 1\otimes 1)\\
&\qquad = (1\otimes\sigma_{MS})(\phi\otimes 1)(m\otimes 1\otimes 1)\\
&\qquad = (1\otimes\sigma_{MS})(\phi(m\otimes 1)\otimes 1)\\
&\qquad = (1\otimes\sigma_{MS})\left(\sum_{\alpha=1}^{\beta} s_\alpha \otimes m_\alpha \otimes 1\right)\\
&\qquad = \sum_{\alpha=1}^{\beta} s_\alpha \otimes 1 \otimes m_\alpha,
\end{aligned}$$

which finally yields

$$\sum_{\alpha=1}^{\beta} s_\alpha \otimes \phi(m_\alpha \otimes 1) = \sum_{\alpha=1}^{\beta} s_\alpha \otimes 1 \otimes m_\alpha.$$

Given such a descent datum ϕ on an S-module M, we define ξ to be the composite

$$M \xrightarrow{\ \eta_M\ } M\otimes_R S \xrightarrow{\ \phi\ } S\otimes_R M \xrightarrow{\ \sigma_{SM}\ } M\otimes_R S$$

where $\eta_M(m) = m \otimes 1$. With the previous notation

$$\xi(m) = \sigma_{SM}\phi(m\otimes 1) = \sigma_{SM}\left(\sum_{\alpha=1}^{\beta} s_\alpha \otimes m_\alpha\right) = \sum_{\alpha=1}^{\beta} m_\alpha \otimes s_\alpha.$$

We shall prove that (M, ξ) is a $\mathbb{T}$-algebra.

First of all, ξ is S-linear. Indeed, given $m \in M$, $t \in S$,

$$\begin{aligned}\xi(mt) &= \sigma_{SM}\phi(mt \otimes 1) = \sigma_{SM}\left(\sum_{\alpha=1}^{\beta} s_\alpha t \otimes m_\alpha\right)\\ &= \sum_{\alpha=1}^{\beta} m_\alpha \otimes s_\alpha t = \left(\sum_{\alpha=1}^{\beta} m_\alpha \otimes s_\alpha\right) t\\ &= \xi(m) \cdot t.\end{aligned}$$

The first condition for a $\mathbb{T}$-coalgebra is just (see 4.1.2)

$$(\varepsilon_M \circ \xi)(m) = \varepsilon_M\left(\sum_{\alpha=1}^{\beta} m_\alpha \otimes s_\alpha\right) = \sum_{\alpha=1}^{\beta} m_\alpha s_\alpha = m.$$

To prove the second condition for a $\mathbb{T}$-coalgebra, observe first that the pentagonal condition for a descent datum yields in particular, with the previous notation,

$$\sum_{\alpha=1}^{\beta} s_\alpha \otimes \xi(m_\alpha) = \sum_{\alpha=1}^{\beta} s_\alpha \otimes m_\alpha \otimes 1.$$

It is then easy to perform the various computations.

$$\begin{aligned}(\xi \otimes 1)\xi(m) &= (\xi \otimes 1)\left(\sum_{\alpha=1}^{\beta} m_\alpha \otimes s_\alpha\right)\\ &= \sum_{\alpha=1}^{\beta} \xi(m_\alpha) \otimes s_\alpha = \sum_{\alpha=1}^{\beta} m_\alpha \otimes 1 \otimes s_\alpha\\ &= \sum_{\alpha=1}^{\beta} \mu_M(m_\alpha \otimes s_\alpha) = \mu_M\left(\sum_{\alpha=1}^{\beta} m_\alpha \otimes s_\alpha\right)\\ &= \mu_M \xi(m).\end{aligned}$$

Thus (M, ξ) is indeed a $\mathbb{T}$-coalgebra.

Choose another S-module $\overline{M}$, a descent datum $\overline{\phi}\colon \overline{M} \otimes_R S \longrightarrow S \otimes_R \overline{M}$ and the corresponding $\mathbb{T}$-coalgebra $(\overline{M}, \overline{\xi})$. If $g\colon (M, \phi) \longrightarrow (\overline{M}, \overline{\phi})$ is a morphism of descent data, let us prove that $g\colon (M, \xi) \longrightarrow (\overline{M}, \overline{\xi})$ is a morphism of $\mathbb{T}$-coalgebra, which will finally yield a functor

$$\tau\colon \mathsf{Des}(f) \longrightarrow \mathsf{Mod}_S^{\mathbb{T}}.$$

We keep the notation $\phi(m \otimes 1) = \sum_{\alpha=1}^{\beta} s_\alpha \otimes m_\alpha$ and write in the same way $\overline{\phi}(g(m) \otimes 1) = \sum_{\gamma=1}^{\delta} \overline{s}_\gamma \otimes \overline{m}_\gamma$. The assumption on g thus implies

$$\sum_{\gamma=1}^{\delta} \overline{s}_\gamma \otimes \overline{m}_\gamma = \sum_{\alpha=1}^{\beta} s_\alpha \otimes g(m_\alpha).$$

This easily yields

$$\begin{aligned}\overline{\xi}(g\otimes 1)(m) &= \overline{\xi}(g(m)\otimes 1) = \sum_{\gamma=1}^{\delta} \overline{m}_\gamma \otimes \overline{s}_\gamma \\ &= \sum_{\alpha=1}^{\beta} g(m_\alpha)\otimes s_\alpha = (g\otimes 1)\left(\sum_{\alpha=1}^{\beta} m_\alpha \otimes s_\alpha\right) \\ &= (g\otimes 1)(\xi)(m).\end{aligned}$$

Let us now check that the functors θ and τ are reciprocal equivalences of categories (see 3.4.3, volume 1). Since the statement on the morphisms is obvious, it suffices to consider the case of objects. Let us start with a $\mathbb{T}$-coalgebra (M,ξ); we put $\theta(M,\xi) = (M,\phi)$ and $\tau(M,\phi) = (M,\overline{\xi})$; we must prove the equality $\xi = \overline{\xi}$. We use the notation already defined in this proof.

$$\begin{aligned}\xi(m) &= \sum_{i=1}^{n} m_i \otimes s_i, \\ \phi(m\otimes 1) &= \sum_{i=1}^{n} s_i \otimes m_i, \\ \overline{\xi}(m) &= \sigma_{SM}\phi(m\otimes 1) \\ &= \sigma_{SM}\left(\sum_{i=1}^{n} s_i\otimes m_i\right) = \sum_{i=1}^{n} m_i \otimes s_i \\ &= \xi(m).\end{aligned}$$

The converse relation is completely analogous. Starting with a descent datum (M,ϕ), we construct $\tau(M,\phi) = (M,\xi)$ and $\theta(M,\xi) = (M,\overline{\phi})$; we must prove the equality $\phi = \overline{\phi}$. With previous notation, we have

$$\begin{aligned}\phi(m\otimes 1) &= \sum_{\alpha=1}^{\beta} s_\alpha \otimes m_\alpha, \\ \xi(m) &= \sum_{\alpha=1}^{\beta} m_\alpha \otimes s_\alpha, \\ \overline{\phi}(m\otimes 1) &= \sum_{\alpha=1}^{\beta} \otimes s_\alpha \otimes m_\alpha = \phi(m\otimes 1),\end{aligned}$$

from which

$$\begin{aligned}\phi(m\otimes s) &= \phi((m\otimes 1)(1\otimes s)) = (\phi(m\otimes 1))(1\otimes s) \\ &= (\overline{\phi}(m\otimes 1))(1\otimes s) = \overline{\phi}((m\otimes 1)(1\otimes s)) \\ &= \overline{\phi}(m\otimes s),\end{aligned}$$

since ϕ and $\overline{\phi}$ are $(S\otimes_R S)$-linear. This proves the stated equivalence.

Finally let us compute the composite of the comparison functor κ of 4.7.2 with our functor τ. This is the functor

$$\tau\circ\kappa\colon \mathsf{Mod}_R \longrightarrow \mathsf{Mod}_S^{\mathbb{T}},\ N\mapsto (N\otimes_R S, \xi_N),$$

where $\xi_N\colon N\otimes_R S \longrightarrow N\otimes_R S\otimes_R S$ is defined by

$$\begin{aligned}\xi_N(n\otimes s) &= (\sigma_{S,N\otimes_R S}\circ\phi_N)(n\otimes s\otimes 1)\\ &= \sigma_{S,N\otimes_R S}(s\otimes n\otimes 1)\\ &= n\otimes 1\otimes s\end{aligned}$$

and an R-linear mapping $g\colon N\longrightarrow\overline{N}$ is mapped to $g\otimes 1_S$. Now observe that the two canonical natural transformations of the adjunction $f_! \dashv f^*$ are given by

$$\begin{aligned}\alpha_N\colon N &\longrightarrow f^*f_!(N) = N\otimes_R S,\quad n\mapsto n\otimes 1,\\ \beta_M\colon f_!f^*(M) = M\otimes_R S &\longrightarrow M,\quad m\otimes s\mapsto ms,\end{aligned}$$

for $N\in\mathsf{Mod}_R$ and $M\in\mathsf{Mod}_S$. Therefore

$$(\tau\circ\kappa)(N) = \big(f_!(N), f_!(\alpha_N)\big),$$

which is precisely the action of the comparison functor described in 4.2.1. The same remark holds for morphisms, since $(\tau\circ\kappa)(g) = f_!(g)$. $\square$

To get a characterization of effective descent morphisms, let us recall two classical definitions.

Definition 4.7.5 *Let M be a module over a commutative ring R with unit. M is flat when the functor*

$$-\otimes_R M\colon \mathsf{Mod}_R\longrightarrow\mathsf{Mod}_R$$

preserves monomorphisms.

Definition 4.7.6 *Let R be a commutative ring with a unit. A monomorphism $i\colon A\longrightarrow B$ of R-modules is pure when, for every R-module M, the canonical mapping*

$$i\otimes 1\colon A\otimes_R M\longrightarrow B\otimes_R M$$

is still a monomorphism.

One should probably emphasize the difference between purity and flatness. In both cases one is interested in getting a monomorphism of the type

$$i\otimes 1\colon A\otimes_R M\longrightarrow B\otimes_R M.$$

When this is the case for all monomorphisms i then M is flat; i is pure when this is the case for all modules M.

Lemma 4.7.7 *Let $f\colon R\longrightarrow S$ be a homomorphism of commutative rings with unit. If f is a pure monomorphism of R-modules, then*

$$-\otimes_R S\colon \mathsf{Mod}_R\longrightarrow\mathsf{Mod}_S$$

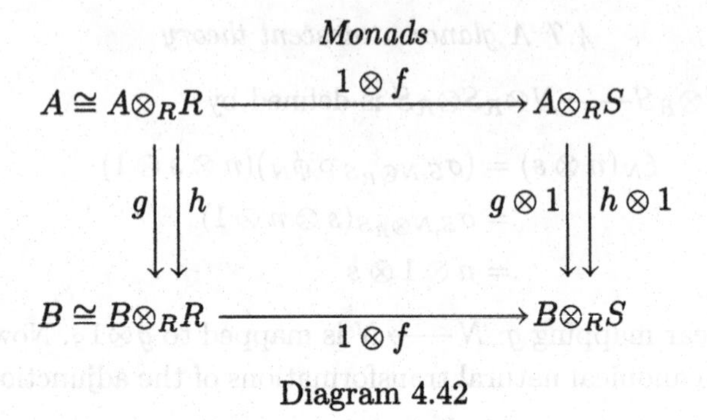

Diagram 4.42

(1) is a faithful functor,
(2) reflects isomorphisms.

Proof Since Mod_R and Mod_S are abelian (see 1.4.6.a), to prove that $-\otimes_R S$ reflects isomorphisms, it suffices to prove that it reflects both monomorphisms and epimorphisms (see 1.5.1). And this will be the case as long as $-\otimes_R S$ is faithful (see 1.7.6). So consider two morphisms $g, h\colon A \rightrightarrows B$ of R-modules such that $g \otimes 1_S = h \otimes 1_S$. Diagram 4.42 is commutative and, by assumption, the horizontal arrows are monomorphisms. Therefore since

$$(1 \otimes f)g = (g \otimes 1)(1 \otimes f) = (h \otimes 1)(1 \otimes f) = (1 \otimes f)h$$

we deduce $g = h$. □

A classical theorem in descent theory is then:

Theorem 4.7.8 *Let $f\colon R \longrightarrow S$ be a homomorphism of commutative rings with units. Suppose that:*
(1) f is a pure monomorphism of R-modules;
(2) S is a flat R-module.
Under these conditions, f is an effective descent morphism.

Proof With 4.7.4 in mind, it suffices to prove the comonadicity of the functor $-\otimes_R S\colon \mathsf{Mod}_R \longrightarrow \mathsf{Mod}_S$. To do it, we apply 4.4.4.

We know already that $-\otimes_R S = f_!$ has a right adjoint f^*; in particular $-\otimes_R S$ is right exact (see 3.2.1, volume 1, and 1.11.2, this volume). By 3.9.3, the morphism $f\colon S \longrightarrow R$ induces an algebraic functor $\mathsf{Mod}_S \longrightarrow \mathsf{Mod}_R$, thus the functor $-\otimes_R S\colon \mathsf{Mod}_S \longrightarrow \mathsf{Mod}_R$ preserves monomorphisms since, by assumption, the composite

$$\mathsf{Mod}_R \xrightarrow{-\otimes_R S} \mathsf{Mod}_S \longrightarrow \mathsf{Mod}_R$$

does, and the algebraic functor reflects limits (see 3.4.1). Thus $-\otimes_R S$ is exact (see 1.11.4) and therefore preserves all equalizers (see 1.11.2).

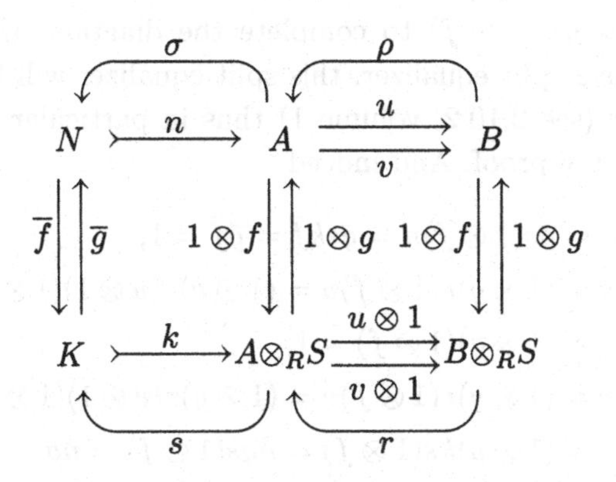

Diagram 4.43

On the other hand Mod_R has all equalizers (see 1.4.6.a). Therefore the dual conditions of (2)(a) and (2)(c) in 4.4.4 are already satisfied and, by 4.7.7.(2)(b), is satisfied as well. □

In 4.7.8, the purity of f is in fact a necessary condition for f being an effective descent morphism (see exercise 4.8.12); the flatness of S is not a necessary condition (see exercise 4.8.13).

Theorem 4.7.9 *Let $f\colon R \longrightarrow S$ be a homomorphism of commutative rings with units. If f admits an R-linear retraction, f is an effective descent morphism.*

Proof Again with 4.7.4 in mind, it suffices to prove the comonadicity of the functor $-\otimes_R S\colon \mathsf{Mod}_R \longrightarrow \mathsf{Mod}_S$. We know already that $-\otimes_R S = f_!$ admits f^* as a right adjoint.

If f admits a retraction $g\colon S \longrightarrow R$, then from $g \circ f = 1_R$ we deduce $(g \otimes 1_M) \circ (f \otimes 1_M) = 1_{R \otimes M}$ for every R-module M. In particular each $f \otimes 1_M$ is a monomorphism and thus f is pure. By 4.7.7, this implies that $-\otimes_R S$ reflects isomorphisms.

It remains to check condition (2)(c) of 4.4.4. Let us consider two morphisms $u, v\colon A \rightrightarrows B$ in Mod_R whose images by $-\otimes_R S$ have a split equalizer. This yields diagram 4.43 in Mod_R where k is the split equalizer of $(u \otimes 1, v \otimes 1)$, i.e.

$$(u \otimes 1) \circ k = (v \otimes 1) \circ k, \quad s \circ k = 1, \quad r \circ (u \otimes 1) = 1, \quad r \circ (v \otimes 1) = k \circ s.$$

If n is the equalizer of (u, v), the commutativity of the squares on the right implies the existence of factorizations $\overline{f}, \overline{g}$ through the equalizer, with $k\overline{f} = (1 \otimes f)n$ and $n\overline{g} = (1 \otimes g)k$. It remains to put $\sigma = \overline{g}s(1 \otimes f)$

and $\rho = (1 \otimes g)r(1 \otimes f)$ to complete the diagram. If we prove that the top line is a split equalizer, this split equalizer will be preserved by every functor (see 2.10.2, volume 1) thus in particular by $-\otimes_R S$; this will conclude the proof. And indeed

$$\begin{aligned} \sigma n &= \overline{g}s(1 \otimes f)n = \overline{g}sk\overline{f} = \overline{g}\overline{f} = 1, \\ \rho u &= (1 \otimes g)r(1 \otimes f)u = (1 \otimes g)r(u \otimes 1)(1 \otimes f) \\ &= (1 \otimes g)(1 \otimes f) = 1, \\ \rho v &= (1 \otimes g)r(1 \otimes f)v = (1 \otimes g)r(v \otimes 1)(1 \otimes f) \\ &= (1 \otimes g)ks(1 \otimes f) = n\overline{g}s(1 \otimes f) = n\sigma. \end{aligned}$$

□

4.8 Exercises

4.8.1 Describe explicitly the monad on Set admitting the category of real vector spaces as category of algebras.

4.8.2 Let $\mathbb{T} = (T, \varepsilon, \mu)$ be a monad over Set. Prove that the category $\mathsf{Set}^{\mathbb{T}}$ of $\mathbb{T}$-algebras is equivalent (in a bigger universe) to the full subcategory of $\mathsf{Fun}(\mathsf{Set}^*_{\mathbb{T}}, \mathsf{Set})$ whose objects are the product preserving functors.

4.8.3 Let $\mathscr{C}$ be a finitely complete category. Show that there is a coincidence (up to equivalences of categories) between:

(1) the localizations of $\mathscr{C}$ (see 3.5.5, volume 1);

(2) the categories of $\mathbb{T}$-algebras for the idempotent and left exact monads $\mathbb{T}$ on $\mathscr{C}$ (i.e. the idempotent monads $\mathbb{T} = (T, \varepsilon, \mu)$) such that T preserves finite limits).

4.8.4 Let $\mathscr{C}$ be a complete and cocomplete category and $\mathbb{T} = (T, \varepsilon, \mu)$ a monad on $\mathscr{C}$ with finite rank. Prove that the category $\mathscr{C}^{\mathbb{T}}$ of $\mathbb{T}$-algebras is cocomplete (compare with 4.3.6).

4.8.5 In a category $\mathscr{C}$, a pair of morphisms $f, g \colon R \rightrightarrows A$ induces a relation on each set $\mathscr{C}(X, A)$ of morphisms: the relation constituted of all the pairs $(f \circ x, g \circ x)$ for $x \colon X \longrightarrow R$. Prove that all those relations are reflexive if and only if f, g are two epimorphisms with a common section. Such a pair (f, g) is called a reflexive pair.

4.8.6 In the adjoint lifting theorem (see 4.5.5), prove that the two morphisms ω_D and $FL(\xi)$ admit $FL(\xi_D)$ as a common section. Conclude that the assumption on $\mathscr{A}$ can be weakened by assuming just the existence of coequalizers of reflexive pairs (see 4.8.5).

4.8.7 In the situation of 4.5.9, prove that the functor Q is full and faithful if and only if each morphism λ_C is an epimorphism in $\mathscr{C}$.

4.8.8 Let $\mathscr{C}$ be a well-powered category and $\mathbb{T} = (T, \varepsilon, \mu)$ a monad on $\mathscr{C}$. Prove that the category $\mathscr{C}^{\mathbb{T}}$ of $\mathbb{T}$-algebras is well-powered.

4.8.9 Let CompLat be the category of complete lattices and $(\bigwedge\bigvee)$-preserving maps. Prove that the corresponding underlying set functor U: CompLat $\longrightarrow$ Set is not monadic. [Hint: U does not have a left adjoint, although it preserves limits; the "free complete lattice" on three generators turns out already to be a proper class, not a set.]

4.8.10 Let $\mathbb{T}$ be a monad on the category of sets. Show that $\mathbb{T}$ has finite rank iff the free algebra $F(\mathbf{1})$ on the singleton has the following property: for every set X and every morphism of $\mathbb{T}$-algebras $g\colon F(\mathbf{1}) \longrightarrow F(X)$, g factors through some $F(i)\colon F(Y) \longrightarrow F(X)$, with $i\colon Y \longrightarrow X$ the inclusion of a finite subset.

4.8.11 In the situation of 4.7.1, prove that given an $(S \otimes_R S)$-linear mapping $\phi\colon M \otimes_R S \longrightarrow S \otimes_R M$ satisfying the two given conditions, ϕ is necessarily an isomorphism and thus a descent datum.

4.8.12 A homomorphism $f\colon R \longrightarrow S$ of commutative rings with unit is called a descent morphism when the comparison functor defined in 4.7.2 $\mathsf{Mod}_R \longrightarrow \mathsf{Des}(f)$ is full and faithful. Prove that f is a descent morphism if and only if f is a pure monomorphism in the category of R-modules.

4.8.13 Consider a homomorphism $f\colon R \longrightarrow S$ of commutative rings with unit. The mapping $\binom{1_R}{f}\colon R \longrightarrow R \times S$ is another homomorphism of commutative rings with unit and it admits a retraction. Show that n general, $R \times S$ is not a flat R-module. [Hint: choose $R = \mathbb{Z}$ and $S = \mathbb{Z}/2\mathbb{Z}$.] Compare with 4.7.8 and 4.7.9.

5

Accessible categories

In chapter 11, we have studied those algebraic theories which can be described via "finite powers": a theory was a category $\mathcal{T}$ with objects $T^0, T^1, \ldots, T^n, \ldots$ and a model was a functor $\mathcal{T} \longrightarrow \mathsf{Set}$ preserving finite powers.

Replacing "finite powers" by "finite limits" leads to the notion of a locally finitely presentable category of models. A theory is now a small category $\mathscr{T}$ with finite limits while a $\mathscr{T}$-model is a functor $\mathscr{T} \longrightarrow \mathsf{Set}$ preserving finite limits.

In section 6.3, volume 1, we observed that in the absence of finite limits, the notion of a "flat functor" was a good substitute for the notion of a functor preserving finite limits. Therefore a further generalization consists in considering a theory as being just a small category $\mathscr{T}$, while a $\mathscr{T}$-model is a flat functor $\mathscr{T} \longrightarrow \mathsf{Set}$. This leads to what is called an "accessible" category of models.

In fact, the accessible categories admit an elegant axiomatic description: they are those categories with a "good" set of generators. And locally presentable categories are just those accessible categories which are cocomplete (or, equivalently, complete). In particular, the existence of generators makes the accessible categories a fruitful context in which to develop categorical constructions which, in general, require smallness conditions.

Finally let us mention that if the categories of limit preserving functors are locally presentable, the categories of limit–colimit preserving functors turn out to be accessible. We present in fact a more flexible approach in terms of general cones and cocones: what is called the theory of sketches. A sketch is a small category provided with a set of distinguished cones and a set of distinguished cocones; a model of this sketch is a set-valued functor transforming the distinguished cones into limit cones and the distinguished cocones into colimit cocones. The categories of models of sketches are exactly the accessible categories.

5.1 Presentable objects in a category

In section 3.8, we studied those models M of an algebraic theory $\mathcal{T}$ which are finitely presentable. This means that M can be described via finitely many generators and finitely many relations (see 3.8.1). But in 3.8.14 we proved this to be equivalent to the preservation of filtered colimits by the representable functor

$$\mathsf{Mod}_{\mathcal{T}}(M,-)\colon \mathsf{Mod}_{\mathcal{T}} \longrightarrow \mathsf{Set}.$$

Clearly, such a definition generalizes easily to the case of an arbitrary category. For the sake of generality, we give it with respect to an arbitrary regular cardinal α (see 6.4) and not just with respect to $\aleph_0$.

Definition 5.1.1 *An object $M \in \mathcal{M}$ of a category $\mathcal{M}$ is α-presentable, for some regular cardinal α, when the representable functor*

$$\mathcal{M}(M,-)\colon \mathcal{M} \longrightarrow \mathsf{Set}$$

preserves α-filtered colimits. An object M is presentable when it is α-presentable for some regular cardinal α.

Clearly, the study of α-presentable objects will be pertinent just when $\mathcal{M}$ has α-filtered colimits. Trivially, one has

Proposition 5.1.2 *Let $\mathcal{M}$ be a category and $\alpha < \beta$ two regular cardinals. Every α-presentable object is also β-presentable.*

Proof This is simply because every β-filtered colimit is α-filtered. □

Here is a useful technical characterization of α-presentable objects.

Proposition 5.1.3 *Let $\mathcal{M}$ be a category and α a regular cardinal. An object $M \in \mathcal{M}$ is α-presentable precisely when, for every α-filtered colimit $L = \mathrm{colim}_{i \in I} M_i$ in $\mathcal{M}$, the following two conditions are satisfied:*

(1) every morphism $m\colon M \longrightarrow L$ factors through one of the canonical morphisms $s_i\colon M_i \longrightarrow L$ of the colimit;

(2) when two morphisms $m_1, m_2\colon M \longrightarrow M_i$ are such that $s_i \circ m_1 = s_i \circ m_2$, for some canonical morphism $s_i\colon M_i \longrightarrow L$ of the colimit, there exists a morphism $s_{ij}\colon M_i \longrightarrow M_j$ in the original diagram such that $s_{ij} \circ m_1 = s_{ij} \circ m_2$.

Proof There exists in Set a canonical mapping

$$\operatorname*{colim}_{i\in I} \mathcal{M}(M, M_i) \longrightarrow \mathcal{M}(M, L)$$

obtained as a factorization through the colimit of the canonical cocone

$$\mathcal{M}(M, s_i)\colon \mathcal{M}(M, M_i) \longrightarrow \mathcal{M}(M, L).$$

The object M is α-presentable when this canonical mapping is bijective, for every α-filtered colimit $L = \operatorname{colim}_{i\in I} M_i$. The first condition expresses the surjectivity of the mapping and the second condition expresses its injectivity. □

Clearly, when the canonical morphisms $s_i\colon M_i \longrightarrow L$ are monomorphisms, the second condition of 5.1.3 vanishes.

Proposition 5.1.4 *Let $\mathcal{M}$ be a category and α a regular cardinal. In $\mathcal{M}$, an α-colimit of α-presentable objects, if it exists, is again an α-presentable object.*

Proof Consider a functor $F\colon \mathcal{J} \longrightarrow \mathcal{M}$, where $\mathcal{J}$ is a category with strictly less than α morphisms and each object $F(J)$ is α-presentable in $\mathcal{M}$. Suppose the colimit of F exists. Given an α-filtered colimit $L = \operatorname{colim}_{i\in I} M_i$ in $\mathcal{M}$, one has by 2.9.5 and 6.4.5, volume 1,

$$\begin{aligned}\mathcal{M}\big(\operatorname*{colim}_{J\in\mathcal{J}} F(J), \operatorname*{colim}_{i\in I} M_i\big) &\cong \lim\nolimits_{J\in\mathcal{J}} \mathcal{M}\big(F(J), \operatorname*{colim}_{i\in I} M_i\big)\\ &\cong \lim\nolimits_{J\in\mathcal{J}} \operatorname*{colim}_{i\in I} \mathcal{M}\big(F(J), M_i\big)\\ &\cong \operatorname*{colim}_{i\in I} \lim\nolimits_{J\in\mathcal{J}} \mathcal{M}\big(F(J), M_i\big)\\ &\cong \operatorname*{colim}_{i\in I} \mathcal{M}\big(\operatorname*{colim}_{J\in\mathcal{J}} F(J), M_i\big).\end{aligned}$$

□

5.2 Locally presentable categories

We refer to section 4.5, volume 1, as far as generators are concerned.

Definition 5.2.1 *A category $\mathcal{M}$ is locally α-presentable, for a regular cardinal α, when:*

(1) $\mathcal{M}$ is cocomplete;

(2) $\mathcal{M}$ has a set $(G_i)_{i\in I}$ of strong generators;

(3) each generator G_i is α-presentable.

A category is locally presentable when it is locally α-presentable for some regular cardinal α. The locally $\aleph_0$-presentable categories are also called locally finitely presentable.

Let us start with some examples.

Examples 5.2.2

5.2.2.a If $\mathcal{T}$ is an algebraic theory as in 3.3.1, the corresponding category $\mathsf{Mod}_{\mathcal{T}}$ of models is locally finitely presentable. Indeed $\mathsf{Mod}_{\mathcal{T}}$ is cocomplete (see 3.4.5) and the finitely generated free models constitute a family of dense, thus strong generators (see 3.8.10, this volume, 4.5.5 and 4.3.6, volume 1). They are finitely presentable by 3.8.14.

5.2.2.b Let $\mathscr{T}$ be a small category. The category $\mathsf{Fun}(\mathscr{T}, \mathsf{Set})$ of set-valued functors on $\mathscr{T}$ is locally finitely presentable. Indeed it is cocomplete (see 2.15.4, volume 1) and the representable functors constitute a family of dense, thus strong generators (see 4.5.17.b, 4.5.5 and 4.3.6, volume 1). Given $T \in \mathscr{T}$, the functor on $\mathsf{Fun}(\mathscr{T}, \mathsf{Set})$ represented by $\mathscr{T}(T, -)$ is just the evaluation at T (see the Yoneda lemma, 1.3.3, volume 1), thus it preserves all colimits (see 2.15.1, volume 1) and in particular filtered colimits.

5.2.2.c Let $\mathscr{T}$ be a small category with finite limits. The corresponding category $\mathsf{Lex}(\mathscr{T}, \mathsf{Set})$ of finite limit preserving functors $\mathscr{T} \longrightarrow \mathsf{Set}$ is locally finitely presentable. Indeed it is cocomplete (see 6.2.4) and the representable functors constitute a family of dense, thus strong generators (same argument as in example b, since the representable functors are left exact). Again given $T \in \mathscr{T}$, the functor on $\mathsf{Lex}(\mathscr{T}, \mathsf{Set})$ represented by $\mathscr{T}(T, -)$ is just the evaluation at T (see the Yoneda lemma again), thus it preserves filtered colimits (see 6.2.2, volume 1).

5.2.2.d Following the lines of section 6.4, volume 1, example 5.2.2.c generalizes immediately to the case of a regular cardinal α. If $\mathscr{T}$ is a small category with α-limits, the category α-$\mathsf{Lex}(\mathscr{T}, \mathsf{Set})$ of α-limit preserving functors $\mathscr{T} \longrightarrow \mathsf{Set}$ is locally α-presentable.

5.2.2.e The category Ban_1 of real Banach spaces and linear contractions is locally $\aleph_1$-presentable. We know that Ban_1 is cocomplete (see 2.8.6, volume 1) and admits $\mathbb{R}$ as a strong generator (see 4.5.17.e, volume 1). It remains to prove that $\mathsf{Ban}_1(\mathbb{R}, -)$, i.e. the unit ball functor, preserves $\aleph_1$-filtered colimits. Let us thus consider an $\aleph_1$-filtered diagram in Ban_1 and the corresponding colimit $L = \operatorname{colim}_{i \in I} B_i$ computed in the category of real vector spaces; since this colimit is filtered, it is computed as in the category of sets (see 3.4.2, volume 1). Define a norm on L by

$$\|l\| = \inf\{\|b\| \mid b \in B_i,\ i \in I,\ [b] = l\}$$

for every element $l \in L$, seen as an equivalence class of elements in the B_i's. If $\|l\| = 0$, for every $n > 0$, $n \in \mathbb{N}$, we find $b_n \in B_{i_n}$ such that $[b_{i_n}] = l$ and $\|b_{i_n}\| < \frac{1}{n}$, where $[b_{i_n}]$ denotes the equivalence class of b_{i_n} in the colimit. By $\aleph_1$-filteredness, we get an index i such that all the elements b_{i_n} are mapped onto the same element $b \in B_i$; then $\|b\| \leq \|b_{i_n}\| < \frac{1}{n}$ for every n, thus $\|b\| = 0$ and $b = 0$; this implies $l = [b] = 0$. From this it follows easily that L has been provided with a norm. The space L is complete for this norm. Indeed given a Cauchy sequence $(l_k)_{k\in\mathbb{N}}$ in L, let us write $l_k = [b_k]$ with $b_k \in B_k$. Again by $\aleph_1$-filteredness we can choose all the b_k's in a same B_i. Given $\varepsilon > 0$, we have $\|l_k - l_m\| < \varepsilon$ for k, m sufficiently large; this implies that fixing k and m, $\|b_k - b_m\| < \varepsilon$ in some B_j and, again by $\aleph_1$-filteredness, we can choose j_ε independently of k, m. Now letting ε run through the numbers $\frac{1}{n}$, $n \neq 0$, $n \in \mathbb{N}$, we get a corresponding family of j_ε's; by $\aleph_1$-filteredness, we can map the corresponding B_{j_ε} into a unique B_j where the image of the elements b_k now constitute a Cauchy sequence. This yields a limit in B_j and thus in L. It is now straightforward to verify that L is indeed the colimit of the original diagram in Ban_1. Just by construction, the open unit ball of L is the filtered colimit (computed in Set) of the open unit balls of the B_i's. Now if $\|l\| = 1$ for some element $l \in L$, write l as $l = \lim(1 - \frac{1}{n})l$ and combine the argument concerning the open unit ball with a limit argument, again by $\aleph_1$-filteredness.

5.2.2.f The category Cat of small categories and functors is locally finitely presentable. Indeed, it is cocomplete (see 5.1.7) and the one-arrow category $\mathbf{2}$ is a strong generator (see 4.5.17.h, volume 1). Now a filtered colimit $\mathscr{L} = \operatorname{colim}_{i\in I} \mathscr{C}_i$ of small categories is easily computed: as far as the sets of objects are concerned, $|\mathscr{L}|$ is just the colimit $|\mathscr{L}| = \operatorname{colim}_{i\in I} |\mathscr{C}|_i$ in Set; next if $L, L' \in \mathscr{L}$, $\mathscr{L}(L, L')$ is the filtered colimit of the sets $\mathscr{C}_i(L_i, L_i')$ where $L = [L_i]$ and $L' = [L_i']$. Since giving a functor $\mathbf{2} \longrightarrow \mathscr{C}$ is just giving an arrow of $\mathscr{C}$, the construction of the colimit $\mathscr{L}$ indicates that $\mathsf{Cat}(\mathbf{2}, -)$ indeed preserves filtered colimits.

The aim of this section is to prove that every locally α-presentable category $\mathscr{M}$ is of the type α-$\mathsf{Lex}(\mathscr{T}, \mathsf{Set})$, for some small category $\mathscr{T}$ with α-limits. But let us first observe the following.

Proposition 5.2.3 *Let $\alpha < \beta$ be regular cardinals. Every locally α-presentable category is also locally β-presentable.*

Proof Follows from 5.1.2. □

Lemma 5.2.4 *Let α be a regular cardinal and $\mathscr{M}$ a locally α-presentable category. With the notation of 5.2.1, write $\mathscr{G}$ for the full subcategory of*

$\mathscr{M}$ generated by the generators $(G_i)_{i\in I}$. The full closure $\mathscr{P}$ of $\mathscr{G}$ in $\mathscr{M}$ under α-colimits exists and has the following properties:

(1) $\mathscr{P}$ is equivalent to a small category;

(2) $\mathscr{P}$ has α-colimits computed as in $\mathscr{M}$;

(3) every object in $\mathscr{P}$ is α-presentable in $\mathscr{M}$.

Proof It suffices to define $\mathscr{P}$ by transfinite induction:

- $\mathscr{P}_0 = \mathscr{G}$;
- $\mathscr{P}_{\beta+1}$ is the full subcategory of $\mathscr{M}$ whose objects can be obtained as colimit of a functor $\mathscr{J} \longrightarrow \mathscr{P}_\beta$, with $\mathscr{J}$ a category with strictly less than α arrows;
- $\mathscr{P}_\gamma = \bigcup_{\beta<\gamma}\mathscr{P}_\beta$ for a limit ordinal γ.

Then $\mathscr{P}$ is defined to be $\mathscr{P}_\alpha$.

Let us prove conditions (1) and (3) of the statement by induction on the ordinal β.

- $\mathscr{G}$ is small since I is a set and by assumption, each object G_i is α-presentable;
- up to equivalence, there is just a set of categories $\mathscr{J}$ with strictly less than α arrows, thus just a set of functors F: $\mathscr{J} \longrightarrow \mathscr{P}_\beta$, assuming $\mathscr{P}_\beta$ to be small; this implies the smallness of $\mathscr{P}_{\beta+1}$; condition (3) holds for $\mathscr{P}_{\beta+1}$ just by 5.1.4;
- the case of a limit ordinal is obvious.

Let us prove condition (2) of the statement. For a functor F: $\mathscr{J} \longrightarrow \mathscr{P}_\alpha$ where $\mathscr{J}$ has strictly less than α arrows, each arrow $F(j)$ is in some $\mathscr{P}_{\alpha_j}$ for $\alpha_j < \alpha$. By regularity of α, there exists $\beta < \alpha$ such that $\alpha_j < \beta$ for each j. Therefore the colimit of F is already in $\mathscr{P}_{\beta+1}$, still with $\beta+1 < \alpha$.

Observe that every α-cocomplete replete subcategory $\mathscr{Q}$, $\mathscr{G} \subseteq \mathscr{Q} \subseteq \mathscr{M}$, contains $\mathscr{P}_0$ and, by transfinite induction, every $\mathscr{P}_\beta$. So $\mathscr{P}$ is indeed the closure of $\mathscr{G}$ under α-colimits. □

Lemma 5.2.5 *Under the conditions of lemma 5.2.4:*

(1) for every object $M \in \mathscr{M}$, the category $\mathscr{P}/M$ is α-filtered;

(2) the objects of $\mathscr{P}$ constitute in $\mathscr{M}$ a dense family of generators.

Proof The objects of $\mathscr{P}/M$ are thus the pairs (P,p) with $P \in \mathscr{P}$ and p: $P \longrightarrow M$; an arrow f: $(P,p) \longrightarrow (Q,q)$ is a morphism f: $P \longrightarrow Q$ such that $q \circ f = p$. We write Γ: $\mathscr{P}/M \longrightarrow \mathscr{M}$ for the functor applying (P,p) on P and f on f (see 4.5.4, volume 1). The category $\mathscr{P}/M$ is

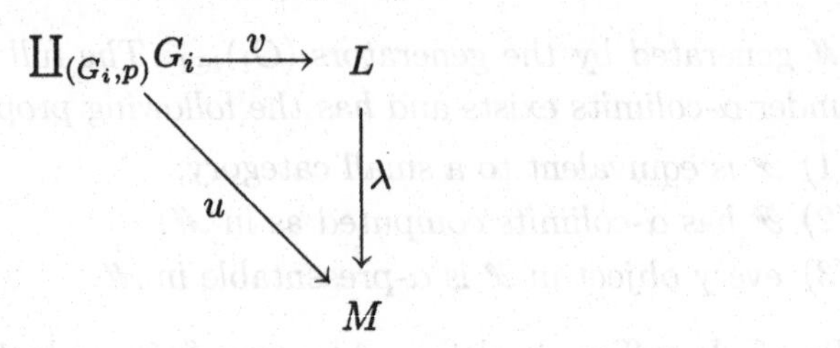

Diagram 5.1

certainly α-filtered since $\mathscr{P}$ has α-colimits (see 5.2.4). We must prove that the colimit of Γ is just $(p\colon P \longrightarrow M)_{(P,p)\in\mathscr{P}/M}$.

Let us consider the colimit $\big(L, s_{(P,p)}\big)_{(P,p)\in\mathscr{P}/M}$ of Γ in $\mathscr{M}$. Since the morphisms $p\colon \Gamma(P,p) \longrightarrow M$ constitute a cocone on Γ, we get a unique factorization $\lambda\colon L \longrightarrow M$ such that $\lambda \circ s_{(P,p)} = p$ for every $(P,p) \in \mathscr{P}/M$. We must prove that λ is an isomorphism.

We consider first diagram 5.1 where the coproduct is indexed by all the pairs (G_i, p) with $i \in I$, $G_i \in \mathscr{G}$ and $p\colon G_i \longrightarrow M$. Writing $\sigma_{(G_i,p)}\colon G_i \longrightarrow \coprod G_i$ for the canonical morphisms of the coproduct, u is the unique morphism such that $u \circ \sigma_{(G_i,p)} = p$ and v is the unique morphism such that $v \circ \sigma_{(G_i,p)} = s_{(G_i,p)}$. In particular,

$$\lambda \circ v \circ \sigma_{(G_i,p)} = \lambda \circ s_{(G_i,p)} = p = u \circ \sigma_{(G_i,p)},$$

which yields $\lambda \circ v = u$. By assumption, u is a strong epimorphism (see 4.5.3, volume 1), thus λ is a strong epimorphism as well (see 4.3.6, volume 1). So by 4.3.6, volume 1, again, it remains to prove that λ is a monomorphism.

Let us consider two morphisms $x, y\colon X \rightrightarrows L$ in $\mathscr{M}$ such that $\lambda \circ x = \lambda \circ y$. Since the G_i's constitute a family of generators, it suffices to prove that for each $i \in I$ and each $z\colon G_i \longrightarrow X$, $x \circ z = y \circ z$. But the colimit $\big(L, s_{(P,p)}\big)_{(P,p)\in\mathscr{P}/M}$ is α-filtered, since $\mathscr{P}/M$ is α-filtered, and G_i is α-presentable. Therefore $x \circ z$ and $y \circ z$ factor through some term of the colimit (see 5.1.3) and, by filteredness of the colimit, there is no restriction in supposing it is the same term. So let us fix $(P,p) \in \mathscr{P}/M$ and $x', y'\colon G_i \longrightarrow P$ such that $s_{(P,p)} \circ x' = x \circ z$ and $s_{(P,p)} \circ y' = y \circ z$. This yields

$$p \circ x' = \lambda \circ s_{(P,p)} \circ x' = \lambda \circ x \circ z, \quad p \circ y' = \lambda \circ s_{(P,p)} \circ y' = \lambda \circ y \circ z,$$

from which we get two morphisms in $\mathscr{P}/M$,

$$x'\colon (G_i, \lambda \circ x \circ z) \longrightarrow (P,p), \quad y'\colon (G_i, \lambda \circ y \circ z) \longrightarrow (P,p).$$

Therefore, since $\lambda \circ x = \lambda \circ y$,

$$x \circ z = s_{(P,p)} \circ x' = s_{(G_i,\lambda xz)} = s_{(G_i,\lambda yz)} = s_{(P,p)} \circ y' = y \circ z. \qquad \square$$

Lemma 5.2.6 *Under the conditions of 5.2.4, $\mathscr{P}$ is the full subcategory of α-presentable objects in $\mathscr{M}$.*

Proof By 5.2.4, it remains to prove that every α-presentable object $M \in \mathscr{M}$ is in $\mathscr{P}$. Using the notation of 5.2.5, we write M as a filtered colimit $M = \operatorname{colim}_{(P,p)} P$ where (P,p) runs through $\mathscr{P}/M$. By 5.1.4, the identity $1_M\colon M \longrightarrow M$ factors through some term of the colimit, yielding $(P,p) \in \mathscr{P}/M$ and $m\colon M \longrightarrow P$ such that $p \circ m = 1_M$. Thus M is a retract of P, i.e. (see 6.5.4) the coequalizer of the two morphisms $1_P, m \circ p\colon P \rightrightarrows P$. Since by 5.2.4 $\mathscr{P}$ is stable under α-colimits, M itself is in $\mathscr{P}$. $\square$

Theorem 5.2.7 *Let $\mathscr{M}$ be a locally α-presentable category, for some regular cardinal α. There exists a small α-complete category $\mathscr{T}$ such that $\mathscr{M}$ is equivalent to the category α-$\mathsf{Lex}(\mathscr{T}, \mathsf{Set})$ of set-valued α-left-exact functors. The dual of $\mathscr{T}$ is itself equivalent to the full subcategory of $\mathscr{M}$ generated by the α-presentable objects.*

Proof We use the notations of the previous lemmas and, by 5.2.4, choose a set Q of objects of $\mathscr{M}$ such that each object of $\mathscr{P}$ is isomorphic to some object of Q. We write $\mathscr{Q}$ for the full subcategory of $\mathscr{M}$ generated by the elements of Q; $\mathscr{T}$ is defined as the dual of $\mathscr{Q}$. We know by 5.2.5 that the objets of $\mathscr{Q}$ constitute a dense family of generators in $\mathscr{M}$, thus the functor

$$Y\colon \mathscr{M} \longrightarrow \mathsf{Fun}(\mathscr{Q}^*, \mathsf{Set}), \quad M \mapsto \mathscr{M}(-, M)$$

is full and faithful (see 4.5.14, volume 1). But each representable functor $\mathscr{M}(-, M)\colon \mathscr{M} \longrightarrow \mathsf{Set}$ transforms colimits into limits (see 2.9.5, volume 1); since $\mathscr{P}$ is stable in $\mathscr{M}$ for α-colimits (see 5.2.4), each functor $\mathscr{M}(-, M)\colon \mathscr{Q} \longrightarrow \mathsf{Set}$ transforms α-colimits into α-limits. Thus we have already defined a full and faithful functor

$$Y\colon \mathscr{M} \longrightarrow \alpha\text{-}\mathsf{Lex}(\mathscr{T}, \mathsf{Set}), \quad M \mapsto \mathscr{M}(-, M).$$

To get an equivalence of categories, it remains to prove that every α-left-exact functor $F\colon \mathscr{T} \longrightarrow \mathsf{Set}$ is isomorphic to $Y(M)$, for some object $M \in \mathscr{M}$. But Y preserves α-filtered colimits since, given a α-filtered colimit $L = \operatorname{colim}_{j \in J} M_j$ in $\mathscr{M}$,

$$Y(L)(Q) = \mathscr{M}(Q, L) = \mathscr{M}(Q, \operatorname*{colim}_{j\in J} M_j)$$
$$= \operatorname*{colim}_{j\in J} \mathscr{M}(Q, M_j) = \operatorname*{colim}_{j\in J} Y(M_j)(Q)$$

just because each $Q \in \mathscr{Q}$ is α-presentable; thus $Y(L) = \operatorname{colim}_{j\in J} Y(M_j)$ since α-filtered colimits are computed pointwise in $\alpha\text{-Lex}(\mathscr{T}, \mathsf{Set})$ (see 6.4.9, volume 1). Now an α-left-exact functor $F\colon \mathscr{T} \longrightarrow \mathsf{Set}$ is the α-filtered colimit $F\colon \operatorname{colim}_{(T,t)} \mathscr{T}(T, -)$, where (T,t) runs through the category of elements of F (see 054.7). Considering the composite

$$\mathsf{Elts}(F)^* \longrightarrow \mathscr{T}^* \cong \mathscr{Q} \subseteq \mathscr{M},$$

we get a functor defined on a small α-filtered category; write $\big(L, s_{(T,t)}\big)$ for its colimit. We have

$$Y(L) \cong Y\big(\operatorname*{colim}_{(T,t)} T\big) \cong \operatorname*{colim}_{(T,t)} Y(T) \cong \operatorname*{colim}_{(T,t)} \mathscr{M}(-, T)$$
$$\cong \operatorname*{colim}_{(T,t)} \mathscr{Q}(-, T) \cong \operatorname*{colim}_{(T,t)} \mathscr{T}(T, -)$$
$$\cong F$$

since T itself is an object of $\mathscr{Q}$. □

Corollary 5.2.8 *Let α be a regular cardinal. In a locally α-presentable category:*

(1) small limits and colimits exist;

(2) α-limits commute with α-filtered colimits;

(3) α-filtered colimits are universal.

Proof By 5.2.7 and 6.4.9, volume 1, conditions (1), (2) and (3) hold. □

Corollary 5.2.9 *Let α be a regular cardinal and $\mathscr{T}$ an α-complete category. In the category $\alpha\text{-Lex}(\mathscr{T}, \mathsf{Set})$ of α-left-exact functors, the α-presentable objects are exactly the representable functors.*

Proof The representable functors constitute a family of strong (and even dense) generators (see 5.2.2.d); this family is closed under α-colimits since the Yoneda embedding

$$Y\colon \mathscr{T} \longrightarrow \alpha\text{-Lex}(\mathscr{T}, \mathsf{Set}),\quad T \mapsto \mathscr{T}(T, -)$$

transforms α-limits into α-colimits (see 6.4.9, volume 1). Therefore the "minimality" of $\mathscr{P}$ in lemma 5.2.4 yields $\mathscr{P} = \mathscr{G} = Y(\mathscr{T})$; thus by 5.2.6 every α-presentable object is a representable functor. □

Proposition 5.2.10 *Let $\mathcal{M}$ be a locally presentable category. For every object $M \in \mathcal{M}$ there exists a regular cardinal α_M such that M is α_M-presentable.*

Proof Suppose $\mathcal{M}$ is locally α-presentable; we use the notation of 5.2.7. Since $\mathcal{Q}$ is small, the category $\mathcal{Q}/M$ is small; let us choose a regular cardinal α_M larger than α and such that $\mathcal{Q}/M$ has strictly less than α_M arrows. By 5.1.2, each object $Q \in \mathcal{Q}$ is α-presentable, thus α_M-presentable. So M is a α_M-colimit of α_M-presentable objects (see 5.2.4), thus M is α_M-presentable (see 5.1.4). □

5.3 Accessible categories

Bearing in mind lemma 5.2.5, we make the following definition:

Definition 5.3.1 *A category $\mathcal{M}$ is α-accessible, for some regular cardinal α, when:*
(1) $\mathcal{M}$ has α-filtered colimits;
(2) $\mathcal{M}$ has a set $(G_i)_{i \in I}$ of dense generators;
(3) writing $\mathcal{G}$ for the full subcategory of $\mathcal{M}$ generated by the generators G_i, the category $\mathcal{G}/M$ is α-filtered for every object $M \in \mathcal{M}$;
(4) each generator G_i is α-presentable.
A category is accessible if it is α-accessible for some regular cardinal α.

Clearly, various equivalent formulations of definition 5.3.1 could be given. Here is a useful "weakening" of it.

Proposition 5.3.2 *A category $\mathcal{M}$ is α-accessible, for some regular cardinal α, if and only if:*
(1) $\mathcal{M}$ has α-filtered colimits;
(2) there exists a set $(G_i)_{i \in I}$ of α-presentable objects such that each object $M \in \mathcal{M}$ is the colimit of some α-filtered diagram in the full subcategory $\mathcal{G}$ generated by the G_i's.

Proof Obviously, the conditions of 5.3.1 imply those of 5.3.2. Conversely, consider a functor $F\colon \mathcal{J} \longrightarrow \mathcal{G}$, with $\mathcal{J}$ a small α-filtered category and $(M, s_J)_{J \in \mathcal{J}}$ the colimit of F in $\mathcal{M}$. Let us prove that $\mathcal{G}/M$ is α-filtered.

By 5.1.3, every arrow $f\colon G_i \longrightarrow M$, with $i \in I$, factors through some $s_J\colon F(J) \longrightarrow M$, $J \in \mathcal{J}$. Thus, given a α-family (G_k, f_k) of objects of $\mathcal{G}/M$, the α-filteredness of $\mathcal{J}$ allows us to factor all the f_k's through the same $s_J\colon F(J) \longrightarrow M$, yielding morphisms $(G_k, f_k) \longrightarrow (F(J), s_J)$

in $\mathcal{G}/M$. In an analogous way if $g_k\colon (G,f) \longrightarrow (G',f')$ is an α-family of morphisms in $\mathcal{G}/M$, let us factor f' through some s_J via a morphism $g\colon G' \longrightarrow F(J)$. By the second condition of proposition 5.1.3, two morphisms $g \circ g_k, g \circ g_{k'}\colon G \rightrightarrows F(J)$ can be coequalized by some $F(j)\colon F(J) \longrightarrow F(J)'$, just because $s_J \circ g_k = f = s_J \circ g_{k'}$. Using once more the α-filteredness of $\mathcal{J}$, we can even choose a single morphism j having the required property for all pairs (k,k'). But this finally yields a morphism $F(j) \circ g\colon (G',f') \longrightarrow (F(J'), s_{J'})$ in $\mathcal{G}/M$ coequalizing all the morphisms g_k. This proves the α-filteredness of $\mathcal{G}/M$, thus condition 5.3.1.(3). But our argument proves also that $\left(F(J), s_J\right)_{i \in I}$ is cofinal in $\mathcal{G}/M$ (see 2.11.2, volume 1), implying condition 5.3.1.(2). □

As a first example we get

Proposition 5.3.3 *An accessible category is locally presentable if and only if it is cocomplete.*

Proof A locally presentable category is cocomplete by definition and accessible by 5.2.5 and 5.2.4. Conversely a cocomplete accessible category is locally presentable because a family of dense generators is also a family of regular (see 4.5.5, volume 1) and thus strong (see 4.3.6, volume 1) generators. □

Thus the two notions of locally presentable and accessible category differ essentially in the amount of colimits whose existence is required. The fundamental example of an accessible category is described in the following proposition.

Proposition 5.3.4 *Let $\mathcal{T}$ be a small category and α a regular cardinal. The category α-$\mathsf{Flat}(\mathcal{T}, \mathsf{Set})$ of α-flat functors $\mathcal{T} \longrightarrow \mathsf{Set}$ is α-accessible.*

Proof By 6.4.14, volume 1, α-$\mathsf{Flat}(\mathcal{T}, \mathsf{Set})$ has α-filtered colimits computed pointwise. By definition of an α-flat functor $F\colon \mathcal{T} \longrightarrow \mathsf{Set}$ (see 6.4.10, volume 1), its category of elements is α-cofiltered. By 4.5.17.b, volume 1, F is thus the α-filtered colimit of the representable functors above it, in $\mathsf{Fun}(\mathcal{T}, \mathsf{Set})$. Since the representable functors are α-flat (see 6.4.12, volume 1) and α-filtered colimits are computed pointwise in the categories $\mathsf{Fun}(\mathcal{T}, \mathsf{Set})$ and α-$\mathsf{Flat}(\mathcal{T}, \mathsf{Set})$ (see 2.15.2 and 6.4.14 of volume 1), the representable functors finally constitute a set of dense generators for α-$\mathsf{Flat}(\mathcal{T}, \mathsf{Set})$; as we observed, condition (3) of 5.3.1 is satisfied as well. By the Yoneda lemma the functor on α-$\mathsf{Flat}(\mathcal{T}, \mathsf{Set})$ represented by a representable functor $\mathcal{T}(T, -)$ is just the evaluation at T; this preserves α-filtered colimits since those are computed pointwise. So the representable functors are α-presentable in α-$\mathsf{Flat}(\mathcal{T}, \mathsf{Set})$. □

The example of α-flat functors is in fact generic, as attested by the following theorem.

Theorem 5.3.5 *Let α be a regular cardinal and $\mathscr{M}$ an α-accessible category. There exists a small category $\mathscr{T}$ such that $\mathscr{M}$ is equivalent to the category α-$\mathsf{Flat}(\mathscr{T}, \mathsf{Set})$ of α-flat functors $\mathscr{T} \longrightarrow \mathsf{Set}$.*

Proof We use the notation of 5.3.1 and define $\mathscr{T}$ as the dual $\mathscr{T} = \mathscr{G}^*$ of $\mathscr{G}$. Since the G_i's constitute a family of dense generators, the functor

$$Y\colon \mathscr{M} \longrightarrow \mathsf{Fun}(\mathscr{G}^*; \mathsf{Set}), \quad M \mapsto \mathscr{M}(-, M)$$

is full and faithful (see 4.5.14, volume 1). Now the category of elements of the functor $\mathscr{M}(-, M)\colon \mathscr{G} \longrightarrow \mathsf{Set}$, by the Yoneda lemma, is equivalent to the category $\mathscr{G}/M$; thus it is α-filtered by assumption. This proves that each functor $\mathscr{M}(-, M)\colon \mathscr{G}^* \longrightarrow \mathsf{Set}$ is α-flat (see 6.4.10, volume 1) so that we already have a full and faithful functor

$$Y\colon \mathscr{M} \longrightarrow \alpha\text{-}\mathsf{Flat}(\mathscr{T}, \mathsf{Set}), \quad M \mapsto \mathscr{M}(-, M).$$

To get an equivalence of categories, it remains to prove that every α-flat functor $F\colon \mathscr{T} \longrightarrow \mathsf{Set}$ is isomorphic to $Y(M)$, for some $M \in \mathscr{M}$. But the Yoneda embedding preserves α-filtered colimits since, given an α-filtered colimit $L = \operatorname{colim}_{j \in J} M_j$ in $\mathscr{M}$ and an object $G_i \in \mathscr{G}$,

$$\begin{aligned} Y(L)(G_i) &= \mathscr{M}(G_i, L) = \mathscr{M}\big(G_i, \operatorname*{colim}_{j \in J} M_j\big) \\ &= \operatorname*{colim}_{j \in J} \mathscr{M}(G_i, M_j) = \operatorname*{colim}_{j \in J} Y(M_j)(G_i), \end{aligned}$$

just because each $G_i \in \mathscr{G}$ is α-presentable; thus $Y(L) = \operatorname{colim}_{j \in J} Y(M_j)$ since α-filtered colimits are computed pointwise in α-$\mathsf{Flat}(\mathscr{T}, \mathsf{Set})$; see 6.4.14, volume 1. Now an α-flat functor $F\colon \mathscr{T} \longrightarrow \mathsf{Set}$ can be written as an α-filtered colimit $F = \operatorname{colim}_{(T,t)} \mathscr{T}(T, -)$, where (T, t) runs through the category of elements of F; see 6.4.13, volume 1. Considering the composite

$$\mathsf{Elts}(F)^* \longrightarrow \mathscr{T}^* \cong \mathscr{G} \subseteq \mathscr{M},$$

we get a functor defined on a small α-filtered category; write $(L, s_{(T,t)})$ for its colimit. We have

$$\begin{aligned} Y(L) &\cong Y\big(\operatorname*{colim}_{(T,t)} T\big) \cong \operatorname*{colim}_{(T,t)} Y(T) \cong \operatorname*{colim}_{(T,t)} \mathscr{M}(-, T) \\ &\cong \operatorname*{colim}_{(T,t)} \mathscr{G}(-, T) \cong \operatorname*{colim}_{(T,t)} \mathscr{T}(T, -) \\ &\cong F, \end{aligned}$$

since T itself is an object of $\mathscr{G}$. □

Proposition 5.3.6 *An accessible category is Cauchy complete.*

Proof Consider the category $\mathcal{I}$ with one single object $*$, the identity arrow on $*$ and an idempotent arrow $e\colon *\longrightarrow *$, thus $e \circ e = e$. Since $e \circ 1_* = e \circ e$, it follows immediately that $\mathcal{I}$ is α-filtered for every regular cardinal α. Therefore, given an accessible category $\mathcal{M}$ and a functor $F\colon \mathcal{I} \longrightarrow \mathcal{M}$, the colimit of F always exists. This implies the splitting of idempotents in $\mathcal{M}$; see 6.5.4, volume 1. □

Proposition 5.3.7 *Let $\mathcal{M}$ be an accessible category. For every object $M \in \mathcal{M}$, there exists a regular cardinal α_M such that M is α_M-presentable.*

Proof Suppose $\mathcal{M}$ is α-accessible; we use the notation of 5.3.1. For $M \in \mathcal{M}$, the category $\mathcal{G}/M$ is small; let us choose a regular cardinal α_M larger than α and such that $\mathcal{G}/M$ has strictly less than α_M arrows. By 5.1.2, each object $G_i \in \mathcal{G}$ is α_M-presentable. So M is a α_M-colimit of α_M-presentable objects, since the G_i's constitute a family of dense generators; thus M is α_M-presentable (see 5.1.4). □

Proposition 5.3.8 *Let $\mathcal{M}$ be a α-accessible category. The full subcategory of α-presentable objects is equivalent to a small category.*

Proof We use the notation of 5.3.1. Let $M \in \mathcal{M}$ be a α-presentable object; by definition of α-accessibility, we can write M as a small α-filtered colimit $M \cong \operatorname{colim}_{(G,g)\in\mathcal{G}/M} G$. The identity arrow $1_M\colon M \longrightarrow M$ factors through some term of this colimit, by α-presentability of M (see 5.1.3); this yields an index $(G, g) \in \mathcal{G}/M$ and an arrow $f\colon M \longrightarrow G$ such that $g \circ f = 1_M$. By 6.5.4, volume 1, M is the equalizer of the pair of morphisms $1_M, f \circ g\colon G \rightrightarrows G$. Since $\mathcal{G}$ is small, there is just a set of such pairs and thus (up to isomorphisms) just a set of α-presentable objects. □

Proposition 5.3.9 *Let $\mathcal{M}$ be a α-accessible category. With the notation of 5.3.1, the family of functors*

$$\mathcal{M}(G_i, -)\colon \mathcal{M} \longrightarrow \mathsf{Set}$$

collectively reflects isomorphisms (see 4.5.7, volume 1).

Proof Given $f\colon A \longrightarrow B$ in $\mathcal{M}$, suppose that each $\mathcal{M}(G_i, f)$ is an isomorphism. Considering the full and faithful embedding of 4.5.14, volume 1,

$$\Gamma\colon \mathcal{M} \longrightarrow \mathsf{Fun}(\mathcal{G}^*, \mathsf{Set}),\ M \mapsto \mathcal{M}(-, M),$$

we have that $\mathcal{M}(G_i, f) = \Gamma(f)_{G_i}$ is an isomorphism for each $G_i \in \mathcal{G}$. Thus $\Gamma(f)$ is an isomorphism, as is f, since Γ is full and faithful. □

Proposition 5.3.9 could be interpreted as the fact that the objects $G_i \in \mathcal{G}$ constitute a family of strong generators (see 4.5.13, volume 1).

5.4 Raising the degree of accessibility

When a problem involves several accessible categories, it is clearly useful (and often essential) to be able to consider them all as α-accessible, for the same regular cardinal α. In the case of locally presentable categories, this can be achieved in an obvious way: given a family $(\alpha_i)_{i \in I}$ of regular cardinals, just choose a regular cardinal α bigger than all the α_i; by 5.2.3, every locally α_i-presentable category is also α-presentable.

The case of accessible categories is more subtle. If $\alpha < \beta$ are regular cardinals, every β-filtered colimit is certainly α-filtered and every α-presentable object is β-presentable (see 5.1.2). Therefore, considering the various conditions of definition 5.3.1, satisfied for a fixed regular cardinal α:

- condition (1) holds for every regular cardinal $\beta > \alpha$;
- condition (2) is independent of α;
- condition (3) holds for every regular cardinal $\beta < \alpha$;
- condition (4) holds for every regular cardinal $\beta > \alpha$.

Thus the obvious way to modify the degree of accessibility varies in opposite directions for conditions (1) and (4) and for condition (3).

The aim of this section is to show that nevertheless the degree of accessibility of a category can always be increased to arbitrarily large but well chosen regular cardinals. This is essentially a game on cardinal arithmetic, which leads to the final conclusion expressed in theorem 5.4.5 and corollary 5.4.8.

Definition 5.4.1 *Let α, β be regular cardinals. The cardinal α is sharply less than β (which we write $\alpha \triangleleft \beta$) when $\alpha < \beta$ and, for every set X of cardinality strictly less than β, there exists a cofinal subset of cardinality strictly less than β in the poset of those subsets of X with cardinality strictly less than α.*

We shall write $\mathcal{P}_\alpha(X)$ for the poset of those subsets of X with cardinality strictly less than α. The condition of definition 5.4.1 means thus the existence of a subset $\mathcal{Q} \subseteq \mathcal{P}_\alpha(X)$, with the cardinality of $\mathcal{Q}$ strictly less than β, and

$$\forall U \in \mathcal{P}_\alpha(X)\ \exists V \in \mathcal{Q}\ U \subseteq V.$$

We shall prove that when α is sharply less than β, every α-accessible category is also β-accessible.

Lemma 5.4.2 *Let $\alpha \triangleleft \beta$ be two regular cardinals. For every category $\mathscr{C}$ with strictly less than β arrows, there exists a subset of cardinality strictly less than β in the poset of those subcategories of $\mathscr{C}$ with strictly less than α arrows.*

Proof Every infinite set X has the same cardinal as its set of finite subsets. Since the subcategory $\mathscr{X}$ generated by a set X of arrows is given by all the finite composites of arrows in X, it follows that if X has cardinality strictly less than α, so does the subcategory $\mathscr{X}$ generated by X.

Under the conditions of the statement, let us choose $\mathcal{Q}$, a cofinal subset of cardinality strictly less than β in the poset $\mathcal{P}_\alpha\big(\mathsf{Ar}(\mathscr{C})\big)$, where $\mathsf{Ar}(\mathscr{C})$ is the set of arrows of $\mathscr{C}$. Each subcategory of $\mathscr{C}$ with strictly less than α arrows is contained in some $X \in \mathcal{Q}$, thus in the subcategory $\mathscr{X}$ generated by X, which still has strictly less than α arrows. Thus the set $\overline{\mathcal{Q}}$ of the subcategories $\mathscr{X}$ generated by the elements $X \in \mathcal{Q}$ satisfies the conditions of the statement. □

Lemma 5.4.3 *Let $\alpha \triangleleft \beta$ be two regular cardinals. If $\mathscr{A}$ is an α-filtered category, every subcategory $\mathscr{B} \subseteq \mathscr{A}$ with cardinality strictly less than β is contained in an α-filtered subcategory $\mathscr{C}$ whose cardinality is still strictly less than β.*

Proof We shall construct $\mathscr{C}$ by transfinite induction.

- $\mathscr{C}_0 = \mathscr{B}$, thus $\#\mathscr{C}_0 < \beta$.
- Suppose that $\mathscr{C}_\lambda$ is defined, for some ordinal λ, with the condition $\#\mathscr{C}_\lambda < \beta$. Since $\alpha \triangleleft \beta$, we choose a cofinal subset $\mathcal{Q}$, with $\#\mathscr{A} < \beta$, in the poset $\mathsf{Cat}_\alpha(\mathscr{C}_\lambda)$ of those subcategories of $\mathscr{C}_\lambda$ with strictly less than α arrows (see 5.4.2). Since $\mathscr{A}$ is α-filtered, for every $\mathscr{X} \in \mathcal{Q}$ we can choose a cocone $(s_X\colon X \longrightarrow A_\mathscr{X})_{X \in \mathscr{X}}$ on $\mathscr{X}$ in $\mathscr{A}$; see 6.4.4, volume 1. We define $\mathscr{C}_{\lambda+1}$ as the subcategory generated by $\mathscr{C}_\lambda$ and all the morphisms s_X, for all $X \in \mathscr{X}$ and $\mathscr{X} \in \mathcal{Q}$. Since the cardinality of $\mathcal{Q}$ and of each $\mathscr{X}$ is strictly less than β, $\#\mathscr{C}_{\lambda+1} < \beta$ by regularity of β.

- If $\mu \leq \beta$ is a limit ordinal, we define $\mathscr{C}_\mu = \bigcup_{\lambda<\mu}\mathscr{C}_\lambda$; by regularity of β, $\#\mathscr{C}_\mu < \beta$. Putting $\mathscr{C} = \mathscr{C}_\beta$, we have thus constructed a subcategory containing $\mathscr{B}$ and with cardinality strictly less than β.

To prove that $\mathscr{C}$ is α-filtered, let us consider a category $\mathscr{J}$ with strictly less than α morphisms and a functor $F: \mathscr{J} \longrightarrow \mathscr{C}$. For each arrow $j \in \mathscr{J}$, F_j is in some $\mathscr{C}_\lambda$, with $\lambda < \beta$. By regularity of β, all arrows F_j are in some $\mathscr{C}_\lambda$, for a unique λ, since $\#\beta < \alpha < \beta$. Therefore there is a cocone on F in $\mathscr{C}_{\lambda+1}$, thus in $\mathscr{C}$; see 6.4.4, volume 1. □

Lemma 5.4.4 *Let $\alpha \triangleleft \beta$ be two regular cardinals. If $\mathscr{A}$ is an α-filtered category, the poset of α-filtered subcategories of $\mathscr{A}$ with cardinality strictly less than β is β-filtered.*

Proof If $(\mathscr{B}_i)_{i \in I}$ is a family of α-filtered subcategories of $\mathscr{A}$, with $\#I < \beta$ and for every $i \in I$, $\#\mathscr{B}_i < \beta$, then by regularity of β the subcategory $\mathscr{B}$ generated by the union of the $\mathscr{B}_i$'s still has cardinality strictly less than β. One concludes the proof by 5.4.3. □

Theorem 5.4.5 *Let $\alpha \triangleleft \beta$ be two regular cardinals. Every α-accessible category is also β-accessible.*

Proof We use the notation of 5.3.1. Certainly $\mathscr{M}$ has β-filtered colimits since it has α-filtered colimits.

Let us now consider all the objects $H \in \mathscr{M}$ which can be written as the colimit of a functor

$$\mathscr{X} \subseteq \mathscr{G}/H \xrightarrow{\quad\Gamma\quad} \mathscr{M}$$

where $\mathscr{X}$ is an α-filtered subcategory of cardinality strictly less than β and Γ is the usual forgetful functor; we write $\mathscr{H}$ for the corresponding full subcategory of $\mathscr{M}$. By 5.1.4, every object $H \in \mathscr{H}$ is β-presentable. But since $\mathscr{G}$ is a dense family of generators, each object $H \in \mathscr{H}$ is in fact the colimit of the forgetful functor $\mathscr{G}/H \longrightarrow \mathscr{M}$; see 4.5.4, volume 1. Therefore H is completely determined (up to isomorphism) by a diagram of objects G_i whose cardinality is strictly less than β. Since $\mathscr{G}$ is small, this implies that $\mathscr{H}$ itself is equivalent to a small category. Write $\mathscr{K}$ for a small full subcategory of $\mathscr{H}$, equivalent to $\mathscr{H}$ and containing $\mathscr{G}$. We shall prove that $\mathscr{K}$ is a subcategory exhibiting the β-accessibility of $\mathscr{M}$, via lemma 5.3.2.

Given an object $M \in \mathscr{M}$, consider the situation

$$\mathscr{G}/M \subseteq \mathscr{K}/M \xrightarrow{\quad\Gamma\quad} \mathscr{M}$$

where Γ is the usual forgetful functor (see 3.5.4, volume 1). Let us consider a cocone $t_{(K,k)}: \Gamma(K,k) \longrightarrow N$ on Γ; its restriction to $\mathscr{G}/M$ yields a unique factorization $m: M \longrightarrow N$ such that $m \circ k = t_{(K,k)}$, for every $K \in \mathscr{G}$, just because $\mathscr{G}$ is a family of dense generators. To get the relation $m \circ k = t_{(K,k)}$ for every $K \in \mathscr{K}$, it suffices to prove $m \circ k \circ l = t_{(K,k)} \circ l$ for every $i \in I$ and every $l: G_i \longrightarrow K$; and indeed

$$m \circ k \circ l = t_{(G_i, k \circ l)} = t_{(K,k)} \circ l$$

since $l: (G_i, k \circ l) \longrightarrow (K,k)$ is a morphism of $\mathscr{K}/M$. Thus $\mathscr{K}$ is a family of dense generators which, as already observed, are β-presentable.

Let us prove now that every object $M \in \mathscr{M}$ is a β-filtered colimit of objects in $\mathscr{K}$. $\mathscr{G}/M$ is α-filtered by assumption. For every α-filtered subcategory $\mathscr{X} \subseteq \mathscr{G}/M$ with $\#\mathscr{X} < \beta$, the colimit of the functor

$$\mathscr{X} \subseteq \mathscr{G}/M \xrightarrow{\Gamma} \mathscr{M}$$

is (up to an isomorphism) some object $K_{\mathscr{X}}$ of $\mathscr{K}$. The various inclusions $\mathscr{X} \subseteq \mathscr{Y} \subseteq \mathscr{G}/M$ yield corresponding factorizations

$$K_{\mathscr{X}} \longrightarrow K_{\mathscr{Y}} \xrightarrow{s_{\mathscr{Y}}} M$$

in $\mathscr{M}$. By 5.4.4, we obtain in this way a β-filtered diagram in $\mathscr{K}$ together with a cocone $(s_{\mathscr{X}}: K_{\mathscr{X}} \longrightarrow M)_{\mathscr{X}}$ on this diagram. This cocone is in fact a colimit one, just by the associativity argument of 2.5.7, volume 1, where "finite" is now replaced by "strictly less than α". Thus M is indeed the colimit of a β-filtered diagram in $\mathscr{K}$. □

So theorem 5.4.5 shows that $\triangleleft$ is a good condition for increasing the degree of accessibility. One can even prove that for two regular cardinals α, β, $\alpha \triangleleft \beta$ iff every α-accessible category is β-accessible (see 5.7.10). But we shall not need this result.

Now the problem which remains is to prove that given a regular cardinal α, or even a family $(\alpha_i)_{i \in I}$ of regular cardinals, it is always possible to find a regular cardinal β such that $\alpha \triangleleft \beta$, or $\alpha_i \triangleleft \beta$ for every $i \in I$. This is just a small game on cardinal arithmetic, at some level which requires of course the axiom of choice. Let us recall some classical notation and results which can be found in most books on set theory and cardinal arithmetic. All Greek letters denote cardinals.

- α^+ stands for the successor cardinal of α;
- for α infinite, α^+ is always a regular cardinal;
- $\alpha^{<\beta}$ stands for $\sum_{\gamma<\beta} \alpha^\gamma$;
- if $\beta \geq 2$, $\alpha \leq \alpha^{<\beta}$;

- if β is a regular cardinal,

$$(\alpha^{<\beta})^\gamma = \begin{cases} \alpha^{<\beta} & \text{if } \gamma < \beta, \\ \alpha^\gamma & \text{if } \gamma \geq \beta; \end{cases}$$

- if X is a set with infinite cardinal α and $\beta \leq \alpha$,

$$\alpha^\beta = \#\{Y \mid Y \subseteq X,\ \#Y = \beta\};$$

- if X is a set with infinite cardinal α and $\beta \leq \alpha$,

$$\#\mathcal{P}_\beta(X) = \#\{Y \mid Y \subseteq X,\ \#Y < \beta\} = \sum_{\gamma<\beta} \alpha^\gamma = \alpha^{<\beta};$$

- an infinite set has the same cardinality as its set of finite subsets, i.e. $\alpha^{<\aleph_0} = \alpha$ for every infinite cardinal α;
- if $\alpha_i < \beta_i$ for every $i \in I$, then $\sum_{i\in I} \alpha_i < \prod_{i\in I} \beta_i$;
- if $\alpha \leq \beta$ with β infinite, $\alpha + \beta = \beta$;
- if α is infinite, $\alpha = \alpha^2$.

The notation $\#Z$ stands clearly for the cardinality of the set Z.

Lemma 5.4.6 *Let $\alpha, \beta, \lambda, \mu$ be cardinals with α, β regular. If $\alpha < \beta$ and for every $\lambda < \alpha$, $\mu < \beta$ one has $\mu^\lambda < \beta$, then $\alpha \lhd \beta$.*

Proof Consider a set X with cardinality $\mu < \beta$. One has $\#\mathcal{P}_\alpha(X) = \mu^{<\alpha}$; but since β is regular, our assumption implies $\mu^{<\alpha} = \sum_{\lambda<\alpha} \mu^\lambda < \beta$. So $\mathscr{P}_\alpha(X)$ itself has cardinality strictly less than β and, certainly, is cofinal in itself. □

Proposition 5.4.7 *Let β be a regular cardinal. There exists a regular cardinal $\overline{\beta} > \beta$ such that, for every regular cardinal $\alpha < \beta$, one has $\alpha \lhd \overline{\beta}$.*

Proof Let us put $\overline{\beta} = (2^{<\beta})^+$; as a successor, this is a regular cardinal. We use lemma 5.4.6 to prove that $\alpha < \beta$ implies $\alpha \lhd \overline{\beta}$, for a regular cardinal α. We choose cardinals $\lambda < \alpha$ and $\mu < \overline{\beta}$; we must check that $\mu^\lambda < \overline{\beta}$. Since $\mu < (2^{<\beta})^+$, we have $\mu \leq 2^{<\beta}$ and thus

$$\mu^\lambda \leq (2^{<\beta})^\lambda = 2^{<\beta} < (2^{<\beta})^+ = \overline{\beta}$$

since $\lambda < \alpha < \beta$. □

Corollary 5.4.8 *Let $(\alpha_i)_{i\in I}$ be a family of regular cardinals. There exists a regular cardinal β such that, for every $i \in I$, $\alpha_i \lhd \beta$.*

Proof Choose a regular cardinal γ such that $\alpha_i < \gamma$, for every $i \in I$. Put $\beta = \overline{\gamma}$ and conclude the proof by 5.4.7. □

5.5 Functors with rank

We shall now study those functors which are compatible with the notion of accessibility.

Definition 5.5.1 *A functor $F\colon \mathscr{A} \longrightarrow \mathscr{B}$ has rank α, for some regular cardinal α, when F preserves α-filtered colimits. It has rank when it has rank α for some regular cardinal α.*

Clearly, the study of functors with rank α can be interesting only in the case where α-filtered colimits exist in the corresponding categories.

Proposition 5.5.2 *Let $\alpha \leq \beta$ be regular cardinals. Every functor with rank α has also rank β.*

Proof Every β-filtered colimit is α-filtered. □

The following propositions characterize functors with rank in the case of accessible categories.

Proposition 5.5.3 *Let $\mathscr{A}$ be an accessible category. A set-valued functor $F\colon \mathscr{A} \longrightarrow \mathsf{Set}$ has rank if and only if it is a small colimit of representable functors.*

Proof Suppose first that we can write $F \cong \operatorname{colim}_{i\in I} \mathscr{A}(A_i, -)$ where the corresponding diagram is small. Each object A_i has some rank α_i; choosing a regular cardinal α such that $\alpha \geq \alpha_i$ for each i, 5.1.2 implies that each A_i is α-presentable. Then each functor $\mathscr{A}(A_i, -)$ preserves α-filtered colimits, and thus so does the colimit F by 2.12.1, volume 1:

$$\begin{aligned} F\big(\operatorname*{colim}_{j\in J} B_j\big) &\cong \operatorname*{colim}_{i\in I} \mathscr{A}\big(A_i, \operatorname*{colim}_{j\in J} B_j\big) \\ &\cong \operatorname*{colim}_{i\in I} \operatorname*{colim}_{j\in J} \mathscr{A}(A_i, B_j) \\ &\cong \operatorname*{colim}_{j\in J} \operatorname*{colim}_{i\in I} \mathscr{A}(A_i, B_j) \\ &\cong \operatorname*{colim}_{j\in J} F(B_j) \end{aligned}$$

for an α-filtered colimit $\operatorname{colim}_{j\in J} B_j$. Thus F has rank α.

Conversely suppose F has rank β and $\mathscr{A}$ is γ-accessible. By 5.5.2, 5.4.8 and 4.4.5, volume 1, we can choose a regular cardinal α such that F has rank α and $\mathscr{A}$ is α-accessible. Using the notation of 5.3.1, we consider the restriction of F to $\mathscr{G}$ and the corresponding category of elements,

$$\mathsf{Elts}(F\circ i) \xrightarrow{\ \phi\ } \mathscr{G} \rightarrowtail^{\ i\ } \mathscr{A} \xrightarrow{\ F\ } \mathsf{Set}.$$

Writing Y for the contravariant Yoneda embedding,

$$\mathsf{Elts}(F\circ i) \xrightarrow{\ \phi\ } \mathscr{G} \xrightarrow{\ Y\ } \mathsf{Fun}(\mathscr{A}, \mathsf{Set}),$$

we can consider the colimit of this composite, since $\mathsf{Elts}(F \circ i)$ is small, due to the smallness of $\mathscr{G}$. By the Yoneda lemma, an object $(G, g) \in \mathsf{Elts}(F \circ i)$ determines a corresponding natural transformation $\overline{g}$: $\mathscr{A}(G, -) \Rightarrow F$; this yields a factorization

$$\varphi\colon \operatorname*{colim}_{(G,g)} \mathscr{A}(G, -) \Rightarrow F$$

and it suffices to prove it is an isomorphism. Each functor $\mathscr{A}(G, -)$ preserves α-filtered colimits since G is α-presentable, thus the colimit also preserves α-filtered colimits (by same argument as in the first part of the proof); on the other hand F preserves α-filtered colimits. By α-accessibility of $\mathscr{A}$ every object $A \in \mathscr{A}$ can be presented as the α-filtered colimit of the functor $\mathscr{G}/A \longrightarrow \mathscr{A}$. Since both $\operatorname{colim}_{(G,g)} \mathscr{A}(G, -)$ and F preserve this α-filtered colimit, the isomorphism $\operatorname{colim}_{(G,g)} \mathscr{A}(G, A) \cong F(A)$ will hold as soon as the isomorphism $\operatorname{colim}_{(G,g)} \mathscr{A}(G, G') \cong F(G')$ holds for every object $G' \in \mathscr{G}$. But this is equivalent to

$$(F \circ i)(G') = \operatorname*{colim}_{(G,g) \in \mathsf{Elts}(F \circ i)} \mathscr{G}(G, G')$$

which is precisely the content of 2.15.6, volume 1. □

Proposition 5.5.4 *Let F: $\mathscr{A} \longrightarrow \mathscr{B}$ be a functor between accessible categories $\mathscr{A}, \mathscr{B}$. Then F has rank iff for every object $B \in \mathscr{B}$, the composite functor $\mathscr{B}(B, F-)$: $\mathscr{A} \longrightarrow \mathsf{Set}$ has rank.*

Proof Suppose F has rank α and fix $B \in \mathscr{B}$; from 5.3.6 B is β-presentable for some regular cardinal β. By 5.5.2 and 5.1.2, we choose a regular cardinal γ such that F has rank γ and B is γ-presentable. Then both functors F and $\mathscr{B}(B, -)$ preserve γ-filtered colimits and thus so does the composite $\mathscr{B}(B, F-)$.

Conversely, by 5.4.5 and 5.4.8 choose a regular cardinal α such that $\mathscr{A}$ and $\mathscr{B}$ are α-accessible. Consider a family $B_i \in \mathscr{B}$, $i \in I$, of objects of $\mathscr{B}$ exhibiting the α-accessibility of $\mathscr{B}$, as in 5.3.1. By assumption, each composite $\mathscr{B}(B_i, F)$ preserves α_i-filtered colimits for some regular cardinal α_i. Choose β a regular cardinal bigger than α and each α_i; then $\mathscr{A}$ and $\mathscr{B}$ have β-filtered colimits and the functors $\mathscr{B}(B_i, -)$, $\mathscr{B}(B_i, F-)$ preserve them. To conclude that F preserves β-filtered colimits, it remains to show that the family $\mathscr{B}(B_i, -)$ of functors reflects β-filtered colimits or, equivalently, by the argument of 2.9.7, volume 1, reflects isomorphisms. This is the case by 5.3.8. □

Proposition 5.5.3, together with 5.5.4, shows that a functor between accessible categories has rank precisely when it satisfies good smallness conditions. Here is another striking smallness condition satisfied by those functors (see 3.3.2, volume 1).

Proposition 5.5.5 *Every functor with rank $F: \mathscr{A} \longrightarrow \mathscr{B}$ between accessible categories satisfies the solution set condition at every object $B \in \mathscr{B}$; see 3.3.2, volume 1.*

Proof Fix $B \in \mathscr{B}$, which is β-presentable for some regular cardinal β (see 5.3.6). By 5.1.2, 5.5.2, 5.4.5 and 5.4.8 choose a regular cardinal α such that B is α-presentable, $\mathscr{A}, \mathscr{B}$ are α-accessible and F has rank α. With the notation of 5.3.1, write $\mathscr{G}$ for the small full subcategory of $\mathscr{A}$ exhibiting the α-accessibility of $\mathscr{A}$ as in 5.3.1.

Given $A \in \mathscr{A}$ and $b: B \longrightarrow FA$, write $A \cong \operatorname{colim}_{i \in I} G_i$ where $G_i \in \mathscr{G}$ and the colimit is α-filtered (see 5.3.1). By choice of the cardinal α, $F(A) \cong \operatorname{colim}_{i \in I} F(G_i)$ so that the morphism b factors through some $F(s_i): F(G_i) \longrightarrow F(A)$, where s_i is a canonical morphism of the colimit (see 5.1.3). Thus $\mathscr{G}$ is a solution set for B. □

With 3.3.3, volume 1, in mind, the next result is in a way related to the previous one.

Proposition 5.5.6 *If a functor $F: \mathscr{A} \longrightarrow \mathscr{B}$ between accessible categories has a left adjoint, it has rank.*

Proof Choose first a regular cardinal α such that both $\mathscr{A}$ and $\mathscr{B}$ are α-accessible (see 5.4.5 and 5.4.8). Using the notation of 5.3.1, write $\mathscr{G}$ for the small full subcategory of $\mathscr{B}$ exhibiting the α-accessibility of $\mathscr{B}$.

Writing G for the left adjoint to F, each object $G(B)$, $B \in \mathscr{G}$, is α_B-presentable for some regular cardinal α_B. We fix a regular cardinal β bigger than α and each α_B. We shall prove that F has rank β.

Let $A = \operatorname{colim}_{i \in I} A_i$ be a β-filtered colimit in $\mathscr{A}$. Since each $B \in \mathscr{G}$ and the corresponding $G(B) \in \mathscr{A}$ are β-presentable (see 5.1.2), one deduces from the adjunction $G \dashv F$ (see 3.1.5, volume 1):

$$\begin{aligned}
\mathscr{B}\big(B, F(\operatorname*{colim}_{i \in I} A_i)\big) &\cong \mathscr{A}\big(G(B), \operatorname*{colim}_{i \in I} A_i\big) \\
&\cong \operatorname*{colim}_{i \in I} \mathscr{A}\big(G(B), A_i\big) \\
&\cong \operatorname*{colim}_{i \in I} \mathscr{B}\big(B, F(A_i)\big) \\
&\cong \mathscr{B}\big(B, \operatorname*{colim}_{i \in I} F(A_i)\big).
\end{aligned}$$

Since this isomorphism holds for every $B \in \mathscr{G}$, one gets the isomorphism $F(\operatorname{colim}_{i \in I} A_i) \cong \operatorname{colim}_{i \in A_i} F(A_i)$; see 5.3.8. □

Putting together the previous results, we get the following form of the adjoint functor theorem (see 3.3.3, volume 1).

Theorem 5.5.7 *Let $F: \mathscr{A} \longrightarrow \mathscr{B}$ be a functor between two locally presentable categories. The following conditions are equivalent:*
(1) F has a left adjoint;
(2) F has rank and preserves small limits.

Proof Locally presentable categories are accessible, complete and cocomplete (see 5.3.2 and 5.2.8).

If F has a left adjoint G, F preserves small limits (see 3.2.2, volume 1) and has rank (see 5.5.6). Conversely if F has rank it satisfies the solution set condition (see 5.5.4) and thus has a left adjoint (see 3.3.3, volume 1). □

Using the same type of arguments, we can now strengthen proposition 5.3.3.

Theorem 5.5.8 *Let $\mathscr{A}$ be an accessible category. The following conditions are equivalent:*
(1) $\mathscr{A}$ is locally presentable;
(2) $\mathscr{A}$ is cocomplete;
(3) $\mathscr{A}$ is complete.

Proof The equivalence between (1) and (2) is established by 5.3.3, while the implication (1) $\Rightarrow$ (3) is contained in 5.2.8. It remains to prove (3) $\Rightarrow$ (2).

Given a small category $\mathscr{I}$, we consider the functor

$$\Delta: \mathscr{A} \longrightarrow \mathsf{Fun}(\mathscr{I}, \mathscr{A})$$

mapping an object A to the constant functor on A; we must prove that Δ has a left adjoint (see 3.2.3, volume 1). But $\mathsf{Fun}(\mathscr{I}, \mathscr{A})$ is complete (see 2.15.2, volume 1) and since limits in $\mathsf{Fun}(\mathscr{I}, \mathscr{A})$ are computed pointwise, Δ preserves limits. By 3.3.3, volume 1, it remains to prove that Δ satisfies the solution set condition. This could be deduced from 5.5.5 via the proof of the result that given a small category $\mathscr{I}$, $\mathsf{Fun}(\mathscr{I}, \mathscr{A})$ is still accessible (see 5.7.5). The direct proof we give here is a crucial step in the proof that $\mathsf{Fun}(\mathscr{I}, \mathscr{A})$ is accessible.

Thus we consider a small category $\mathscr{I}$, a functor $F: \mathscr{I} \longrightarrow \mathscr{A}$, an object $A \in \mathscr{A}$ and a natural transformation $\theta: F \Rightarrow \Delta(A)$. Each $F(I)$, $I \in \mathscr{I}$, is α_I-presentable for some regular cardinal α_I. Applying 5.4.5 and 5.4.8, we raise the degree of accessibility of $\mathscr{A}$ to some regular cardinal α, strictly bigger than all the α_I and the cardinality of the set of arrows of

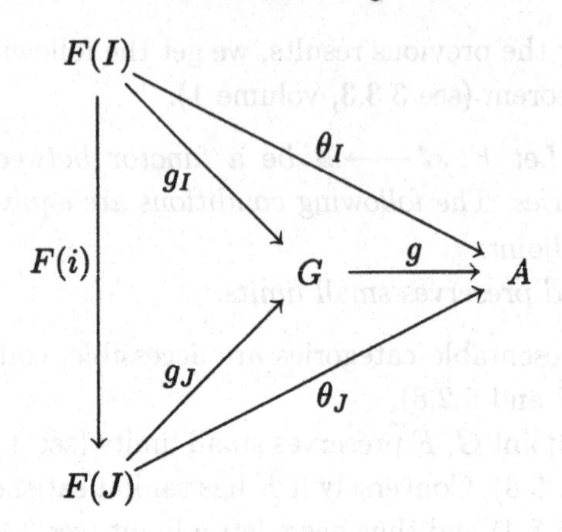

Diagram 5.2

$\mathscr{I}$. Using the notation of 5.3.1, we consider the full subcategory $\mathscr{G} \subseteq \mathscr{A}$ exhibiting the α-accessibility of $\mathscr{A}$. We write $A \cong \operatorname{colim}_{(G,g)} G$, where $(G,g) \in \mathscr{G}/A$, for the corresponding α-filtered colimit concerning A. We recall that α-filtered colimits in $\mathsf{Fun}(\mathscr{I}, \mathscr{A})$ are computed pointwise (see 2.15.1, volume 1).

For every object $I \in \mathscr{I}$, the morphism $\theta_I\colon F(I) \longrightarrow A$ factors through some canonical morphism $s_{(G,g)}\colon G \longrightarrow A$, since $F(I)$ is α-presentable (see 5.1.3). Since $\#\mathscr{I} < \alpha$, by α-filteredness we can assume that the same index (G,g) can be used for all objects $I \in \mathscr{I}$. Now given a morphism $i\colon I \longrightarrow J$ in $\mathscr{I}$, we have diagram 5.2 where $g \circ g_I = \theta_I$ and $g \circ g_J = \theta_J$. There is *a priori* no reason to have $g_J \circ F(i) = g_I$. But since

$$g \circ g_J \circ F(i) = \theta_J \circ F(i) = \theta_I = g \circ g_I,$$

5.1.3 implies the existence of some $f\colon (G,g) \longrightarrow (G',g')$ such that the relation $f \circ g_J \circ F(i) = f \circ g_I$ holds. Since $\#\mathscr{I} < \alpha$, by α-filteredness we can again assume that the same (G',g') can be used for all arrows $i \in \mathscr{I}$. Putting $\theta'_I = f \circ g_I$, we have thus defined a natural transformation $\theta'\colon F \Rightarrow \Delta(G')$ such that $\Delta(g') \circ \theta' = \theta$. This proves that $\mathscr{G}$ is a solution set condition for Δ at F. □

To conclude this section, let us come back to the case of monads with rank, already studied in section 4.6.

Theorem 5.5.9 *Let $\mathscr{A}$ be a locally presentable category and $\mathbb{T} = (T, \varepsilon, \mu)$ a monad with rank on $\mathscr{A}$ (i.e. the functor $T\colon \mathscr{A} \longrightarrow \mathscr{A}$ has rank). Under these conditions, the category $\mathscr{A}^{\mathbb{T}}$ of $\mathbb{T}$-algebras is locally presentable as well.*

Proof We write $U\colon \mathscr{A}^{\mathbb{T}} \longrightarrow \mathscr{A}$ for the forgetful functor and F for its left adjoint (see 4.1.4). The very difficult part of the proof is to show the cocompleteness of $\mathscr{A}$, but this has already been done in 4.3.6, since a locally presentable category is both complete and cocomplete (see 5.2.8).

Choose now, by 1.3.3 and 1.5.2, volume 1, a regular cardinal α such that $\mathscr{A}$ is locally α-presentable and T has rank α. By 5.2.8, U preserves α-filtered colimits. Let us choose a family $(G_i)_{i\in I}$ of strong generators of $\mathscr{A}$, each G_i being α-presentable.

The functor $\mathscr{A}^{\mathbb{T}}(F(G_i), -)\colon \mathscr{A}^{\mathbb{T}} \longrightarrow \mathsf{Set}$ is, via the adjunction $F \dashv U$, isomorphic to the functor $\mathscr{A}(G_i, U-)$. But $\mathscr{A}(G_i, -)$ preserves α-filtered colimits since G_i is α-presentable and U preserves α-filtered colimits by 5.2.8. Thus the composite $\mathscr{A}(G_i, U-) \cong \mathscr{A}^{\mathbb{T}}(F(G_i), -)$ preserves α-filtered colimits and $F(G_i)$ is thus α-presentable in $\mathscr{A}^{\mathbb{T}}$.

Moreover U reflects isomorphisms (see 4.1.4) as well as the family of functors $\mathscr{A}(G_i, -)$, $i \in I$, since the G_i's constitute a family of strong generators in $\mathscr{A}$; see 4.5.10, volume 1. Therefore the family of functors $\mathscr{A}(G_i, U-) \cong \mathscr{A}^{\mathbb{T}}(F(G_i), -)$, $i \in I$, reflects isomorphisms and $\big(F(G_i)\big)_{i\in I}$ is thus a family of strong generators in $\mathscr{A}^{\mathbb{T}}$. $\square$

5.6 Sketches

In section 5.3, we have characterized the accessible categories as being the categories of α-flat functors $\mathscr{T} \longrightarrow \mathsf{Set}$, for some small category $\mathscr{T}$. In this section, we shall investigate another characterization: the accessible categories are also the categories of models for some "sketch".

Definition 5.6.1 *A sketch is a triple* $\mathbb{S} = (\mathscr{T}, \mathscr{P}, \mathscr{I})$ *where:*

(1) $\mathscr{T}$ *is a small category;*

(2) $\mathcal{P}$ *is a set of cones on functors* $F\colon \mathscr{D} \longrightarrow \mathscr{T}$*, defined on small categories* $\mathscr{D}$*;*

(3) $\mathcal{I}$ *is a set of cocones on functors* $F\colon \mathscr{D} \longrightarrow \mathscr{T}$*, defined on small categories* $\mathscr{D}$*.*

Definition 5.6.2 *Let* $\mathbb{S} = (\mathscr{T}, \mathcal{P}, \mathcal{I})$ *be a sketch. A model of* $\mathbb{S}$ *is a functor* $M\colon \mathscr{T} \longrightarrow \mathsf{Set}$ *with the following properties:*

(1) if $\big(p_D\colon T \longrightarrow F(D)\big)_{D\in\mathscr{D}}$ *is a cone of* $\mathcal{P}$ *on the functor* $F\colon \mathscr{D} \longrightarrow \mathscr{T}$*, then* $\big(M(p_D)\colon M(T) \longrightarrow MF(D)\big)_{D\in\mathscr{D}}$ *is the limit of* $M \circ F$*;*

(2) *if* $\left(s_D\colon F(D)\longrightarrow T\right)_{D\in\mathscr{D}}$ *is a cocone of* $\mathcal{I}$ *on the functor* $F\colon\mathscr{D}\longrightarrow\mathscr{T}$, *then* $\left(M(s_D)\colon MF(D)\longrightarrow M(T)\right)_{D\in\mathscr{D}}$ *is the colimit of* $M\circ F$.

A morphism of $\mathbb{S}$*-models is just a natural transformation. We write* $\mathsf{Mod}_{\mathbb{S}}$ *for the category of* $\mathbb{S}$*-models.*

The definition of a sketch $\mathbb{S}=(\mathscr{T},\mathcal{P},\mathcal{I})$ could be generalized in a rather straightforward way by replacing the category $\mathscr{T}$ by a graph $\mathscr{G}$, together with a set $\mathcal{C}$ of commutativity conditions (see 5.1.1 and 5.1.5, volume 1). But applying 5.1.6, volume 1, one observes immediately that replacing the pair $(\mathscr{G},\mathcal{C})$ by the induced category $\mathscr{T}$ does not change the category of models.

On the other hand, definition 5.6.1 could have been strengthened by requiring that all the cones in $\mathcal{P}$ be already limit cones and all the cocones in $\mathcal{I}$ be colimit cones. This again does not change the range of categories of models, as corollary 5.6.7 shows.

We shall prove that the category of models of a sketch is accessible. For this, we need a lemma that people trained in logic will recognize as a form of the downward Löwenheim–Skolem theorem.

Lemma 5.6.3 *Let* $\mathbb{S}=(\mathscr{T},\mathcal{P},\mathcal{I})$ *be a sketch. There exists a regular cardinal* α *such that for every model* $M\in\mathsf{Mod}_{\mathbb{S}}$ *and every family of elements* $\left(x_o\in M(T_i)\right)_{i\in I}$, *with* $\#I<\alpha$, *there exists a submodel* $N\subseteq M$ *containing all the elements* x_i *and such that* $\#\left(\coprod_{T\in\mathscr{T}}NT\right)<\alpha$.

Proof Let us choose first a regular cardinal β which is strictly larger than all the following:

- the cardinality of the set of arrows of $\mathscr{T}$;
- the cardinality of the set $\mathcal{P}$;
- the cardinality of the set $\mathcal{I}$;
- the cardinality of the set of arrows of each small category $\mathscr{D}$ which appears in the definition of a cone of $\mathcal{P}$;
- the cardinality of the set of arrows of each small category $\mathscr{D}$ which appears in the definition of a cocone of $\mathcal{I}$.

We define α to be the regular cardinal $\alpha=(\beta^{<\beta})^+$; see section 5.4. Clearly, this definition makes sense because of all the smallness conditions in the definition of a sketch.

Now let $M\colon\mathscr{T}\longrightarrow\mathsf{Set}$ be a model of $\mathbb{S}$. A family $(X_T)_{T\in\mathscr{T}}$ of subsets $X_T\subseteq M(T)$ constitutes the values of some submodel $M'\subseteq M$, i.e. $X_T=M'(T)$, precisely when the following conditions are satisfied.

(1) *Functoriality condition:* for every arrow $t\colon T' \longrightarrow T$ of $\mathscr{T}$ and element $x \in X_{T'}$, $M(t)(x) \in X_T$.

(2) *Limit condition* (see 2.8, volume 1, for the description of limits in Set): for every cone $\left(p_D\colon T \longrightarrow F(D)\right)_{D\in\mathscr{D}}$ in $\mathcal{P}$ and every compatible family of elements $\left(x_D \in X_{F(D)}\right)_{D\in\mathscr{D}}$, the unique element $x \in M(T)$ such that $M(p_D)(x) = x_D$ for each $D \in \mathscr{D}$ belongs to X_T; here compatibility means $(MF)(d)(x_D) = x_{D'}$ for every $d\colon D \longrightarrow D'$ in $\mathscr{D}$).

(3) *Colimit condition* (see 2.8 and 2.4.6.b, volume 1, for the description of colimits in Set): for every cocone $\left(s_D\colon F(D) \longrightarrow T\right)_{T\in\mathscr{T}}$ in $\mathcal{I}$:

  (a) every element $x \in X_T$ has the form $M(s_D)(y)$ for some $D \in \mathscr{D}$ and $y \in X_{F(D)}$;

  (b) if two elements $x \in X_{F(D)}$ and $y \in X_{F(D')}$ are such that $M(s_D)(x) = M(s_{D'})(y)$, there exists a finite sequence $D = D_0, \dots, D_n = D'$ of objects of $\mathscr{D}$ and a corresponding sequence of elements $x_i \in X_{F(D_i)}$, $x_0 = x$, $x_n = y$ such that for each pair (x_i, x_{i+1}) of consecutive elements, there exists $d\colon D_i \longrightarrow D_{i+1}$ in $\mathscr{D}$ such that $(MF)(d)(x_i) = x_{i+1}$ or there exists $d\colon D_{i+1} \longrightarrow D_i$ in $\mathscr{D}$ such that $(MF)(d)(x_{i+1}) = x_i$.

For the sake of brevity, a sequence $x = x_0, \dots, x_i, \dots, x_n = y$ of elements as in (3)(b) will be referred to as a finite sequence of elements exhibiting the equivalence of x and y.

It is quite straightforward to realize conditions (1) to (3). For this we introduce various constructions, where $M\colon \mathscr{T} \longrightarrow \mathsf{Set}$ is thus a model of $\mathbb{S}$ and $X_T \subseteq M(T)$ for each $T \in \mathscr{T}$.

- *Construction f:* define X^f_T, for $T \in \mathscr{T}$, by the formula

  $$X^f_T = \left\{ M(t)(x) \,\middle|\, t\colon T' \longrightarrow T \text{ in } \mathscr{T},\ \ x \in X_{T'} \right\}.$$

  Clearly, $X_T \subseteq X^f_T$, just by choosing $t = 1_T$.

- *Construction l:* for each cone $\left(p_D\colon T \longrightarrow F(D)\right)_{D\in\mathscr{D}}$ in $\mathcal{P}$ and each compatible family $\left(x_D \in X_{F(D)}\right)_{D\in\mathscr{D}}$, add to X_T the unique element $x \in M(T)$ such that $M(p_D)(x) = x_D$ for each $D \in \mathscr{D}$; write X^l_T for this bigger set containing X_T.

- *Construction c^1:* for each cocone $\left(s_D\colon F(D) \longrightarrow T\right)_{D\in\mathscr{D}}$ and each element $x \in X_T$, choose one index $D \in \mathscr{D}$ and one element $y \in (MF)(D)$ such that $M(s_D)(y) = x$; add this element y to $X_{F(D)}$; write $X^{c^1}_{T'}$ for the bigger set containing $X_{T'}$ obtained at $T' \in \mathscr{T}$.

- *Construction c^2:* for each cocone $\left(s_D: F(D) \longrightarrow T\right)_{D\in\mathscr{D}}$ and each pair of elements $x \in X_{F(D)}$, $y \in X_{F(D')}$ such that $M(s_D)(x) = M(s_{D'})(y)$, choose one sequence $x = x_0, \ldots, x_i, \ldots, x_n = y$ of elements $x_i \in (MF)(D_i)$ exhibiting the equivalence between x and y and add each x_i to each $X_{F(D_i)}$; write $X_{T'}^{c^2}$ for the bigger set containing $X_{T'}$ obtained at $T' \in \mathscr{T}$.

The crucial point is now to verify that all the previous constructions are compatible with the choice of the regular cardinal α; more precisely

$$\left(\sum_{T\in\mathscr{T}} \#X_T < \alpha\right) \Rightarrow \left(\sum_{T\in\mathscr{T}} \#X_T^* < \alpha\right)$$

where $*$ stands for any of the four constructions. Let us thus assume $\#\sum_{T\in\mathscr{T}} X_T < \alpha$ and prove the corresponding conclusion for each of the four constructions. Observe that "$< \alpha$" is equivalent to "$\leq \beta^{<\beta}$" by definition of α.

Construction f: Given $T \in \mathscr{T}$, there are strictly less than α arrows $t: T' \longrightarrow T$ and strictly less than α elements in each $X_{T'}$; thus $\#X_T^f < \alpha$ by regularity of α.

Construction l: For each cone $\left(p_D: T \longrightarrow F(D)\right)_{D\in\mathscr{D}}$ in $\mathcal{P}$, $\#\mathscr{D} < \beta$ and $\#X_{F(D)} \leq \beta^{<\beta}$, for every $D \in \mathscr{D}$. The number of compatible families is thus bounded by $(\beta^{<\beta})^{\#\mathscr{D}} = \beta^{<\beta}$ (see 5.4), so is strictly less than α. But $\#X_T < \alpha$ and there are strictly less than $\beta < \alpha$ cones to consider; by regularity of α, this implies $\#X_T^l < \alpha$ and thus also $\sum_{T\in\mathscr{T}} \#X_T^l < \alpha$ since $\#\mathscr{T} < \alpha$.

Construction c^1: For each cocone $\left(s_D: F(D) \longrightarrow T\right)_{D\in\mathscr{D}}$ in $\mathcal{I}$ one element (at most) is added to the disjoint union $\coprod_{T\in\mathscr{T}} X_T$; as the number of elements added is strictly less than $\beta \leq \beta^{<\beta}$,

$$\sum_{T\in\mathscr{T}} \#X_T^{c^1} = \#\coprod_{T\in\mathscr{T}} X_T^{c^1} \leq \beta^{<\beta} + \beta = \beta^{<\beta} < \alpha.$$

Construction c^2: For each cocone $\left(s_D: F(D) \longrightarrow T\right)_{D\in\mathscr{D}}$ in $\mathcal{I}$, there are at most $(\beta^{<\beta})^2 = \beta^{<\beta}$ pairs (x, y) to consider and for each pair of this kind, finitely many elements are added to the disjoint union $\coprod_{T\in\mathscr{T}} X_T$, i.e. strictly less than $\aleph_0$ elements. Thus for a given cocone, at most $(\beta^{<\beta}) \cdot (\aleph_0) \leq (\beta^{<\beta})(\beta^{<\beta}) = \beta^{<\beta}$ elements are added. So we start with strictly less than α elements for each cocone in a family $\mathcal{I}$ with $\#\mathcal{I} < \beta < \alpha$; by regularity of α, we end up with strictly less than α elements.

We are now ready to prove our statement. Let us choose a family

$(x_i \in M(T_i))_{i\in I}$ in an $\mathbb{S}$-model $M: \mathscr{T} \longrightarrow \mathsf{Set}$, with $\#\mathcal{I} < \alpha$. We define

$$X_T = \{x_i \mid i \in I,\ x_i \in M(T)\}$$

for every $T \in \mathscr{T}$. By assumption, $\sum_{T\in\mathscr{T}} \#X_T \leq \#\mathcal{I} < \alpha$. We construct now a transfinite sequence of families $(X_T^\lambda \subseteq M(T))_{T\in\mathscr{T}}$:

(1) $X_T^0 = X_T$;
(2) if $(X_T^\lambda)_{T\in\mathscr{T}}$ is defined, $(X_T^{\lambda+1})_{T\in\mathscr{T}}$ is obtained by successively applying the four constructions f, l, c^1, c^2 to $(X_T^\lambda)_{T\in\mathscr{T}}$;
(3) if λ is a limit ordinal, we define $X_T^\lambda = \bigcup_{\nu<\lambda} X_T^\nu$ for every $T \in \mathscr{T}$.

We shall prove that the required submodel $N \subseteq M$ is given by $N(T) = X_T^{\beta^{<\beta}}$, for every $T \in \mathscr{T}$; as usual, we identify a cardinal with the smallest ordinal of this cardinality, which is thus a limit ordinal.

First of all, let us prove by transfinite induction that $\sum_{T\in\mathscr{T}} \#N(T) < \alpha$. This is true at the level 0 of the construction and we have already observed that the four constructions f, l, c^1, c^2 respect boundedness by α. Now if $\lambda \leq \beta^{<\beta}$ is a limit ordinal, then $\lambda < \alpha$ and assuming that $\sum_{T\in\mathscr{T}} \#X_T^\nu < \alpha$ for every $\nu < \lambda$, we get $\sum_{T\in\mathscr{T}} \#X_T^\lambda$ by regularity of α.

By construction, N contains the original family $(x_i)_{i\in I}$ and it remains to check that N is a submodel of M. By the description we have made of a submodel, see conditions (1), (2), (3)(a), (3)(b), this is equivalent to proving the stability of N under each of the four constructions f, l, c^1, c^2.

Construction f: If $t: T' \longrightarrow T$ is an arrow of $\mathscr{T}$ and $x \in N(T')$, then $x \in X_{T'}^\lambda$ for some $\lambda < \beta^{<\beta}$ and thus $M(t)(x) \in X_T^{\lambda+1} \subseteq N(T)$.

Construction l: Given a cone $(p_D: T \longrightarrow F(D))_{D\in\mathscr{D}}$ in $\mathcal{P}$ and a compatible family of elements $(x_D \in (NF)(D))_{D\in\mathscr{D}}$, each element x_D belongs to some $X_{F(D)}^{\lambda_D}$, for $\lambda_D < \beta^{<\beta}$. But since $\#\mathscr{D} < \beta$

$$\sum\nolimits_{D\in\mathscr{D}} \lambda_D < \prod\nolimits_{D\in\mathscr{D}} \beta^{<\beta} = (\beta^{<\beta})^{\#\mathscr{D}} = \beta^{<\beta};$$

see section 5.4. This implies that we can choose $\lambda < \beta^{<\beta}$ with $\lambda_D \leq \lambda$ for all $D \in \mathscr{D}$. But the unique element $x \in M(T)$ such that $M(p_D)(x) = x_D$ lies in $X_T^{\lambda+1} \subseteq N(T)$.

Construction c^1: Given a cocone $(s_D: F(D) \longrightarrow T)_{D\in\mathscr{D}}$ of $\mathcal{I}$, an element $x \in N(T)$ belongs to some X_T^λ, for $\lambda < \beta^{<\beta}$. Therefore we get $D \in \mathscr{D}$ and $y \in X_{F(D)}^{\lambda+1} \subseteq (NF)(D)$ such that $M(s_D)(y) = x$.

Construction c^2: Given a cocone $(s_D: F(D) \longrightarrow T)_{D\in\mathscr{D}}$ and two elements $x \in (NF)(D)$, $y \in (NF)(D')$ such that $M(s_D)(x) = M(s_{D'})(y)$, we have $x \in X^{\lambda'}_{F(D)}$ and $y \in X^{\lambda''}_{(FD')}$, for $\lambda' < \beta^{<\beta}$ and $\lambda'' < \beta^{<\beta}$. Writing λ for the bigger of the two ordinals λ', λ'', we have $x \in X^{\lambda}_{F(D)}$ and $y \in X^{\lambda}_{F(D')}$. Therefore in $\left(X_T^{\lambda+1}\right)_{T\in\mathscr{T}}$, and thus in N, there is a finite sequence exhibiting the equivalence of x and y. □

Proposition 5.6.4 *The category of models of a sketch is accessible.*

Proof Let $\mathbb{S} = (\mathscr{T}, \mathcal{P}, \mathcal{I})$ be a sketch. We consider the regular cardinal α constructed in lemma 5.6.3 and prove that the category $\mathsf{Mod}_{\mathbb{S}}$ of models is α-accessible.

The category $\mathsf{Mod}_{\mathbb{S}}$ is a full subcategory of $\mathsf{Fun}(\mathscr{T}, \mathsf{Set})$; let us prove first that $\mathsf{Mod}_{\mathbb{S}}$ is closed in $\mathsf{Fun}(\mathscr{T}, \mathsf{Set})$ under α-filtered colimits. Consider for this a α-filtered small category $\mathscr{F}$ and a functor $F\colon \mathscr{F} \longrightarrow \mathsf{Mod}_{\mathbb{S}}$; let us write $\left(s_X\colon F(X) \longrightarrow L\right)_{X\in\mathscr{F}}$ for the colimit of F in $\mathsf{Fun}(\mathscr{T}, \mathsf{Set})$, which is computed pointwise by 2.15.2, volume 1.

Choose a cone $\left(p_D\colon T \longrightarrow G(D)\right)_{D\in\mathscr{D}}$ on $G\colon \mathscr{D} \longrightarrow \mathscr{T}$, in the set $\mathcal{P}$ of distinguished cones. For every object $X \in \mathscr{F}$, the composite

$$\mathscr{D} \xrightarrow{\quad G \quad} \mathscr{T} \xrightarrow{\quad F(X) \quad} \mathsf{Set}$$

admits

$$\Big(F(X)(p_D)\colon F(X)(T) \longrightarrow F(X)\big(G(D)\big)\Big)_{D\in\mathscr{D}}$$

as a limit, just because $F(X)$ is a model of $\mathbb{S}$. But $\mathscr{D}$ is an α-category and $\mathscr{F}$ is α-filtered; therefore by 6.4.5, volume 1, one gets the isomorphism

$$\operatorname*{colim}_{X\in\mathscr{F}}\Big(\lim_{D\in\mathscr{D}} F(X)\big(G(D)\big)\Big) \cong \lim_{D\in\mathscr{D}}\Big(\operatorname*{colim}_{X\in\mathscr{F}} F(X)\big(G(D)\big)\Big).$$

But since the colimit L of the functor F in $\mathsf{Fun}(\mathscr{T}, \mathsf{Set})$ is computed pointwise,

$$\operatorname*{colim}_{X\in\mathscr{F}} F(X)\big(G(D)\big) \cong L\big(G(D)\big).$$

On the other hand since $F(X)$ is a model of $\mathbb{S}$,

$$\lim_{D\in\mathscr{D}} F(X)\big(G(D)\big) \cong F(X)(T).$$

Thus the previous isomorphism yields

$$\begin{aligned}
L(T) &\cong \operatorname*{colim}_{X\in\mathscr{F}} F(X)(T) \\
&\cong \operatorname*{colim}_{X\in\mathscr{F}}\Big(\lim_{D\in\mathscr{D}} F(X)\big(G(D)\big)\Big) \\
&\cong \lim_{D\in\mathscr{D}}\Big(\operatorname*{colim}_{X\in\mathscr{F}} F(X)\big(G(D)\big)\Big) \\
&\cong \lim_{D\in\mathscr{D}} L\big(G(D)\big)
\end{aligned}$$

which is the required condition. The proof concerning a cocone of $\mathcal{I}$ is completely analogous, applying 2.12.1, volume 1, instead of 6.4.5, volume 1.

The previous argument shows that L is indeed a model of $\mathbb{S}$, proving that $\mathsf{Mod}_{\mathbb{S}}$ has α-filtered colimits computed pointwise.

We must now define a set $(M_i: \mathscr{T} \longrightarrow \mathsf{Set})_{i \in I}$ of $\mathbb{S}$-models exhibiting the α-accessibility of $\mathsf{Mod}_{\mathbb{S}}$ as in lemma 5.3.2. We choose those models M such that the set $\coprod_{T \in \mathscr{T}} M(T)$ has strictly less than α elements. The elements of this set are in bijection with the objects of the category $\mathsf{Elts}(M)$ (see 1.6.4, volume 1) and since $\mathscr{T}$ itself has strictly less than α arrows, so has $\mathsf{Elts}(M)$, by the regularity of α. By 2.15.6, volume 1, we know that each M is the colimit of the functor

$$\mathsf{Elts}(M) \xrightarrow{\ \phi\ } \mathscr{T} \xrightarrow{\ Y\ } \mathsf{Fun}(\mathscr{T}, \mathsf{Set})$$

where ϕ is the forgetful functor and Y the Yoneda embedding. Since there is – up to isomorphisms – just a set of categories with strictly less than α arrows, this implies that – up to isomorphisms – there is just a set of such functors M with $\#\coprod_{T \in \mathscr{T}} M(T) < \alpha$; more precisely, choose one functor M in each isomorphism class. We write $\mathscr{G}$ for the full subcategory of $\mathsf{Mod}_{\mathbb{S}}$ generated by those models M.

By the Yoneda lemma, the representable functor $\mathscr{T}(T, -)$ represents on $\mathsf{Fun}(\mathscr{T}, \mathsf{Set})$ the evaluation functor at T. Since colimits are computed pointwise in $\mathsf{Fun}(\mathscr{T}, \mathsf{Set})$ (see 2.15.2, volume 1), this functor represented by $\mathscr{T}(T, -)$ preserves all colimits. In particular each functor $\mathscr{T}(T, -)$ is α-presentable in $\mathsf{Fun}(\mathscr{T}, \mathsf{Set})$ and thus so is every functor obtained as a α-colimit of representable functors (see 5.1.4). By 2.15.6, volume 1, again, this shows that each of our models $M \in \mathscr{G}$ is α-presentable in $\mathsf{Fun}(\mathscr{T}, \mathsf{Set})$. But α-filtered colimits are computed in the same way in $\mathsf{Mod}_{\mathbb{S}}$ and in $\mathsf{Fun}(\mathscr{T}, \mathsf{Set})$, since $\beta < \alpha$; thus each $M \in \mathscr{G}$ is α-presentable in $\mathsf{Mod}_{\mathbb{S}}$.

Finally we shall prove that each model $M \in \mathsf{Mod}_{\mathbb{S}}$ is the α-filtered union (and thus colimit) of its submodels $M' \subseteq M$ which are (up to an isomorphism) in $\mathscr{G}$. This will yield the conclusion by 5.3.2. First of all, by 5.6.3, every element $x \in M(T)$, $T \in \mathscr{T}$, belongs to a submodel $M' \in \mathscr{G}$, which proves already that the union is M itself. Moreover, given a family $(M_k \subseteq M)_{k \in K}$ of submodels, with $\#K < \alpha$, the union $M' = \bigcup_{k \in K} M_k$ in $\mathsf{Fun}(\mathscr{T}, \mathsf{Set})$ is still such that $\#\coprod_{T \in \mathscr{T}} M'(T) < \alpha$, by regularity of α. Thus by 5.6.3 again, M' is contained in a submodel $M'' \in \mathscr{G}$. This proves the α-filteredness of the union. □

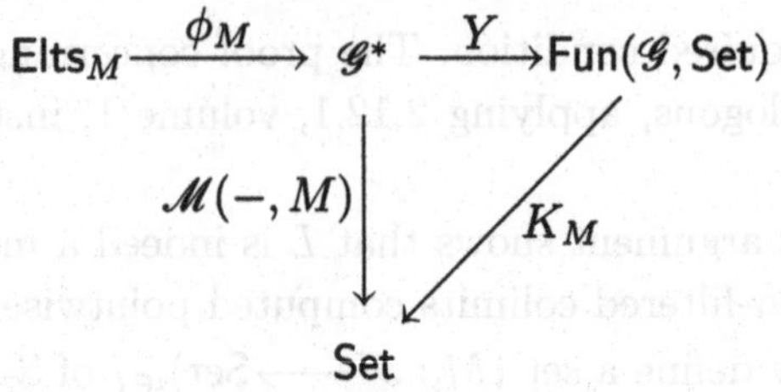

$$\begin{array}{ccc} \mathsf{Elts}_M \xrightarrow{\ \phi_M\ } & \mathscr{G}^* & \xrightarrow{\ Y\ } \mathsf{Fun}(\mathscr{G}, \mathsf{Set}) \\ {\scriptstyle \mathscr{M}(-,M)} & \Big\downarrow & \swarrow {\scriptstyle K_M} \\ & \mathsf{Set} & \end{array}$$

Diagram 5.3

The rest of this section is now devoted to proving the converse of 5.6.4, namely that every accessible category is the category of models of some sketch. The core of the proof is the following lemma which is a classifying topos-like theorem (see 4.1.7, volume 3).

Lemma 5.6.5 *Let $\mathscr{M}$ be an α-accessible category; write $\mathscr{G}$ for the full subcategory of α-presentable objects. Then $\mathscr{M}$ is equivalent to the category of those functors $\mathsf{Fun}(\mathscr{G}, \mathsf{Set}) \longrightarrow \mathsf{Set}$ which preserve α-limits and small colimits.*

Proof By 5.3.8, $\mathscr{G}$ is equivalent to a small category, so that there is no restriction in supposing in the proof that $\mathscr{G}$ is a small full subcategory of $\mathscr{M}$ containing exactly one object in each isomorphism class of α-presentable objects.

For each object $M \in \mathscr{M}$, we can consider the Kan extension K_M of the functor

$$\mathscr{M}(M,-)\colon \mathscr{G}^* \longrightarrow \mathsf{Set}, \quad G \mapsto \mathscr{M}(G,M)$$

along the Yoneda embedding, as in diagram 5.3. Let us write Elts_M for the category of elements of $\mathscr{M}(-,M)$ and $\phi_M\colon \mathsf{Elts}_M \longrightarrow \mathscr{G}$ for the corresponding forgetful functor. Applying 3.8.1, volume 1, and the formula for computing Kan extensions pointwise (see 3.7.2, volume 1), we get $K_M(F) \cong \operatorname{colim} F \circ \phi_M$, for every functor $F\colon \mathscr{G} \longrightarrow \mathsf{Set}$. The interchange property for colimits (see 2.12.1, volume 1) and the previous formula imply immediately that K_M preserves all small colimits. On the other hand $\mathscr{M}(-,M)$ is α-flat (see proof of 5.3.5), thus K_M preserves α-limits (see 6.4.13, volume 1).

Let us also recall that $K_M \circ Y \cong \mathscr{M}(-,M)$ since Y is full and faithful (see 3.7.3, volume 1). Given two objects $M, M' \in \mathscr{M}$, the definition of a Kan extension (see 3.7.1, volume 1) and the proof of 5.3.5 yield the

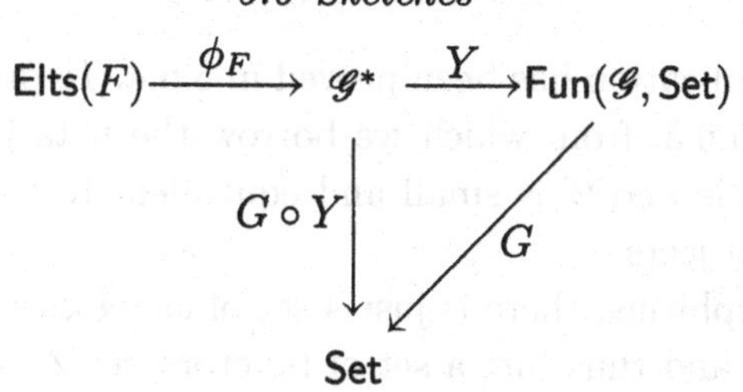

Diagram 5.4

isomorphisms

$$\begin{aligned}\mathsf{Nat}(K_M, K_{M'}) &\cong \mathsf{Nat}\big(\mathscr{M}(M,-), K_{M'} \circ Y\big)\\ &\cong \mathsf{Nat}\big(\mathscr{M}(-,M), \mathscr{M}(-,M')\big)\\ &\cong \mathscr{M}(M,M').\end{aligned}$$

This proves already that $\mathscr{M}$ is equivalent to the category of functors K_M and natural transformations between them.

It remains to prove that each functor G: $\mathsf{Fun}(\mathscr{G}, \mathsf{Set}) \longrightarrow \mathsf{Set}$ which preserves α-limits and small colimits is isomorphic to some K_M. Let us consider diagram 5.4 where F is some functor F: $\mathscr{G} \longrightarrow \mathsf{Set}$, $\mathsf{Elts}(F)$ is its category of elements and ϕ_F the corresponding forgetful functor. We know that $F \cong \operatorname{colim} Y \circ \phi_F$ (see 2.15.6, volume 1), while the Kan extension $\mathrm{Lan}_Y G$ of G along Y is given by the formula (see 3.7.2, volume 1)

$$\begin{aligned}(\mathrm{Lan}_Y G)(F) &\cong \operatorname{colim} G \circ Y \circ \phi_F \cong G(\operatorname{colim} Y \circ \phi_F)\\ &\cong G(F),\end{aligned}$$

since G preserves small colimits. Thus $\mathrm{Lan}_Y G$ is just G, which preserves α-limits by assumption. Therefore $G \circ Y$ is α-flat (see 6.4.13, volume 1) and thus isomorphic to some $\mathscr{M}(-, M)$ by 5.3.5. So G is isomorphic to K_M. □

Lemma 5.6.5 can be interpreted as the fact that the accessible category $\mathscr{M}$ is equivalent to the category of models of a "large sketch" $\mathbb{S} = (\mathscr{T}, \mathcal{P}, \mathcal{I})$, where $\mathscr{T}$ is the (large) category $\mathsf{Fun}(\mathscr{G}, \mathsf{Set})$, $\mathcal{P}$ is the class of all α-cones and $\mathcal{I}$ is the class of all small cocones. It remains to reduce sufficiently the size of those classes to end up eventually with just sets.

Theorem 5.6.6 *A category is accessible if and only if it is the category of models of some sketch.*

Proof One implication has been proved in 5.6.4. To prove the converse, we start from 5.6.5, from which we borrow the notation. In particular $\mathscr{M}$ is α-accessible and $\mathscr{G}$ is small and equivalent to the subcategory of α-presentable objects.

Up to isomorphisms, there is just a set of categories with strictly less than α arrows, and thus just a set of functors $H\colon \mathscr{D} \longrightarrow \mathscr{G}^*$ defined on such categories $\mathscr{D}$. For every functor of this kind, we choose one "distinguished" limit cone $\big(p_D\colon L \longrightarrow YH(D)\big)_{D\in\mathscr{D}}$ on $Y \circ H$ in $\mathsf{Fun}(\mathscr{G}, \mathsf{Set})$,

$$\mathscr{D} \xrightarrow{\ H\ } \mathscr{G}^* \xrightarrow{\ Y\ } \mathsf{Fun}(\mathscr{G}, \mathsf{Set}),$$

where Y is the Yoneda embedding.

Our category $\mathscr{T}$ is the full subcategory of $\mathsf{Fun}(\mathscr{G}, \mathsf{Set})$ whose objects are the functors $\mathscr{G}(G, -)$, for $G \in \mathscr{G}$, and the objects L, chosen as distinguished limits of the α-diagrams of representable functors. Clearly, $\mathscr{T}$ is small.

As set $\mathscr{P}$ of cones, we choose for each α-diagram of representable functors the corresponding distinguished limit cone $\big(p_D\colon L \longrightarrow YH(D)\big)_{D\in\mathscr{D}}$.

As set of cocones, we choose, for each object $F \in \mathscr{T}$, the canonical cocone $\mathscr{G}/F$. In other words we consider the category $\mathsf{Elts}(F)$ of elements of F and put in $\mathcal{I}$ the canonical colimit cone of

$$\mathsf{Elts}(F) \xrightarrow{\ \phi_F\ } \mathscr{G}^* \xrightarrow{\ Y\ } \mathsf{Fun}(\mathscr{G}, \mathsf{Set})$$

with vertex F (see 2.15.6, volume 1).

By construction of the sketch $\mathbb{S} = (\mathscr{T}, \mathcal{P}, \mathcal{I})$, we have:

(1) $\mathscr{T}$ is a full subcategory of $\mathsf{Fun}(\mathscr{G}, \mathsf{Set})$;
(2) every representable functor belongs to $\mathscr{T}$;
(3) every cone of $\mathcal{P}$ is a α-limit cone in $\mathsf{Fun}(\mathscr{G}, \mathsf{Set})$;
(4) every cocone of $\mathcal{I}$ is a small colimit cone in $\mathsf{Fun}(\mathscr{G}, \mathsf{Set})$;
(5) for every α-category $\mathscr{D}$ and every functor $H\colon \mathscr{D} \longrightarrow \mathscr{G}^*$, $\mathcal{P}$ contains a limit cone on the functor $Y \circ H$;
(6) for every functor $F \in \mathscr{T}$, $\mathcal{I}$ contains a colimit cone on the functor $Y \circ \phi_F$.

We shall prove that $\mathscr{M}$ is equivalent to the category of models of the sketch $\mathbb{S} = (\mathscr{T}, \mathcal{P}, \mathcal{I})$.

By the first condition in the preceding list, we can consider the full embedding $i\colon \mathscr{T} \hookrightarrow \mathsf{Fun}(\mathscr{G}, \mathsf{Set})$. By the second condition, the Yoneda embedding factors as $Y = i \circ Y'$, with $Y'\colon \mathscr{G}^* \longrightarrow \mathscr{T}$. Conditions (3) and (4) imply that given $M \in \mathscr{M}$, $K_M \circ i\colon \mathscr{T} \longrightarrow \mathsf{Set}$ is a model of

$\mathbb{S}$, just because K_M preserves α-limits and small colimits. In the same way a natural transformation $\theta\colon K_M \Rightarrow K_{M'}$ yields another natural transformation $\theta * i\colon K_M \circ i \Rightarrow K_{M'} \circ i$. Combining this with 5.6.5, we have already defined a functor $\varphi\colon \mathscr{M} \longrightarrow \mathsf{Mod}_{\mathbb{S}}$.

Every functor $F\colon \mathscr{G} \longrightarrow \mathsf{Set}$ can be written as $F \cong \operatorname{colim} Y \circ \phi_F$, i.e. as a colimit of representable functors (see 2.15.6, volume 1). Given $M, M' \in \mathscr{M}$, K_M and $K_{M'}$ preserve this colimit so that a natural transformation $\theta\colon K_M \Rightarrow K_{M'}$ is completely determined by its values on the representable functors. Since $\mathscr{T}$ contains the representable functors, θ is thus determined by $\theta * i$, proving the faithfulness of φ.

To prove that φ is full, we must check that every natural transformation $\tau\colon K_M \circ i \Rightarrow K_{M'} \circ i$ has the form $\theta * i$ for some $\theta\colon K_M \Rightarrow K_{M'}$. Again write $F\colon \mathscr{G} \longrightarrow \mathsf{Set}$ as $F \cong \operatorname{colim} Y \circ \phi_F$. By naturality of τ, we get a cocone

$$K_M \circ Y \circ \phi_F(A,a) \xrightarrow{\;\tau_{Y'A}\;} K_{M'} \circ Y \circ \phi_F(A,a) \xrightarrow{\;K_{M'}(\overline{a})\;} K_{M'}(F)$$

where (A,a) runs through $\mathsf{Elts}(F)$ and $\overline{a}$ corresponds with a by the Yoneda lemma. This yields a unique factorization θ_F,

$$K_M(F) \cong K_M(\operatorname{colim} Y \circ \phi_F) \cong \operatorname{colim} K_M \circ Y \circ \phi_F \longrightarrow K_{M'}(F),$$

such that $\theta_F \circ K_M(\overline{a}) = K_{M'}(\overline{a}) \circ \tau_{Y'A}$ for every $(A,a) \in \mathsf{Elts}(F)$. By naturality of τ, $\theta_F = \tau_F$ when $F \in \mathscr{T}$. On the other hand θ is natural since given $\sigma\colon F \Rightarrow G$ in $\mathsf{Fun}(\mathscr{G}, \mathsf{Set})$, the relation $\theta_G \circ K_M(\sigma) = K_{M'}(\sigma) \circ \theta_F$ follows from the Yoneda lemma via the equalities

$$\begin{aligned}
\theta_G \circ K_M(\sigma) \circ K_M(\overline{a}) &= \theta_G \circ K_M\big(\overline{\sigma_A(a)}\big) \\
&= K_{M'}\big(\overline{\sigma_A(a)}\big) \circ \tau_{Y'A} \\
&= K_{M'}(\sigma) \circ K_{M'}(\overline{a}) \circ \tau_{Y'A} \\
&= K_{M'}(\sigma) \circ \theta_F \circ K_M(\overline{a})
\end{aligned}$$

when (A,a) runs through $\mathsf{Elts}(F)$; indeed the morphisms $\overline{a}$ constitute a colimit cone (see 2.15.6, volume 1) preserved by K_M. Thus θ is natural and $\tau = \theta * i$.

To prove that φ is an equivalence, it remains to show that every model $R\colon \mathscr{T} \longrightarrow \mathsf{Set}$ of $\mathbb{S}$ is isomorphic to the restriction of some K_M. Given an $\mathbb{S}$-model $R\colon \mathscr{T} \longrightarrow \mathsf{Set}$, we extend it to a functor $\overline{R}\colon \mathsf{Fun}(\mathscr{G}, \mathsf{Set}) \longrightarrow \mathsf{Set}$ in the following way. A functor $F\colon \mathscr{G} \longrightarrow \mathsf{Set}$ can be written as $F = \operatorname{colim} Y \circ \phi_F$; we define $\overline{R}(F) \cong \operatorname{colim} R \circ Y' \circ \phi_F$. By 3.7.2, volume 1, $\overline{R}(F) \cong \operatorname{Lan}_Y(R \circ Y')(F)$ and thus by 3.8.1, volume 1, one also has $\overline{R}(F) \cong \operatorname{colim} F \circ \phi_{R \circ Y'}$. The interchange property for colimits (see

2.12.1, volume 1) implies immediately that $\overline{R}$ preserves colimits. Now if $F \in \mathscr{T}$, the colimit cone of $Y \circ \phi_F$ is in $\mathcal{I}$ so that

$$\overline{R}(F) \cong \operatorname{colim} R \circ Y' \circ \phi_F \cong R(\operatorname{colim} Y' \circ \phi_F) \cong R(F).$$

This proves that $\overline{R}$ extends R.

Let us prove now that $\overline{R}$ preserves α-limits. Since $\overline{R}$ is the Kan extension of $R \circ Y'$ along Y, this is equivalent to proving that this composite $R \circ Y'$ is α-flat. The proof is analogous to the implication (1) $\Rightarrow$ (3) in 6.4.13 (i.e. in 6.3.6) volume 1. Since we work on $\mathscr{G}^*$, we must prove the α-filteredness of $\mathsf{Elts}(R \circ Y')$. Given an α-category $\mathscr{D}$ and a functor $H\colon \mathscr{D} \longrightarrow \mathsf{Elts}(R \circ Y')$, write $HD = (A_D, a_D)$. The limit $(p_D\colon L \longrightarrow Y A_D)_{D \in \mathscr{D}}$ of $Y \circ \phi_{R \circ Y'} \circ H$ is a cone in $\mathcal{P}$, by assumption. Thus

$$\begin{aligned}\lim_{D\in\mathscr{D}} RY(A_D) &\cong R\big(\lim_{D\in\mathscr{D}} Y(A_D)\big) \cong \overline{R}\big(\lim_{D\in\mathscr{D}} Y(A_D)\big)\\ &\cong \operatorname{colim}\big(\lim_{D\in\mathscr{D}} Y(A_D)\big) \circ \phi_{R\circ Y'}.\end{aligned}$$

Since $(a_D)_{D\in\mathscr{D}}$ is a compatible family in $\lim_{D\in\mathscr{D}} RY(A_D)$, it is represented by some element in some term of the colimit. This means the existence of an object $(A, a) \in \mathsf{Elts}(R \circ Y')$ and a compatible family of morphisms $\big(f_D \in \mathscr{G}(A_D, A)\big)_{D\in\mathscr{D}}$. In other words, we have found a cocone $\big(f_D\colon (A_D, a_D) \longrightarrow (A, a)\big)_{D\in\mathscr{D}}$ on H.

Finally $\overline{R}$ preserves all small colimits and all α-limits, proving by 5.6.5 that it is isomorphic to some K_M. And since the restriction of $\overline{R}$ to $\mathscr{T}$ is just R, this ends the proof that φ is an equivalence of categories. □

Corollary 5.6.7 *Every sketch has a category of models equivalent to the category of models of a sketch* $\mathbb{S} = (\mathscr{T}, \mathcal{P}, \mathcal{I})$, *where* $\mathcal{P}$ *is a set of limit cones and* $\mathcal{I}$ *is a set of colimit cones.*

Proof The category of models of a sketch is accessible (see 5.6.4) and an α-accessible category can be presented as the category of models for a sketch $\mathbb{S} = (\mathscr{T}, \mathcal{P}, \mathcal{I})$ where $\mathcal{P}$ is a family of α-limit cones and $\mathcal{I}$ a family of small colimits (see proof of 5.6.6). □

Corollary 5.6.8 *Consider a sketch* $\mathbb{S} = (\mathscr{T}, \mathcal{P}, \mathcal{I})$. *If* $\mathcal{P}$ *or* $\mathcal{I}$ *are empty, the category of models of* $\mathbb{S}$ *is locally presentable.*

Proof We treat the case $\mathcal{P}$ empty; the other is analogous, using colimits instead of limits. By 5.6.6 and 5.5.8, it suffices to observe that $\mathsf{Mod}_{\mathbb{S}}$ is stable in $\mathsf{Fun}(\mathscr{T}, \mathsf{Set})$ under small limits, which follows immediately from 2.12.1, volume 1. □

Theorem 5.6.6, disregarding the fact that it is written in terms of equivalent conditions, does not give any information on the degree of accessibility. For example, consider a small category $\mathscr{T}$ with finite limits and colimits. The category of those functors $F\colon \mathscr{T} \longrightarrow \mathsf{Set}$ preserving finite limits and colimits is accessible, as attested by 5.6.4. But what is its degree of accessibility? The cardinal α constructed in 5.6.3 is so big that it does not give much precise information. Another example: start with a κ-accessible category; as in 5.6.6, present it as the category of models of a sketch; then using this sketch, construct the cardinal α as in 5.6.3; α is in general much bigger than κ, while the accessible category which is considered is still the same.

5.7 Exercises

5.7.1 Let $\mathscr{M}$ be a category such that both $\mathscr{M}$ and its dual $\mathscr{M}^*$ are locally presentable; show that $\mathscr{M}$ is equivalent to a poset.

5.7.2 Show that the category of Hilbert spaces and linear contractions is $\aleph_0$-accessible and equivalent to its dual (compare with 5.7.1).

5.7.3 Show that the category of sets and partial bijections is $\aleph_0$-accessible and equivalent to its dual (compare with 5.7.1). By a partial bijection from A to B we mean a relation R from A to B which is the graph of a bijection $f'\colon A' \longrightarrow B'$, with $A' \subseteq A$ and $B' \subseteq B$.

5.7.4 Show that every Cauchy complete small category is accessible. [Hint: take α infinite and larger than the cardinality of the category and prove α^{++}-accessibility.]

5.7.5 Let $\mathscr{I}$ be a small category and $\mathscr{M}$ an accessible category. Show that the category $\mathsf{Fun}(\mathscr{I}, \mathscr{M})$ of functors and natural transformations is accessible.

5.7.6 Let $\mathscr{M}$ be an accessible category and α a regular cardinal. Show that the category of α-presentable objects of $\mathscr{M}$ is equivalent to a small category.

5.7.7 Show that an accessible category with pushouts is co-well-powered.

5.7.8 Prove that the relation $\alpha \triangleleft \beta$ between regular cardinals is transitive.

5.7.9 Consider regular cardinals $\alpha \triangleleft \beta$ and an α-accessible category $\mathscr{M}$. Prove that an object $M \in \mathscr{M}$ is β-presentable iff it is a α-filtered colimit of α-presentable objects over a diagram of size strictly less than β.

5.7.10 Given regular cardinals α, β, prove that one has $\alpha \triangleleft \beta$ iff every α-accessible category is β-accessible.

5.7.11 Let $\mathscr{A}, \mathscr{B}$ be accessible categories with $\mathscr{A}$ small. Prove that if $\mathscr{B}$ is complete (resp. cocomplete), the category of functors with rank from $\mathscr{A}$ to $\mathscr{B}$ is complete (resp. cocomplete).

5.7.12 Prove that the category of small accessible categories and functors with rank is complete.

5.7.13 If $\mathscr{A}, \mathscr{B}, \mathscr{C}$ are accessible categories and $F\colon \mathscr{A} \longrightarrow \mathscr{C}$, $G\colon \mathscr{B} \longrightarrow \mathscr{C}$ are functors with rank, prove that the comma category (F, G) is accessible.

5.7.14 Let $\mathbb{S}_1 = (\mathscr{T}_1, \mathcal{P}_1, \mathcal{I}_1)$ and $\mathbb{S}_2 = (\mathscr{T}_2, \mathcal{P}_2, \mathcal{I}_2)$ be sketches. A functor $F\colon \mathscr{T}_1 \longrightarrow \mathscr{T}_2$ is a morphism of sketches when it takes each cone of $\mathcal{P}_1$ to a cone of $\mathcal{P}_2$ and each cocone of $\mathcal{I}_1$ to a cocone of $\mathcal{I}_2$. For such a morphism of sketches, prove that composition with F yields a functor with rank $\mathsf{Mod}_{\mathbb{S}_1} \longrightarrow \mathsf{Mod}_{\mathbb{S}_2}$.

5.7.15 If $\mathbb{S} = (\mathscr{T}, \mathcal{P}, \mathcal{I})$ is a sketch, a model of $\mathbb{S}$ in some category $\mathscr{C}$ is a functor $F\colon \mathscr{T} \longrightarrow \mathscr{C}$ mapping the cones of $\mathcal{P}$ to limit cones and the cocones of $\mathcal{I}$ on colimit cocones. Prove that the models of $\mathbb{S}$ in a locally presentable category constitute an accessible category.

6
Enriched category theory

A category $\mathscr{A}$ consists of:

- a class of objects $|\mathscr{A}|$;
- for every pair A, B of objects, a set $\mathscr{A}(A, B)$ of arrows;
- for every triple A, B, C of objects, a composition law

$$\mathscr{A}(A,B) \times \mathscr{A}(B,C) \longrightarrow \mathscr{A}(A,C),$$

these data satisfying the identity and associativity axioms.

In 7.1.1, volume 1, we have seen that in the case of a 2-category, the sets $\mathscr{A}(A, B)$ of arrows turn out to be provided with an additional structure: that of a small category, while the composition respects this structure in each variable. In 1.2.1, an analogous situation has been encountered for preadditive categories: the sets $\mathscr{A}(A, B)$ are now provided with the structure of an abelian group while the composition respects this structure in each variable. We shall say that a 2-category is "enriched over the category of small categories" while a preadditive category is "enriched over the category of abelian groups".

Preserving a structure in each variable is something which can be easily expressed for a mapping $X \times Y \longrightarrow Z$, as long as it makes sense to fix an element of X or Y. This contradicts the very spirit of category theory, where objects are basic entities. Now observe that when X, Y, Z are small categories, being functorial in each variable is just being globally functorial (see 1.6, volume 1). And when X, Y, Z are abelian groups, the biadditive mappings $X \times Y \longrightarrow Z$ on the product are in bijection with the group homomorphisms $X \otimes Y \longrightarrow Z$ on the tensor product.

In this chapter we shall introduce first the so-called "symmetric monoidal closed categories", which are the best substitutes for **Set** for developing enriched category theory. Then we shall indicate how to enrich

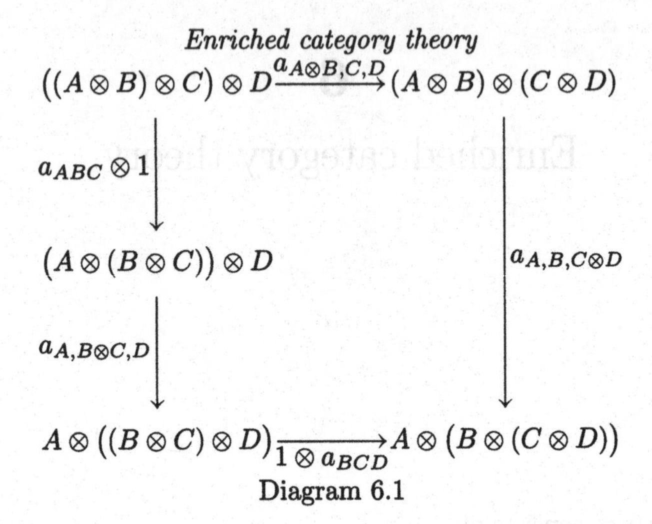

Diagram 6.1

over them the basic concepts and results of category theory.

It is a deliberate choice, in this chapter, to reduce the proofs to the necessary constructions and the key arguments. We hope this will clarify the spirit of enriched category theory, even if the conscientious reader is left with a large amount of routine calculations.

6.1 Symmetric monoidal closed categories

Definition 6.1.1 *A monoidal category $\mathscr{V}$ consists in giving:*

(1) a category $\mathscr{V}$;

(2) a bifunctor $\otimes$: $\mathscr{V}\times\mathscr{V}\longrightarrow\mathscr{V}$, called the tensor product – we write $A\otimes B$ for the image under $\otimes$ of the pair (A,B);

(3) an object $I\in\mathscr{V}$, called the unit;

(4) for every triple A,B,C of objects, an "associativity" isomorphism

$$a_{ABC}\colon (A\otimes B)\otimes C\longrightarrow A\otimes (B\otimes C);$$

(5) for every object A, a "left unit" isomorphism

$$l_A\colon I\otimes A\longrightarrow A;$$

(6) for every object A, a "right unit" isomorphism

$$r_A\colon A\otimes I\longrightarrow A.$$

These data must satisfy the following requirements:

(1) the morphisms a_{ABC} are natural in A,B,C;

(2) the morphisms l_A are natural in A;

(3) the morphisms r_A are natural in A;

(4) diagram 6.1 is commutative for every quadruple of objects A, B, C, D (associativity coherence);

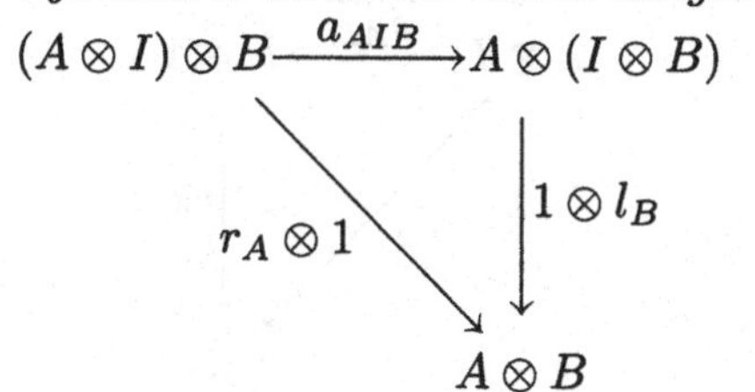

Diagram 6.2

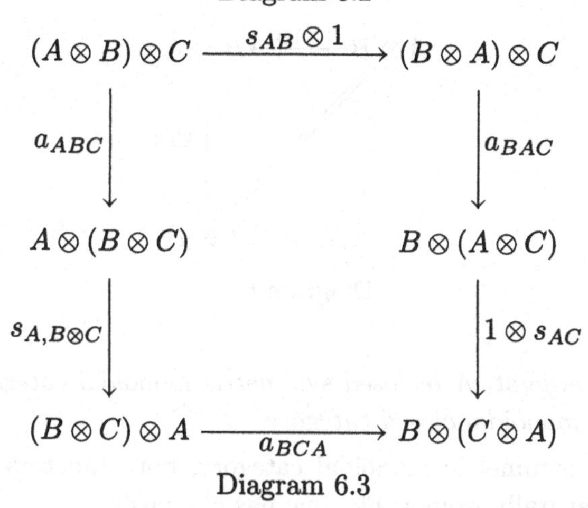

Diagram 6.3

(5) *diagram 6.2 is commutative for every pair A, B of objects (unit coherence)*

Definition 6.1.2 *With the notation of 6.1.1, a monoidal category is symmetric when, moreover, an isomorphism*

$$s_{AB}: A \otimes B \longrightarrow B \otimes A$$

is given for every pair A, B of objects. These isomorphisms must be such that:

(1) *the morphisms s_{AB} are natural in A, B;*
(2) *diagram 6.3 is commutative for every triple A, B, C of objects (associativity coherence);*
(3) *diagram 6.4 is commutative for every object A (unit coherence);*
(4) *diagram 6.5 is commutative (symmetry axiom) for every pair A, B of objects.*

Definition 6.1.3 *With the notation of 6.1.1, a monoidal category $\mathscr{V}$ is biclosed when, for each object $B \in \mathscr{V}$, both functors*

$$- \otimes B: \mathscr{V} \longrightarrow \mathscr{V}, \quad B \otimes -: \mathscr{V} \longrightarrow \mathscr{V}$$

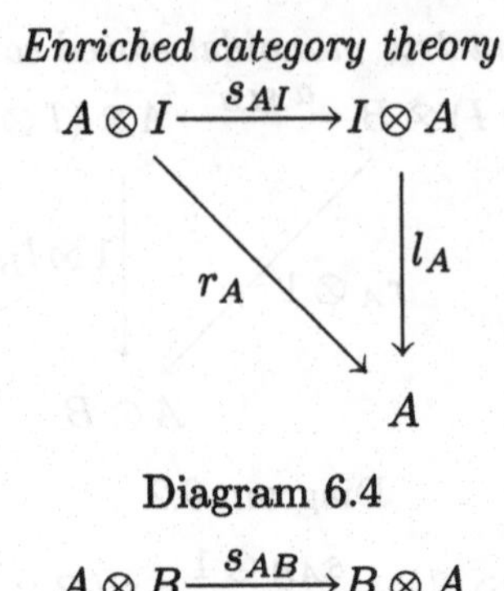

Diagram 6.4

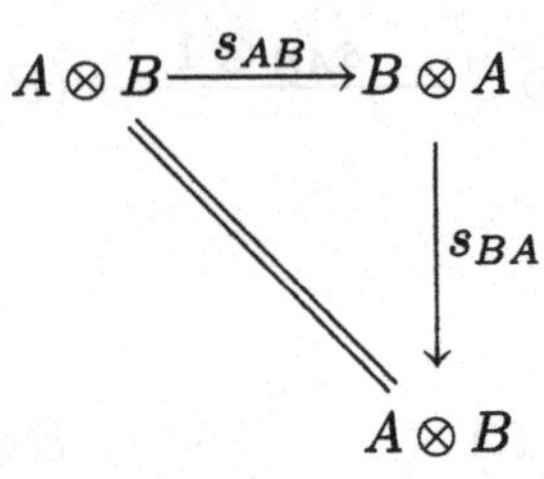

Diagram 6.5

have a right adjoint. A biclosed symmetric monoidal category is called a symmetric monoidal closed category.

Since in a symmetric monoidal category, both functors $- \otimes B$ and $B \otimes -$ are naturally isomorphic, one has obviously

Proposition 6.1.4 *A symmetric monoidal category $\mathscr{V}$ is closed if and only if, for each object $B \in \mathscr{V}$, the functor $- \otimes B \colon \mathscr{V} \longrightarrow \mathscr{V}$ has a right adjoint.* □

Let us immediately introduce a crucial example.

Definition 6.1.5 *A category $\mathscr{V}$ is cartesian closed when it admits all finite products and, for every object $B \in \mathscr{V}$, the functor $- \times B \colon \mathscr{V} \longrightarrow \mathscr{V}$ has a right adjoint, generally written $(-)^B \colon \mathscr{V} \longrightarrow \mathscr{V}$.*

Proposition 6.1.6 *Every cartesian closed category is symmetric monoidal closed, with the cartesian product as a tensor product.*

Proof The existence of the required isomorphisms and the various coherence conditions follow immediately from the universal property of products (as in 2.1.6, volume 1). The unit is just the terminal object. □

In the rest of this chapter, we shall most often reduce our attention to the case of symmetric monoidal closed categories. As long as this can be done at no extra cost, we shall nevertheless give definitions or state results in a more general context.

Let us now introduce some more notation. Given a monoidal category $\mathscr{V}$, the functor $\mathscr{V}(I,-)\colon \mathscr{V} \longrightarrow \mathsf{Set}$ represented by the unit I is generally called the "forgetful functor" or "underlying set functor". Sometimes we just write it U. We make a strong point of the fact that this forgetful functor is in general not faithful: for example, when $\mathscr{V}$ is the cartesian closed category of small categories, I is the one point category and thus $\mathscr{V}(I,A)$ is – up to isomorphism – just the set of objects of the small category A. And clearly a functor is not determined by its action on the objects.

When $\mathscr{V}$ is a symmetric monoidal closed category, we write

$$[B,-]\colon \mathscr{V} \longrightarrow \mathscr{V}, \quad C \mapsto [B,C]$$

for the right adjoint to the functor $-\otimes B\colon \mathscr{V} \longrightarrow \mathscr{V}$. In particular the isomorphisms

$$\mathscr{V}(B,B) \cong \mathscr{V}\big(I \otimes B, B\big) \cong \mathscr{V}\big(I,[B,B]\big)$$

yield a "unit" morphism $u_B\colon I \longrightarrow [B,B]$ corresponding with the identity on B. In an analogous way the isomorphisms

$$\mathscr{V}(C,C) \cong \mathscr{V}\big(C \otimes I, C\big) \cong \mathscr{V}\big(C,[I,C]\big)$$

yield a morphism $i_C\colon C \longrightarrow [I,C]$ corresponding with the identity on C. It is also useful to consider the "evaluation morphisms"

$$\mathrm{ev}_{AB}\colon [A,B] \otimes A \longrightarrow B$$

corresponding by adjunction with the identity on $[A,B]$ and the "composition morphisms"

$$c_{ABC}\colon [A,B] \otimes [B,C] \longrightarrow [A,C]$$

corresponding by adjunction with the composite of diagram 6.6.

Proposition 6.1.7 *On a symmetric monoidal closed category $\mathscr{V}$, we get a bifunctor*

$$[-,-]\colon \mathscr{V}^* \times \mathscr{V} \longrightarrow \mathscr{V}, \quad (A,B) \mapsto [A,B]$$

whose composite with the forgetful functor $\mathscr{V}(I,-)\colon \mathscr{V} \longrightarrow \mathsf{Set}$ is just

$$\mathscr{V}^* \times \mathscr{V} \longrightarrow \mathsf{Set}, \quad (A,B) \mapsto \mathscr{V}(A,B).$$

Proof Given $f\colon A \longrightarrow A'$, $[f,B]\colon [A',B] \longrightarrow [A,B]$ corresponds by adjunction with the composite

$$[A',B] \otimes A \xrightarrow{1 \otimes f} [A',B] \otimes A' \xrightarrow{\mathrm{ev}_{A'B}} B.$$

$$\begin{array}{c} [A,B]\otimes[B,C]\otimes A \\ \downarrow \cong \\ [A,B]\otimes A\otimes[B,C] \\ \downarrow \mathrm{ev}_{AB}\otimes 1 \\ B\otimes[B,C] \\ \downarrow \cong \\ [B,C]\otimes B \\ \downarrow \mathrm{ev}_{BC} \\ C \end{array}$$

Diagram 6.6

It is routine to check the functoriality.

On the other hand the isomorphisms

$$\mathscr{V}(I,[B,C]) \cong \mathscr{V}(I\otimes B,C) \cong \mathscr{V}(B,C)$$

prove the second assertion. □

Proposition 6.1.8 *In a symmetric monoidal closed category $\mathscr{V}$:*

(1) the morphisms $u_B\colon I \longrightarrow [B,B]$ are natural in B;
(2) the morphisms $i_C\colon C \longrightarrow [I,C]$ are isomorphisms;
(3) the morphisms $i_C\colon C \longrightarrow [I,C]$ are natural in C;
(4) diagram 6.7 and diagram 6.8 are commutative, for all objects A, B, C, D in $\mathscr{V}$.

Proof The inverse of the morphism i_C is the composite

$$[I,C] \xrightarrow{r^{-1}_{[I,C]}} [I,C]\otimes I \xrightarrow{\mathrm{ev}_{IC}} C.$$

The rest of the proof is routine computations left to the reader. □

Examples 6.1.9

6.1.9.a The category Set of sets and mappings is cartesian closed, as observed in 3.1.6.f, volume 1.

6.1.9.b The category Cat of small categories and functors is cartesian closed, as proved in 3.1.6.g, volume 1.

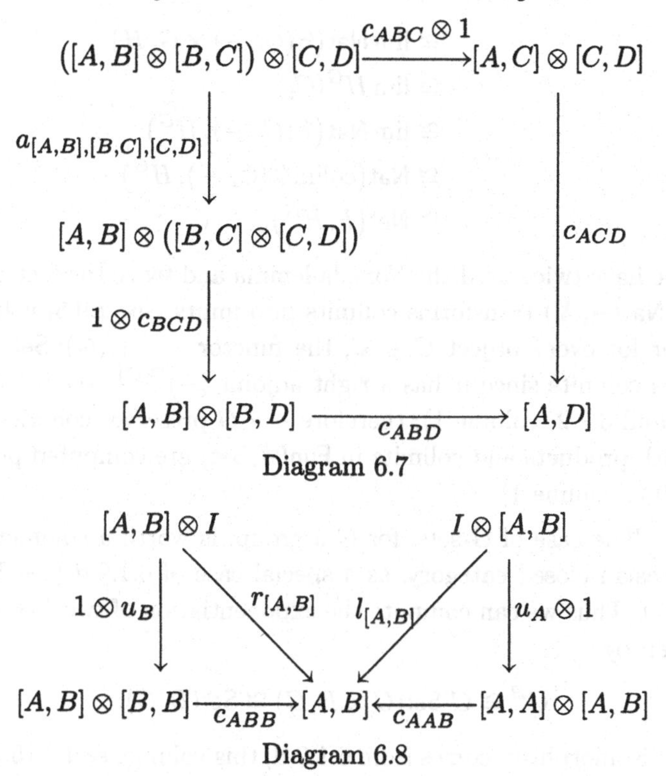

Diagram 6.7

Diagram 6.8

6.1.9.c The category of models of a commutative algebraic theory is symmetric monoidal closed (see 3.10.3); in particular the following are examples of symmetric monoidal closed categories: abelian groups, modules over a commutative ring, pointed sets, G-sets for a commutative group G, ...

6.1.9.d If $\mathscr{C}$ is a small category, the category $\mathsf{Fun}(\mathscr{C}, \mathsf{Set})$ of functors and natural transformations is cartesian closed. Indeed, given two functors $G, H: \mathscr{C} \rightrightarrows \mathsf{Set}$, define

$$H^G: \mathscr{C} \longrightarrow \mathsf{Set}, \quad C \mapsto \mathsf{Nat}(\mathscr{C}(-,C) \times G, H),$$

where this definition extends in an obvious way to the case of morphisms. Given a third functor $F: \mathscr{C} \longrightarrow \mathsf{Set}$, write it as a colimit of representable functors (see 2.15.6, volume 1). The conclusion follows now from the isomorphisms

$$\begin{aligned}\mathsf{Nat}(F \times G, H) &\cong \mathsf{Nat}\Big((\operatorname{colim} \mathscr{C}(C_i, -)) \times G, H\Big)\\ &\cong \mathsf{Nat}\Big(\operatorname{colim}(\mathscr{C}(C_i, -) \times G), H\Big)\end{aligned}$$

$$\begin{aligned}
&\cong \lim \mathsf{Nat}\big(\mathscr{C}(C_i,-)\times G, H\big)\\
&\cong \lim H^G(C_i)\\
&\cong \lim \mathsf{Nat}\big(\mathscr{C}(C_i,-), H^G\big)\\
&\cong \mathsf{Nat}\big(\operatorname{colim}\mathscr{C}(C_i,-), H^G\big)\\
&\cong \mathsf{Nat}\big(F, H^G\big),
\end{aligned}$$

where we have twice used the Yoneda lemma and twice the fact that the functor $\mathsf{Nat}(-,X)$ transforms colimits into limits (see 2.9.5, volume 1); moreover for every object $C \in \mathscr{C}$, the functor $-\times G(C)\colon \mathsf{Set}\longrightarrow\mathsf{Set}$ preserves colimits since it has a right adjoint $(-)^{G(C)}$ (see 6.1.7.a, this volume and 3.2.2, volume 1); therefore $-\times G$ preserves colimits as well since both products and colimits in $\mathsf{Fun}(\mathscr{C},\mathsf{Set})$ are computed pointwise (see 2.15.2, volume 1).

6.1.9.e The case of G-sets, for G a group, is worth a comment. This is a cartesian closed category, as a special case of 6.1.9.d (see 2.15.7.a, volume 1). Thus we can compute the exponentiation of two G-sets B, C; it is given by

$$C^B \cong G\text{-}\mathsf{Set}(G\times B, C) \cong \mathsf{Set}(B,C).$$

The first isomorphism comes from 6.1.9.d, this volume, and 2.15.7.a, volume 1, while the second one is easily proved. Indeed, given a morphism of G-sets $f\colon G\times B\longrightarrow C$, just consider the mapping $\overline{f} = f(1,-)\colon B\longrightarrow C$, where the group has been written additively. This mapping completely determines f since, given $x \in G$, $b \in B$,

$$f(x,b) = f\big(x(1,x^{-1}b)\big) = xf\big(1,x^{-1}b\big) = x\overline{f}\big(x^{-1}b\big).$$

Conversely every mapping $g\colon B\longrightarrow C$ is induced in such a way by a morphism $\widetilde{g}\colon G\times B\longrightarrow G$ of G-sets: just put $\widetilde{g}(x,b) = xg(x^{-1}b)$, which yields of course $\widetilde{g}(1,b) = g(b)$. This $\widetilde{g}$ is a morphism of G-sets since, given $y \in G$,

$$\widetilde{g}(yx,yb) = yxg\big(x^{-1}y^{-1}yb\big) = yxg\big(x^{-1}b\big) = y\widetilde{g}(x,b).$$

6.1.9.f When the group G is commutative, the category of G-sets is symmetric monoidal closed (see 3.10.3) for a structure determined by

$$[A,B] \cong G\text{-}\mathsf{Set}(A,B)$$

for G-sets A, B. But the category is also cartesian closed, for a structure determined by

$$B^A \cong \mathsf{Set}(A,B);$$

see 6.1.7.e. This is an example of a category bearing two distinct symmetric monoidal closed structures.

6.1.9.g The category of topological spaces and continuous mappings can be provided with the structure of a symmetric monoidal closed category: define $[Y, Z]$ as the set of continuous mappings from Y to Z, provided with the pointwise topology; the basic open subsets are

$$\langle y, U\rangle = \{f\colon Y \longrightarrow Z \mid f \text{ continuous}, \ f(y) \in U\}$$

for $y \in Y$ and U open in Z. The corresponding tensor product $X \otimes Y$ of two spaces is the cartesian product of the two sets provided with the final topology for the mappings

$$X \longrightarrow X \times Y, \ x \mapsto (x, y),$$
$$Y \longrightarrow X \times Y \ y \mapsto (x, y),$$

for all elements $x \in X$, $y \in Y$.

6.1.9.h The category Ban_1 of Banach spaces and linear contractions is symmetric monoidal closed for a structure characterized by

$$[B, C] \cong \mathsf{Ban}_\infty(B, C)$$

for two Banach spaces B, C (see 1.2.5.f, volume 1). It is indeed well-known that $\mathsf{Ban}_\infty(B, C)$ is a Banach space when provided with the norm

$$\|f\| = \sup\{\|f(b)\| \mid b \in B, \ \|b\| \leq 1\}.$$

The tensor product $A \otimes B$ of two Banach spaces is the so-called "projective tensor product" of A, B.

6.1.9.i If $\mathscr{C}$ is a small category, the category $\mathsf{Fun}(\mathscr{C}, \mathscr{C})$ of endofunctors (i.e. functors from the category to itself) and natural transformations is monoidal (but not symmetric, in general) when choosing the composition of two functors as their tensor product.

6.1.9.j Every category with finite products is monoidal symmetric, when choosing the cartesian product as tensor product. This structure is closed just when the category is cartesian closed.

6.1.9.k If R is a ring, the category of R-R-bimodules and left–right-R-linear mappings is monoidal biclosed, for the structure given by $L \otimes M \cong L \otimes_R M$. The right adjoint to $- \otimes M$ is given by $[M, -]_r$ where

$$[M, N]_r = \{f\colon M \longrightarrow N \mid f \text{ right-}R\text{-linear }\},$$

while the right adjoint to $L \otimes -$ is given by $[L, -]_l$ where

$$[L, N]_l = \{f \colon L \longrightarrow N \mid f \text{ left-}R\text{-linear}\}.$$

When R is not commutative, this is an asymmetric example.

6.1.9.1 A $\wedge$-semi-lattice $(A, \leq)$ with top element 1, viewed as a category, admits the element $a \wedge b$ as product of the elements $a, b \in A$ (see 2.1.7.b, volume 1). This category is cartesian closed as long as, for every pair b, c of elements, there exists an element $b \Rightarrow c$ such that

$$a \wedge b \leq c \quad \text{iff} \quad a \leq b \Rightarrow c.$$

In particular, every Heyting algebra (see 1.2.1, volume 3) is a cartesian closed category.

6.2 Enriched categories

When a symmetric monoidal closed category is fixed, we show how to enrich over it the notions of category, functor, natural transformation and distributor. We use freely the notation of section 6.1.

Definition 6.2.1 *Let $\mathscr{V}$ be a monoidal category. A $\mathscr{V}$-category $\mathscr{C}$ consists in the following data:*

(1) a class $|\mathscr{C}|$ of "objects";

(2) for every pair $A, B \in |\mathscr{C}|$ of objects, an object $\mathscr{C}(A, B)$ of $\mathscr{V}$;

(3) for every triple $A, B, C \in |\mathscr{C}|$ of objects, a "composition" morphism in $\mathscr{V}$,

$$c_{ABC} \colon \mathscr{C}(A, B) \otimes \mathscr{C}(B, C) \longrightarrow \mathscr{C}(A, C);$$

(4) for every object $A \in |\mathscr{C}|$, a "unit" morphism in $\mathscr{V}$,

$$u_A \colon I \longrightarrow \mathscr{C}(A, A).$$

These data must satisfy the following conditions:

(1) given objects $A, B, C, D, E \in \mathscr{C}$, diagram 6.9 commutes (associativity axiom);

(2) given objects $A, B \in \mathscr{C}$, diagram 6.10 commutes (unit axiom).

When $|\mathscr{C}|$ is a set, the $\mathscr{V}$-category $\mathscr{C}$ is called a small $\mathscr{V}$-category.

As a first obvious result we get

Proposition 6.2.2 *Let $\mathscr{V}$ be a symmetric monoidal category. With the notation of 6.2.1, the $\mathscr{V}$-category $\mathscr{C}$ gives rise to a "dual" $\mathscr{V}$-category $\mathscr{C}^*$ defined by:*

(1) $|\mathscr{C}^*| = |\mathscr{C}|$;

(2) $\mathscr{C}^*(A, B) = \mathscr{C}(B, A)$ for $A, B \in \mathscr{C}$;

$$\begin{array}{ccc}
(\mathscr{C}(A,B)\otimes\mathscr{C}(B,C))\otimes\mathscr{C}(C,D) & \xrightarrow{c_{ABC}\otimes 1} & \mathscr{C}(A,C)\otimes\mathscr{C}(C,D) \\
\downarrow{\scriptstyle a_{\mathscr{C}(A,B),\mathscr{C}(B,C),\mathscr{C}(C,D)}} & & \\
\mathscr{C}(A,B)\otimes(\mathscr{C}(B,C)\otimes\mathscr{C}(C,D)) & & \downarrow{\scriptstyle c_{ACD}} \\
\downarrow{\scriptstyle 1\otimes c_{BCD}} & & \\
\mathscr{C}(A,B)\otimes\mathscr{C}(B,D) & \xrightarrow[c_{ABD}]{} & \mathscr{C}(A,D)
\end{array}$$

Diagram 6.9

$$\begin{array}{ccccc}
I\otimes\mathscr{C}(A,B) & \xrightarrow{l_{\mathscr{C}(A,B)}} & \mathscr{C}(A,B) & \xleftarrow{r_{\mathscr{C}(A,B)}} & \mathscr{C}(A,B)\otimes I \\
\downarrow{\scriptstyle u_A\otimes 1} & & \downarrow{\scriptstyle 1_{\mathscr{C}(A,B)}} & & \downarrow{\scriptstyle 1\otimes u_B} \\
\mathscr{C}(A,A)\otimes\mathscr{C}(A,B) & \xrightarrow[c_{AAB}]{} & \mathscr{C}(A,B) & \xleftarrow[c_{ABB}]{} & \mathscr{C}(A,B)\otimes\mathscr{C}(B,B)
\end{array}$$

Diagram 6.10

(3) $c^*_{ABC} = c_{CBA} \circ s_{\mathscr{C}(B,A),\mathscr{C}(C,B)}$ *for* $A, B, C \in \mathscr{C}^*$*;*
(4) $u^*_A = u_A$ *for* $A \in \mathscr{C}^*$*.* □

Definition 6.2.3 *Let* $\mathscr{V}$ *be a monoidal category. Given* $\mathscr{V}$*-categories* $\mathscr{A}, \mathscr{B}$*, a* $\mathscr{V}$*-functor* $F\colon \mathscr{A} \longrightarrow \mathscr{B}$ *consists in giving:*
(1) for every object $A \in \mathscr{A}$*, an object* $F(A) \in \mathscr{B}$*;*
(2) for every pair $A, A' \in \mathscr{A}$ *of objects, a morphism in* $\mathscr{V}$*,*

$$F_{AA'}\colon \mathscr{A}(A,A') \longrightarrow \mathscr{B}(F(A), F(A'));$$

in such a way that the following axioms hold:
(1) for all objects $A, A', A'' \in \mathscr{A}$*, diagram 6.11 commutes (composition axiom);*
(2) for every object $A \in \mathscr{A}$*, diagram 6.12 commutes (unit axiom).*

Definition 6.2.4 *Let* $\mathscr{V}$ *be a monoidal category. Let* $\mathscr{A}, \mathscr{B}$ *be two* $\mathscr{V}$*-categories and* $F, G\colon \mathscr{A} \rightrightarrows \mathscr{B}$ *two* $\mathscr{V}$*-functors. A* $\mathscr{V}$*-natural transformation* $\alpha\colon F \Rightarrow G$ *consists in giving, for every object* $A \in \mathscr{A}$*, a morphism*

$$\alpha_A\colon I \longrightarrow \mathscr{B}(F(A), G(A))$$

in $\mathscr{V}$ *such that diagram 6.13 commutes, for all objects* $A, A' \in \mathscr{A}$

$$\begin{array}{ccc}
\mathscr{A}(A,A')\otimes\mathscr{A}(A',A'') & \xrightarrow{c_{AA'A''}} & \mathscr{A}(A,A'') \\
{\scriptstyle F_{AA'}\otimes F_{A'A''}}\downarrow & & \downarrow{\scriptstyle F_{A,A''}} \\
\mathscr{B}(FA,FA')\otimes\mathscr{B}(FA',FA'') & \xrightarrow[c_{FA,FA',FA''}]{} & \mathscr{B}(FA,FA'')
\end{array}$$

Diagram 6.11

$$\begin{array}{ccc}
I & \xrightarrow{u_A} & \mathscr{A}(A,A) \\
 & {\scriptstyle u_{FA}}\searrow & \downarrow{\scriptstyle F_{AA}} \\
 & & \mathscr{B}(FA,FA)
\end{array}$$

Diagram 6.12

$$\begin{array}{ccc}
 & \mathscr{A}(A,A') & \\
{\scriptstyle l^{-1}_{\mathscr{A}(A,A')}}\swarrow & & \searrow{\scriptstyle r^{-1}_{\mathscr{A}(A,A')}} \\
I\otimes\mathscr{A}(A,A') & & \mathscr{A}(A,A')\otimes I \\
{\scriptstyle \alpha_A\otimes G_{AA'}}\downarrow & & \downarrow{\scriptstyle F_{AA'}\otimes\alpha_{A'}} \\
\mathscr{B}(FA,GA)\otimes\mathscr{B}(GA,GA') & & \mathscr{B}(FA,FA')\otimes\mathscr{B}(FA',GA') \\
{\scriptstyle c_{FA,GA,GA'}}\searrow & & \swarrow{\scriptstyle c_{FA,FA',GA'}} \\
 & \mathscr{B}(FA,GA') &
\end{array}$$

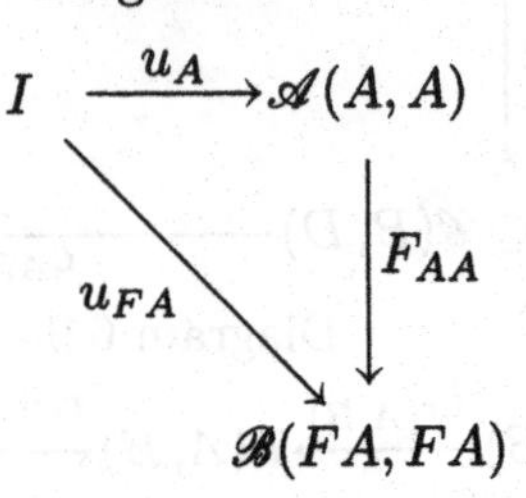

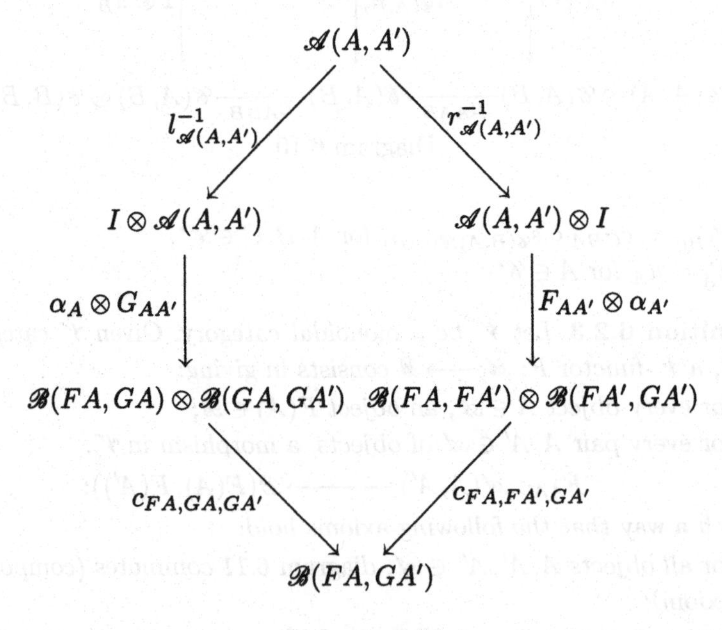

Diagram 6.13

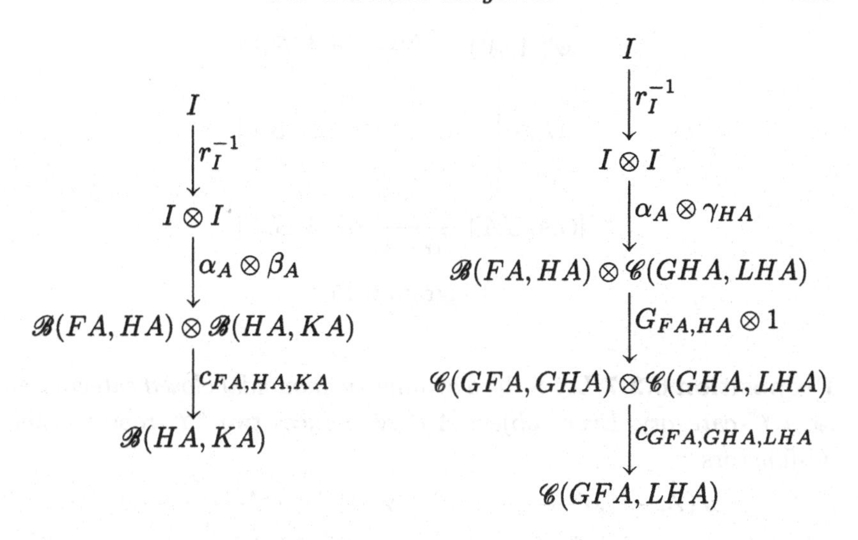

Diagram 6.14

As expected, one gets the following result.

Proposition 6.2.5 *Let $\mathscr{V}$ be a monoidal category. The small $\mathscr{V}$-categories, together with the $\mathscr{V}$-functors and the $\mathscr{V}$-natural transformations, constitute a 2-category written $\mathscr{V}$-*Cat.

Proof If $F\colon \mathscr{A} \longrightarrow \mathscr{B}$ and $G\colon \mathscr{B} \longrightarrow \mathscr{C}$ are $\mathscr{V}$-functors, the composite $G \circ F\colon \mathscr{A} \longrightarrow \mathscr{C}$ is defined by $(G \circ F)(A) = G\big(F(A)\big)$ for $A \in \mathscr{A}$, while $(G \circ F)_{AA'}$, is the composite

$$\mathscr{A}(A, A') \xrightarrow{F_{AA'}} \mathscr{B}\big(F(A), F(A')\big) \xrightarrow{G_{F(A),F(A')}} \mathscr{C}\big(GF(A), GF(A')\big).$$

If $H, K\colon \mathscr{A} \rightrightarrows \mathscr{B}$ are other $\mathscr{V}$-functors and $\alpha\colon F \Rightarrow H$, $\beta\colon H \Rightarrow K$ are $\mathscr{V}$-natural transformations, the composite $\beta \circ \alpha\colon F \Rightarrow K$ is defined by the first composite of diagram 6.14, where A runs through $\mathscr{A}$. Consider yet another $\mathscr{V}$-functor $L\colon \mathscr{B} \longrightarrow \mathscr{C}$ and another $\mathscr{V}$-natural transformation $\gamma\colon G \Rightarrow L$. The $\mathscr{V}$-natural transformation $\gamma * \alpha\colon G \circ F \Rightarrow L \circ H$ is defined by the second composite of diagram 6.14, where A runs through $\mathscr{A}$.

It is routine to check the required commutativities. □

Proposition 6.2.6 *If $\mathscr{V}$ is a symmetric monoidal closed category, $\mathscr{V}$ itself can be provided with the structure of a $\mathscr{V}$-category.*

Proof See 6.1.8. □

$$\begin{array}{ccc}\mathscr{A}(A,A') & \xrightarrow{F_{AA'}} & [FA,FA'] \\ {\scriptstyle G_{AA'}}\downarrow & & \downarrow{\scriptstyle [1,\alpha_{A'}]} \\ [GA,GA'] & \xrightarrow[{[\alpha_A,1]}]{} & [FA,GA']\end{array}$$

Diagram 6.15

Proposition 6.2.7 *Let $\mathscr{V}$ be a symmetric monoidal closed category and $\mathscr{A}$ a $\mathscr{V}$-category. Every object $A \in \mathscr{A}$ induces two "$\mathscr{V}$-representable" $\mathscr{V}$-functors*

$$\mathscr{A}(A,-): \mathscr{A} \longrightarrow \mathscr{V}, \quad \mathscr{A}(-,A): \mathscr{A}^* \longrightarrow \mathscr{V},$$

whose images under the forgetful functor U of 6.4.4 are just the ordinary functors on $U(\mathscr{A})$ represented by A.

Proof The object $B \in \mathscr{A}$ is mapped to the object $\mathscr{A}(A,B) \in \mathscr{V}$ by $\mathscr{A}(A,-)$. Given another object $C \in \mathscr{A}$, the composite

$$\mathscr{A}(B,C)\otimes\mathscr{A}(A,B) \xrightarrow{s_{\mathscr{A}(A,B),\mathscr{A}(B,C)}} \mathscr{A}(A,B)\otimes\mathscr{A}(B,C) \xrightarrow{c_{ABC}} \mathscr{A}(A,C)$$

corresponds by adjunction with a morphism

$$\mathscr{A}(B,C) \longrightarrow [\mathscr{A}(A,B),\mathscr{A}(A,C)]$$

making $\mathscr{A}(A,-)$ a $\mathscr{V}$-functor.

The case of $\mathscr{A}(-,A)$ is analogous and the statement concerning the forgetful functor holds just by definition of $\mathscr{A}(-,A)$ and $\mathscr{A}(A,-)$. □

Finally let us observe that, between $\mathscr{V}$-valued $\mathscr{V}$-functors, the notion of $\mathscr{V}$-natural transformations takes a more normal form.

Proposition 6.2.8 *Let $\mathscr{V}$ be a symmetric monoidal closed category. If $\mathscr{A}$ is a $\mathscr{V}$-category and $F,G\colon \mathscr{A} \rightrightarrows \mathscr{V}$ are $\mathscr{V}$-functors, giving a $\mathscr{V}$-natural transformation $\alpha\colon F \Rightarrow G$ is equivalent to giving a family of morphisms $\alpha_A\colon FA \longrightarrow GA$ in $\mathscr{V}$, for $A \in \mathscr{A}$, in such a way that diagram 6.15 commutes for all objects $A, A' \in \mathscr{A}$.*

Proof This reduces immediately to definition 6.2.4 via the isomorphisms

$$\mathscr{V}\Big(I, \big[F(A),G(A)\big]\Big) \cong \mathscr{V}\Big(I \otimes F(A), G(A)\Big) \cong \mathscr{V}\Big(F(A), G(A)\Big). \quad \square$$

$$
\begin{array}{c}
(\mathscr{A}\otimes\mathscr{B})\big((A,B),(A',B')\big)\otimes(\mathscr{A}\otimes\mathscr{B})\big((A',B'),(A'',B'')\big) \\
\Big\| \\
\mathscr{A}(A,A')\otimes\mathscr{B}(B,B')\otimes\mathscr{A}(A',A'')\otimes\mathscr{B}(B',B'') \\
\Big\downarrow \cong \\
\mathscr{A}(A,A')\otimes\mathscr{A}(A',A'')\otimes\mathscr{B}(B,B')\otimes\mathscr{B}(B',B'') \\
\Big\downarrow c_{AA'A''}\otimes c_{BB'B''} \\
\mathscr{A}(A,A'')\otimes\mathscr{B}(B,B'') \\
\Big\| \\
(\mathscr{A}\otimes\mathscr{B})\big((A,B),(A'',B'')\big)
\end{array}
$$

Diagram 6.16

We still want to introduce the notion of a $\mathscr{V}$-distributor. For this we must be able to handle $\mathscr{V}$-bifunctors. This will follow from the next result, where the reader will notice that symmetry is essential.

Proposition 6.2.9 *Let $\mathscr{V}$ be a symmetric monoidal category. The category of small $\mathscr{V}$-categories and $\mathscr{V}$-functors is itself provided with the structure of a symmetric monoidal category.*

Proof Consider two small $\mathscr{V}$-categories $\mathscr{A},\mathscr{B}$. One gets a new $\mathscr{V}$-category $\mathscr{A}\otimes\mathscr{B}$ by putting:

- $|\mathscr{A}\otimes\mathscr{B}| = |\mathscr{A}|\times|\mathscr{B}|$;
- $(\mathscr{A}\otimes\mathscr{B})\big((A,B),(A',B')\big) = \mathscr{A}(A,A')\otimes\mathscr{B}(B,B')$ for $A,A'\in\mathscr{A}$ and $B,B'\in\mathscr{B}$;
- if $A,A',A''\in\mathscr{A}$ and $B,B',B''\in\mathscr{B}$ the composition map of $\mathscr{A}\otimes\mathscr{B}$ is given as in diagram 6.16, where the isomorphism is constructed from the associativity and symmetry isomorphisms;
- if $A\in\mathscr{A}, B\in\mathscr{B}$, the corresponding unit map of $\mathscr{A}\otimes\mathscr{B}$ is given by

$$I \xrightarrow{r_I^{-1}} I\otimes I \xrightarrow{u_A\otimes u_B} \mathscr{A}(A,A)\otimes\mathscr{B}(B,B).$$

The rest is now routine calculations: the unit for the tensor product of $\mathscr{V}$-categories is the $\mathscr{V}$-category $\mathscr{I}$ with a single object $*$ and $\mathscr{I}(*,*)=I$. □

$$\begin{array}{c}\coprod_{B',B''\in\mathscr{B}}\psi(C,B'')\otimes\mathscr{B}(B'',B')\otimes\phi(B',A)\\ u\Big\downarrow\Big\downarrow v\\ \coprod_{B\in\mathscr{B}}\psi(C,B)\otimes\phi(B,A)\\ \Big\downarrow q_{CA}\\ (\psi\circ\phi)(C,A)\end{array}$$

Diagram 6.17

Definition 6.2.10 *Let $\mathscr{V}$ be a symmetric monoidal closed category. By a $\mathscr{V}$-distributor $\phi\colon \mathscr{A} \longrightarrow\!\!\!\!\!\circ\!\!\!\rightarrow \mathscr{B}$ between small $\mathscr{V}$-categories, we mean a $\mathscr{V}$-functor $\mathscr{B}^* \otimes \mathscr{A} \longrightarrow \mathscr{V}$. By a morphism of $\mathscr{V}$-distributors, we mean a $\mathscr{V}$-natural transformation between the corresponding $\mathscr{V}$-functors.*

The previous definition makes sense by 6.2.9 and 6.2.6. But considering the composite of two distributors requires the cocompleteness of $\mathscr{V}$, as in 7.8.2, volume 1.

Proposition 6.2.11 *Let $\mathscr{V}$ be a cocomplete symmetric monoidal closed category. The small $\mathscr{V}$-categories, the $\mathscr{V}$-distributors and their morphisms constitute a bicategory.*

Proof The only difficult point is to define the composite of two $\mathscr{V}$-distributors $\phi\colon \mathscr{A} \longrightarrow\!\!\!\!\!\circ\!\!\!\rightarrow \mathscr{B}$ and $\psi\colon \mathscr{B} \longrightarrow\!\!\!\!\!\circ\!\!\!\rightarrow \mathscr{C}$. Thus ϕ and ψ are $\mathscr{V}$-functors

$$\phi\colon \mathscr{B}^* \otimes \mathscr{A} \longrightarrow \mathscr{V}, \quad \psi\colon \mathscr{C}^* \otimes \mathscr{B} \longrightarrow \mathscr{V},$$

which yields in particular morphisms in $\mathscr{V}$,

$$\phi_{BAB'A'}\colon \mathscr{B}^*(B,B') \otimes \mathscr{A}(A,A') \longrightarrow [\phi(B,A),\phi(B',A')],$$
$$\psi_{CBC'B'}\colon \mathscr{C}^*(C,C') \otimes \mathscr{B}(B,B') \longrightarrow [\psi(C,B),\psi(C',B')],$$

for $A,A' \in \mathscr{A}$, $B,B' \in \mathscr{B}$, $C,C' \in \mathscr{C}$.

Given $A \in \mathscr{A}$ and $C \in \mathscr{C}$, we define $(\psi\circ\phi)(C,A)$ as the coequalizer in diagram 6.17 (see 7.8.2, volume 1), where u, v are defined in the following way. Fixing $B', B'' \in \mathscr{B}$, we consider first the two morphisms

$$\overline{\phi}_{B'AB''A'}\colon \mathscr{B}(B'',B') \otimes \mathscr{A}(A,A') \otimes \phi(B',A) \longrightarrow \phi(B'',A'),$$
$$\overline{\psi}_{CB''C'B'}\colon \psi(C,B'') \otimes \mathscr{C}(C',C) \otimes \mathscr{B}(B'',B') \longrightarrow \psi(C',B'),$$

corresponding by adjunction with $\phi_{B'AB''A'}$ and $\psi_{CB''C'B'}$. The morphisms u, v are then the unique factorizations through the first coproduct of the composites of diagram 6.18, where $s_{B'}, s_{B''}$ are canonical

$$
\begin{array}{c}
\psi(C,B'')\otimes\mathscr{B}(B'',B')\otimes\phi(B',A)\\
\Big\downarrow 1\otimes r^{-1}\otimes 1\\
\psi(C,B'')\otimes\mathscr{B}(B'',B')\otimes I\otimes\phi(B',A)\\
\Big\downarrow 1\otimes 1\otimes u_A\otimes 1\\
\psi(C,B'')\otimes\mathscr{B}(B'',B')\otimes\mathscr{A}(A,A)\otimes\phi(B',A)\\
\Big\downarrow 1\otimes\overline{\phi}_{B'AB''A}\\
\psi(C,B'')\otimes\phi(B'',A)\\
\Big\downarrow s_{B''}\\
\coprod_{B\in\mathscr{B}}\psi(C,B)\otimes\phi(B,A)
\end{array}
$$

$$
\begin{array}{c}
\psi(C,B'')\otimes\mathscr{B}(B'',B')\otimes\phi(B',A)\\
\Big\downarrow 1\otimes l^{-1}\otimes 1\\
\psi(C,B'')\otimes I\otimes\mathscr{B}(B'',B')\otimes\phi(B',A)\\
\Big\downarrow 1\otimes u_A\otimes 1\otimes 1\\
\psi(C,B'')\otimes\mathscr{C}(C,C)\otimes\mathscr{B}(B'',B')\otimes\phi(B',A)\\
\Big\downarrow \overline{\psi}_{CB''CB'}\\
\psi(C,B')\otimes\phi(B',A)\\
\Big\downarrow s_{B'}\\
\coprod_{B\in\mathscr{B}}\psi(C,B)\otimes\phi(B,A)
\end{array}
$$

Diagram 6.18

morphisms of the second coproduct and, for simplicity, the associativity isomorphisms have been omitted.

Given $A, A' \in \mathscr{A}$ and $C, C' \in \mathscr{C}$, it remains to define morphisms

$$(\psi\circ\phi)_{CAC'A'}\colon \mathscr{C}^*(C,C')\otimes\mathscr{A}(A,A')\longrightarrow\big[(\psi\circ\phi)(C,A),(\psi\circ\phi)(C',A')\big].$$

By adjunction, this is equivalent to constructing a morphism

$$x\colon \mathscr{C}(C',C)\otimes\mathscr{A}(A,A')\otimes(\psi\circ\phi)(C,A)\longrightarrow(\psi\circ\phi)(C',A').$$

$$\begin{array}{c}
\mathscr{C}(C',C)\otimes\mathscr{A}(A,A')\otimes\psi(C,B)\otimes\phi(B,A)\\
\Big\downarrow\cong\\
\psi(C,B)\otimes\mathscr{C}(C',C)\otimes\mathscr{A}(A,A')\otimes\phi(B,A)\\
\Big\downarrow\cong\\
\psi(C,B)\otimes\mathscr{C}(C',C)\otimes I\otimes I\otimes\mathscr{A}(A,A')\otimes\phi(B,A)\\
\Big\downarrow 1\otimes 1\otimes u_B\otimes u_B\otimes 1\otimes 1\\
\psi(C,B)\otimes\mathscr{C}(C',C)\otimes\mathscr{B}(B,B)\otimes\mathscr{B}(B,B)\otimes\mathscr{A}(A,A')\otimes\phi(B,A)\\
\Big\downarrow \overline{\psi}_{CBC'B}\otimes\overline{\phi}_{BABA'}\\
\psi(C',B)\otimes\phi(B,A')\\
\Big\downarrow s_B\\
\coprod_{B\in\mathscr{B}}\psi(C',B)\otimes\phi(B,A')\\
\Big\downarrow q_{C'A'}\\
(\psi\circ\phi)(C',A')
\end{array}$$

Diagram 6.19

But for every object $V\in\mathscr{V}$, the functor $V\otimes -$ admits $[V,-]$ as a right adjoint (see 6.1.4), therefore it preserves coequalizers (see 3.2.2, volume 1). Thus constructing x is equivalent to constructing

$$y\colon \mathscr{C}(C',C)\otimes\mathscr{A}(A,A')\otimes\Big(\coprod_{B\in\mathscr{B}}\psi(C,B)\otimes\phi(B,A)\Big)\longrightarrow(\psi\circ\phi)(C',A')$$

such that $y\circ u = y\circ v$. Using the fact that each functor $V\otimes -$ preserves coproducts as well (see 3.2.2, volume 1), constructing y reduces to constructing a suitable family of morphisms

$$y_B\colon \mathscr{C}(C',C)\otimes\mathscr{A}(A,A')\otimes\psi(C,B)\otimes\phi(B,A)\longrightarrow(\psi\circ\phi)(C',A').$$

Still avoiding writing the associativity isomorphisms, these are the composites of diagram 6.19.

The proof is now a list of lengthy but straightforward computations left to the reader. □

$$\begin{array}{c} \mathcal{V}-\mathsf{Nat}(F,G) \\ \big\downarrow k \\ \prod_{A\in\mathcal{A}}\mathcal{B}(FA,GA) \\ u \downarrow\downarrow v \\ \prod_{A',A''\in\mathcal{A}}[\mathcal{A}(A',A''),\mathcal{B}(FA',GA'')] \end{array}$$

Diagram 6.20

$$\begin{array}{ccc} (\prod_A\mathcal{B}(FA,GA))\otimes\mathcal{A}(A',A'') & & \mathcal{A}(A',A'')\otimes(\prod_A\mathcal{B}(FA,GA)) \\ \big\downarrow p_{A'}\otimes G_{A'A''} & & \big\downarrow F_{A',A''}\otimes p_{A''} \\ \mathcal{B}(FA',GA')\otimes\mathcal{B}(GA',GA'') & & \mathcal{B}(FA',FA'')\otimes\mathcal{B}(FA'',GA'') \\ \big\downarrow c_{FA',GA',GA''} & & \big\downarrow c_{FA',FA'',GA''} \\ \mathcal{B}(FA',GA'') & & \mathcal{B}(FA',GA'') \end{array}$$

Diagram 6.21

6.3 The enriched Yoneda lemma

We enrich first some of the constructions of section 6.2, using completeness assumptions on $\mathcal{V}$.

Proposition 6.3.1 *Let $\mathcal{V}$ be a complete symmetric monoidal closed category. Given two $\mathcal{V}$-categories $\mathcal{A},\mathcal{B}$, with $\mathcal{A}$ small, the category of $\mathcal{V}$-functors $\mathcal{A}\longrightarrow\mathcal{B}$ and $\mathcal{V}$-natural transformations can be provided with the structure of a $\mathcal{V}$-category, written $\mathcal{V}[\mathcal{A},\mathcal{B}]$.*

Proof Given two $\mathcal{V}$-functors $F,G\colon\mathcal{A}\rightrightarrows\mathcal{B}$, we must define the object $\mathcal{V}\text{-}\mathsf{Nat}(F,G)\in\mathcal{V}$ of $\mathcal{V}$-natural transformations from F to G. It is given by the equalizer of diagram 6.20, where it remains to define u and v. Fixing $A',A''\in\mathcal{A}$, the composites of diagram 6.21 correspond by adjunction (and symmetry) with morphisms

$$\prod_A\mathcal{B}(F(A),G(A))\rightrightarrows\big[\mathcal{A}(A',A''),\mathcal{B}(F(A'),G(A''))\big]$$

which we choose as u,v followed by the projection $p_{(A',A'')}$. This defines u,v and the rest is lengthy but routine computation. □

Observe that in the case $\mathcal{V}=\mathsf{Set}$, we have defined $\mathcal{V}\text{-}\mathsf{Nat}(F,G)$ as the set of those families $(\alpha_A\colon FA \longrightarrow GA)_{A\in\mathscr{A}}$ such that the mappings

$$\mathscr{A}(A',A'') \longrightarrow \mathscr{B}(F(A'),G(A'')),\ \ f \mapsto G(f)\circ\alpha_{A'},$$
$$\mathscr{A}(A',A'') \longrightarrow \mathscr{B}(F(A'),G(A'')),\ \ f \mapsto \alpha_{A''}\circ F(f)$$

coincide for all choices $A', A'' \in \mathscr{A}$. This indeed defines the set of natural transformations from F to G.

Corollary 6.3.2 *Let $\mathscr{V}$ be a complete symmetric monoidal closed category. The category of small $\mathscr{V}$-categories and $\mathscr{V}$-functors is itself provided with the structure of a symmetric monoidal closed category.*

Proof The symmetric monoidal structure has been described in 6.2.9 while the closedness follows from 6.3.1. □

When $\mathscr{V}$ is a complete and cocomplete symmetric monoidal closed category, putting together propositions 6.2.11 and 6.3.1 yields that small $\mathscr{V}$-categories, $\mathscr{V}$-distributors and morphisms of $\mathscr{V}$-distributors constitute what should be called a $\mathscr{V}$-bicategory, a gadget whose precise definition is left to the reader.

The next lemma, together with 6.2.8, will simplify a good deal various proofs involving objects of $\mathscr{V}$-natural transformations.

Lemma 6.3.3 *Let $\mathscr{V}$ be a complete symmetric monoidal closed category. Given a small $\mathscr{V}$-category $\mathscr{A}$ and two $\mathscr{V}$-functors $F, G\colon \mathscr{A} \rightrightarrows \mathscr{V}$, an object $N \in \mathscr{V}$ is isomorphic to the object of $\mathscr{V}$-natural transformations $\mathscr{V}\text{-}\mathsf{Nat}(F,G)$ if and only if, for every object $V \in \mathscr{V}$, there is a bijective correspondence between*

(1) the morphisms $V \longrightarrow N$,

(2) the $\mathscr{V}$-natural transformations $F \Rightarrow [V, G-]$.

Proof With the notation of 6.3.1, a morphism $f\colon V \longrightarrow \mathscr{V}\text{-}\mathsf{Nat}(F,G)$ is induced by a morphism $f'\colon V \longrightarrow \coprod_{A\in\mathscr{A}} \big[F(A), G(A)\big]$ equalizing u, v, thus by a family of morphisms $f_A\colon V \longrightarrow \big[F(A), G(A)\big]$. This is equivalent to giving morphisms $V \otimes F(A) \longrightarrow G(A)$ and thus, by symmetry and adjunction, to giving morphisms $g_A\colon F(A) \longrightarrow \big[V, G(A)\big]$. By 6.2.8, the $\mathscr{V}$-naturality of the family $(g_A)_{A\in\mathscr{A}}$ reduces to the commutativity of diagram 6.22, where the first vertical morphism, expressing the action of the composite $\mathscr{V}$-functor $[V, G-]\colon \mathscr{A} \longrightarrow \mathscr{V}$, corresponds by adjunction and symmetry with

$$[V,GA]\otimes\mathscr{A}(A,A') \xrightarrow{1\otimes G_{AA'}} [V,GA]\otimes[GA,GA'] \xrightarrow{c_{V,GA,GA'}} [V,GA'].$$

The required commutativity is thus equivalent by adjunction to that of diagram 6.23, where the last two arrows correspond by adjunction and

$$\begin{array}{ccc}
\mathscr{A}(A,A') & \xrightarrow{F_{AA'}} & [FA,FA'] \\
{\scriptstyle [V,G]_{AA'}}\downarrow & & \downarrow{\scriptstyle [1,g_{A'}]} \\
[[V,GA],[V,GA']] & \xrightarrow[{[g_A,1]}]{} & [FA,[V,GA']]
\end{array}$$

Diagram 6.22

$$\begin{array}{ccc}
V & \xrightarrow{f_{A'}} & [FA',GA'] \\
{\scriptstyle f_A}\downarrow & & \downarrow \\
[FA,GA] & \longrightarrow & [\mathscr{A}(A,A'),[FA,GA']]
\end{array}$$

Diagram 6.23

symmetry with the composites of diagram 6.24. But this last commutativity is itself equivalent to the relation $u \circ f' = v \circ f'$, which concludes the proof by 1.9.5 and 1.5.2, volume 1. □

The previous lemma indicates in particular how to define the object of $\mathscr{V}$-natural transformations $\mathscr{V}\text{-}\mathsf{Nat}(F,G)$ between two $\mathscr{V}$-functors $F, G\colon \mathscr{A} \rightrightarrows \mathscr{V}$, when $\mathscr{A}$ is not necessarily small and $\mathscr{V}$ is not necessarily complete.

Definition 6.3.4 *Let $\mathscr{V}$ be a symmetric monoidal closed category. Given a $\mathscr{V}$-category $\mathscr{A}$ and two $\mathscr{V}$-functors $F, G\colon \mathscr{A} \rightrightarrows \mathscr{V}$, an object N of $\mathscr{V}$ is called the object of $\mathscr{V}$-natural transformations from F to G, and one writes $N \cong \mathscr{V}\text{-}\mathsf{Nat}(F,G)$, if for all $V \in \mathscr{V}$ there exist bijections, natural in the variable $V \in \mathscr{V}$, between*

(1) the set of morphisms $V \longrightarrow N$,

(2) the class of $\mathscr{V}$-natural transformations $F \Rightarrow [V, G-]$.

Theorem 6.3.5 (Enriched Yoneda lemma)
Let $\mathscr{V}$ be a symmetric monoidal closed category and $\mathscr{A}$ a small $\mathscr{V}$-category. For every object $A \in \mathscr{A}$ and every $\mathscr{V}$-functor $F\colon \mathscr{A} \longrightarrow \mathscr{V}$, the object of $\mathscr{V}$-natural transformations from $\mathscr{A}(A,-)$ to F exists and there is an isomorphism in $\mathscr{V}$,

$$\mathscr{V}\text{-}\mathsf{Nat}(\mathscr{A}(A,-),F) \cong F(A),$$

$$\begin{array}{ccc}
[FA,GA]\otimes\mathscr{A}(A,A') & & \mathscr{A}(A,A')\otimes[FA',GA'] \\
\downarrow 1\otimes G_{AA'} & & \downarrow F_{AA'}\otimes 1 \\
[FA,GA]\otimes[GA,GA'] & & [FA,FA']\otimes[FA',GA'] \\
\downarrow c_{FA,GA,GA'} & & \downarrow c_{FA,FA',GA'} \\
[FA,GA'] & & [FA,GA']
\end{array}$$

Diagram 6.24

$$\begin{array}{ccc}
\mathscr{A}(B,C) & \xrightarrow{\mathscr{A}(A,-)} & [\mathscr{A}(A,B),\mathscr{A}(A,C)] \\
{\scriptstyle [V,F-]}\downarrow & & \downarrow{\scriptstyle [1,\varphi_C]} \\
[[V,FB],[V,FC]] & \xrightarrow[{[\varphi_B,1]}]{} & [\mathscr{A}(A,B),[V,FC]]
\end{array}$$

Diagram 6.25

which is $\mathscr{V}$-natural both in F and in A.

Proof We refer to 6.3.4. Given an object $V\in\mathscr{V}$ and a morphism $f\colon V\longrightarrow F(A)$, we must construct a $\mathscr{V}$-natural transformation

$$\varphi\colon\mathscr{A}(A,-)\longrightarrow[V,F-],$$

i.e. a compatible family of morphisms $\varphi_B\colon\mathscr{A}(A,B)\longrightarrow[V,F(B)]$ for B in $\mathscr{B}$. They correspond by adjunction with the composites

$$\mathscr{A}(A,B)\otimes V\xrightarrow{F_{A,B}\otimes f}[F(A),F(B)]\otimes F(A)\xrightarrow{\mathrm{ev}_{F(A),F(B)}}F(B).$$

The $\mathscr{V}$-naturality of φ means the commutativity of diagram 6.25 for $B,C\in\mathscr{A}$; this is equivalent, by adjunction and symmetry, to the equality of the composites of diagram 6.26, which holds by $\mathscr{V}$-functoriality of F.

Conversely given a $\mathscr{V}$-natural transformation $\varphi\colon\mathscr{A}(A,-)\longrightarrow[V,F-]$, we get the composite

$$V\xrightarrow{\overline{\varphi}_A}[\mathscr{A}(A,A),F(A)]\xrightarrow{[u_A,1]}[I,F(A)]\cong F(A)$$

where $\overline{\varphi}_A$ corresponds with φ_A by symmetry and adjunction.

It is now straightforward to show from the definitions that we have defined reciprocal and natural constructions. □

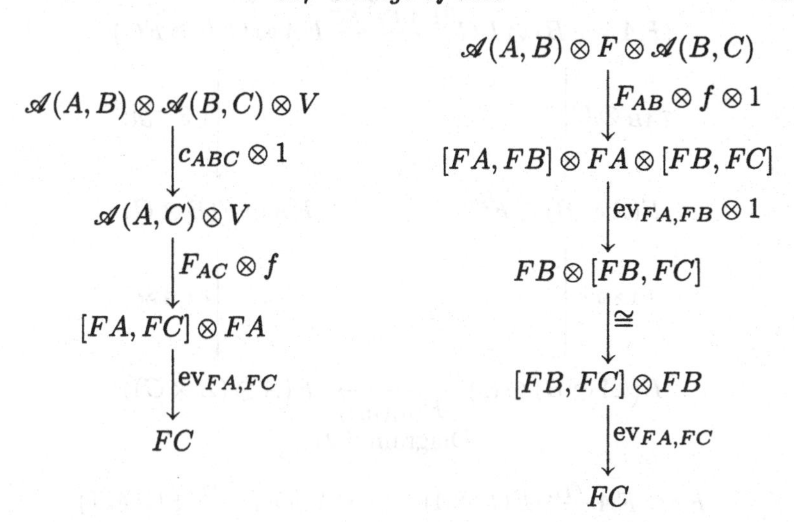

Diagram 6.26

Corollary 6.3.6 *Let $\mathscr{V}$ be a complete symmetric monoidal closed category. For every small $\mathscr{V}$-category $\mathscr{A}$, the mapping*

$$Y\colon \mathscr{A} \longrightarrow \mathscr{V}\text{-}\mathsf{Fun}(\mathscr{A}^*,\mathscr{V}),\ \ A\mapsto \mathscr{A}(-,A)$$

can be extended to a $\mathscr{V}$-functor called the $\mathscr{V}$-Yoneda-embedding.

Proof It remains to construct, for $A, A' \in \mathscr{A}$

$$Y_{AA'}\colon \mathscr{A}(A,A') \longrightarrow \mathscr{V}\text{-}\mathsf{Nat}(\mathscr{A}(-,A),\mathscr{A}(-,A')).$$

But by the $\mathscr{V}$-Yoneda-lemma (see 6.3.4), we have indeed such an isomorphism $Y_{AA'}$. Checking the axioms is now routine. □

6.4 Change of base

We study the effect of a "morphism of monoidal categories" on the corresponding enriched categories, functors and natural transformations. As a special case, we obtain the ordinary categories, functors, ... associated with enriched ones.

Definition 6.4.1 *A morphism $F\colon \mathscr{V} \longrightarrow \mathscr{W}$ of monoidal categories consists in the following data:*

(1) a functor $F\colon \mathscr{V} \longrightarrow \mathscr{W}$;

(2) for each pair (A,B) of objects of $\mathscr{V}$, a morphism of $\mathscr{W}$,

$$\tau_{AB}\colon F(A)\otimes F(B) \longrightarrow F(A\otimes B);$$

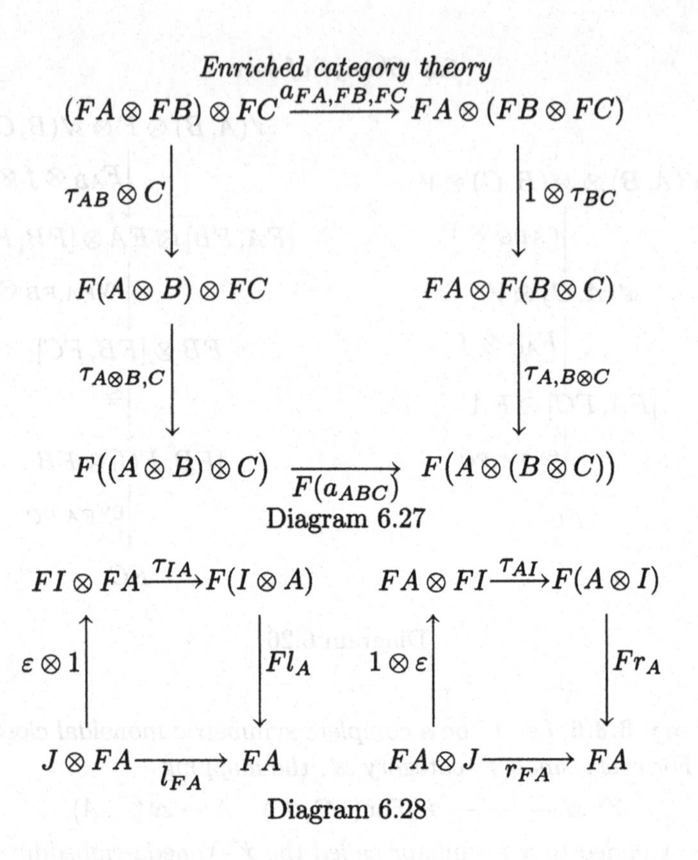

Diagram 6.27

Diagram 6.28

(3) a morphism $\varepsilon\colon J \longrightarrow F(I)$ of $\mathscr{W}$, where I stands for the unit of $\mathscr{V}$ and J for the unit of $\mathscr{W}$.

These data must satisfy the following conditions:

(1) the morphisms τ_{AB} are natural in A, B;

(2) diagram 6.27 commutes for all objects A, B, C of $\mathscr{V}$ (associativity condition);

(3) both parts of diagram 6.28 commute for each object A of $\mathscr{V}$ (unit conditions).

When $\mathscr{V}, \mathscr{W}$ are symmetric monoidal categories, F is a morphism of symmetric monoidal categories when moreover

(4) diagram 6.29 commutes, for every pair A, B of objects of $\mathscr{V}$ (symmetry condition).

Proposition 6.4.2 *For every monoidal category $\mathscr{V}$, the forgetful functor $\mathscr{V}(I, -)\colon \mathscr{V} \longrightarrow$ Set is a morphism of monoidal categories, when Set is provided with its structure of cartesian closed category. If $\mathscr{V}$ is symmetric, this is a morphism of symmetric monoidal categories.*

$$\begin{array}{ccc} FA \otimes FB & \xrightarrow{s_{FA,FB}} & FB \otimes FA \\ \Big\downarrow{\scriptstyle \tau_{AB}} & & \Big\downarrow{\scriptstyle \tau_{BA}} \\ F(A \otimes B) & \xrightarrow[F s_{AB}]{} & F(B \otimes A) \end{array}$$

Diagram 6.29

$$\begin{array}{c} F\big(\mathscr{A}(A,B)\big) \otimes F\big(\mathscr{A}(B,C)\big) \\ \Big\downarrow{\scriptstyle \tau} \\ F\big(\mathscr{A}(A,B) \otimes \mathscr{A}(B,C)\big) \\ \Big\downarrow{\scriptstyle F c_{ABC}} \\ F\big(\mathscr{A}(A,C)\big) \end{array}$$

Diagram 6.30

Proof Given morphisms $a: I \longrightarrow A$ and $b: I \longrightarrow B$ in $\mathscr{V}$, the composite

$$I \xrightarrow{\quad a \quad} A \xrightarrow{\quad r_A^{-1} \quad} A \otimes I \xrightarrow{\quad 1 \otimes b \quad} A \otimes B$$

can be written $\tau_{AB}(a,b)$, which yields a mapping

$$\tau_{A,B}: \mathscr{V}(I,A) \times \mathscr{V}(I,B) \longrightarrow \mathscr{V}(I, A \otimes B).$$

On the other hand the identity on I yields a mapping

$$\varepsilon: \{*\} \longrightarrow \mathscr{V}(I,I), \quad * \mapsto 1_I.$$

It is routine to check the conditions of 6.4.1. □

Proposition 6.4.3 *Let $F: \mathscr{V} \longrightarrow \mathscr{W}$ be a morphism of monoidal categories. F induces a 2-functor $\varphi: \mathscr{V}$-Cat$\longrightarrow \mathscr{W}$-Cat.*

Proof Let $\mathscr{A}$ be a $\mathscr{V}$-category. The $\mathscr{W}$-category $\varphi(\mathscr{A})$ is defined in the following way, with the notations of 6.4.1:

- the objects of $\varphi(\mathscr{A})$ are those of $\mathscr{A}$;
- for $A, B \in \mathscr{A}$, $\varphi(\mathscr{A})(A,B) = F\big(\mathscr{A}(A,B)\big)$;
- for $A, B, C \in \mathscr{A}$, the corresponding composition morphism of $\varphi(\mathscr{A})$ is the composite of diagram 6.30;

- for $A \in \mathscr{A}$, the corresponding unit morphism of $\varphi(\mathscr{A})$ is the following composite:

$$J \xrightarrow{\ \varepsilon\ } F(I) \xrightarrow{F(u_A)} F(\mathscr{A}(A,A)).$$

If $G\colon \mathscr{A} \longrightarrow \mathscr{B}$ is a $\mathscr{V}$-functor, the $\mathscr{W}$-functor $\varphi(G)\colon \varphi(\mathscr{A}) \longrightarrow \varphi(\mathscr{B})$ has the same action as G on the objects, while the morphism $\big(\varphi(G)\big)_{AA'}$, for $A, A' \in \mathscr{A}$, is just $F(G_{AA'})$.

If $H\colon \mathscr{A} \longrightarrow \mathscr{B}$ is another $\mathscr{V}$-functor and $\alpha\colon G \Rightarrow H$ is a $\mathscr{V}$-natural transformation, we get a $\mathscr{W}$-natural transformation $\varphi(\alpha)\colon \varphi(G) \Rightarrow \varphi(H)$ by defining $\varphi(\alpha)_A$ as the following composite, for $A \in \mathscr{A}$:

$$J \xrightarrow{\ \varepsilon\ } F(I) \xrightarrow{F(\alpha_A})F\Big(\mathscr{A}\big(G(A), H(A)\big)\Big).$$

The rest is routine computations. □

Corollary 6.4.4 *For every monoidal category, there exists a "forgetful" 2-functor*

$$U\colon \mathscr{V}\text{-}\mathsf{Cat} \longrightarrow \mathsf{Cat}.$$

Proof Just apply 6.4.3 to $\mathscr{V}(I,-)$; see 6.4.2. □

If $\mathscr{A}$ is a $\mathscr{V}$-category, the category $U(\mathscr{A})$ obtained from 6.4.4 is often referred to as the ordinary category underlying $\mathscr{A}$. One has:

- $\big|U(\mathscr{A})\big| = |\mathscr{A}|$;
- $\big(U(\mathscr{A})\big)(A,B) = \mathscr{V}\big(I, \mathscr{A}(A,B)\big)$;
- the composite of $f\colon I \longrightarrow \mathscr{A}(A,B)$, $g\colon I \longrightarrow \mathscr{A}(B,C)$ is given by

$$I \xrightarrow{r_I^{-1}} I \otimes I \xrightarrow{f \otimes g} \mathscr{A}(A,B) \otimes \mathscr{A}(B,C) \xrightarrow{c_{ABC}} \mathscr{A}(A,C).$$

Observe also that a given $f\colon A \longrightarrow B$ in $U(\mathscr{A})$ yields a morphism

$$\mathscr{A}(X,f)\colon \mathscr{A}(X,A) \longrightarrow \mathscr{A}(X,B)$$

in $\mathscr{V}$: this is just the composite

$$\mathscr{A}(X,A) \cong \mathscr{A}(X,A) \otimes I \xrightarrow{1 \otimes f} \mathscr{A}(X,A) \otimes \mathscr{A}(A,B) \xrightarrow{c_{XAB}} \mathscr{A}(X,B).$$

An analogous definition holds for $\mathscr{A}(f,X)$.

Proposition 6.4.5 *Let $F\colon \mathscr{V} \longrightarrow \mathscr{W}$ be a morphism of symmetric monoidal categories, where $\mathscr{V}$ and $\mathscr{W}$ are in fact symmetric monoidal closed categories. Under these conditions there exist morphisms in $\mathscr{W}$,*

$$\sigma_{AB}\colon F[A,B] \longrightarrow [F(A), F(B)],$$

for every pair A, B of objects of $\mathscr{V}$, these morphisms satisfying the following conditions:

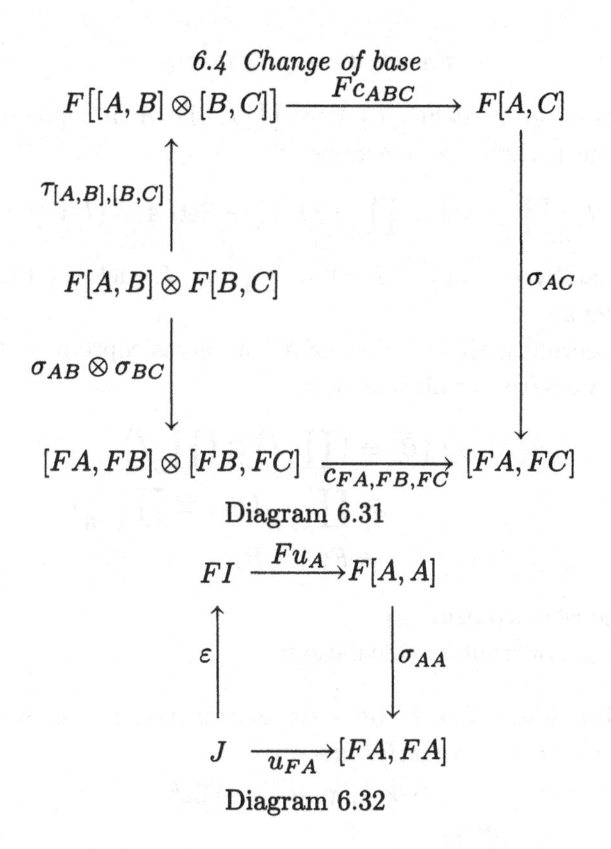

Diagram 6.31

Diagram 6.32

(1) *the morphisms σ_{AB} are natural in A, B;*

(2) *diagram 6.31 commutes, for all objects A, B, C of $\mathscr{V}$ (composition condition);*

(3) *diagram 6.32 commutes, for every object A of $\mathscr{V}$ (unit condition)*

Proof The σ_{AB} correspond by adjunction with the following composite

$$F[A,B]\otimes F(A) \xrightarrow{\tau_{[A,B],A}} F[[A,B]\otimes A] \xrightarrow{F(\mathrm{ev}_{AB})} F(B).$$

The rest is routine computation. □

Proposition 6.4.6 *Let $\mathscr{V}$ be a symmetric monoidal closed category with coproducts. The functor*

$$\mathscr{V}(I,-)\colon \mathscr{V} \longrightarrow \mathsf{Set}$$

has a left adjoint $F\colon \mathsf{Set} \longrightarrow \mathscr{V}$. This adjoint functor is itself a morphism of symmetric monoidal categories with the additional fact that the required natural morphisms

$$\tau_{AB}\colon F(A)\otimes F(B) \longrightarrow F(A\times B),\ \ \varepsilon\colon I \longrightarrow F(1)$$

are isomorphisms.

Proof For a set A, define $F(A) = \coprod_A I$, the A-th copower of I. By 3.2.2, volume 1, there are bijections

$$\mathscr{V}\left(\coprod_A I, V\right) \cong \prod_A \mathscr{V}(I,V) \cong \mathsf{Set}(A, \mathscr{V}(I,V)).$$

Observe that by definition, $F(1) = \coprod_1 I = I$, yielding the required isomorphism ε.

Since tensoring with an object of $\mathscr{V}$ preserves coproducts (see 3.2.2, volume 1), we have, for all sets A, B,

$$\begin{aligned} F(A)\otimes F(B) &\cong \left(\coprod_A I\right)\otimes\left(\coprod_B I\right) \\ &\cong \coprod_{A\times B} I\otimes I \cong \coprod_{A\times B} I \\ &\cong F(A\times B), \end{aligned}$$

yielding the isomorphism τ_{AB}.

The rest is now routine calculations. □

Proposition 6.4.7 *Let $\mathscr{V}$ be a symmetric monoidal closed category with coproducts. The forgetful functor*

$$U\colon \mathscr{V}\text{-}\mathsf{Cat} \longrightarrow \mathsf{Cat}$$

of 6.4.4 has a left adjoint.

Proof If $\mathscr{C}$ is a small category, define a $\mathscr{V}$-category $\overline{\mathscr{C}}$ as follows:

- $|\overline{\mathscr{C}}| = |\mathscr{C}|$;
- $\overline{\mathscr{C}}(A,B) = F(\mathscr{C}(A,B))$, where F is left adjoint to $\mathscr{V}(I,-)$ (see 6.4.6) and $A, B \in \mathscr{C}$;
- $c_{ABC}\colon \overline{\mathscr{C}}(A,B)\otimes\overline{\mathscr{C}}(B,C) \longrightarrow \overline{\mathscr{C}}(A,C)$, for $A, B, C \in \mathscr{C}$, is the composite

  $$F(\mathscr{C}(A,B))\otimes F(\mathscr{C}(B,C)) \cong F(\mathscr{C}(A,B)\times\mathscr{C}(B,C)) \longrightarrow F(\mathscr{C}(A,C))$$

  where the isomorphism is that of 6.4.6 and the arrow is the image under F of the ordinary composition in $\mathscr{C}$;
- $u_A\colon I \longrightarrow \overline{\mathscr{C}}(A,A)$, for $A \in \mathscr{C}$ is the composite

  $$I \cong F(\mathbf{1}) \longrightarrow F(\mathscr{C}(A,A))$$

  where the isomorphism is that of 6.4.6 and the arrow is the image under F of the mapping applying $* \in \mathbf{1}$ on 1_A.

It is just routine to check that $\overline{\mathscr{C}}$ is indeed a $\mathscr{V}$-category.

It is easy to define a functor $\gamma\colon \mathscr{C} \longrightarrow U(\overline{\mathscr{C}})$:

- $\gamma(C) = C$ for $C \in \mathscr{C}$;
- if $f\colon A \longrightarrow B$ is a morphism of $\mathscr{C}$, $\gamma(f)$ is the arrow

$$s_f\colon I \longrightarrow \coprod_{\mathscr{C}(A,B)} I \cong F(I).$$

The definitions of c_{ABC} and u_A imply immediately that γ is a functor.

Now given a functor $G\colon \mathscr{C} \longrightarrow U(\mathscr{A})$, where $\mathscr{A}$ is a $\mathscr{V}$-category (not necessarily small), we get the required unique $\mathscr{V}$-factorization $\overline{G}\colon \overline{\mathscr{C}} \longrightarrow \mathscr{A}$ in the following way:

- $\overline{G}(C) = G(C)$ for $C \in \mathscr{C}$;
- $\overline{G}_{AB}\colon \overline{\mathscr{C}}(A,B) \longrightarrow \mathscr{A}\big(\overline{G}(A), \overline{G}(B)\big)$ must thus be a morphism of the type $F\big(\mathscr{C}(A,B)\big) \longrightarrow \mathscr{A}\big(G(A), G(B)\big)$; it is the one corresponding by adjunction with the action of G,

$$\mathscr{C}(A,B) \longrightarrow U\Big(\mathscr{A}\big(G(A), G(B)\big)\Big) \cong \mathscr{V}\Big(I, \mathscr{A}\big(G(A), G(B)\big)\Big).$$

The details follow immediately from 6.4.6. □

Proposition 6.4.8 *Let $\mathscr{V}$ be a symmetric monoidal closed category with coproducts. Consider a small category $\mathscr{C}$ and the corresponding free $\mathscr{V}$-category $\overline{\mathscr{C}}$ (see 6.4.7). For every $\mathscr{V}$-category $\mathscr{A}$, there is an isomorphism of categories between*

- *the category of functors and natural transformations from $\mathscr{C}$ to $U(\mathscr{A})$;*
- *the category of $\mathscr{V}$-functors and $\mathscr{V}$-natural transformations from $\overline{\mathscr{C}}$ to $\mathscr{A}$.*

Proof The last part of the proof of 6.4.7 has been developed for an arbitrary $\mathscr{V}$-category $\mathscr{A}$. This already yields a bijective correspondence between

- the functors $G\colon \mathscr{C} \longrightarrow U(\mathscr{A})$,
- the $\mathscr{V}$-functors $\overline{G}\colon \overline{\mathscr{C}} \longrightarrow \mathscr{A}$.

We must prove now that given functors $G, H\colon \mathscr{C} \rightrightarrows U\mathscr{A}$ with corresponding $\mathscr{V}$-functors $\overline{G}, \overline{H}\colon \overline{\mathscr{C}} \rightrightarrows \mathscr{A}$, there is a bijective correspondence between

- the natural transformations $G \Rightarrow H$,
- the $\mathscr{V}$-natural transformations $\overline{G} \Rightarrow \overline{H}$.

$$\begin{array}{ccc}
\mathscr{C}(C,D) & \xrightarrow{G_{CD}} & U\mathscr{A}(GC,GD) \\
{\scriptstyle H_{CD}}\downarrow & & \downarrow{\scriptstyle U\mathscr{A}(1,\alpha_D)} \\
U\mathscr{A}(HC,HD) & \xrightarrow[U\mathscr{A}(\alpha_C,1)]{} & U\mathscr{A}(GC,HD)
\end{array}$$

Diagram 6.33

A natural transformation $\alpha\colon G \Rightarrow H$ is by definition a family of morphisms $\alpha_C\colon G(C) \longrightarrow H(C)$ in $U\mathscr{A}$, thus by 6.2.4 a family of morphisms $\alpha_C\colon I \longrightarrow \mathscr{A}(G(C), H(C))$ in $\mathscr{V}$, with diagram 6.33 commutative for all C, D in $\mathscr{C}$. But since

$$U\Big(\mathscr{A}(G(C), G(D))\Big) \cong \mathscr{V}\Big(I, \mathscr{A}(G(C), G(D))\Big),$$

we can transform the diagram by the adjunction of 6.4.6 and get equivalently the commutativity of diagram 6.34, which is precisely the requirement for $\alpha\colon \overline{G} \Rightarrow \overline{H}$ being $\mathscr{V}$-natural (see 6.2.4). □

6.5 Tensors and cotensors

Definition 6.5.1 *Let $\mathscr{V}$ be a symmetric monoidal closed category, $\mathscr{A}$ a $\mathscr{V}$-category and $A \in \mathscr{A}, V \in \mathscr{V}$ two objects.*

- *The cotensor of V and A exists if there is an object $[V, A] \in \mathscr{A}$ together with isomorphisms*

  $$\mathscr{A}\big(B, [V, A]\big) \cong \big[V, \mathscr{A}(B, A)\big]$$

  in $\mathscr{V}$ which are $\mathscr{V}$-natural in $B \in \mathscr{A}$. We say $\mathscr{A}$ is cotensored when $[V, A]$ exists for all objects $V \in \mathscr{V}, A \in \mathscr{A}$.
- *The tensor of V and A exists if there is an object $V \otimes A \in \mathscr{A}$ together with isomorphisms*

  $$\mathscr{A}(V \otimes A, B) \cong \big[V, \mathscr{A}(A, B)\big]$$

  in $\mathscr{V}$ which are $\mathscr{V}$-natural in $B \in \mathscr{A}$. We say $\mathscr{A}$ is tensored when $V \otimes A$ exists for all objects $V \in \mathscr{V}, A \in \mathscr{A}$.

Example 6.5.2

In the case $\mathscr{V} = \mathsf{Set}$ and $\mathscr{A}$ an ordinary category with products and coproducts, one has

$$\mathscr{A}\Big(B, \textstyle\prod_{i \in I} A\Big) \cong \textstyle\prod_{i \in I} \mathscr{A}(B, A) \cong \mathsf{Set}\big[I, \mathscr{A}(B, A)\big],$$

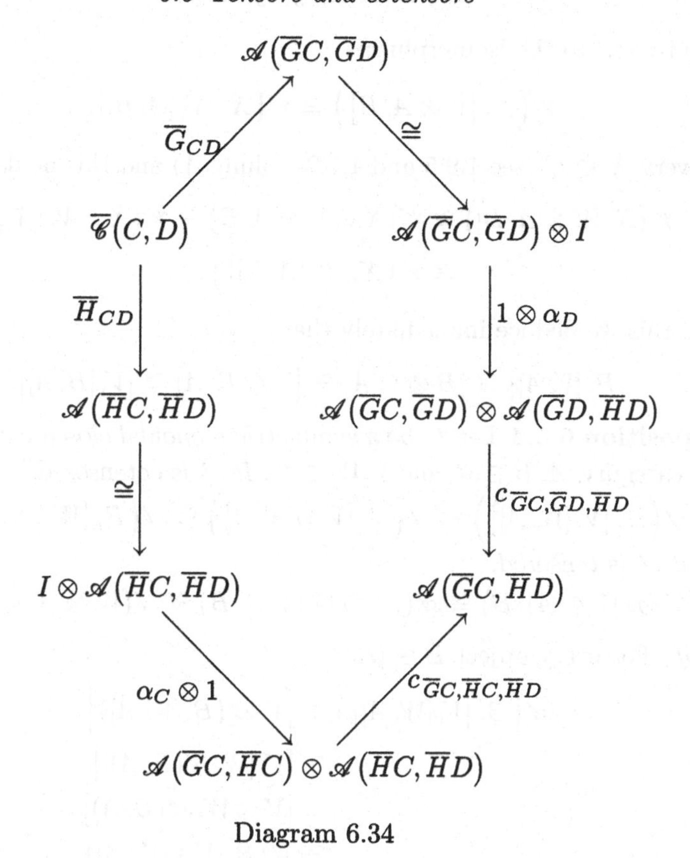

Diagram 6.34

$$\mathscr{A}\left(\coprod_{i\in I} A, B\right) \cong \prod_{i\in I} \mathscr{A}(A,B) \cong \mathsf{Set}\big[I, \mathscr{A}(A,B)\big],$$

from which one concludes that the cotensor $[I, A]$ is just the I-th power of A and the tensor $I \otimes A$ is the I-th copower of A, for $A, B \in \mathscr{A}$ and $I \in \mathsf{Set}$.

The following proposition indicates in particular that no notational confusion at all arises when $\mathscr{A} = \mathscr{V}$ in 6.5.1.

Proposition 6.5.3 *Let $\mathscr{V}$ be a symmetric monoidal closed category. Then $\mathscr{V}$ is both a tensored and cotensored $\mathscr{V}$-category, with the tensor $V \otimes A$ and the cotensor $[V, A]$ as in 6.1.7.*

Proof Indeed the isomorphism

$$[V \otimes A, B] \cong \big[V, [A, B]\big]$$

is equivalent to the isomorphisms

$$\mathscr{V}\Big(X, \big[V \otimes A, B\big]\Big) \cong \mathscr{V}\Big(X, \big[V, [A, B]\big]\Big)$$

for every $X \in \mathscr{V}$ (see 1.9.5 and 1.5.2, volume 1) and this holds since

$$\mathscr{V}(X, [V \otimes A, B]) \cong \mathscr{V}(X \otimes V \otimes A, B) \cong \mathscr{V}(X \otimes V, [A, B]) \cong \mathscr{V}\Big(X, \big[V, [A, B]\big]\Big).$$

From this we deduce immediately that

$$[B, [V, A]] \cong [B \otimes V, A] \cong [V \otimes B, A] \cong [V, [B, A]]. \qquad \square$$

Proposition 6.5.4 *Let $\mathscr{V}$ be a symmetric monoidal closed category, $\mathscr{A}$ a $\mathscr{V}$-category, $A, B \in \mathscr{A}$ and $V, W \in \mathscr{V}$. If $\mathscr{A}$ is cotensored,*

$$\mathscr{A}\Big(B, \big[V, [W, A]\big]\Big) \cong \mathscr{A}\Big(B, \big[V \otimes W, A\big]\Big) \cong \mathscr{A}\Big(B, \big[W, [V, A]\big]\Big),$$

and if $\mathscr{A}$ is tensored,

$$\mathscr{A}(V \otimes (W \otimes A), B) \cong \mathscr{A}((V \otimes W) \otimes A, B) \cong \mathscr{A}(W \otimes (V \otimes A), B).$$

Proof For every object $B \in \mathscr{A}$,

$$\begin{aligned}\mathscr{A}\Big(B, \big[V, [W, A]\big]\Big) &\cong \Big[V, \mathscr{A}(B, [W, A])\Big] \\ &\cong [V, [W, \mathscr{A}(B, A)]] \\ &\cong [V \otimes W, \mathscr{A}(B, A)] \\ &\cong \mathscr{A}(B, [V \otimes W, A])\end{aligned}$$

from which follows the first statement.

An analogous proof holds for tensors. $\square$

Proposition 6.5.5 *Let $\mathscr{V}$ be a symmetric monoidal closed category and $V \in \mathscr{V}$. Let $\mathscr{A}$ be a $\mathscr{V}$-category. If $\mathscr{A}$ is cotensored, the correspondence*

$$[V, -]\colon \mathscr{A} \longrightarrow \mathscr{A}, \quad A \mapsto [V, A]$$

induces a $\mathscr{V}$-functor and if $\mathscr{A}$ is tensored, the correspondence

$$V \otimes -\colon \mathscr{A} \longrightarrow \mathscr{A}, \quad A \mapsto V \otimes A$$

induces a $\mathscr{V}$-functor as well.

Proof Suppose $\mathscr{A}$ is cotensored. For $A, B \in \mathscr{A}$, the composite

$$I \xrightarrow{u_{\mathscr{A}(A,B)}} [\mathscr{A}(A, B), \mathscr{A}(A, B)] \cong \mathscr{A}\Big(A, \big[\mathscr{A}(A, B), B\big]\Big)$$

induces a morphism $\mathrm{ev}_{AB}\colon A \longrightarrow \big[\mathscr{A}(A, B), B\big]$ in $U(\mathscr{A})$. The $\mathscr{V}$-functoriality of $[V, -]$ is then expressed by morphisms

$$\mathscr{A}(A, B) \longrightarrow \mathscr{A}([V, A], [V, B])$$

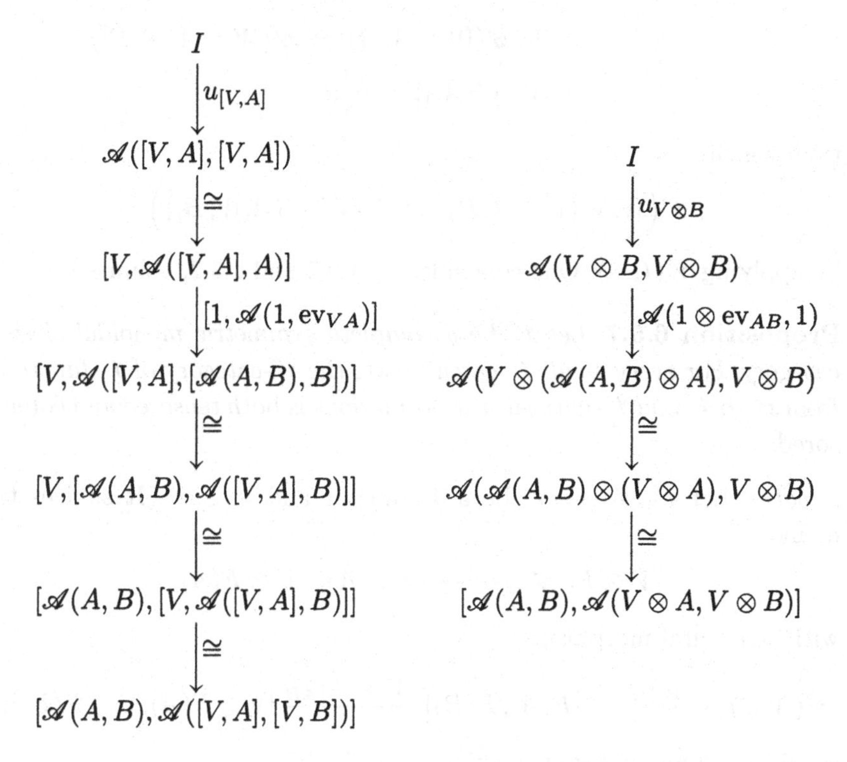

Diagram 6.35

corresponding by adjunction with the first composites of diagram 6.35.

In the same way the $\mathscr{V}$-functoriality of $V \otimes -$ is expressed by morphisms

$$\mathscr{A}(A,B) \longrightarrow \mathscr{A}(V \otimes A, V \otimes B)$$

which correspond by adjunction with the second composites of diagram 6.35 (see 6.5.2). □

Proposition 6.5.6 *Let $\mathscr{V}$ be a symmetric monoidal closed category. If $\mathscr{A}$ is a tensored and cotensored $\mathscr{V}$-category, for every $V \in \mathscr{V}$ the two $\mathscr{V}$-functors*

$$[V,-]\colon \mathscr{A} \longrightarrow \mathscr{A}, \quad V \otimes -\colon \mathscr{A} \longrightarrow \mathscr{A}$$

induce isomorphisms

$$\mathscr{A}(V \otimes A, B) \cong \mathscr{A}(A, [V,B])$$

which are $\mathscr{V}$-natural in $A, B \in \mathscr{A}$.

Proof For every $W \in \mathscr{V}$, (see 6.5.12),

$$[W, \mathscr{A}(V \otimes A, B)] \cong \mathscr{A}(W \otimes (V \otimes A), B) \cong \mathscr{A}(V \otimes (W \otimes A), B)$$

$$\cong [V, \mathscr{A}(W \otimes A, B)] \cong \mathscr{A}(W \otimes A, [V, B])$$
$$\cong \Big[W, \mathscr{A}\big(A, [V, B]\big)\Big],$$

from which

$$\mathscr{V}\Big(W, \mathscr{A}(V \otimes A, B)\Big) \cong \mathscr{V}\Big(W, \mathscr{A}\big(A, [V, B]\big)\Big)$$

by applying $\mathscr{V}(I, -)$. One concludes by 1.9.5 and 1.5.2, volume 1. □

Proposition 6.5.7 *Let $\mathscr{V}$ be a complete symmetric monoidal closed category. For every small $\mathscr{V}$-category $\mathscr{A}$, the $\mathscr{V}$-category of $\mathscr{V}$-functors from $\mathscr{A}$ to $\mathscr{V}$ and $\mathscr{V}$-natural transformations is both tensored and cotensored.*

Proof Let $F\colon \mathscr{A} \longrightarrow \mathscr{V}$ be a $\mathscr{V}$-functor and $V \in \mathscr{V}$. It suffices to define

$$V \otimes F\colon \mathscr{A} \longrightarrow \mathscr{A}, \quad A \mapsto V \otimes F(A)$$

with structural morphisms

$$\mathscr{A}(A, B) \xrightarrow{F_{AB}} [F(A), F(B)] \xrightarrow{(V \otimes -)_{AB}} [V \otimes F(A), V \otimes F(B)];$$

see 6.5.5. In an analogous way

$$[V, F]\colon \mathscr{A} \longrightarrow \mathscr{V}, \quad A \mapsto [V, F(A)]$$

has the structural morphisms (see 6.5.5)

$$\mathscr{A}(A, B) \xrightarrow{F_{AB}} [F(A), F(B)] \xrightarrow{[V, -]_{AB}} \Big[[V, F(A)], [V, F(B)]\Big].$$

It is routine to check that those definitions fit our needs (see 6.3.3). □

The notion of cotensor allows in particular a generalization of 6.3.3.

Lemma 6.5.8 *Let $\mathscr{V}$ a complete symmetric monoidal closed category, $\mathscr{A}$ a small $\mathscr{V}$-category and $\mathscr{B}$ a cotensored $\mathscr{V}$-category. Given two $\mathscr{V}$-functors $F, G\colon \mathscr{A} \rightrightarrows \mathscr{B}$, an object $N \in \mathscr{V}$ is isomorphic to the object of $\mathscr{V}$-natural transformations $\mathscr{V}$-$\mathsf{Nat}(F, G)$ if and only if, for every object $V \in \mathscr{V}$, there is a bijective correspondence between*

(1) the morphisms $V \longrightarrow N$,
(2) the $\mathscr{V}$-natural transformations $F \Rightarrow [V, G-]$.

Proof The proof of 6.3.3 can almost be repeated. With the notation of 6.3.1, a morphism $f\colon V \longrightarrow \mathscr{V}$-$\mathsf{Nat}(F, G)$ is induced by a morphism $V \longrightarrow \prod_{A \in \mathscr{A}} \mathscr{B}(F(A), G(A))$, thus equivalently by a family of morphisms $f_A\colon V \longrightarrow \mathscr{B}(F(A), G(A))$. By cotensorization of $\mathscr{B}$, this is equivalent to giving a compatible family of morphisms

$$I \longrightarrow \Big[V, \mathscr{B}\big(F(A), G(A)\big)\Big] \cong \mathscr{B}\Big(F(A), \big[V, G(A)\big]\Big),$$

thus finally to giving a $\mathscr{V}$-natural transformation $F \Rightarrow [V, G-]$. One concludes again by 1.9.5 and 1.5.2, volume 1. □

6.6 Weighted limits

Enriching the notion of limit with respect to some base category $\mathscr{V}$ is much more subtle than the straightforward generalizations of section 6.2. Indeed, given a functor $F\colon \mathscr{A} \longrightarrow \mathscr{B}$, the limit of F is defined using the notion of "cone on F": this is a pair (B, α) where $B \in \mathscr{B}$ is an object and $\alpha\colon \Delta_B \Rightarrow F$ is a natural transformation, with Δ_B the constant functor on the object B, i.e. the composite

$$\mathscr{A} \longrightarrow \mathbf{1} \xrightarrow{\ \delta_B\ } \mathscr{B}$$

where $\mathbf{1}$ is the terminal category (one single object $*$ and just the identity on it) and δ_B is defined by $\delta_B(*) = B$.

When $\mathscr{V}$ is, for example, a symmetric monoidal closed category and F is a $\mathscr{V}$-functor, for every object $B \in \mathscr{B}$ a $\mathscr{V}$-functor $\delta_B\colon \mathscr{I} \longrightarrow \mathscr{B}$ is easily defined on the unit $\mathscr{V}$-category $\mathscr{I}$ (see 6.2.10): just put $\delta_B(*) = B$, where $*$ is the unique object of $\mathscr{I}$, and choose the morphism

$$\mathscr{I}(*, *) = I \xrightarrow{\ u_B\ } \mathscr{B}(B, B)$$

as $(\delta_B)_{**}$. But there is no way to find a canonical $\mathscr{V}$-functor $\mathscr{A} \longrightarrow \mathscr{I}$, except when $I \in \mathscr{V}$ is the terminal object (for example, when $\mathscr{V}$ is cartesian closed; see 6.1.6).

The first observation for enriching the notion of limit is the fact that in many fundamental results of the previous chapters, we had to deal with a situation like in diagram 6.36, where F, G are functors and $\mathsf{Elts}(G)$ is the category of elements of G; the limit considered was that of $F \circ \phi_G$ (see 2.15.6, 3.3.1, 3.7.2, ..., volume 1). Restricting one's attention to limits of the form $\lim(F \circ \phi_G)$ is not at all a restriction since, choosing for G the constant functor on the singleton, the category of elements of G is just $\mathscr{A}$ itself and $\phi_G = 1_{\mathscr{A}}$, so that we recapture the limit of F.

But the key observation for enriching the notion of limit is the following fact

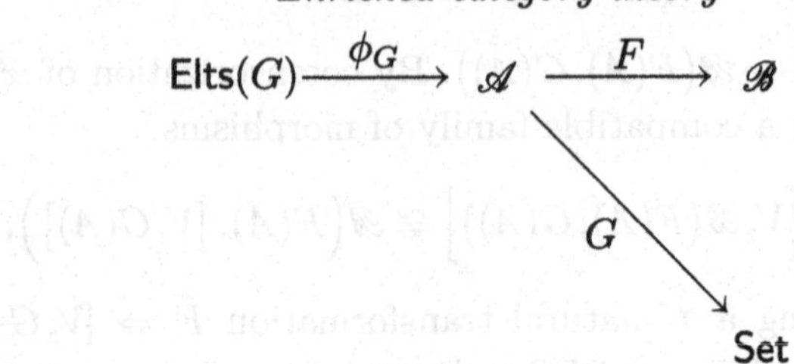

Diagram 6.36

Lemma 6.6.1 *In the situation which has just been described, there exist natural bijections in $B \in \mathscr{B}$,*

$$\mathsf{Nat}(\Delta_B, F \circ \phi_G) \cong \mathsf{Nat}(G, \mathscr{B}(B, F-))$$

where Δ_B: $\mathsf{Elts}(G) \longrightarrow \mathscr{B}$ is the constant functor on the object $B \in \mathscr{B}$.

Proof Given a cone $\big(q_{(A,a)}: B \longrightarrow (F \circ \phi_G)(A,a)\big)_{(A,a)}$, we define a natural transformation α: $G \Rightarrow \mathscr{B}(B, F-)$ by putting $\alpha_A(a) = q_{(A,a)}$, for $a \in G(A)$; it is natural since given f: $(A,a) \longrightarrow (A', a')$ in $\mathsf{Elts}(G)$

$$\mathscr{B}\big(B, F(f)\big) \circ \alpha_A(a) = F(f) \circ q_{(A',a')} = \alpha_{A'}(a') = \alpha_{A'} \circ G(f)(a).$$

Now, given a natural transformation α: $G \Rightarrow \mathscr{B}(B, F-)$, we define $q_{(A,a)}$: $B \longrightarrow F(A)$ by putting $q_{(A,a)} = \alpha_A(a)$. This is a cone on $F \circ \phi_G$ since given f: $(A,a) \longrightarrow (A',a')$ in $\mathsf{Elts}(G)$, the naturality of α implies

$$\begin{aligned} F(f) \circ q_{(A,a)} &= F(f) \circ \alpha_A = \mathscr{B}\big(B, F(f)\big) \circ \alpha_A(a) \\ &= \alpha_{A'} \circ G(f)(a) = \alpha_{A'}(a') = q_{(A',a')}. \end{aligned}$$

The fact that we have defined reciprocal bijections is obvious from the definitions. The naturality in B is immediate since, given b: $B' \longrightarrow B$ and the previous cone $q_{(A,a)}$, the composite cone is $q_{(A,a)} \circ b$ which is applied on β: $G \Rightarrow \mathscr{B}(B', F-)$ defined by

$$\beta_A(a) = q_{(A,a)} \circ b = \alpha_A(a) \circ b = \mathscr{B}\big(b, F(A)\big) \circ \alpha_A(a). \qquad \square$$

Corollary 6.6.2 *In the situation of the beginning of this section, the limit of the functor $F \circ \phi_G$ exists iff there is an object $L \in \mathscr{B}$ together with bijections, natural in $B \in \mathscr{B}$,*

$$\mathsf{Nat}\big(G, \mathscr{B}(B, F-)\big) \cong \mathscr{B}(B, L).$$

Proof Just apply 6.6.1 and the fact that the limit exists when there is $L \in \mathscr{B}$ and natural bijections

$$\mathsf{Nat}(\Delta_B, F \circ \phi_G) \cong \mathscr{B}(B, L);$$

see 3.2.3, volume 1. □

Choosing $B = L$ in the previous formula yields the "canonical" natural transformation $\lambda\colon G \Rightarrow \mathscr{B}(L, F-)$ associated with the limit: it has the universal property that given $\mu\colon G \Rightarrow \mathscr{B}(B, F-)$, there exists a unique morphism $b\colon B \longrightarrow L$ such that $\mu = \mathscr{B}(b, F-) \circ \lambda$.

Definition 6.6.3 *Let $\mathscr{V}$ be a symmetric monoidal closed category. Given $\mathscr{V}$-functors $F\colon \mathscr{A} \longrightarrow \mathscr{B}$, $G\colon \mathscr{A} \longrightarrow \mathscr{V}$, the $\mathscr{V}$-limit of F weighted by G exists when:*

(1) for every $B \in \mathscr{B}$, the object $\mathscr{V}\text{-}\mathsf{Nat}\big(G, \mathscr{B}(B, F-)\big)$ of $\mathscr{V}$-natural transformations exists;

(2) there exists an object $L \in \mathscr{B}$ and isomorphisms in $\mathscr{V}$,

$$\lambda_B\colon \mathscr{V}\text{-}\mathsf{Nat}\big(G, \mathscr{B}(B, F-)\big) \cong \mathscr{B}(B, L),$$

which are $\mathscr{V}$-natural in B.

We write in general $\lim_G F$ for this weighted limit, when it exists. When $\lim_G F$ exists for all choices of F, G, with $\mathscr{A}$ small, $\mathscr{B}$ is said to be $\mathscr{V}$-complete.

As proved in 6.3.1, when $\mathscr{A}$ is small and $\mathscr{V}$ is complete, the required objects of $\mathscr{V}$-natural transformations exist in any case.

The $\mathscr{V}$-naturality in $B \in \mathscr{B}$ is worth a comment. We have a first $\mathscr{V}$-functor $\mathscr{B}(-, L)\colon \mathscr{B}^* \longrightarrow \mathscr{V}$ and, when $\mathscr{A}$ is small, a second one obtained as the composite

$$\mathscr{B}^* \xrightarrow{\ Y\ } \mathscr{V}\text{-}\mathsf{Fun}(\mathscr{B}, \mathscr{V}) \xrightarrow{\ F^*\ } \mathscr{V}\text{-}\mathsf{Fun}(\mathscr{A}, \mathscr{V}) \xrightarrow{\ (G,-)\ } \mathscr{V},$$

where $Y(B) \cong \mathscr{B}(B, -)$ is the $\mathscr{V}$-Yoneda-embedding (see 6.3.4), F^* is composition with F and $(G, -)$ is the $\mathscr{V}$-functor represented by G. What is required is the existence of a $\mathscr{V}$-natural isomorphism λ between those two $\mathscr{V}$-functors. Applying 6.2.8, this can be expressed by isomorphisms λ_B as in 6.6.3 and the commutativity of some diagrams. Due to the first requirement in 6.6.3, these last commutativities make sense even when $\mathscr{A}$ is not small, and it is what we mean by the $\mathscr{V}$-naturality in $B \in \mathscr{B}$.

The notion of weighted $\mathscr{V}$-colimit is just dual ... which implies in particular that the weighting $\mathscr{V}$-functor G is now contravariant!

Definition 6.6.4 *Let $\mathscr{V}$ be a symmetric monoidal closed category. Given $\mathscr{V}$-functors $F\colon \mathscr{A} \longrightarrow \mathscr{B}$ and $G\colon \mathscr{A}^* \longrightarrow \mathscr{V}$, the $\mathscr{V}$-colimit of F weighted by G exists when:*

(1) for every $B \in \mathscr{B}$, the object $\mathscr{V}\text{-}\mathsf{Nat}\big(G, \mathscr{B}(F-, B)\big)$ of $\mathscr{V}$-natural transformations exists;

$$\begin{array}{c} \mathscr{A}(C,A)\otimes\mathscr{A}(A,B)\otimes\mathscr{A}(B,D) \\ \Big\downarrow 1\otimes c_{ABD} \\ \mathscr{A}(C,A)\otimes\mathscr{A}(A,D) \\ \Big\downarrow c_{CAD} \\ \mathscr{A}(C,D) \end{array}$$

Diagram 6.37

(2) there exists an object $L \in \mathscr{B}$ and isomorphisms in $\mathscr{V}$,

$$\lambda_B\colon \mathscr{V}\text{-}\mathsf{Nat}\big(G, \mathscr{B}(F-, B)\big) \cong \mathscr{B}(L, B),$$

which are $\mathscr{V}$-natural in B.

We write in general $\mathrm{colim}_G F$ *for this weighted colimit, when it exists. When* $\mathrm{colim}_G F$ *exists for all choices of F, G, with $\mathscr{A}$ small, $\mathscr{B}$ is said to be $\mathscr{V}$-cocomplete.*

In order to specify two important particular cases of weighted limits, we need some lemmas related with the tensor product of $\mathscr{V}$-categories, as described in 6.2.9.

Lemma 6.6.5 *Let $\mathscr{V}$ be a symmetric monoidal closed category. For every small $\mathscr{V}$-category $\mathscr{A}$ there is a $\mathscr{V}$-functor*

$$\mathscr{A}\colon \mathscr{A}^* \otimes \mathscr{A} \longrightarrow \mathscr{V}$$

with values $\mathscr{A}(A, B)$ on the objects.

Proof Given objects $(A, B), (C, D) \in \mathscr{A}^* \otimes \mathscr{A}$, it remains to define the required morphism

$$\mathscr{A}_{ABCD}\colon \mathscr{A}^*(A, C) \otimes \mathscr{A}(B, D) \longrightarrow [\mathscr{A}(A, B), \mathscr{A}(C, D)].$$

By adjunction and symmetry, this corresponds with the composite of diagram 6.37. The rest is routine computation. □

Lemma 6.6.6 *Let $\mathscr{V}$ be a symmetric monoidal closed category. The unit $\mathscr{V}$-category $\mathscr{I}$ is $\mathscr{V}$-isomorphic to its $\mathscr{V}$-dual.*

Proof Indeed $|\mathscr{I}| = \{*\} = |\mathscr{I}|^*$ and $\mathscr{I}(*,*) = I = \mathscr{I}^*(*,*)$, from which we get the result. □

Lemma 6.6.7 *Let $\mathscr{V}$ be a symmetric monoidal closed category, $\mathscr{B}$ a $\mathscr{V}$-category and $\mathscr{I}$ the unit $\mathscr{V}$-category. The category of $\mathscr{V}$-functors $\mathscr{I} \longrightarrow \mathscr{B}$ and $\mathscr{V}$-natural transformations can be provided with the structure of a $\mathscr{V}$-category, $\mathscr{V}$-isomorphic to $\mathscr{B}$.*

Proof A $\mathscr{V}$-functor $F\colon \mathscr{I} \longrightarrow \mathscr{B}$ is the choice of an object $F(*) \in \mathscr{B}$ together with a morphism $\varphi\colon \mathscr{I}(*,*) \longrightarrow \mathscr{B}\big(F(*), F(*)\big)$. But the second axiom for a $\mathscr{V}$-functor implies that $\varphi = u_{F(*)}$, so that giving a $\mathscr{V}$-functor $F\colon \mathscr{I} \longrightarrow \mathscr{B}$ is equivalent to giving an object $F(*) \in \mathscr{B}$.

Next considering the construction of $\mathscr{V}\text{-}\mathsf{Nat}(F,G)$ in 6.3.1 we realize that both products reduce to a single factor since $\mathscr{I}$ is a singleton. Moreover $\mathscr{A}(A', A'') \cong \mathscr{I}(*,*) \cong I$, from which u and v are just the identity on $\mathscr{B}(F*, G*)$. In particular the equalizer k exists and it is the identity on $\mathscr{B}\big(F(*), G(*)\big)$. Thus $\mathscr{V}\text{-}\mathsf{Nat}(F,G) \cong \mathscr{B}\big(F(*), G(*)\big)$, yielding the conclusion. □

With the notations of 6.6.7, we write $\delta_B\colon \mathscr{I} \longrightarrow \mathscr{B}$ for the $\mathscr{V}$-functor corresponding with the object $B \in \mathscr{B}$. By 6.6.6, δ_B is both $\mathscr{V}$-covariant and $\mathscr{V}$-contravariant.

Definition 6.6.8 *Let $\mathscr{V}$ be a symmetric monoidal closed category and $\mathscr{A}, \mathscr{B}$ two $\mathscr{V}$-categories.*

- *By the end $\int_{A\in\mathscr{A}} F(A,A)$ of a $\mathscr{V}$-functor $F\colon \mathscr{A}^* \otimes \mathscr{A} \longrightarrow \mathscr{B}$, we mean the $\mathscr{V}$-limit of F weighted by $\mathscr{A}\colon \mathscr{A}^* \otimes \mathscr{A} \longrightarrow \mathscr{V}$, when this exists.*
- *By the coend $\int^{A\in\mathscr{A}} F(A,A)$ of a $\mathscr{V}$-functor $F\colon \mathscr{A} \otimes \mathscr{A}^* \longrightarrow \mathscr{B}$, we mean the $\mathscr{V}$-colimit of F weighted by $\mathscr{A}\colon \mathscr{A}^* \otimes \mathscr{A} \longrightarrow \mathscr{V}$, when this exists.*

Proposition 6.6.9 *Let $\mathscr{V}$ be a symmetric monoidal closed category and $\mathscr{A}$ a $\mathscr{V}$-category. Given $\mathscr{V}$-functors $F, G\colon \mathscr{A} \rightrightarrows \mathscr{V}$, the following conditions are equivalent:*

(1) the object of $\mathscr{V}$-natural transformations $\mathscr{V}\text{-}\mathsf{Nat}(F,G)$ exists;

(2) the end of the $\mathscr{V}$-functor

$$T\colon \mathscr{A}^* \otimes \mathscr{A} \longrightarrow \mathscr{V}, \quad (A,B) \mapsto \big[F(A), G(B)\big]$$

exists;

(3) the weighted $\mathscr{V}$-limit $\lim_F G$ exists.

When these conditions are realized,

$$\mathscr{V}\text{-}\mathsf{Nat}(F,G) = \int_{A\in\mathscr{A}} T(A,A) = \lim_F G.$$

Proof Observe first that T organizes itself into a $\mathscr{V}$-functor. The structural morphisms

$$(\mathscr{A}^* \times \mathscr{A})\big((A,B),(C,D)\big) \longrightarrow \Big[\big[F(A), G(B)\big], \big[F(C), G(D)\big]\Big]$$

correspond by adjunction with the composites of diagram 6.38, where the second arrow itself corresponds by adjunction and symmetry with the composite of diagram 6.39.

$$\begin{array}{c} \mathscr{A}(C,A)\otimes\mathscr{A}(B,D) \\ \downarrow F_{CA}\otimes G_{BD} \\ [FC,FA]\otimes[GB,GD] \\ \downarrow \\ [[FA,GB],[FC,GD]] \end{array}$$

Diagram 6.38

$$\begin{array}{c} [FC,FA]\otimes[FA,GB]\otimes[GB,GD]\otimes FC \\ \downarrow c_{FC,FA,GB}\otimes 1\otimes 1 \\ [FC,GB]\otimes[GB,GD]\otimes FC \\ \downarrow c_{FC,GB,GD}\otimes 1 \\ [FC,GD]\otimes FC \\ \downarrow \mathrm{ev}_{FC,GD} \\ GD \end{array}$$

Diagram 6.39

For each $V\in\mathscr{V}$, giving a $\mathscr{V}$-natural transformation $\alpha\colon \mathscr{A}\Rightarrow[V,T-]$ is giving a family of morphisms

$$\alpha_{AB}\colon \mathscr{A}(A,B)\longrightarrow\Big[V,\big[F(A),G(B)\big]\Big]$$

which is $\mathscr{V}$-natural in A,B; see 6.2.8. On the other hand giving a $\mathscr{V}$-natural transformation $\beta\colon F\Rightarrow[V,G-]$ is giving a family of morphisms

$$\beta_C\colon F(C)\longrightarrow[V,G(C)]$$

which is $\mathscr{V}$-natural in C.

Given α, the composite

$$I\xrightarrow{\ u_C\ }\mathscr{A}(C,C)\xrightarrow{\ \alpha_{CC}\ }\Big[V,\big[F(C),G(C)\big]\Big]$$

defines a morphism $V\longrightarrow[F(C),G(C)]$ in $\mathscr{V}$ which, by adjunction, corresponds with a morphism $V\otimes F(C)\longrightarrow G(C)$ and finally with a morphism $\beta_C\colon FC\longrightarrow[V,GC]$. Conversely given β, one constructs α_{AB} as the composite in diagram 6.40.

$$\begin{array}{c} \mathscr{A}(A,B) \\ \downarrow F_{AB} \\ [F(A),F(B)] \\ \downarrow [1,\beta_B] \\ \big[F(A),[V,G(B)]\big] \\ \downarrow \cong \\ \big[V,[F(A),G(B)]\big] \end{array}$$

Diagram 6.40

It is immediate that the correspondences $\alpha \mapsto \beta$, $\beta \mapsto \alpha$ are reciprocal. Observe that given $W \in \mathscr{V}$, this implies also the existence of natural bijections between

- the $\mathscr{V}$-natural transformations $\mathscr{A} \Rightarrow \big[W,[V,T-]\big]$,
- the $\mathscr{V}$-natural transformations $F \Rightarrow \big[W,[V,G-]\big]$,

just because $\big[W,[V,-]\big] \cong \big[W \otimes V,-\big]$; see 6.5.4. Therefore by 6.3.3 we get the isomorphism

$$\mathscr{V}\text{-Nat}\big(\mathscr{A},[V,T-]\big) \cong \mathscr{V}\text{-Nat}\big(F,[V,G-]\big).$$

This concludes the proof of (2) $\Leftrightarrow$ (3) since the end of T exists when

$$\mathscr{V}\text{-Nat}\big(\mathscr{A},[V,T-]\big) \cong \left[V,\int_{A\in\mathscr{A}} T(A,A)\right]$$

while the weighted $\mathscr{V}$-limit $\lim_F G$ exists when

$$\mathscr{V}\text{-Nat}\big(F,[V,G-]\big) \cong [V,\lim_F G].$$

The case of $\mathscr{V}\text{-Nat}(F,G)$ is treated by the same remark as above; it exists when we have bijections

$$\mathscr{V}\text{-Nat}\big(F,[V,G-]\big) \cong \mathscr{V}\big(V,\mathscr{V}\text{-Nat}(F,G)\big).$$

Replacing V by $V \otimes W$ yields by 6.5.4 bijections

$$\mathscr{V}\text{-Nat}\Big(F,\big[W,[V,G-]\big]\Big) \cong \mathscr{V}\Big(W,\big[V,\mathscr{V}\text{-Nat}(F,G)\big]\Big)$$

and thus by 6.3.4 isomorphisms

$$\mathscr{V}\text{-Nat}\big(F,[V,G-]\big) \cong \big[V,\mathscr{V}\text{-Nat}(F,G)\big]. \qquad \square$$

Proposition 6.6.10 *Let $\mathscr{V}$ be a complete symmetric monoidal closed category, $\mathscr{A}$ a $\mathscr{V}$-category and $A \in \mathscr{A}$, $V \in \mathscr{V}$ two objects.*

- *The cotensor $[V, A]$ of the objects V, A exists if and only if the $\mathscr{V}$-limit of $\delta_A \colon \mathscr{I} \longrightarrow \mathscr{A}$ weighted by $\delta_V \colon \mathscr{I} \longrightarrow \mathscr{V}$ exists; when this is the case, both objects are isomorphic.*
- *The tensor $V \otimes A$ of the objects V, A exists if and only if the $\mathscr{V}$-colimit of $\delta_A \colon \mathscr{I} \longrightarrow \mathscr{A}$ weighted by $\delta_V \colon \mathscr{I} \longrightarrow \mathscr{V}$ exists; when this is the case, both objects are isomorphic.*

Proof The $\mathscr{V}$-functor $\mathscr{A}(B, \delta_A -) \colon \mathscr{I} \longrightarrow \mathscr{A}$ is just the $\mathscr{V}$-functor $\delta_{\mathscr{A}(B,A)}$, when $B \in \mathscr{A}$. Therefore by 6.6.7

$$\mathscr{V}\text{-}\mathsf{Nat}\big(\delta_V, \mathscr{A}(B, \delta_A)\big) \cong \mathscr{V}\text{-}\mathsf{Nat}\big(\delta_V, \delta_{\mathscr{A}(B,A)}\big) \cong \big[V, \mathscr{A}(B, A)\big].$$

The existence of the weighted limit $\lim_{\delta_V} \delta_A$ thus reduces to the existence of $L \in \mathscr{A}$ together with $\mathscr{V}$-natural isomorphisms $\big[V, \mathscr{A}(B, A)\big] \cong \mathscr{A}(B, L)$, which is just the definition of the cotensor $[V, A]$; see 6.5.1. An analogous proof holds for tensors. □

Proposition 6.6.11 *Let $\mathscr{V}$ be a complete symmetric monoidal closed category. For every $\mathscr{V}$-category $\mathscr{A}$ and every object $A \in \mathscr{A}$:*

(1) the $\mathscr{V}$-functor $\mathscr{A}(A, -) \colon \mathscr{A} \longrightarrow \mathscr{V}$ preserves all weighted $\mathscr{V}$-limits;
(2) the $\mathscr{V}$-functor $\mathscr{A}(-, A) \colon \mathscr{A} \longrightarrow \mathscr{V}$ transforms weighted $\mathscr{V}$-colimits into weighted $\mathscr{V}$-limits.

Proof Consider a small $\mathscr{V}$-category $\mathscr{D}$ and two $\mathscr{V}$-functors $F \colon \mathscr{D} \longrightarrow \mathscr{A}$, $G \colon \mathscr{D} \longrightarrow \mathscr{V}$. Suppose the weighted $\mathscr{V}$-limit $\lim_G F$ exists and is given by $L \in \mathscr{A}$ together with $\mathscr{V}$-natural isomorphisms

$$\mathscr{V}\text{-}\mathsf{Nat}\big(G, \mathscr{A}(B, F-)\big) \cong \mathscr{A}(B, L)$$

for $B \in \mathscr{A}$. For every $V \in \mathscr{V}$, the isomorphisms (see 6.5.7)

$$\begin{aligned} \mathscr{V}\text{-}\mathsf{Nat}\Big(G, \big[V, \mathscr{A}(A, F-)\big]\Big) &\cong \Big[V, \mathscr{V}\text{-}\mathsf{Nat}\big(G, \mathscr{A}(A, F-)\big)\Big] \\ &\cong \big[V, \mathscr{A}(A, L)\big] \end{aligned}$$

show that $\mathscr{A}(A, L) = \lim_G \mathscr{A}(A, F-)$.

The second assertion holds by duality. □

Proposition 6.6.12 *Let $\mathscr{V}$ be a complete symmetric monoidal closed category, V an object of $\mathscr{V}$ and $\mathscr{A}$ a $\mathscr{V}$-category. If $\mathscr{A}$ is cotensored, the functor $[V, -] \colon \mathscr{A} \longrightarrow \mathscr{A}$ preserves weighted $\mathscr{V}$-limits. If $\mathscr{A}$ is tensored, the functor $V \otimes - \colon \mathscr{A} \longrightarrow \mathscr{A}$ preserves weighted $\mathscr{V}$-colimits.*

Proof Let $\mathscr{D}$ be a small $\mathscr{V}$-category and $F\colon \mathscr{D} \longrightarrow \mathscr{A}$, $G\colon \mathscr{D} \longrightarrow \mathscr{V}$ two $\mathscr{V}$-functors. If $\lim_G F$ exists, we have an object $L \in \mathscr{A}$ together with $\mathscr{V}$-natural isomorphisms

$$\mathscr{V}\text{-Nat}\big(G, \mathscr{A}(A, F-)\big) \cong \mathscr{A}(A, L)$$

for $A \in \mathscr{A}$. If $\mathscr{A}$ is tensored, applying 6.5.7 we get

$$\begin{aligned}\mathscr{V}\text{-Nat}\Big(G, \mathscr{A}\big(A, [V, F-]\big)\Big) &\cong \mathscr{V}\text{-Nat}\Big(G, \big[V, \mathscr{A}(A, F-)\big]\Big)\\ &\cong \Big[V, \mathscr{V}\text{-Nat}\big(G, \mathscr{A}(A, F-)\big)\Big]\\ &\cong \big[V, \mathscr{A}(A, L)\big]\\ &\cong \mathscr{A}\big(A, [V, L]\big),\end{aligned}$$

from which $[V, L] \cong \lim_G [V, F-]$. □

With view to proving an existence theorem for weighted $\mathscr{V}$-limits, let us make clear the relation between ordinary limits and weighted limits.

Lemma 6.6.13 *Let $\mathscr{V}$ be a complete and cocomplete symmetric monoidal closed category. Consider, with the notation of 6.4.7,*

- *a cotensored $\mathscr{V}$-category $\mathscr{A}$,*
- *a small category $\mathscr{C}$ and the corresponding free $\mathscr{V}$-category $\overline{\mathscr{C}}$,*
- *a functor $F\colon \mathscr{C} \longrightarrow U(\mathscr{A})$ with corresponding $\mathscr{V}$-functor $\overline{F}\colon \overline{\mathscr{C}} \longrightarrow \mathscr{A}$,*
- *the constant functor $\Delta_I\colon \mathscr{C} \longrightarrow \mathscr{V}$ on $I \in \mathscr{V}$ and the corresponding $\mathscr{V}$-functor $\overline{\Delta_I}\colon \overline{\mathscr{C}} \longrightarrow \mathscr{V}$.*

If, for every $A \in \mathscr{A}$, the functor $\mathscr{A}(A, -)\colon U(\mathscr{A}) \longrightarrow \mathscr{V}$ preserves limits, the following conditions are equivalent:

(1) the limit of F exists;

(2) the $\mathscr{V}$-limit of $\overline{F}$ weighted by $\overline{\Delta_I}$ exists.

Moreover, the limit object is the same in both cases.

Proof The $\mathscr{V}$-limit of $\overline{F}$ weighted by $\overline{\Delta_I}$ exists if we can find $L \in \mathscr{A}$ together with $\mathscr{V}$-natural isomorphisms

$$\mathscr{V}\text{-Nat}\big(\overline{\Delta_I}, \mathscr{A}(A, \overline{F}-)\big) \cong \mathscr{A}(A, L)$$

for $A \in \mathscr{A}$. By 6.3.3, this is equivalent to having a bijective correspondence between

- the $\mathscr{V}$-natural transformations $\overline{\Delta_I} \Rightarrow \big[V, \mathscr{A}(A, F-)\big]$,
- the morphisms $V \longrightarrow \mathscr{A}(A, L)$

for every object $V \in \mathscr{V}$.

A $\mathscr{V}$-natural transformation $\alpha\colon \overline{\Delta_I} \Rightarrow [V, \mathscr{A}(A, \overline{F}-)]$, corresponds with a compatible family of morphisms

$$\overline{\alpha}_B\colon I \otimes V \longrightarrow \mathscr{A}(A, F(B)).$$

Via the isomorphisms

$$\begin{aligned}\mathscr{V}\Big(I \otimes V, \mathscr{A}(A, F(B))\Big) &\cong \mathscr{V}\Big(I, \big[V, \mathscr{A}(A, F(B))\big]\Big)\\ &\cong \mathscr{V}\Big(I, \mathscr{A}\big(A, [V, F(B)]\big)\Big)\\ &\cong U(\mathscr{A})\Big(A, \big[V, F(B)\big]\Big)\end{aligned}$$

this corresponds with a cone

$$\widetilde{\alpha}_B\colon A \longrightarrow [V, FB]$$

on $[V, F-]$ in $U(\mathscr{A})$.

On the other hand via the isomorphisms

$$\begin{aligned}\mathscr{V}(V, \mathscr{A}(A, L)) \cong \mathscr{V}(I \otimes V, \mathscr{A}(A, L)) &\cong \mathscr{V}\Big(I, [V, \mathscr{A}(A, L)]\Big)\\ \cong \mathscr{V}\Big(I, \mathscr{A}(A, [V, L])\Big) &\cong U(\mathscr{A})(A, [V, L])\end{aligned}$$

a morphism $V \longrightarrow \mathscr{A}(A, L)$ in $\mathscr{V}$ corresponds with sone $A \longrightarrow [V, L]$ in $U(\mathscr{A})$.

As a conclusion, the existence of the $\mathscr{V}$-limit of $\overline{F}$ weighted by $\overline{\Delta_I}$ is equivalent to the existence of natural bijections between

- the cones of vertex A on $[V, F-]$,
- the morphisms $A \longrightarrow [V, L]$ in $U(\mathscr{A})$.

for all $A \in \mathscr{A}$, $V \in \mathscr{V}$. And since $[I, -]$ is just the identity functor on $\mathscr{A}$, this means finally that the ordinary limit of F exists in $U(\mathscr{A})$ and is preserved by each functor $[V, -]\colon U(\mathscr{A}) \longrightarrow U(\mathscr{A})$.

It remains to prove that every functor $[V, -]\colon U(\mathscr{A}) \longrightarrow U(\mathscr{A})$ preserves limits. This follows from the bijections

$$\begin{aligned}\mathsf{Nat}\big(\Delta_A, [V, F-]\big) &\cong \mathsf{Nat}\big(\Delta_V, \mathscr{A}(A, F-)\big)\\ &\cong \mathscr{V}(V, \mathscr{A}(A, L)) \cong U(\mathscr{A})(A, [V, L])\end{aligned}$$

since $\mathscr{A}$ is cotensored and $\mathscr{A}(A, -)\colon U(\mathscr{A}) \longrightarrow \mathscr{V}$ preserves limits, by assumption. □

$$\begin{array}{ccc}
\mathscr{D}(C,D) & \xrightarrow{\mathscr{A}(L,F-)_{CD}} & [\mathscr{A}(L,FC),\mathscr{A}(L,FD)] \\
G_{CD}\downarrow & & \downarrow [\alpha_C,1] \\
[GC,GD] & \xrightarrow[{[1,\alpha_D]}]{} & [GC,\mathscr{A}(L,FD)]
\end{array}$$

Diagram 6.41

The reader should convince himself that, in 6.6.12, the assumption "$\mathscr{A}(A,-)\colon U(\mathscr{A})\longrightarrow\mathscr{V}$ preserves limits" does not follow from 6.6.11 when $\mathscr{A}$ is not supposed to be $\mathscr{V}$-complete.

Theorem 6.6.14 *Let $\mathscr{V}$ be a complete and cocomplete symmetric monoidal closed category. A $\mathscr{V}$-category $\mathscr{A}$ admits all weighted $\mathscr{V}$-limits iff:*
(1) $\mathscr{A}$ is cotensored;
(2) the underlying category $U(\mathscr{A})$ is complete;
(3) the functors $\mathscr{A}(A,-)\colon U(\mathscr{A})\longrightarrow\mathscr{V}$ preserve (ordinary) limits.
A $\mathscr{V}$-category $\mathscr{A}$ admits all weighted $\mathscr{V}$-colimits iff:
(1) $\mathscr{A}$ is tensored;
(2) the underlying category $U(\mathscr{A})$ is cocomplete;
(3) the functors $\mathscr{A}(-,A)\colon U(\mathscr{A})\longrightarrow\mathscr{V}$ transform (ordinary) colimits into limits.

Proof We prove the first statement; the second one follows by duality. If $\mathscr{A}$ admits all weighted $\mathscr{V}$-limits, it is cotensored (see 6.6.10) and $U(\mathscr{A})$ is complete (see 6.6.13). Moreover, using the notations of 6.6.13 together with 6.6.11,

$$\begin{aligned}
\mathscr{A}(A,\lim F) &\cong \mathscr{A}\big(A,\lim_{\overline{\Delta_I}}\overline{F}\big)\\
&\cong \lim_{\overline{\Delta_I}}\mathscr{A}\big(A,\overline{F}-\big)\cong \lim_{\overline{\Delta_I}}\overline{\mathscr{A}(A,F-)}\cong \lim\mathscr{A}(A,F-)
\end{aligned}$$

which proves that $\mathscr{A}(A,-)\colon U(\mathscr{A})\longrightarrow\mathscr{V}$ preserves limits.

Conversely, assume that $\mathscr{A}$ is cotensored and $U(\mathscr{A})$ is complete. Choose a small $\mathscr{V}$-category $\mathscr{D}$ and $\mathscr{V}$-functors $F\colon\mathscr{D}\longrightarrow\mathscr{A}$, $G\colon\mathscr{D}\longrightarrow\mathscr{V}$. For an object $A\in\mathscr{A}$, we shall first construct an object $L\in\mathscr{A}$ and a "universal" $\mathscr{V}$-natural transformation $\alpha\colon G\Rightarrow\mathscr{A}(L,F-)$, that is a family of morphisms $\alpha_D\colon G(D)\longrightarrow\mathscr{A}\big(L,F(D)\big)$, $D\in\mathscr{D}$, such that the square in diagram 6.41 commutes (see 6.2.8). But giving $\alpha_D\colon G(D)\longrightarrow\mathscr{A}\big(L,F(D)\big)$ in $\mathscr{V}$ is equivalent to giving $\beta_D\colon L\longrightarrow\big[G(D),F(D)\big]$ in $U(\mathscr{A})$, by definition of a cotensor. The $\mathscr{V}$-

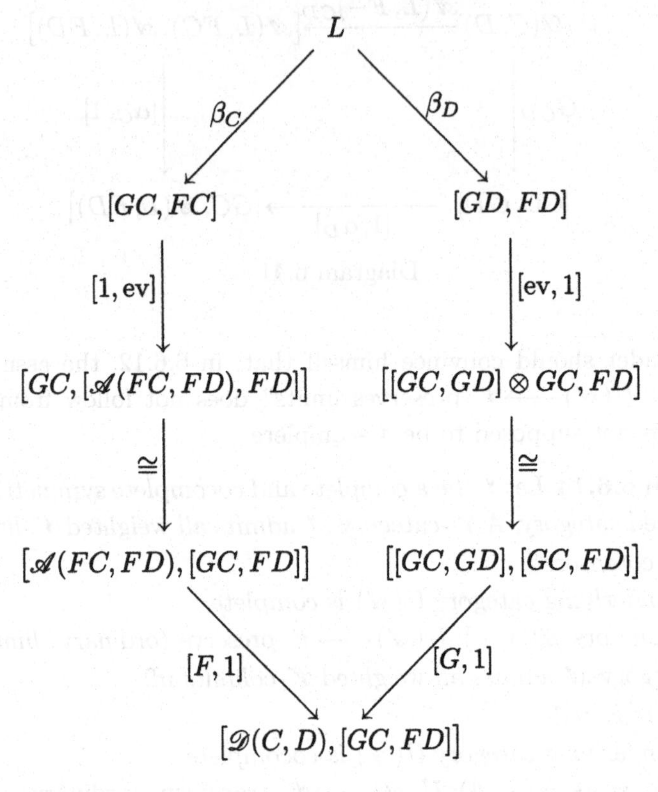

Diagram 6.42

naturality of α, in terms of the morphisms β_D, means the commutativity of diagram 6.42 for all objects $C, D \in \mathscr{D}$. Defining $\big(\beta_D\colon L \longrightarrow [G(D), F(D)]\big)_{D\in\mathscr{D}}$ to be the corresponding limit in $U(\mathscr{A})$ of the global diagram obtained when C, D run through $\mathscr{D}$, we get by adjunction the required $\mathscr{V}$-natural transformation $\alpha\colon G \Rightarrow \mathscr{A}(L, F-)$ with the property that, for every $\mathscr{V}$-natural transformation $\gamma\colon G \Rightarrow \mathscr{A}(A, F-)$ with $A \in \mathscr{A}$, there exists a unique morphism $a\colon A \longrightarrow L$ in $U(\mathscr{A})$ such that $\gamma = \mathscr{A}(a, 1)\circ\alpha$. In other words, we have got natural bijections

$$\mathscr{V}\text{-}\mathsf{Nat}\big(G, \mathscr{A}(A, F-)\big) \cong U(\mathscr{A})(A, L).$$

Fix now an object $V \in \mathscr{V}$ and, in the previous argument, replace F by the composite $\mathscr{V}$-functor $[V, F-]\colon \mathscr{D} \longrightarrow \mathscr{A}$. One now gets a limit $\big(\beta_D^V\colon L^V \longrightarrow [G(D), [V, F(D)]]\big)_{D\in\mathscr{D}}$ and natural bijections

$$\mathscr{V}\text{-}\mathsf{Nat}\Big(G, \mathscr{A}\big(A, [V, F-]\big)\Big) \cong U(\mathscr{A})\big(A, L^V\big).$$

Let us prove that $L^V \cong [V, L]$. In fact the diagram defining L is built up from a family of pairs

$$[GC, FC] \longrightarrow [\mathscr{D}(C,D), [GC, FD]] \longleftarrow [GD, FD]$$

and the diagram defining L^V is built up from a family of pairs

$$[GC, [V, FC]] \longrightarrow \Big[\mathscr{D}(C,D), [GC, [V, FD]]\Big] \longleftarrow [GD, [V, FD]].$$

By definition of cotensors, this last diagram is isomorphic to

$$[V, [GC, FC]] \longrightarrow \Big[V, [\mathscr{D}(C,D), [GC, FD]]\Big] \longleftarrow [V, [GD, FD]]$$

and it is immediately clear that this is just the diagram obtained from that defining L after application of $[V, -]$. To prove the isomorphism $L^V \cong [V, L]$, it remains thus to prove that $[V, -]$ preserves the limit defining L.

But the very last argument in the proof of 6.6.12 was precisely that the functor $[V, -]$ preserves limits as long as every representable functor $\mathscr{A}(A, -)\colon U(\mathscr{A}) \longrightarrow \mathscr{V}$ does, which is one of our assumption.

Finally we have got natural bijections

$$\mathscr{V}\text{-}\mathsf{Nat}\Big(G, \mathscr{A}(A, [V, F-])\Big) \cong U(\mathscr{A})(A, [V, L]) \cong \mathscr{V}\big(V, \mathscr{A}(A, L)\big)$$

and we can conclude by 6.3.3 that there exists an isomorphism in $\mathscr{V}$,

$$\mathscr{V}\text{-}\mathsf{Nat}\big(G, \mathscr{A}(A, F-)\big) \cong \mathscr{A}(A, L). \qquad \square$$

Corollary 6.6.15 *Let $\mathscr{V}$ be a complete and cocomplete symmetric monoidal closed category and $\mathscr{A}$, $\mathscr{B}$ two $\mathscr{V}$-categories, $\mathscr{A}$ being $\mathscr{V}$-complete. A $\mathscr{V}$-functor $F\colon \mathscr{A} \longrightarrow \mathscr{B}$ preserves all weighted $\mathscr{V}$-limits iff*

(1) F preserves cotensors,

(2) F preserves ordinary limits.

Proof The necessity of the conditions follows immediately from 6.6.10 and 6.6.13. The construction of the weighted $\mathscr{V}$-limit in the proof of 6.6.14 shows that the conditions are also sufficient. $\square$

Corollary 6.6.16 *Let $\mathscr{V}$ be a complete and cocomplete symmetric monoidal closed category and $\mathscr{A}$ a tensored and cotensored category. Then:*

- *$\mathscr{A}$ is $\mathscr{V}$-complete iff $U(\mathscr{A})$ is complete;*
- *$\mathscr{A}$ is $\mathscr{V}$-cocomplete iff $U(\mathscr{A})$ is cocomplete.*

Proof Applying the functor $\mathscr{V}(I,-)$ to the isomorphism of 6.5.6, we get for $A, B \in \mathscr{A}$ and $V \in \mathscr{V}$

$$U(\mathscr{A})(V \otimes A, B) \cong U(\mathscr{A})(A, [V, B])$$

which proves the adjunction $V\otimes - \dashv [V,-]$ (see 3.1.5, volume 1) and thus the facts that $[V,-]\colon U(\mathscr{A}) \longrightarrow \mathscr{A}$ preserves limits and $V\otimes -\colon \mathscr{A} \longrightarrow \mathscr{A}$ preserves colimits. Reversing the argument at the end of the proof of 6.6.13 yields, with the same notation,

$$\begin{aligned}\mathsf{Nat}(\Delta_V, \mathscr{A}(A, F-)) &\cong \mathsf{Nat}(\Delta_A, [V, F-]) \\ &\cong U(\mathscr{A})(A, [V, L]) \cong \mathscr{V}(V, \mathscr{A}(A, L))\end{aligned}$$

since the functor $[V,-]$ preserves limits. But this means precisely that the functor $\mathscr{A}(A,-)\colon U(\mathscr{A}) \longrightarrow \mathscr{V}$ preserves limits. In the same way one proves that $\mathscr{A}(-,A)\colon U(\mathscr{A}) \longrightarrow \mathscr{V}$ transforms colimits into limits. □

Let us now generalize 2.15.6, volume 1, to the case of $\mathscr{V}$-functors. First of all observe that

Proposition 6.6.17 *Let $\mathscr{V}$ be a complete and cocomplete symmetric monoidal closed category. For every small $\mathscr{V}$-category $\mathscr{A}$, the $\mathscr{V}$-category of $\mathscr{V}$-functors $\mathscr{A} \longrightarrow \mathscr{V}$ and $\mathscr{V}$-natural transformations is $\mathscr{V}$-complete and $\mathscr{V}$-cocomplete.*

Proof By 6.6.16 and 6.5.7, it suffices to prove that a pointwise limit or colimit of $\mathscr{V}$-functors is still a $\mathscr{V}$-functor. We do it for limits; the case of colimits is analogous.

Consider a diagram $(F_D)_{D\in\mathscr{D}}$ of $\mathscr{V}$-functors and $\mathscr{V}$-natural transformations indexed by a small category $\mathscr{D}$. By 6.2.8, it makes sense to define in the following way a $\mathscr{V}$-functor $L\colon \mathscr{A} \longrightarrow \mathscr{V}$. It maps the object A on $\lim_{D\in\mathscr{D}} F_D(A)$ while the structural morphisms

$$\mathscr{A}(A, B) \longrightarrow \lim_{D\in\mathscr{D}} \big[\lim_{D\in\mathscr{D}} F_D(A), F_D(B)\big]$$

can equivalently be given by morphisms

$$\mathscr{A}(A, B) \longrightarrow \big[\lim_{D\in\mathscr{D}} F_D(A), \lim_{D\in\mathscr{D}} F_D(B)\big]$$

and these are just the factorizations through the limit of the composites

$$\mathscr{A}(A, B) \xrightarrow{\ F_D\ } \big[F_D(A), F_D(B)\big] \xrightarrow{\ [p_D, 1]\ } \big[\lim_{D\in\mathscr{D}} F_D(A), F_D(B)\big].$$

It is straightforward to check that L is indeed the required limit.

In the previous argument we have used the fact that the functors $[V,-]\colon \mathscr{V} \longrightarrow \mathscr{V}$ preserve limits, which is clear since they admit the functors $-\otimes V\colon \mathscr{V} \longrightarrow \mathscr{V}$ as left adjoints (see 3.2.2, volume 1). The proof in the case of colimits requires the fact that the functors $[-,V]\colon \mathscr{V} \longrightarrow \mathscr{V}$ transform colimits into limits. This is the case by 6.6.11 and 6.6.15. $\square$

Theorem 6.6.18 *Let $\mathscr{V}$ be a complete and cocomplete symmetric monoidal closed category. For a small $\mathscr{V}$-category $\mathscr{A}$, consider the $\mathscr{V}$-Yoneda-embedding*

$$Y\colon \mathscr{A}^* \longrightarrow \mathscr{V}\text{-}\mathsf{Fun}(\mathscr{A},\mathscr{V}),\quad A \mapsto \mathscr{A}(A,-).$$

For every $\mathscr{V}$-functor $F\colon \mathscr{A} \longrightarrow \mathscr{V}$, the isomorphism $F \cong \operatorname{colim}_F Y$ holds.

Proof For every $\mathscr{V}$-functor $G\colon \mathscr{A} \longrightarrow \mathscr{V}$ we must find an isomorphism

$$\mathscr{V}\text{-}\mathsf{Nat}\Big(F(-),\mathscr{V}\text{-}\mathsf{Nat}\big(Y(-),G\big)\Big) \cong \mathscr{V}\text{-}\mathsf{Nat}(F,G).$$

This is an immediate consequence of the $\mathscr{V}$-Yoneda-lemma (see 6.3.5) which implies $\mathscr{V}\text{-}\mathsf{Nat}\big(Y(-),G\big) \cong G$. $\square$

Examples 6.6.19

6.6.19.a Choose $\mathscr{V} = \mathsf{Cat}$, the cartesian closed category of small categories (see 6.1.9.b). A $\mathscr{V}$-category is just a 2-category (see 7.1.1, volume 1) and a $\mathscr{V}$-functor is just a 2-functor (see 7.2.1, volume 1). Given a 2-functor $F\colon \mathscr{A} \longrightarrow \mathscr{B}$ with $\mathscr{A}$ small, the $\mathscr{V}$-limit of F weighted by the constant functor $\Delta_1\colon \mathscr{A} \longrightarrow \mathscr{V}$ on the terminal object of Cat is just the 2-limit of F (see 7.4.5, volume 1).

Examples 6.6.20

6.6.20.a Pseudo-limits and lax limits can also be presented as special cases of weighted Cat-limits. Let us indicate this with an example. In a 2-category $\mathscr{A}$, the lax equalizer of two morphisms $f,g\colon A \rightrightarrows B$ is a morphism $k\colon K \longrightarrow A$ together with a 2-cell $\kappa\colon f\circ k \Rightarrow g\circ k$, these data being 2-universal for these requirements (see 7.6.1, volume 1). Write $\mathscr{I}$ for the weighting category with two objects X,Y and two non-identity arrows $u,v\colon X \rightrightarrows Y$; $F\colon \mathscr{I} \longrightarrow \mathscr{A}$ is the 2-functor defined by $F(u) = f$, $F(v) = g$. Write $\mathbf{1}$ for the category with a single object 0 and just the identity on it; write $\mathbf{2}$ for the category with two objects $1,2$ and one single non-identity arrow $t\colon 1 \longrightarrow 2$. The functor $G\colon \mathscr{I} \longrightarrow \mathsf{Cat}$ maps X to $\mathbf{1}$, Y to $\mathbf{2}$, u to $G(u)\colon \mathbf{1} \longrightarrow \mathbf{2}$ defined by $G(u)(0) = 1$ and v on $G(v)\colon \mathbf{1} \longrightarrow \mathbf{2}$ defined by $G(v)(0) = 2$. For every object $C \in \mathscr{A}$, a 2-natural transformation $\alpha\colon G \Rightarrow \mathscr{A}(C,F-)$ consists in giving

- a functor $\alpha_X\colon \mathbf{1} \longrightarrow \mathscr{A}(C,A)$, i.e. a morphism $h\colon C \longrightarrow A$ in $\mathscr{A}$,
- a functor $\alpha_Y\colon \mathbf{2} \longrightarrow \mathscr{A}(C,B)$, i.e. two morphisms $l,m\colon C \rightrightarrows B$ in $\mathscr{A}$ and a 2-cell $\lambda\colon l \Rightarrow m$,

and these data must satisfy the requirements $f \circ h = l$, $g \circ h = m$. In other words giving α is just giving $h\colon C \longrightarrow A$ together with a 2-cell $\lambda\colon f \circ h \Rightarrow g \circ h$. Giving such a 2-universal α is thus indeed giving the lax equalizer of f, g.

6.7 Enriched adjunctions

The use of weighted limits and colimits now makes possible the direct generalization of most results of category theory in the context of enriched category theory. We shall limit our attention to the case of adjoint functors and Kan extensions.

Definition 6.7.1 *Let $\mathscr{V}$ be a symmetric monoidal category and $\mathscr{A}, \mathscr{B}$ two $\mathscr{V}$-categories. Two $\mathscr{V}$-functors $F\colon \mathscr{A} \longrightarrow \mathscr{B}$, $G\colon \mathscr{B} \longrightarrow \mathscr{A}$ are $\mathscr{V}$-adjoint, G left adjoint to F and F right adjoint to G, when there exist isomorphisms in $\mathscr{V}$*

$$\mathscr{A}\big(G(B),A\big) \cong \mathscr{B}\big(B,F(A)\big),$$

which are $\mathscr{V}$-natural in $A \in \mathscr{A}$, $B \in \mathscr{B}$.

Proposition 6.7.2 *Let $\mathscr{V}$ be a symmetric monoidal closed category and $F\colon \mathscr{A} \longrightarrow \mathscr{B}$ a $\mathscr{V}$-functor. The following conditions are equivalent:*

(1) F has a left $\mathscr{V}$-adjoint $G\colon \mathscr{B} \longrightarrow \mathscr{A}$;

(2) for every $B \in \mathscr{B}$, there exists $G(B) \in \mathscr{A}$ together with isomorphisms

$$\mathscr{A}\big(G(B),A\big) \cong \mathscr{B}\big(B,F(A)\big)$$

which are $\mathscr{V}$-natural in $A \in \mathscr{A}$.

Proof We make G a $\mathscr{V}$-functor by observing that the isomorphism

$$\mathscr{V}\Big(I, \mathscr{A}\big(G(B),G(B)\big)\Big) \cong \mathscr{V}\Big(I, \mathscr{B}\big(B,FG(B)\big)\Big)$$

yields a morphism $\eta_B\colon B \longrightarrow FG(B)$ in the underlying category $U(\mathscr{B})$, η_B corresponding with $u_{G(B)}$. The composite

$$\mathscr{B}(C,B) \xrightarrow{\mathscr{B}(1,\eta_B)} \mathscr{B}\big(C,FG(B)\big) \cong \mathscr{A}\big(G(C),G(B)\big)$$

provides G with the structure of a $\mathscr{V}$-functor.

It remains to prove the naturality in B of the isomorphisms

$$\theta_{AB}\colon \mathscr{A}\big(G(B),A\big) \cong \mathscr{B}\big(B,F(A)\big).$$

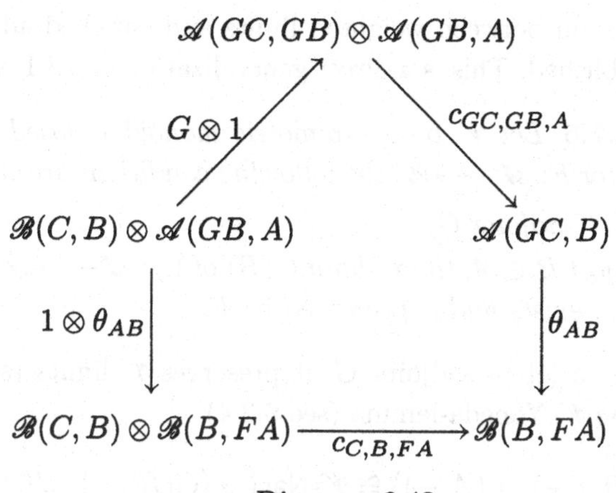

Diagram 6.43

By 6.2.8 this is equivalent, by adjunction, to the commutativity of diagram 6.43, which follows immediately from the definition of the $\mathscr{V}$-functor structure of G. □

Proposition 6.7.3 *Let $\mathscr{V}$ be a symmetric monoidal closed category. If a $\mathscr{V}$-functor $F\colon \mathscr{A}\longrightarrow\mathscr{B}$ has a left $\mathscr{V}$-adjoint, it preserves all weighted $\mathscr{V}$-limits.*

Proof Choose two $\mathscr{V}$-functors $H\colon \mathscr{D}\longrightarrow\mathscr{A}$, $K\colon \mathscr{D}\longrightarrow\mathscr{V}$ defined on a $\mathscr{V}$-category $\mathscr{D}$. Suppose the weighted limit $L\cong \lim_K H$ exists, yielding $\mathscr{V}$-natural isomorphisms in $A\in\mathscr{A}$,

$$\mathscr{V}\text{-}\mathsf{Nat}\big(K,\mathscr{A}(A,H-)\big)\cong\mathscr{A}(A,L).$$

Write G for the left adjoint to F.

We must find isomorphisms, $\mathscr{V}$-natural in $B\in\mathscr{B}$,

$$\mathscr{V}\text{-}\mathsf{Nat}\big(K,\mathscr{B}(B,FH-)\big)\cong\mathscr{B}\big(B,F(L)\big);$$

putting $A=G(B)$, this is immediate since $\mathscr{B}(B,F-)\cong\mathscr{A}\big(G(B),H-\big)$ and $\mathscr{B}\big(B,F(L)\big)\cong\mathscr{A}\big(G(B),L\big)$. □

Proposition 6.7.4 *Let $\mathscr{V}$ be a symmetric monoidal closed category.*

- *A $\mathscr{V}$-category $\mathscr{A}$ is tensored iff every $\mathscr{V}$-functor $\mathscr{A}(A,-)\colon \mathscr{A}\longrightarrow\mathscr{V}$, for $A\in\mathscr{A}$, has a left $\mathscr{V}$-adjoint.*
- *A $\mathscr{V}$-category $\mathscr{A}$ is cotensored iff every $\mathscr{V}$-functor $\mathscr{A}(-,A)\colon \mathscr{A}^*\longrightarrow\mathscr{V}$, for $A\in\mathscr{A}$, has a left $\mathscr{V}$-adjoint.*

Proof This is immediate from 6.5.1 and 6.7.2. □

A general relation between weighted limits and enriched adjunctions can now be established. This is a direct generalization of 3.3.1, volume 1.

Proposition 6.7.5 *Let $\mathscr{V}$ be a symmetric monoidal closed category. Given a $\mathscr{V}$-functor $F: \mathscr{A} \longrightarrow \mathscr{B}$, the following conditions are equivalent:*

(1) F has a left $\mathscr{V}$-adjoint G;

(2) for every object $B \in \mathscr{B}$, the $\mathscr{V}$-limit $G(B)$ of $1_{\mathscr{A}}: \mathscr{A} \longrightarrow \mathscr{A}$ weighted by $\mathscr{B}(B, F-)$ exists and is preserved by F.

Proof If F has a left $\mathscr{V}$-adjoint G, it preserves $\mathscr{V}$-limits (see 6.7.2). Moreover, by the $\mathscr{V}$-Yoneda-lemma (see 6.3.4)

$$\begin{aligned}\mathscr{V}\text{-Nat}(\mathscr{B}(B,F-),\mathscr{A}(A,-)) &\cong \mathscr{V}\text{-Nat}\Big(\mathscr{A}\big(G(B),-\big),\mathscr{A}(A,-)\Big)\\ &\cong \mathscr{A}\big(A,G(B)\big).\end{aligned}$$

Thus $G(B) = \lim_{\mathscr{B}(B,F-)} 1_{\mathscr{A}}$.

Conversely the assumptions mean

$$\mathscr{V}\text{-Nat}(\mathscr{B}(B,F-),\mathscr{A}(A,-)) \cong \mathscr{A}\big(A,G(B)\big)$$
$$\mathscr{V}\text{-Nat}(\mathscr{B}(B,F-),\mathscr{B}(C,F-)) \cong \mathscr{B}\big(C,FG(B)\big)$$

for $A \in \mathscr{A}$ and $B, C \in \mathscr{B}$. In particular

$$\mathscr{A}\big(G(B),G(B)\big) \cong \mathscr{V}\text{-Nat}\Big(\mathscr{B}(B,F-),\mathscr{A}\big(G(B),-\big)\Big),$$
$$\begin{aligned}\mathscr{V}\text{-Nat}(\mathscr{B}(B,F-),\mathscr{B}(B,F-)) &\cong \mathscr{B}\big(B,FG(B)\big)\\ &\cong \mathscr{V}\text{-Nat}\Big(\mathscr{A}\big(G(B),-\big),\mathscr{B}(B,F-)\Big),\end{aligned}$$

with the last isomorphism coming from the $\mathscr{V}$-Yoneda-lemma. Applying $\mathscr{V}(I,-)$, the identities on $\mathscr{B}(B,F-)$ correspond with $\mathscr{V}$-natural transformations

$$\alpha_A: \mathscr{B}(B,F-) \Rightarrow \mathscr{A}\big(G(B),-\big), \quad \beta_A: \mathscr{A}\big(G(B),-\big) \Rightarrow \mathscr{B}(B,F-),$$

and it is a straightforward matter to check that $\beta_A = (\alpha_A)^{-1}$. One concludes the proof by 6.7.2. □

Proposition 6.7.5 leaves us with the problem of finding sufficient conditions for the existence of the weighted limits $\lim_{\mathscr{B}(B,F-)} 1_{\mathscr{A}}$ defined on the $\mathscr{V}$-category $\mathscr{A}$, which is generally large. In practice, it is often more efficient to reduce the problem to one of the classical adjoint functor theorems (see section 3.3, volume 1). This is possible via the following theorem.

Theorem 6.7.6 *Let $\mathscr{V}$ be a symmetric monoidal closed category, $\mathscr{A}$ a cotensored $\mathscr{V}$-category, $\mathscr{B}$ an arbitrary $\mathscr{V}$-category and $F: \mathscr{A} \longrightarrow \mathscr{B}$ a $\mathscr{V}$-functor. Then F has a left $\mathscr{V}$-adjoint functor if and only if*

(1) F preserves cotensors,
(2) the underlying functor $U(F): U(\mathscr{A}) \longrightarrow U(\mathscr{B})$ has a left adjoint.

Proof If F has a $\mathscr{V}$-left adjoint G, F preserves cotensors by 6.7.3 and 6.6.10. Moreover applying $\mathscr{V}(I, -)$ to the isomorphisms

$$\mathscr{A}\big(G(B), A\big) \cong \mathscr{B}\big(B, F(A)\big)$$

yields bijections

$$U(\mathscr{A})\big(G(B), A\big) \cong U(\mathscr{B})\big(B, F(A)\big)$$

proving that the ordinary functor underlying G is left adjoint to the ordinary functor underlying F.

Conversely, suppose $G: U(\mathscr{B}) \longrightarrow U(\mathscr{A})$ is a functor left adjoint to the functor $U(F): U(\mathscr{A}) \longrightarrow U(\mathscr{B})$. By 6.7.2, it remains to prove the existence of isomorphisms

$$\mathscr{A}\big(G(B), A\big) \cong \mathscr{B}\big(B, F(A)\big)$$

$\mathscr{V}$-natural in $\mathscr{A}$. The adjunction $G \dashv U(F)$ yields two natural transformations $\eta: 1_{U\mathscr{B}} \Rightarrow U(F) \circ G$ and $\varepsilon: G \circ U(F) \Rightarrow 1_{\mathscr{A}}$; see 3.1.5, volume 1. In particular we have a composite, $\mathscr{V}$-natural in A,

$$\mathscr{A}\big(G(B), A\big) \xrightarrow{F_{G(B),A}} \mathscr{B}\big(FG(B), F(A)\big) \xrightarrow{\mathscr{B}(\eta_B, 1)} \mathscr{B}\big(B, F(A)\big).$$

Constructing its inverse

$$\mathscr{B}\big(B, F(A)\big) \longrightarrow \mathscr{A}\big(G(B), A\big)$$

is equivalent, since $\mathscr{A}$ is cotensored, to constructing a morphism in $U(\mathscr{A})$

$$G(B) \longrightarrow \Big[\mathscr{B}\big(B, F(A)\big), A\Big].$$

Applying the adjunction $G \dashv U(F)$, this is still equivalent to constructing a morphism in $U(\mathscr{B})$

$$B \longrightarrow F\Big[\mathscr{B}\big(B, F(A)\big), A\Big].$$

Since F preserves cotensors, this reduces to constructing

$$B \longrightarrow \Big[\mathscr{B}\big(B, F(A)\big), F(A)\Big],$$

i.e. to constructing a morphism in $\mathscr{V}$,

$$\mathscr{B}(B, F(A)) \longrightarrow \mathscr{B}(B, F(A)),$$

which we choose to be the identity.

It is now straightforward computation to verify that we have defined reciprocal isomorphisms. $\square$

Let us conclude with the case of enriched Kan extensions (see section 3.7, volume 1).

Theorem 6.7.7 *Let $\mathscr{V}$ be a complete symmetric monoidal closed category. Consider a $\mathscr{V}$-functor $F: \mathscr{A} \longrightarrow \mathscr{B}$, between two small $\mathscr{V}$-categories $\mathscr{A}, \mathscr{B}$, and a $\mathscr{V}$-cocomplete and cotensored $\mathscr{V}$-category $\mathscr{C}$. Under these conditions, composing with F*

$$F^*: \mathscr{V}\text{-Fun}(\mathscr{B}, \mathscr{C}) \longrightarrow \mathscr{V}\text{-Fun}(\mathscr{A}, \mathscr{C}),$$

yields a $\mathscr{V}$-functor F^* admitting a $\mathscr{V}$-left-adjoint functor written Lan_F. *For a $\mathscr{V}$-functor $G: \mathscr{A} \longrightarrow \mathscr{C}$, the $\mathscr{V}$-functor $\mathrm{Lan}_F G$ is called the left $\mathscr{V}$-Kan extension of G along F.*

Proof Now F^* maps a $\mathscr{V}$-functor $H: \mathscr{B} \longrightarrow \mathscr{C}$ to $H \circ F$. This yields a $\mathscr{V}$-functor with structural morphisms

$$F^*_{HH'}: \mathscr{V}\text{-Nat}(H, H') \longrightarrow \mathscr{V}\text{-Nat}(H \circ F, H' \circ F)$$

defined in the following way. By the $\mathscr{V}$-Yoneda-lemma

$$\mathscr{V}\text{-Nat}([-, M], [-, N]) \cong [M, N]$$

for all $M, N \in \mathscr{V}$ (see 6.3.5), thus applying $\mathscr{V}(I, -)$ indicates that defining a morphism $M \longrightarrow N$ is equivalent to defining a $\mathscr{V}$-natural transformation $[-, M] \Rightarrow [-, N]$, thus for every object $V \in \mathscr{V}$ a corresponding morphism $[V, M] \longrightarrow [V, N]$. In our case, M, N are objects of natural transformations so that the problem reduces, by 6.5.8, to defining mappings

$$\mathscr{V}\text{-Nat}\big(H, [V, H'-]\big) \longrightarrow \mathscr{V}\text{-Nat}\big(H \circ F, [V, H' \circ F-]\big).$$

But since $[V, H'-] \circ F = [V, H' \circ F-]$, this is just composition with F. The $\mathscr{V}$-naturality is obvious. This completes the definition of F^*.

To prove the existence of a $\mathscr{V}$-left-adjoint to F^*, choose a $\mathscr{V}$-functor $G: \mathscr{A} \longrightarrow \mathscr{C}$. We must define a $\mathscr{V}$-functor $\mathrm{Lan}_F G: \mathscr{B} \longrightarrow \mathscr{C}$ and $\mathscr{V}$-natural isomorphisms

$$\mathscr{V}\text{-Nat}(\mathrm{Lan}_F G, H) \cong \mathscr{V}\text{-Nat}(G, H \circ F)$$

$$\begin{array}{c}\mathscr{B}(F-,B)\otimes\mathscr{B}(B,B') \\ \Big\downarrow c_{F-,B,B'} \\ \mathscr{B}(F-,B') \\ \Big\downarrow \gamma_{B'} \\ \mathscr{C}(G-,\operatorname{colim}_{\mathscr{B}(F-,B')}G)\end{array}$$

Diagram 6.44

where H runs through the $\mathscr{V}$-functors $H\colon \mathscr{B}\longrightarrow\mathscr{C}$; see 6.7.2. Given B in $\mathscr{B}$, we define $(\mathrm{Lan}_F G)(B)$ as the colimit of G weighted by $\mathscr{B}(F-,B)$. To make this a $\mathscr{V}$-functor, it remains to construct the structural morphisms

$$\mathscr{B}(B,B')\longrightarrow\mathscr{C}\big(\operatorname{colim}_{\mathscr{B}(F-,B)}G,\operatorname{colim}_{\mathscr{B}(F-,B')}G\big).$$

Considering the isomorphisms

$$\begin{aligned}&\mathscr{V}\text{-}\mathsf{Nat}\big(\mathscr{B}(F-,B'),\mathscr{C}(G-,\operatorname{colim}_{\mathscr{B}(F-,B')}G)\big)\\ &\qquad\qquad\cong\mathscr{C}\big(\operatorname{colim}_{\mathscr{B}(F-,B')}G,\operatorname{colim}_{\mathscr{B}(F-,B')}G\big)\end{aligned}$$

and applying $\mathscr{V}(I,-)$, the identity on the right-hand side corresponds with a $\mathscr{V}$-natural transformation

$$\gamma_{B'}\colon \mathscr{B}(F-,B')\Rightarrow\mathscr{C}(G-,\operatorname{colim}_{\mathscr{B}(F-,B')}G).$$

On the other hand, by 6.6.11

$$\begin{aligned}&\mathscr{C}\big(\operatorname{colim}_{\mathscr{B}(F-,B)}G,\operatorname{colim}_{\mathscr{B}(F-,B')}G\big)\\ &\qquad\qquad\cong\lim_{\mathscr{B}(F-,B)}\mathscr{C}\big(G-,\operatorname{colim}_{\mathscr{B}(F-,B')}G\big).\end{aligned}$$

By definition of a weighted limit, constructing the required structural morphisms is thus equivalent to constructing $\mathscr{V}$-natural transformations

$$\mathscr{B}(F-,B)\Rightarrow\Big[\mathscr{B}(B,B'),\mathscr{C}\big(G-,\operatorname{colim}_{\mathscr{B}(F-,B')}G\big)\Big].$$

These correspond by adjunction with the composites of diagram 6.44. The reader will check that this indeed provides $\mathrm{Lan}_F G$ with the structure of a $\mathscr{V}$-functor.

It remains to prove the isomorphism

$$\mathscr{V}\text{-}\mathsf{Nat}(\mathrm{Lan}_F G,H)\cong\mathscr{V}\text{-}\mathsf{Nat}(G,H\circ F).$$

$$\begin{array}{c}\mathscr{B}(F-,B)\\ \Big\downarrow H\\ \mathscr{C}(HF-,HB)\\ \Big\downarrow [V,-]\\ \mathscr{C}([V,HF-],[V,HB])\\ \Big\downarrow \mathscr{C}(\beta,1)\\ \mathscr{C}(G,[V,HB])\end{array}$$

Diagram 6.45

The argument at the beginning of the proof shows that this is equivalent to finding natural bijections

$$\mathscr{V}\text{-Nat}\big(\mathrm{Lan}_F G,[V,H-]\big)\cong\mathscr{V}\text{-Nat}\big(G,[V,H\circ F-]\big)$$

for $V\in\mathscr{V}$. To do this, observe that the composites

$$I\xrightarrow{u_{FA}}\mathscr{B}\big(F(A),F(A)\big)\xrightarrow{(\lambda_{FA})_A}\mathscr{C}\big(G(A),(\mathrm{Lan}_F G\circ F)(A)\big)$$

define a $\mathscr{V}$-natural transformation $\gamma\colon G\Rightarrow\mathrm{Lan}_F G\circ F$. Given another natural transformation $\alpha\colon\mathrm{Lan}_F G\Rightarrow[V,H-]$ we get the composite

$$G\xrightarrow{\gamma}\mathrm{Lan}\,_F G\circ F\xrightarrow{\alpha*1}[V,H-]\circ F\cong[V,H\circ F-].$$

Now given $\beta\colon G\Rightarrow[V,H\circ F-]$, the definition of $\mathrm{Lan}_F G$ yields isomorphisms

$$\mathscr{V}\text{-Nat}\big(\mathscr{B}(F-,B),\mathscr{C}(G-,C)\big)\cong\mathscr{C}\big((\mathrm{Lan}_F G)(B),C\big).$$

Putting $C=[V,H(B)]$, the composite of diagram 6.45 defines a $\mathscr{V}$-natural transformation $\mathscr{B}(F-,B)\Rightarrow\mathscr{C}\big(G,[V,H(B)]\big)$, thus by the previous isomorphism a morphism

$$I\longrightarrow\mathscr{C}\Big((\mathrm{Lan}_F G)(B),\big[V,H(B)\big]\Big).$$

These last morphisms, for all $b\in\mathscr{B}$, define the required $\mathscr{V}$-natural transformation $\mathrm{Lan}_F G\Rightarrow[V,H-]$ which is required and we again leave to the reader the verification that we have indeed defined reciprocal bijections. □

6.8 Exercises

For simplicity, in these exercises, $\mathscr{V}$ always denotes a complete and cocomplete symmetric monoidal closed category.

6.8.1 Let $\mathscr{A}, \mathscr{B}$ be small $\mathscr{V}$-categories and $T\colon (\mathscr{A}\otimes\mathscr{B})^*\otimes(\mathscr{A}\otimes\mathscr{B}) \longrightarrow \mathscr{V}$ a $\mathscr{V}$-functor. Prove the "Fubini formula"

$$\int_{A\in\mathscr{A}}\int_{B\in\mathscr{B}} T(A,B,A,B) \cong \int_{B\in\mathscr{B}}\int_{A\in\mathscr{A}} T(A,B,A,B).$$

6.8.2 Let $\mathscr{A}, \mathscr{B}$ be $\mathscr{V}$-categories: $\mathscr{A}$ is small and $\mathscr{B}$ is $\mathscr{V}$-complete. Prove that the $\mathscr{V}$-category $\mathscr{V}\text{-}\mathsf{Fun}(\mathscr{A},\mathscr{B})$ of $\mathscr{V}$-functors and $\mathscr{V}$-natural transformations is $\mathscr{V}$-complete.

6.8.3 Let $\mathscr{V}$ be a small symmetric monoidal category. Prove that $\mathscr{V}$ is a full subcategory of a symmetric monoidal closed category $\mathscr{W}$, the structure of $\mathscr{V}$ being induced by that of $\mathscr{W}$. [Hint: choose $\mathscr{W} = \mathsf{Fun}(\mathscr{V}^*, \mathsf{Set})$.]

6.8.4 Let $\mathscr{A}$ be a $\mathscr{V}$-category. Prove that a $\mathscr{V}$-functor $F\colon \mathscr{A} \longrightarrow \mathscr{V}$ is $\mathscr{V}$-representable iff $\lim_F 1_{\mathscr{A}}$ exists and is preserved by F.

6.8.5 Consider $\overline{\mathbb{R}}_+ = \{r \mid r \in \mathbb{R}, r \geq 0\} \cup \{+\infty\}$ with the reversed poset structure, i.e. there is a morphism $r \longrightarrow s$ when $r \geq s$. Show that $\overline{\mathbb{R}}_+$ is a complete and cocomplete symmetric monoidal closed category if we define

$$r \otimes s = r + s, \quad [s,t] = \max\,\{t - s, 0\};$$

as a matter of convention, $\infty - \infty = 0$.

6.8.6 If (X,d) is a metric space, prove that X together with the distance $d\colon X \times X \longrightarrow \overline{\mathbb{R}}_+$ defines a category enriched in $\overline{\mathbb{R}}_+$.

6.8.7 If (X,d) is a metric space viewed as a $\overline{\mathbb{R}}_+$-category, prove that a $\overline{\mathbb{R}}_+$-functor $(X,d) \longrightarrow \overline{\mathbb{R}}_+$ is just a mapping $f\colon X \longrightarrow \overline{\mathbb{R}}$ such that $[f(x), f(y)] \leq d(x,y)$, $x, y \in X$.

6.8.8 If (X,d) is a metric space and $\mathbf{1}$ is the singleton, show that a pair of $\overline{\mathbb{R}}_+$-adjoint $\overline{\mathbb{R}}_+$-distributors $f\colon \mathbf{1} \,-\!\!\circ\!\!\!\rightarrow (X,d)$, $g\colon (X,d) \,-\!\!\circ\!\!\!\rightarrow \mathbf{1}$, $f \dashv g$ is just a pair of mappings $f, g\colon X \rightrightarrows \overline{\mathbb{R}}_+$ such that

(1) $\bigwedge_{x\in X}\{fx + gx \mid x \in X\} = 0$,
(2) $\forall x, y \in X \quad f(x) + g(y) \geq d(x,y)$.

6.8.9 Let (X,d) be a metric space and $(\overline{X}, \overline{d})$ its Cauchy completion, obtained as the set of equivalence classes of Cauchy sequences in X. Prove that the elements of $\overline{X}$ correspond bijectively with the pairs of adjoint distributors as in 6.8.8. [Hint: given $\overline{x} \in \overline{X}$, define $f\colon X \longrightarrow \overline{\mathbb{R}}_+$ by $f(x) = \overline{d}(\overline{x}, x)$ and put $g = f$; given an adjoint pair f, g choose

$a_n \in X$ such that $f(a_n) + g(a_n) < \frac{1}{n} (n \in \mathbb{N}^*)$ and show this yields a Cauchy sequence.]

6.8.10 Conclude that a metric space (X, d) is complete in the classical sense (i.e. every Cauchy sequence has a limit) iff every $\overline{\mathbb{R}}_+$-distributor $\mathbf{1} \rightarrowtriangle (X, d)$ which has a right $\overline{\mathbb{R}}_+$-adjoint is induced by an $\overline{\mathbb{R}}_+$-functor $\mathbf{1} \longrightarrow (X, d)$; see 7.9.3, volume 1.

7
Topological categories

Categorical methods have proved to be particularly useful when studying various questions in algebraic topology: today, most books on the subject contain a crash course in category theory, which turns out to provide a fruitful setting for handling the required structures. In this topic, the notions of abelian category and exact sequence play a key role.

This chapter is mainly concerned with the description of good categorical settings for developing general topology. And if this deserves a chapter in a book, this is clearly because the most obvious category one could think of – the category of topological spaces and continuous mappings – does not have rich categorical properties (like being regular, monadic, cartesian closed, a topos, ...) which would have made applicable the results of some other chapters of this book.

In topology, one is mainly concerned with the problem of convergence and in particular problems of convergence in spaces of continuous functions. One is particularly interested in situations where "if a sequence of continuous functions converges, the limit is again continuous". In fact, this requires a notion of convergence in the set of all functions (not necessarily continuous)... to express finally the continuity of the limit. Such a "good" situation is thus obtained when the set of continuous functions is closed in the set of all functions, for the corresponding topology inducing the notion of convergence. For example, in calculus, when studying the functions from $\mathbb{R}^n$ to $\mathbb{R}^m$, one puts a special emphasis on uniform convergence on compact subsets ... and proves that if a sequence of continuous functions converges in this sense, the limit is again continuous. This is a striking difference from pointwise convergence, for which a limit of continuous functions is in general not continuous. The previous example uses explicitly the notion of "uniform structure", but topological substitutes exist for defining, for continuous functions between topological spaces, the notion of "uniform convergence on the compact subsets" (via the compact–open topology) or the notion of pointwise convergence (via the pointwise topology).

In categorical terms, one is thus interested in topologizing in a nice way the spaces $\mathcal{C}(X,Y)$ of continuous functions, thus finally in finding a good symmetric monoidal closed structure on the category Top of topological spaces. A (unique) such structure exists: the corresponding notion of convergence on the spaces $\mathcal{C}(X,Y)$ of continuous functions is pointwise convergence... which, unfortunately, is not topologically very interesting. Finally, the very bad thing is the fact that Top is not cartesian closed: indeed, the categorical product in Top is the usual topological product, thus a very good one from the point of view of topology; therefore topologizing $\mathcal{C}(X,Y)$ in such a way to get a right adjoint to the cartesian product would yield an ideal situation! But this is not always possible. In fact it is possible when X is locally compact, in which case the corresponding topology on $\mathcal{C}(X,Y)$ is the expected one, namely the compact–open topology.

So Top is not cartesian closed; we have to live with this! Can one find a good "cartesian closed approximation" of Top? For example by dropping some "bad" topological spaces ... or by adding some new objects or arrows to create the function spaces Y^X which do not exist! In the Hausdorff case, we present an example of each type: the subcategory of compactly generated spaces and the subcategory of compactly continuous mappings ... and prove that these two categories are in fact equivalent!

Finally we define those functors, called topological, which satisfy axiomatically the basic properties of the forgetful functor Top$\longrightarrow$Set. The key notion consists in axiomatizing categorically the notion of "initial topology on a set for a given family of mappings to topological spaces".

7.1 Exponentiable spaces

As indicated in the introduction, one of the first questions asks if the category Top of topological spaces and continuous mappings admits the structure of a symmetric monoidal closed category (see chapter 6).

Proposition 7.1.1 *Suppose the structure of a symmetric monoidal closed category is given on* Top. *For two spaces X, Y:*

(1) the set underlying the tensor product $X \otimes Y$ is necessarily the cartesian product $X \times Y$ of the underlying sets;

$$\begin{array}{ccc} \mathcal{C}(\mathbf{1},[\mathbf{1},X]) & \xrightarrow{\mathcal{C}(1_\mathbf{1},[j,\mathbf{1}])} & \mathcal{C}(\mathbf{1},[I,X]) \\ \cong\downarrow & & \downarrow\cong \\ \mathcal{C}(\mathbf{1}\otimes\mathbf{1},X) & \xrightarrow[\mathcal{C}(1_\mathbf{1}\otimes j,\mathbf{1})]{} & \mathcal{C}(\mathbf{1}\otimes I,X) \end{array}$$

Diagram 7.1

(2) *the set underlying the internal function space* $[X,Y]$ *is necessarily the set* $\mathcal{C}(X,Y)$ *of continuous mappings from* X *to* Y;

(3) *the unit space* I *is necessarily the singleton.*

Proof First of all, observe that the space I cannot be empty. Indeed since $\emptyset$ is the initial object of Top, it is preserved by every functor $-\otimes X$, because such a functor admits $[X,-]$ as a right adjoint (see 3.2.2, volume 1). Therefore $I=\emptyset$ would imply $X\cong I\otimes X\cong\emptyset\otimes X\cong\emptyset$ for every space X.

Since I is not empty, fix an arbitrary (obviously continuous) mapping $i\colon \mathbf{1}\longrightarrow I$ from the singleton $\mathbf{1}$ to I; considering the unique (obviously continuous) mapping $j\colon I\longrightarrow\mathbf{1}$, one has $j\circ i=1_\mathbf{1}$ and thus $\mathbf{1}$ is presented as a retract of I. Every functor, thus in particular $\mathbf{1}\otimes -$, preserves retracts, thus $\mathbf{1}\otimes\mathbf{1}$ is a retract of $\mathbf{1}\otimes I\cong\mathbf{1}$. Therefore $\mathbf{1}\otimes\mathbf{1}$ is the singleton or the empty set.

But $\mathbf{1}\otimes\mathbf{1}$ cannot be empty. Indeed, $\mathbf{1}\otimes\mathbf{1}=\emptyset$ would imply

$$\mathcal{C}(\emptyset,\emptyset)\cong\mathcal{C}(\mathbf{1}\otimes\mathbf{1},\emptyset)\cong\mathcal{C}(\mathbf{1},[\mathbf{1},\emptyset]).$$

Thus $\mathbf{1}_\emptyset$ would correspond with a mapping $k\colon \mathbf{1}\longrightarrow[\mathbf{1},\emptyset]$ and the composite $k\circ j\colon I\longrightarrow[\mathbf{1},\emptyset]$ with a continuous mapping $\mathbf{1}\cong I\otimes\mathbf{1}\longrightarrow\emptyset$ (see 6.1.7), which is a contradiction. So $\mathbf{1}\otimes\mathbf{1}$ is the singleton.

Thus $1_\mathbf{1}\otimes j\colon \mathbf{1}\otimes I\longrightarrow\mathbf{1}\otimes\mathbf{1}$ is a mapping from the singleton to itself, so is the identity. The commutativity of diagram 7.1 indicates that the upper horizontal arrow is a bijection, hence the continuous mapping

$$[j,1_X]\colon [\mathbf{1},X]\longrightarrow[I,X]$$

is bijective for every X. Since this mapping has a continuous retraction, namely $[i,1_X]$, we know $[j,1_X]$ and $[i,1_X]$ are inverse homeomorphisms. This implies immediately that i,j are themselves inverse homeomorphisms: just put $X=I$ and apply 6.1.7. So we have proved that I is just the singleton.

Condition (2) of the statement follows immediately. For two spaces X, Y one has bijections

$$\mathcal{C}(X,Y) \cong \mathcal{C}(\mathbf{1} \otimes X, Y) \cong \mathscr{C}(\mathbf{1}, [X,Y]),$$

proving that the continuous mappings in $\mathcal{C}(X,Y)$ are indeed in bijection with the elements of $[X,Y]$.

It remains to prove condition (1) of the statement. Consider again two spaces X, Y and write X_d for the set X provided with the discrete topology. Notice that in Top, X_d is just the copower $X_d = \coprod_{x \in X} \mathbf{1}$. The identity $1_X \colon X_d \longrightarrow X$ is an epimorphism in Top. The functor $- \otimes Y \colon \mathsf{Top} \longrightarrow \mathsf{Top}$ preserves colimits, because it has a right adjoint $[Y, -]$; see 3.2.2, volume 1. In particular it preserves coproducts and epimorphisms (see 2.9.3, volume 1). One has therefore

$$X_d \otimes Y \cong \Big(\coprod_{x \in X} \mathbf{1}\Big) \otimes Y \cong \coprod_{x \in X} (\mathbf{1} \otimes Y) \cong \coprod_{x \in X} Y \cong X_d \times Y.$$

This yields an epimorphism

$$X_d \times Y \cong X_d \otimes Y \xrightarrow{1_X \otimes 1_Y} X \otimes Y.$$

On the other hand considering the morphisms $\alpha \colon X \longrightarrow \mathbf{1}$, $\beta \colon Y \longrightarrow \mathbf{1}$ one constructs

$$X \otimes Y \xrightarrow{1_X \otimes \beta} X \otimes \mathbf{1} \cong X, \quad X \otimes Y \xrightarrow{\alpha \otimes 1_Y} \mathbf{1} \otimes Y \cong Y,$$

from which one gets a corresponding factorization

$$\begin{pmatrix} 1_X \otimes \beta \\ \alpha \otimes 1_Y \end{pmatrix} \colon X \otimes Y \longrightarrow X \times Y.$$

Let us prove that the composite

$$X_d \times Y \xrightarrow{1_X \otimes 1_Y} X \otimes Y \xrightarrow{\begin{pmatrix} 1_X \otimes \beta \\ \alpha \otimes 1_Y \end{pmatrix}} X \times Y$$

is the identity mapping on $X \times Y$; this will prove that the surjection $1_X \otimes 1_Y$ is also injective and thus bijective. Indeed the composite

$$X_d \times Y \cong X_d \otimes Y \xrightarrow{1_X \otimes 1_Y} X \otimes Y \xrightarrow{1_X \otimes \beta} X \otimes \mathbf{1} \cong X$$

has by definition the form

$$\coprod_{x \in X} Y \longrightarrow X \otimes Y \xrightarrow{1_X \otimes \beta} X \otimes \mathbf{1} \cong X.$$

Its composite with the injection s_X of the coproduct is thus

$$Y \cong \mathbf{1} \otimes Y \xrightarrow{\quad x \otimes 1_Y \quad} X \otimes Y \xrightarrow{\quad 1_X \otimes \beta \quad} X \otimes \mathbf{1} \cong X$$

or equivalently

$$Y \cong \mathbf{1} \otimes Y \xrightarrow{\quad 1_{\mathbf{1}} \otimes \beta \quad} \mathbf{1} \otimes \mathbf{1} \xrightarrow{\quad x \otimes 1_{\mathbf{1}} \quad} X \otimes \mathbf{1} \cong X$$

which is just the constant mapping on x since $\mathbf{1} \otimes \mathbf{1} \cong \mathbf{1}$. An analogous argument holds for the factor Y. □

So, by 7.1.1, a tensor product of two spaces X, Y must be the cartesian product $X \times Y$ of the underlying sets, provided with a convenient topology. The most convenient one is clearly the product topology; being allowed to choose it as a tensor product reduces to proving that Top is cartesian closed (see 6.1.5), which unfortunately ... is false!

Proposition 7.1.2 *The category of topological spaces is not cartesian closed.*

Proof Let us assume that the category Top of topological spaces and continuous mappings is cartesian closed. We shall deduce a contradiction. By 7.1.1 we know that for two topological spaces Y, Z, the exponentiation Z^Y is the set $\mathcal{C}(Y, Z)$ of continuous functions provided with some topology. For this topology, the evaluation map

$$\mathsf{ev}_{YZ}\colon Z^Y \times Y \longrightarrow Z, \quad (f, y) \mapsto f(y)$$

is continuous since it corresponds, by adjunction, with the identity on Z^Y. Let us observe moreover that given a topology $\mathcal{T}$ on $\mathcal{C}(Y, Z)$ such that the evaluation map

$$\mathsf{ev}_{YZ}\colon \big(\mathcal{C}(Y, Z), \mathcal{T}\big) \times Y \longrightarrow Z$$

is continuous, $\mathcal{T}$ is finer than the topology of Z^Y: indeed, by adjunction, the identity map $\mathcal{C}(Y, Z) = Z^Y$ must be continuous. So the topology of Z^Y is the coarsest one making the evaluation map ev_{YZ} continuous.

In particular, let us choose for Z the unit interval $[0, 1]$ of the real line and for Y the space $\mathbb{Q}$ of rational numbers, with the topology induced by that of the real line $\mathbb{R}$. The real line is a completely regular space, i.e. two disjoint closed subsets can be separated by a continuous function to the unit interval. This property is obviously inherited by every subspace, in particular by $\mathbb{Q}$. We have already proved that $[0, 1]^{\mathbb{Q}}$ is the set $\mathcal{C}\big(\mathbb{Q}, [0, 1]\big)$

of continuous functions, provided with the coarsest topology making the evaluation map

$$\text{ev}\colon \mathcal{C}(\mathbb{Q},[0,1]) \times \mathbb{Q} \longrightarrow [0,1], \quad (f,q) \mapsto f(q)$$

continuous. Let us deduce from this that $\mathbb{Q}$ is locally compact, which will be a contradiction.

It is a matter of fact that $\mathbb{Q}$ is not locally compact. Observe indeed that if $r \in \mathbb{R}\setminus\mathbb{Q}$ is an irrational number, the closed interval $[-r,+r]$ of $\mathbb{R}$ determines by intersection with $\mathbb{Q}$ a closed neighbourhood of 0 in $\mathbb{Q}$, which we still write $[-r,+r]$. One has obviously $[-r,+r] = \bigcup_q]-q,+q[$ in $\mathbb{Q}$, where q runs through the rational numbers $0 < q < r$. Since no finite subunion covers $[-r,+r]$, $[-r,+r]$ is not compact in $\mathbb{Q}$. But the local compactness of $\mathbb{Q}$ would imply the existence of a compact neighbourhood V of 0 in $\mathbb{Q}$. As a neighbourhood of 0, V contains a neighbourhood of the form $]-q,+q[$ with $q \in \mathbb{Q}$; choosing $0 < r < q$ with r irrational, the closed subset $[-r,+r]$ of $\mathbb{Q}$ is contained in the compact subset V, thus is itself compact, which is a contradiction.

It remains thus to deduce the local compactness of $\mathbb{Q}$ from the assumption of cartesian closedness of Top. We consider for this the continous function $\delta_0\colon \mathbb{Q} \longrightarrow [0,1]$ given by the constant mapping on 0. For every rational number $q \in \mathbb{Q}$, $\delta_0(q) \in [0,\frac{1}{2}[$ and $[0,\frac{1}{2}[$ is open in $[0,1]$. By continuity of the evaluation map, we can find $U_q \subseteq [0,1]^{\mathbb{Q}}$, $V_q \subseteq \mathbb{Q}$, open neighbourhoods of δ_0 and q such that $\text{ev}(U_q \times V_q) \subseteq [0,\frac{1}{2}[$. The closure $\overline{V_q}$ of V_q is thus a neighbourhood of q and it suffices to prove it is compact.

Let us consider an open covering $\bigcup_{k\in K} W_k$ of $\overline{V_q}$. Putting $W_0 = \complement\overline{V_q}$ (the complement of the closure of V_q), we get an open covering $\bigcup_{l\in L} W_l$ of $\mathbb{Q}$, where $L = K \cup \{0\}$. We shall construct from this a topology $\mathcal{S}$ on $\mathcal{C}(\mathbb{Q},[0,1])$ making the evaluation map continuous. We take as fundamental open subsets all the subsets

$$[C,J] = \left\{ g \in \mathcal{C}(\mathbb{Q},[0,1]) \,\middle|\, g(C) \subseteq J \right\}$$

where C runs through the closed subsets of $\mathbb{Q}$ contained in some W_l, $l \in L$, while J runs through the open subsets of $[0,1]$. To prove that $\mathcal{S}$ makes the evaluation map continuous, consider $g \in \mathcal{C}(\mathbb{Q},[0,1])$, $p \in \mathbb{Q}$ and J an open neighbourhood of $g(p)$ in $[0,1]$. $p \in W_{l_0}$ for some $l_0 \in L$ and $g^{-1}(J)$ is another open neighbourhood of p, since g is continuous. By regularity of $\mathbb{Q}$, there is an open neighbourhood W of p whose closure is contained in $W_{l_0} \cap g^{-1}(J)$

$$p \in W \subseteq \overline{W} \subseteq W_{l_0} \cap g^{-1}(J).$$

This implies in particular $g(\overline{W}) \subseteq J$, thus $g \in [\overline{W}, J]$. It remains to observe that

$$\mathsf{ev}\Big([\overline{W}, J] \times W\Big) \subseteq J$$

which is just obvious.

Let us write $\mathcal{T}$ for the topology of $[0,1]^{\mathbb{Q}}$. It is the coarsest topology making the evaluation map continuous. Since $\mathcal{S}$ is another topology of this kind, $\mathcal{S}$ is finer than $\mathcal{T}$. The $\mathcal{T}$-neighbourhood U_q of δ_0 in $\mathcal{C}(\mathbb{Q}, [0,1])$ thus contains an $\mathcal{S}$-neighbourhood of δ_0. This means that it is possible to find indices $l_1, \ldots, l_n \in L$, closed subsets $C_i \subseteq W_{l_i}$ in $\mathbb{Q}$ and open subsets J_i in $[0,1]$ such that

$$\delta_0 \in [C_1, W_{l_1}] \cap \ldots \cap [C_n, W_{l_n}] \subseteq U_q.$$

Let us prove that $V_q \subseteq C_1 \cup \ldots \cup C_n$.

If $V_q \not\subseteq C_1 \cup \ldots \cup C_n$, fix a point $p \in V_q$, $p \notin C_1 \cup \ldots \cup C_n$. Then $\{p\}$ and $C_1 \cup \ldots \cup C_n$ are disjoint closed subsets in $\mathbb{Q}$; by complete regularity, we can find a continuous function $f\colon \mathbb{Q} \longrightarrow [0,1]$ such that $f(C_1 \cup \ldots \cup C_n) = 0$, $f(p) = 1$. In particular f coincides with δ_0 on each C_i, thus $\delta_0(C_i) = f(C_i) \subseteq W_{l_i}$. This proves $f \in [C_i, W_{l_i}]$ for each $i = 1, \ldots, n$, thus $f \in U_q$. But this is impossible since by definition of U_q, V_q, $f \in U_q$ and $p \in V_q$ imply $0 \leq f(p) < \frac{1}{2}$. Therefore we must have $V_q \subseteq C_1 \cup \ldots \cup C_n$. Since the C_i's are closed, this yields

$$V_q \subseteq \overline{V_q} \subseteq C_1 \cup \ldots \cup C_n \subseteq W_{l_1} \cup \ldots \cup W_{l_n}.$$

Observe finally that if $W_0 = \complement\overline{V_q}$ appears in the right-hand term, dropping it from the union will still produce a finite covering of $\overline{V_q}$, proving finally the compactness of $\overline{V_q}$. □

Since the category of topological spaces is not cartesian closed, it is sensible to make the following definition.

Definition 7.1.3 *A topological space X is exponentiable when the "product functor"*

$$- \times X\colon \mathsf{Top} \longrightarrow \mathsf{Top}$$

has a right adjoint.

We shall not give an explicit characterization of exponentiable spaces, since the corresponding condition is rather technical. But we shall prove an interesting sufficient condition for exponentiability.

Let us make clear that our topological spaces are not required to be Hausdorff. In particular the notion of compactness does not require the Hausdorff axiom. Now in the non-Hausdorff case, the notion of local compactness should be made precise, since various definitions appear in the literature, which are equivalent just in the Hausdorff case.

Definition 7.1.4 *A topological space X is locally compact when every neighbourhood of a point contains a compact neighbourhood of this point.*

Proposition 7.1.5 *A locally compact space is exponentiable. The right adjoint to the functor $-\times X$ is given by the functor mapping a space Y to the set $C(X,Y)$ of continuous functions, topologized with the compact–open topology.*

Proof We recall that the compact open topology on $C(X,Y)$ is that determined by the fundamental open subsets

$$\langle K,U\rangle = \{f\in C(X,Y) \mid f(K)\subseteq U\}$$

where K runs through the compact subsets of X and U runs through the open subsets of Y (our notion of compactness does not require Hausdorffness). Let us write $[X,Y]$ for the set $C(X,Y)$ topologized in this way. It is easy to observe that this construction is functorial in Y. Indeed given a continuous mapping $h\colon Y\longrightarrow Z$, we must observe that composition with h,

$$[X,h]\colon [X,Y]\longrightarrow [X,Z],$$

is continuous. If $K\subseteq X$ is compact and $V\subseteq Z$ is open,

$$\begin{aligned}[K,h]^{-1}\langle K,V\rangle &= \{f\in C(X,Y) \mid hf(K)\subseteq V\}\\ &= \{f\in C(X,Y) \mid f(K)\subseteq h^{-1}(V)\}\\ &= \langle K,h^{-1}(V)\rangle\end{aligned}$$

is open since h is continuous.

Let us consider a continuous mapping $g\colon Z\times X\longrightarrow Y$ and the corresponding function

$$\gamma\colon Z\longrightarrow [X,Y],\quad z\mapsto g(z,-).$$

To prove the continuity of γ, let us choose K compact in X and U compact in Y; then

$$\gamma^{-1}\langle K,U\rangle = \{z\in Z \mid \forall x\in K\ \ g(z,x)\in U\}.$$

Fix a point $z\in\gamma^{-1}\langle K,U\rangle$; we shall construct an open neighbourhood W of z contained in $\gamma^{-1}\langle K,U\rangle$; this will prove that $\gamma^{-1}\langle K,U\rangle$ is open and thus γ is continuous. For every $x\in K$, $g^{-1}(U)$ is an open neighbourhood of (z,x), thus contains an open neighbourhood of the form $W_x \times V_x$. The open subsets V_x cover the compact subset K, thus K is already covered by finitely many of them, let us say $V_{x_1}, \ldots, V_{x_n}$. Putting $W = W_{x_1} \cap \ldots \cap W_{x_n}$ we get an open neighbourhood of z. Choose $z' \in W$. For every $x \in K$ there is an index i such that $x \in V_{x_i}$; on the other hand $z' \in W_{x_i}$, thus $g(z',x) \in U$. This proves that W is contained in $\gamma^{-1}\langle K, U\rangle$.

Conversely consider a continuous mapping $f\colon Z \longrightarrow [X,Y]$ and the corresponding function

$$\varphi\colon Z \times X \longrightarrow Y, \quad \varphi(z,x) = f(z)(x).$$

To prove the continuity of φ, consider elements $x \in X$, $y \in Y$, $z \in Z$ such that $\varphi(z,x) = y$ and an open neighbourhood U of Y; then

$$\varphi^{-1}(U) = \{(z',x') \in Z \times X \,|\, f(z')(x') \in U\}.$$

Since $f(z)$ is continuous, $f(z)^{-1}(U)$ is a neighbourhood of x. By assumption on X, we can choose a compact neighbourhood $K \subseteq f(z)^{-1}(U)$ of x. Since f is continuous

$$f^{-1}\langle K, U\rangle = \{z' \in Z \,|\, \forall x' \in K \;\; f(z')(x') \in U\}$$

is open in Z. One obviously has

$$(z,x) \in f^{-1}\langle K, U\rangle \times K \subseteq \varphi(U),$$

proving that $\varphi^{-1}(U)$ is a neighbourhood of (z,x). Thus φ is continuous.

It is obvious that the previous data define natural inverse bijections. □

Let us conclude this section with describing a symmetric monoidal closed structure on Top. As proved in **Pedicchio and Solimini**, this is in fact the unique symmetric monoidal closed structure existing on Top.

Proposition 7.1.6 *There exists a symmetric monoidal closed structure on* Top *for which the internal hom-functor $[X,Y]$ is given by the space $\mathcal{C}(X,Y)$ of continuous functions provided with the pointwise topology.*

Proof We recall that the pointwise topology is obtained by choosing as fundamental open subsets

$$\langle x, U\rangle = \{f \in \mathcal{C}(X,Y) \,|\, f(x) \in U\}$$

where x runs through the points of X and U runs through the open subsets of Y. Let us write $[X,Y]$ for the set $\mathcal{C}(X,Y)$ topologized in this way. It is easy to observe that this construction is functorial in Y. Given a continuous mapping $h\colon Y \longrightarrow Z$, we must prove that composition with h is continuous:

$$[X,h]\colon [X,Y] \longrightarrow [X,Z].$$

Given $x \in K$ and V open in Z,

$$\begin{aligned}[X,h]^{-1}\langle x,V\rangle &= \{f \in \mathcal{C}(X,Y) \mid hf(x) \in V\}\\ &= \{f \in \mathcal{C}(X,Y) \mid f(x) \in h^{-1}(V)\}\\ &= \langle x, h^{-1}(V)\rangle\end{aligned}$$

is open since h is continuous.

To discover the form of the tensor product topology, consider a continuous function $f\colon Z \longrightarrow [X,Y]$ and the corresponding mapping

$$\varphi\colon Z \times X \longrightarrow Y, \quad (z,x) \mapsto f(z)(x),$$

where $Z \times X$ is the topological product of Z, X in Top. Since f is continuous, fixing $z \in Z$ yields

$$\varphi(z,-)\colon X \longrightarrow Y, \quad x \mapsto f(z)(x),$$

which is the continuous function $f(z)$. Fixing $x \in X$ yields

$$\varphi(-,x)\colon Z \longrightarrow Y, \quad z \mapsto f(z)(x).$$

Given an open subset $U \subseteq Y$

$$\begin{aligned}\varphi(-,x)^{-1}(U) &= \{z \in Z \mid f(z)(x) \in U\}\\ &= \{z \in Z \mid f(z) \in \langle x,U\rangle\}\\ &= f^{-1}\langle x,U\rangle,\end{aligned}$$

which is open in Z, since f is continuous. Thus when f is continuous, φ is continuous in each variable.

Conversely consider a function $\varphi\colon Z \times X \longrightarrow Y$ which is continuous in each variable. In particular, fixing $z \in Z$, the function

$$\varphi(z,-)\colon X \in Y, \quad x \mapsto \varphi(z,x)$$

is continuous, yielding a mapping

$$f\colon Z \longrightarrow [X,Y], \quad z \mapsto \varphi(z,-).$$

This mapping f is itself continuous since, given $x \in X$ and U open in Y,

$$\begin{aligned}f^{-1}\langle x,U\rangle &= \{z \in Z \mid \varphi(z,x) \in U\}\\ &= \varphi(-,x)^{-1}(U),\end{aligned}$$

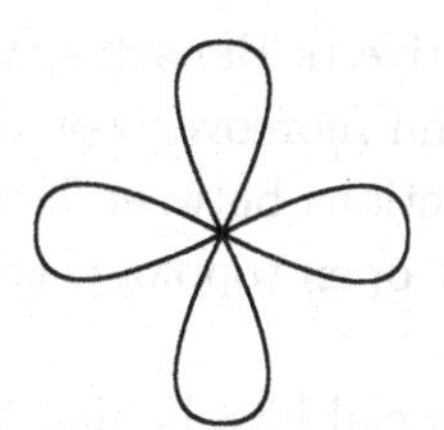

Figure 7.2

which is open in Z since $\varphi(-,x)$ is continuous.

So we have already constructed bijections

$$\mathcal{SC}(Z \times X, Y) \cong \mathcal{C}(Z, [X, Y])$$

where $\mathcal{SC}$ stands for "separately continuous", i.e. continuous in each variable. These bijections are obviously natural and it remains to find natural bijections

$$\mathcal{C}(Z \otimes X, Y) \cong \mathcal{SC}(Z \times X, Y).$$

We know that $Z \otimes X$ must be the product set of Z, X (see 7.1.1). Thus the problem is to find a topology $Z \otimes X$ on the product set $Z \times X$ such that being continuous on $Z \otimes X$ is equivalent to being separately continuous on the topological product. Since a function $f\colon Z \times X \longrightarrow Y$ is separately continuous precisely when the composites

$$Z \xrightarrow{(-,x)} Z \times X \xrightarrow{\quad f \quad} Y, \quad X \xrightarrow{(z,-)} Z \times X \xrightarrow{\quad f \quad} Y$$

are continuous for all $x \in X$, $z \in Z$, the topology of $Z \otimes X$ is just the final topology for all the mappings

$$Z \xrightarrow{(-,x)} Z \times X, \quad X \xrightarrow{(z,-)} Z \times X. \qquad \square$$

The topology of $Z \otimes X$ is sometimes referred to as the topology of "asterisks". In $\mathbb{R}^2$, figure 7.2 is an asterisk, where the boundary is not in U, except the central point: all "horizontal" and "vertical" sections of this asterisk are open subsets of the corresponding real lines. This is thus a neighbourhood of the central point for the topology of asterisks.

7.2 Compactly generated spaces

Proposition 7.1.5 could suggest one restricts one's attention to locally compact spaces in order to find a cartesian closed subcategory of **Top**.

This would be very restrictive (a Banach space is locally compact when it is finite dimensional!) and moreover would not work, since the space $\mathcal{C}(X,Y)$ of continuous functions between locally compact spaces X, Y, provided with the compact open topology, is in general not locally compact.

Let us recall that the crucial idea behind the compact open topology is to describe a topology nicely related with the notion of "uniform convergence on compact subsets". This suggests we weaken the notion of continuity in the following way:

Definition 7.2.1 *Let X, Y be topological spaces. An arbitrary mapping $f: X \longrightarrow Y$ is compactly continuous when its restriction to each compact subset of X is continuous.*

This notion of "compact continuity" will turn out to be stable under a limit process which is uniform on the compact subsets.

Obviously, one has

Proposition 7.2.2

(1) The composite of two compactly continuous functions is compactly continuous.

(2) Every continuous function is compactly continuous.

(3) If X is a locally compact space, a function $f: X \longrightarrow Y$ is compactly continuous if and only if it is continuous.

Proof If $f: X \longrightarrow Y$, $g: Y \longrightarrow Z$ are compactly continuous, for every compact subset $K \subseteq X$ the continuity of $f: K \longrightarrow Y$ implies that $f(K)$ is compact in Y. Since g is continuous on $f(K)$ by assumption, $f \circ g$ is continuous on K.

The second statement is obvious. The third statement reduces to the fact that a function is continuous as long as it is continuous on a neighbourhood of every point. □

One of the essential features of the notion of continuity is its local character: a function is continuous if and only if it is continuous in the neighbourhood of every point. The notion of compact continuity does not share this property ... except if we assume that each compact subset is also locally compact. This property is automatic when the space is Hausdorff. Moreover when studying problems of convergence on compact subsets, it is convenient to assume that the compact subsets are closed: again this is automatic for Hausdorff spaces. For this reason, in this section, we shall restrict our attention to the case of Hausdorff spaces.

It is probably useful to underline the fact that for non-Hausdorff spaces, it does not suffice to replace "compact" by "compact locally compact", or "compact closed", or "compact closed locally compact" to get a direct generalization of the following results. Indeed, while compactness is preserved by continuous mappings – an important fact in what we shall do – closedness and local compactness are not.

Proposition 7.2.3 *The category $\mathcal{CC}$-Comp of compact Hausdorff spaces and compactly continuous mappings is cartesian closed.*

Proof The singleton is obviously a terminal object in $\mathcal{CC}$-Comp. Moreover if X, Y are Hausdorff spaces, their topological product $X \times Y$ is also their product in $\mathcal{CC}$-Comp. Indeed, given two compactly continuous mappings $f\colon Z \longrightarrow X$, $g\colon Z \longrightarrow Y$, the unique factorization

$$\begin{pmatrix} f \\ g \end{pmatrix} : Z \longrightarrow X \times Y$$

in Set is compactly continuous, since given a compact subset $K \subseteq Z$, the universal property of the topological product $X \times Y$ reduces the continuity of $\binom{f}{g}$ on K to the continuity of both f and g on K.

Now given two Hausdorff spaces Y, Z, let us write $\mathcal{CC}(Y, Z)$ for the set of compactly continuous functions from Y to Z provided with the topology whose fundamental open subsets are given by

$$[K, U] = \{f \in \mathcal{CC}(Y, Z) \mid f(K) \subseteq U\}$$

where K is compact in Y and U is open in Z (the "compact open topology" for compactly continuous mappings).

First of all, we must prove that the topology on $\mathcal{CC}(Y, Z)$ is Hausdorff. If $f, g\colon Y \rightrightarrows Z$ are distinct compactly continuous mappings, choose an element $y \in Y$ such that $f(y) \neq g(y)$. Since Z is Hausdorff, choose U, V disjoint open subsets such that $f(y) \in U$, $g(y) \in V$. The singleton $\{y\}$ is compact in Y and one has $f \in \big[\{y\}, U\big]$, $g \in \big[\{y\}, V\big]$. Moreover $\big[\{y\}, U\big]$ and $\big[\{y\}, V\big]$ are disjoint, just because U and V are disjoint.

Next, let us prove that the evaluation mapping

$$\text{ev}\colon \mathcal{CC}(Y, Z) \times Y \longrightarrow Z, \quad (f, y) \mapsto f(y)$$

is compactly continuous, i.e. is continuous on every compact subset P of the product. Since the product is the topological one, the projections map P to compact subsets $C \subseteq \mathcal{CC}(Y, Z)$ and $K \subseteq Y$. Since $P \subseteq C \times K$, it suffices to prove the continuity of the evaluation on $C \times K$. To do this choose $(f, y) \in C \times K$ and an open neighbourhood $U \subseteq Z$ of $f(y)$.

Since $f\colon K \longrightarrow Z$ is continuous, there exists a neighbourhood V of y in K such that $f(V) \subseteq U$. Since K is compact Hausdorff, it is locally compact and we can choose a compact neighbourhood $L \subseteq V$ of y in K; in particular $f(L) \subseteq U$. Now $[L,U] \cap C$ is a neighbourhood of f in C and L is a neighbourhood of y in K. Therefore $([L,U] \cap C) \times L$ is a neighbourhood of (f,y) in $C \times K$ and, by definition, is mapped in U by the evaluation mapping. This proves the continuity of the evaluation on $C \times K$ and thus on P.

Finally consider a Hausdorff space X and a compactly continuous mapping $f\colon X \times Y \longrightarrow Z$. We must still prove the compact continuity of the transposed mapping

$$g\colon X \longrightarrow \mathcal{CC}(Y,Z), \quad x \mapsto f(x,-),$$

that is, the unique mapping such that $\mathsf{ev} \circ (g \times 1_Y) = f$. First of all observe that the mapping g is well-defined, i.e. $f(x,-)$ is compactly continuous for each $x \in X$. Indeed, this is just the composite of the compactly continuous mapping f with the continuous mapping

$$Y \longrightarrow X \times Y, \quad y \mapsto (x,y).$$

To prove that g itself is compactly continuous, consider compact subsets $A \subseteq X$, $K \subseteq Y$ and an open subset $U \subseteq Z$. We must prove that g restricted to A is continuous, i.e.

$$g^{-1}([K,U]) \cap A = \{x \in A \mid \forall y \in K \;\; f(x,y) \in U\}$$

is open in A. Fix $x \in g^{-1}([K,U])$. For every $y \in K$, the continuity of f on the compact subset $A \times K$ implies the existence of two neighbourhoods V_y of x in A and W_y of y in K, such that $f(V_y \times W_y) \subseteq U$. Since K is compact, there exist finitely many elements $y_1, \ldots, y_n \in K$ such that $W_{y_1}, \ldots, W_{y_n}$ already cover K. But $V = V_{y_1} \cap \ldots \cap V_{y_n}$ is still a neighbourhood of x in A and

$$f(V \times K) = \bigcup_{i=1}^{n} f(V \times W_{y_i}) \subseteq \bigcup_{i=1}^{n} f(V_{y_i} \times W_{y_i}) \subseteq U.$$

This implies $x \in V \subseteq g^{-1}([K,U])$, thus $g^{-1}([K,U])$ is a neighbourhood of each of its points x. □

Via proposition 7.2.3, we know how to enlarge the category **Haus** of Hausdorff spaces and continuous mappings in order to get a cartesian closed category, with the exponentiation Z^Y given by the set of arrows from Y to Z, provided with the compact open topology. This is nice, but the reader can argue that he does not want to weaken the notion of continuity to that of compact continuity. Never mind, the category $\mathcal{CC}$-Comp is in fact equivalent to a full subcategory of the category Haus of Hausdorff spaces and continuous mappings!

Proposition 7.2.4 *The canonical inclusion* Haus $\subseteq \mathcal{CC}$-Comp *admits a left adjoint functor which is full and faithful. Thus* $\mathcal{CC}$-Comp *is equivalent to a full subcategory of* Haus.

Proof Given a Hausdorff space $(X, \mathcal{T})$, where $\mathcal{T}$ is the topology on the set X, let us first construct another topology $\mathcal{K}(\mathcal{T})$ on the same set X. A subset $U \subseteq X$ is $\mathcal{K}(\mathcal{T})$-open if and only if for every $\mathcal{T}$-compact subset $K \subseteq X$, $K \cap U$ is $\mathcal{T}$-open in K. Equivalently $C \subseteq X$ is $\mathcal{K}(\mathcal{T})$-closed if and only if for every $\mathcal{T}$-compact subset $K \subset X$, $K \cap C$ is $\mathcal{T}$-closed in K, i.e. $\mathcal{T}$-closed in X since K itself is $\mathcal{T}$-closed in X. Since intersecting with $K \subseteq X$ distributes over arbitrary set-theoretical intersections and unions, $\mathcal{K}(\mathcal{T})$ is a topology on X. By construction of $\mathcal{K}(\mathcal{T})$, every $\mathcal{T}$-open subset is $\mathcal{K}(\mathcal{T})$-open, thus $\mathcal{K}(\mathcal{T})$ is finer than $\mathcal{T}$. In particular $\mathcal{K}(\mathcal{T})$ is a Hausdorff topology and the identity mapping $1_X\colon (X, \mathcal{K}(\mathcal{T})) \longrightarrow (X, \mathcal{T})$ is continuous.

Next, let us prove that the two Hausdorff topologies $\mathcal{T}$ and $\mathcal{K}(\mathcal{T})$ have the same compact subsets and coincide on those compact subsets. Since $\mathcal{K}(\mathcal{T})$ is finer than $\mathcal{T}$, every $\mathcal{K}(\mathcal{T})$-compact subset is certainly $\mathcal{T}$-compact. Conversely if K is $\mathcal{T}$-compact and $\bigcap_{i\in I} C_i \subseteq K$, with each C_i $\mathcal{K}(\mathcal{T})$-closed, we have in fact $\bigcap_{i\in I}(C_i \cap K) \subseteq K$ with each $C_i \cap K$ $\mathcal{T}$-closed, by definition of $\mathcal{K}(\mathcal{T})$. Thus a finite sub-intersection is already contained in K, proving that K is $\mathcal{K}(\mathcal{T})$-compact. Since we know that $\mathcal{K}(\mathcal{T})$ is finer than $\mathcal{T}$, it remains to prove that given K compact and $C \subseteq K$ with C $\mathcal{K}(\mathcal{T})$-closed in K, C is also $\mathcal{T}$-closed in K. But since K is $\mathcal{K}(\mathcal{T})$-closed, C is $\mathcal{K}(\mathcal{T})$-closed in X thus $C = K \cap C$ is $\mathcal{T}$-closed in K, by definition of $\mathcal{K}(\mathcal{T})$. This proves that the identity mapping $1_X\colon (X, \mathcal{T}) \longrightarrow (X, \mathcal{K}(\mathcal{T}))$ is an isomorphism in $\mathcal{CC}$-Comp.

Now we verify the functoriality of the construction mapping a Hausdorff space $(X, \mathcal{T})$ to the corresponding space $\big(X, \mathcal{K}(\mathcal{T})\big)$. Given a compactly continuous mapping $f\colon (X, \mathcal{T}) \longrightarrow (Y, \mathcal{S})$ to another Hausdorff space $(Y, \mathcal{S})$, we must prove that $f\colon \big(X, \mathcal{K}(\mathcal{T})\big) \longrightarrow \big(Y, \mathcal{K}(\mathcal{S})\big)$ is continuous. Given $C \subseteq Y$ which is $\mathcal{K}(\mathcal{S})$-closed, we must prove that $f^{-1}(C)$ is $\mathcal{K}(\mathcal{T})$-closed, i.e. for every $\mathcal{T}$-compact subset $K \subseteq X$, $f^{-1}(C) \cap K$ is $\mathcal{T}$-closed. But since K is $\mathcal{T}$-compact, $f(K)$ is $\mathcal{S}$-compact and

$$f^{-1}(C) \cap K = f^{1}\big(C \cap f(K)\big) \cap K.$$

Since $f(K)$ is $\mathcal{S}$-compact and C is $\mathcal{K}(\mathcal{S})$-closed, $C \cap f(K)$ is $\mathcal{S}$-closed.

Thus $f^{-1}(C \cap f(K)) \cap K$ is $\mathcal{T}$-closed in K, thus in X, by continuity of f on K.

We have thus defined a functor $\mathcal{K}$: $\mathcal{CC}$-Comp $\longrightarrow$ Haus; on the other hand we write ι: Haus $\longrightarrow$ $\mathcal{CC}$-Comp for the canonical (non-full) inclusion. There are natural transformations ε: $\mathcal{K} \circ \iota \Rightarrow \text{id}$, η: $\text{id} \Rightarrow \iota \circ \mathcal{K}$ given by the identity mappings. The triangular identities of 3.1.5, volume 1, are then clearly satisfied, from which adjunction follows. Finally, $\mathcal{K}$ is full and faithful since η is an isomorphism, see 3.4.1, volume 1. □

$\mathcal{CC}$-Comp is thus equivalent to the full subcategory of Haus generated by those spaces $(X, \mathcal{T})$ for which $\mathcal{T} = \mathcal{K}(\mathcal{T})$.

Definition 7.2.5 *A Hausdorff topological space $(X, \mathcal{T})$ is compactly generated when a subset $C \subseteq X$ is closed as long as its intersection with every compact subset $K \subseteq X$ is closed. Such a space is also called a "Kelley space".*

Corollary 7.2.6 *The category of compactly generated spaces and continuous mappings is equivalent to the category of Hausdorff spaces and compactly continuous mappings. It is a cartesian closed, full coreflective subcategory of the category of Hausdorff spaces and continuous mappings.* □

Examples 7.2.7

7.2.7.a *Every locally compact Hausdorff space $(X, \mathcal{T})$ is compactly generated.*

Indeed if $U \subseteq X$ is $\mathcal{K}(\mathcal{T})$-open and $x \in U$, choose a compact $\mathcal{T}$-neighbourhood K of x. But $K \cap U$ is $\mathcal{T}$-open in K, thus a $\mathcal{T}$-neighbourhood of x in K. Since K itself is a $\mathcal{T}$-neighbourhood of x in X, $K \cap U$ and thus U are $\mathcal{T}$-neighbourhoods of x in X. So U is a $\mathcal{T}$-neighbourhood of each of its points, thus is $\mathcal{T}$-open.

7.2.7.b *A Hausdorff space which satisfies the first axiom of countability is compactly generated.*

So we assume that X is a Hausdorff space in which each point has a denumerable basis of neighbourhoods. We consider a subset $S \subseteq X$ which is not closed and construct a compact subset $K \subseteq X$ such that $S \cap K$ is not closed.

Since S is not closed, we choose $x \in \overline{S} \setminus S$. Consider a denumerable basis $(V_n)_{n \in \mathbb{N}}$ of neighbourhoods of x and, for every integer $n \in \mathbb{N}$, a point $x_n \in S \cap V_n$. Consider the subset K constituted of x and the various x_n, $n \in \mathbb{N}$. Since every open covering $(U_i)_{i \in I}$ of K is such that

$x \in U_{i_0}$, for some index i_0 we know K is compact. Since the V_n, $n \in \mathbb{N}$, constitute a basis of neighbourhoods of x, there exists n_0 such that $V_{n_0} \subseteq U_{i_0}$. So U_{i_0} contains x and all the elements x_n, $n \geq n_0$, and it remains to consider finitely many U_i containing the remaining elements $x_0, \ldots, x_{n-1}$. Thus K is indeed compact and $K \cap S$ is just the sequence of elements $(x_n)_{n\in\mathbb{N}}$, which is not closed since x is not in it ($x \notin S$) but belongs to its closure (every neighbourhood of x contains some V_n which contains x_n).

7.2.7.c *Every metric space is compactly generated.*

A metric space is Hausdorff, since two points x, y at a distance d are contained in the open balls $B(x, \frac{d}{3})$, $B(y, \frac{d}{3})$ which are disjoint. It satisfies also the first axiom of countability because the open balls $B(x, \frac{1}{n})$, $n \in \mathbb{N}^*$, constitute a basis of neighbourhoods of x. One gets the conclusion by 7.2.6.b.

Let us conclude this section with a warning. In the category $\mathcal{CC}$-Comp of Hausdorff spaces and compactly continuous mappings, the cartesian closed structure is given by the topological product and the compact–open topology. Going through the equivalence with the category $\mathcal{CG}$-Haus of compactly generated spaces, one gets the corresponding cartesian closed structure on $\mathcal{CG}$-Haus. The cartesian product in $\mathcal{CG}$-Haus is *a priori* no longer the usual topological product. Let us prove that it will be, under an assumption of local compactness.

Lemma 7.2.8 *For a Hausdorff space $(X, \mathcal{T})$, the following conditions are equivalent:*

(1) $(X, \mathcal{T})$ is compactly generated;

(2) $\mathcal{T}$ is the final topology for all the inclusions $K \hookrightarrow X$, where K runs through all the compact subspaces of $(X, \mathcal{T})$;

(3) $\mathcal{T}$ is the final topology for a family $(f_i\colon K_i \longrightarrow X)_{i\in I}$ of mappings, with each K_i a compact Hausdorff space;

(4) $\mathcal{T}$ is the final topology for the class of all continuous mappings $f\colon K \longrightarrow (X, \mathcal{T})$, where K runs through all the compact Hausdorff spaces.

Proof (1) $\Rightarrow$ (2) is just the definition of a compactly generated space. (2) $\Rightarrow$ (3) is obvious. (3) $\Rightarrow$ (4) is obvious since adding already continuous functions to a family does not change the finality of the original topology. To prove (4) $\Rightarrow$ (1), choose $U \subseteq X$ whose intersection with each compact subset $C \subseteq X$ is open in C. For each continuous mapping $f\colon K \longrightarrow X$ with K compact, $f(K)$ is a compact subset of X and

$$f^{-1}(U) = f^{-1}(U \cap f(K))$$

is therefore open in K. By the finality of the topology $\mathcal{T}$, U is open in X. □

Proposition 7.2.9 *Let Y be a locally compact Hausdorff space. In the cartesian closed category of compactly generated spaces, each cartesian product $X \times Y$ is the usual topological product.*

Proof Let X, Z be topological spaces with X compactly generated. We write $X \times Y$ for the topological product and Z^Y for the set of continuous mappings provided with the compact–open topology. By 7.1.5 a mapping $f\colon X \times Y \longrightarrow Z$ is continuous if and only if the corresponding mapping $\overline{f}\colon X \longrightarrow Z^Y$ is continuous. By 7.2.8, $\overline{f}$ is continuous when its restriction to each compact subset $K \subseteq X$ is continuous. Applying 7.1.5 again, we find that f is continuous if and only if its restriction to each subspace $K \times Y$ is continuous, where K runs through the compact subsets of X. Since each point of Y has a compact neighbourhood, the continuity of f on $X \times Y$ is still equivalent to its continuity on each subset $K \times K'$, with $K \subseteq X$ and $K' \subseteq Y$ compact subspaces. In other words the Hausdorff space $X \times Y$ has the final topology for all those inclusions $K \times K' \subseteq X \times Y$. Since each $K \times K'$ is compact, one concludes by 7.2.8.(3) that $X \times Y$ is compactly generated. □

7.3 Topological functors

One typical feature about topological spaces is the possibility of defining initial or final topologies. Given a family $(X_i, \mathcal{T}_i)_{i \in I}$ of topological spaces and given mappings $f_i\colon X \longrightarrow X_i$ from a set X, there exists a "best" topology on X making all the f_i's continuous: this is the "initial topology" for the f_i's (it is generated by all the $f_i(U)$, for all $i \in I$ and U open in X_i). In this example, we were dealing with a discrete diagram $(X_i, \mathcal{T}_i)_{i \in I}$ of topological spaces. Here is the general notion for an arbitrary diagram.

Definition 7.3.1 *Let $U\colon \mathscr{A} \longrightarrow \mathscr{B}$ be a functor. Consider*

(1) a category $\mathscr{D}$ and a functor $H\colon \mathscr{D} \longrightarrow \mathscr{A}$,

(2) a cone $\left(f_D\colon B \longrightarrow UH(D)\right)_{D \in \mathscr{D}}$ on $U \circ H$ in $\mathscr{B}$.

An initial structure for those data is a cone $\left(g_D\colon A \longrightarrow H(D)\right)_{D \in \mathscr{D}}$ on H such that:

(1) $U(A) = B$ and, for all $D \in \mathscr{D}$, $f_D = U(g_D)$;

(2) if $\big(h_D\colon A' \longrightarrow H(D)\big)_{D\in\mathscr{D}}$ is a cone on the functor H whose image $\big(U(h_D)\colon U(A') \longrightarrow UH(D)\big)_{D\in\mathscr{D}}$ under U factors via a morphism $b\colon U(A') \longrightarrow U(A)$ through $\big(f_D\colon U(A) \longrightarrow UH(D)\big)_{D\in\mathscr{D}}$, there exists a unique morphism $a\colon A' \longrightarrow A$ such that $U(a) = b$ and $h_D = g_D \circ a$ for every $D \in \mathscr{D}$.

As usual, the uniqueness of the arrow a implies the uniqueness up to isomorphism of the initial structure.

Definition 7.3.2 *Let $U\colon \mathscr{A} \longrightarrow \mathscr{B}$ be a functor.*

(1) Given a category $\mathscr{D}$, U has initial structures of shape $\mathscr{D}$ if for every functor $H\colon \mathscr{D} \longrightarrow \mathscr{B}$ and every cone on $U \circ H$, there exists a corresponding initial structure.

(2) The functor U has initial structures when it has initial structures of shape $\mathscr{D}$ for every small category $\mathscr{D}$.

(3) The functor U is topological when it has initial structures of shape $\mathscr{D}$ for every category $\mathscr{D}$.

Examples 7.3.3

7.3.3.a The forgetful functor $U\colon \mathsf{Top} \longrightarrow \mathsf{Set}$ is topological. With the notation of 7.3.1, it suffices to define A as the set B provided with the initial topology for all the mappings f_D: the topology generated by all the subsets $f_D^{-1}(V)$, with V open in $H(D)$.

7.3.3.b If **Cat** is the category of small categories and functors, the functor $U\colon \mathsf{Cat} \longrightarrow \mathsf{Set}$ mapping a small category $\mathscr{C}$ to its set of objects has initial structures. With the notation of 7.3.1, the set B is made a category with B as a set of objects by choosing $B(b, b')$ to be the projective limit

$$B(b, b') = \lim H(D)\big(f_D(b), f_D(b')\big).$$

7.3.3.c A functor $H\colon \mathscr{D} \longrightarrow \mathscr{A}$ has a limit precisely when there exists an initial structure along the functor $U\colon \mathscr{A} \longrightarrow \mathbf{1}$ (where $\mathbf{1}$ is the terminal category) for the unique possible cone on $U \circ H$.

The difference between topological functors and functors having initial structures is not easy to grasp intuitively: it is just a question of size of the diagrams involved in the problem. But we have already seen in 2.7.1, volume 1, that requiring a property for all diagrams (possibly large) can have drastic consequences. This is the case for topological functors:

Proposition 7.3.4 *Every topological functor is faithful.*

Proof Consider a topological functor $U: \mathscr{A} \longrightarrow \mathscr{B}$ and two morphisms $u, v: X \rightrightarrows Y$ in $\mathscr{A}$, such that $U(u) = U(v)$. Consider the discrete category $\mathscr{D}$ whose objects are all the arrows d of $\mathscr{A}$. Take $H: \mathscr{D} \longrightarrow \mathscr{A}$ to be the constant functor on Y and define a cone $\big(f_d: B \longrightarrow UH(d)\big)_{d\in\mathscr{D}}$ by putting $B = U(X)$ and $f_d = U(u)$. Write $\big(g_d: A \longrightarrow H(d)\big)_{d\in\mathscr{D}}$ for the corresponding initial structure. Define the cone $\big(h_d: X \longrightarrow H(d)\big)_{d\in\mathscr{D}}$ by

$$h_d = \begin{cases} u & \text{if } d \in \mathscr{A}(X, A) \text{ and } g_d \circ d = v, \\ v & \text{otherwise.} \end{cases}$$

Since $U(u) = U(v)$, the cone $\big(U(h_d)\big)_{d\in\mathscr{D}}$ coincides with the cone $(f_d)_{d\in\mathscr{D}}$, from which there is a unique morphism $a: X \longrightarrow A$ such that $U(a) = 1_{U(X)}$ and $g_d \circ a = h_d$. Putting $d = a$, we get in particular $g_a \circ a = h_a$. If $h_a \neq v$, by definition of h_a we get $h_a = v$, thus a contradiction. Thus $h_a = v$ and so by definition of h_a, $h_a = u$ and finally $u = v = h_a$. □

Since topological functors are faithful, the following lemma is often useful:

Lemma 7.3.5 *Let $F: \mathscr{A} \longrightarrow \mathscr{B}$ be a faithful functor, $H: \mathscr{D} \longrightarrow \mathscr{A}$ a diagram in $\mathscr{A}$ and $\big(f_D: B \longrightarrow FH(D)\big)_{D\in\mathscr{D}}$ a cone on $F \circ H$. There exists an initial structure for these data as long as such a structure exists for the corresponding discrete diagram.*

Proof Let us write $\mathcal{D}$ for the discrete category having the same objects as $\mathscr{D}$. Consider an initial structure $\big(g_D: A \longrightarrow H(D)\big)_{D\in\mathcal{D}}$ for the diagram $H: \mathcal{D} \longrightarrow \mathscr{A}$ and the cone $(f_D)_{D\in\mathscr{D}}$. For every morphism $d: D \longrightarrow D'$ of $\mathscr{D}$,

$$U\big(H(d) \circ g_D\big) = UH(d) \circ U(g_D) = UH(d) \circ f_D = f_{D'} = U(g_{D'}),$$

from which $H(d) \circ g_D = g_{D'}$ since U is faithful. This yields the result. □

Proposition 7.3.6 *Every topological functor is also cotopological.*

Proof By "cotopological", we mean clearly the notion dual to "topological".

A topological functor $U: \mathscr{A} \longrightarrow \mathscr{B}$ is faithful (see 7.3.4), thus by 7.3.5 it suffices to consider a discrete category $\mathscr{D}$, a functor $H: \mathscr{D} \longrightarrow \mathscr{A}$ and a cocone $\big(f_D: FH(D) \longrightarrow B\big)_{D\in\mathscr{D}}$. Let us consider the discrete category $\mathscr{D}'$ whose objects are the pairs (X, b), with $X \in \mathscr{A}$ and $b: B \longrightarrow U(X)$

an arrow such that the composites $b \circ f_D$ have the form $U(a_{D,b})$, with $a_{D,b}\colon H(D) \longrightarrow X$ in $\mathscr{A}$, for each object $D \in \mathscr{D}$. We define a functor $H'\colon \mathscr{D}' \longrightarrow \mathscr{A}$ by putting $H'(X,b) = X$ and a cone $f'_{(X,b)}\colon B \longrightarrow F(X)$ on $F \circ H'$ by putting $f'_{(X,b)} = b$. By assumption, we find an initial structure $\big(g'_{(X,b)}\colon A \longrightarrow X\big)_{(X,B)\in\mathscr{D}'}$. Fixing $D \in \mathscr{D}$, the cone $(a_{D,b})_{(X,b)\in\mathscr{D}'}$ is such that $U(a_{D,b})$ factors through $f'_{(X,b)} = b$ via f_D; therefore we get a unique $g_D\colon H(D) \longrightarrow A$ such that $U(g_D) = f_D$ and $g'_{(X,b)} \circ g_D = a_{D,b}$. This yields the required cocone $\big(g_D\colon H(D) \longrightarrow A\big)_{D\in\mathscr{D}}$ mapped by U to $\big(f_D\colon FH(D) \longrightarrow B\big)_{D\in\mathscr{D}}$.

To prove the universal property of $(g_D)_{D\in\mathscr{D}}$, we must consider a cocone $\big(h_D\colon H(D) \longrightarrow A'\big)_{D\in\mathscr{D}}$ in $\mathscr{A}$ and a morphism $b\colon F(A) \longrightarrow F(A')$ such that $b \circ U(g_D) = U(h_D)$. The pair (A', b) is an object of $\mathscr{D}'$, thus putting $a = g'_{(A',b)}$ we get a morphism $a\colon A \longrightarrow A'$ such that $U(a) = b$ and $a \circ g_D = a_{D,b}$. From the relations

$$U(a \circ g_D) = U(a) \circ U(g_D) = b \circ U(g_D) = U(h_D)$$

and the faithfulness of U, we get $a \circ g_D = h_D$. The uniqueness of a follows again from the faithfulness of U and the relation $U(a) = b$. □

Proposition 7.3.7 *A topological functor has both a right and a left adjoint and these adjoint functors are full and faithful.*

Proof By 7.3.6, it suffices to prove the existence of a full and faithful right adjoint to a topological functor $U\colon \mathscr{A} \longrightarrow \mathscr{B}$. Given an object $B \in \mathscr{B}$, we consider the empty diagram $\mathscr{D}$ (notation of 7.3.1) and the corresponding initial structures for the empty cone of vertex B. This is an object $A \in \mathscr{A}$ such that $U(A) = B$ and for every $A' \in \mathscr{A}$ and $b\colon U(A') \longrightarrow B$, there exists a unique arrow $a\colon A' \longrightarrow A$ such that $U(a) = b$. This is precisely saying that the pair $(A, 1_B)$ is the coreflection of B along U (see 3.1.1, volume 1). So U has a right adjoint functor V (see 3.1.3, volume 1). Since each couniversal morphism $UV(B) \longrightarrow B$ is an isomorphism (the identity on B) the functor V is full and faithful (see 3.4.1, volume 1). □

Following the classical terminology for the topological functor

$$U\colon \mathsf{Top} \longrightarrow \mathsf{Set},\ \ (X,\mathcal{T}) \mapsto X$$

the right adjoint to a topological functor is called the "discrete object functor" and the left adjoint is called the "indiscrete" or "chaotic object functor".

Proposition 7.3.8 *A topological functor creates limits and colimits.*

Proof Consider a topological functor $U: \mathscr{A} \longrightarrow \mathscr{B}$ and an arbitrary functor $H: \mathscr{D} \longrightarrow \mathscr{A}$ such that the composite $U \circ H$ admits in $\mathscr{B}$ a limit $(p_D: L \longrightarrow UH(D))_{D \in \mathscr{D}}$. The initial structure $(g_D: A \longrightarrow H(D))_{D \in \mathscr{D}}$ for these data yields the limit of H in $\mathscr{A}$. Indeed given another cone $(h_D: A' \longrightarrow H(D))_{D \in \mathscr{D}}$ on the functor H, the cone $(U(h_D))_{D \in \mathscr{D}}$ factors uniquely through the limit cone $(p_D)_{D \in \mathscr{D}}$ via an arrow $b: U(A') \longrightarrow L$; this yields a factorization $a: A' \longrightarrow L$ such that $g_D \circ a = h_D$ and $U(a) = b$. The uniqueness of a follows immediately from that of b and the faithfulness of U. The case of colimits follows by duality (see 7.3.6). □

When $U: \mathscr{A} \longrightarrow \mathscr{B}$ is a topological functor, all the objects $A \in \mathscr{A}$ mapped to the same object $B \in \mathscr{B}$ can be thought of as "all the U-topological structures which can be put on the object B". As for ordinary topologies, one has the following result:

Proposition 7.3.9 *Let $U: \mathscr{A} \longrightarrow \mathscr{B}$ be a faithful functor with initial structures. For an object $B \in \mathscr{B}$, the objects $A \in \mathscr{A}$ such that $U(A) = B$ and morphisms $f \in \mathscr{A}$ such that $U(f) = 1_B$ constitute a complete preordered class. When U is a topological functor, all limits (even large) exist in this preordered class.*

Proof If $U(A) = B = U(A')$, by faithfulness of U there is at most one morphism $A \longrightarrow A'$ mapped to 1_B; so we indeed get a preordered class. Now given a family $(A_i)_{i \in I}$ of objects mapped to B, the identity morphisms $(1_B: B \longrightarrow U(A_i))_{i \in I}$ constitute a cone for which there exists a corresponding initial structure $(g_i: A \longrightarrow A_i)_{i \in I}$. The universal property of this last cone means precisely that A is the infimum of the family $(A_i)_{i \in I}$ in the preordered class indicated.

When U is topological, the assumption that I is a set can be dropped (see 7.3.2). □

In the basic case of the topological functor U: Top $\longrightarrow$ Set, on a given set X there is just a set of possible topologies. So the preordered classes of 7.3.9 are in fact complete preordered sets.

Definition 7.3.10 *A functor $U: \mathscr{A} \longrightarrow \mathscr{B}$ is fibre small when, for every object $B \in \mathscr{B}$, the category of those objects $A \in \mathscr{A}$ such that $U(A) = B$ and arrows $f \in \mathscr{A}$ such that $U(f) = 1_B$ is equivalent to a small category.*

The notion of "fibre smallness" is an alternative to the use of large diagrams in the definition of a topological functor (see 7.3.2).

Proposition 7.3.11 *Let $U: \mathscr{A} \longrightarrow \mathscr{B}$ be a functor. The following conditions are equivalent:*
(1) U is topological and fibre small;
(2) U has initial structures and is faithful and fibre small.

Proof By 7.3.4, one has (1) $\Rightarrow$ (2). Conversely, by 7.3.5 it remains to consider a discrete category $\mathscr{D}$, a functor $H: \mathscr{D} \longrightarrow \mathscr{A}$ and a cone $\big(f_D: B \longrightarrow FH(D)\big)_{D \in \mathscr{D}}$ on $F \circ H$ and prove the existence of a corresponding initial structure. For every object $D \in \mathscr{D}$, let us consider the one element cone $f_D: B \longrightarrow FH(D)$ which induces an initial lifting $v_D: A_D \longrightarrow H(D)$. All the objects A_D are in the "fibre" over B, i.e. $U(A_D) = B$; since this "fibre" is equivalent to a small preordered set (by assumption) there is just a set of isomorphism classes of such A_D's; since moreover this "fibre" is complete (by 7.3.9), we can choose A to be the infimum of all the A_D's. For every $D \in \mathscr{D}$, we define g_D to be the composite

$$A \xrightarrow{u_D} A_D \xrightarrow{v_D} H(D)$$

where u_D expresses the relation $A \leq A_D$ in the fibre over B.

Let us prove that $\big(g_D: A \longrightarrow H(D)\big)_{D \in \mathscr{D}}$ is the initial structure associated with the original data. First of all

$$U(g_D) = U(v_D) \circ U(u_D) = f_D \circ 1_B = f_D.$$

Next choose a cone $\big(h_D: A' \longrightarrow H(D)\big)_{D \in \mathscr{D}}$ in $\mathscr{A}$ and, in $\mathscr{B}$, a factorization $b: U(A') \longrightarrow B$ such that $f_D \circ b = U(h_D)$. By universality of v_D, there exists a unique morphism $a_D: A' \longrightarrow A_D$ such that $U(a_D) = b$ and $v_D \circ a_D = h_D$. But the cone $(u_D: A \longrightarrow A_D)_{D \in \mathscr{D}}$ is the initial lifting of the cone $(a_D: A' \longrightarrow A_D)_{D \in \mathscr{D}}$ and the factorization $b: U(A') \longrightarrow B$ yields a unique morphism $a: A' \longrightarrow A$ such that $U(a) = b$ and $u_D \circ a = a_D$. This implies

$$g_D \circ a = v_D \circ u_D \circ a = v_D \circ a_D = h_D$$

from which the result follows immediately. □

7.4 Exercises

7.4.1 Let F: Top $\longrightarrow$ Top be a colimit preserving functor. Prove that at the level of underlying sets, F is just the product with a fixed topological space X.

7.4.2 Let F: Top $\longrightarrow$ Top be a colimit preserving functor. Prove the existence of a unique natural transformation from F to the identity.

7.4.3 Prove that the category of compactly generated spaces is complete and cocomplete.

7.4.4 Prove that every closed subspace of a compactly generated space is itself compactly generated.

7.4.5 Prove that every open subspace of a compactly generated space is itself compactly generated.

7.4.6 Show that the product of uncountably many copies of the real line is not a compactly generated space. [Hint: consider those families $(r_i)_{i \in I}$ for which there exists $n \in \mathbb{N}$ such that for all i, $r_i = n$ or $r_i = 0$, the number of those components for which $r_i = 0$ being less that n.]

7.4.7 An uncountable power of the unit interval $[0,1]$ is compact by the Tychonoff theorem, thus a compactly generated space. Show that the subspace of those families $(r_i)_{i \in I}$ for which r_i is never 0 or 1 is not compactly generated, proving that a subspace of a compactly generated space is in general not compactly generated. [Hint: use 7.4.6.]

7.4.8 Let Z be a T_0, T_1, T_2, regular or completely regular space. Given a space Y, consider the set $\mathcal{C}(Y,Z)$ of continuous functions provided with the compact open topology. Prove that $\mathcal{C}(Y,Z)$ is again, respectively, a T_0, T_1, T_2, regular or completely regular space.

7.4.9 Prove that a directed set of continuous functions from a compact space A to a metric space B converges for the compact open topology on $\mathcal{C}(A,B)$ if and only if it converges uniformly on A.

7.4.10 Prove that a directed set of continuous functions on a compactly generated space A to a uniform space B converges for the compact open topology on $\mathcal{C}(A,B)$ if and only if it converges uniformly on the compact subsets of A.

7.4.11 Let Y, Z be topological spaces. If the corresponding evaluation map $\mathcal{C}(Y,Z) \times Y \longrightarrow Z$ is continuous, when $\mathcal{C}(Y,Z)$ is provided with the compact open topology, prove for every topological space X the existence of natural bijections

$$\mathsf{Top}(X \times Y, Z) \cong \mathsf{Top}(X, \mathcal{C}(Y,Z)).$$

7.4.12 Let $U: \mathscr{A} \longrightarrow \mathsf{Set}$ be a topological functor. Prove that in $\mathscr{A}$ every arrow factors as a regular epimorphism followed by a monomorphism.

7.4.13 Let $U: \mathscr{A} \longrightarrow \mathsf{Set}$ be a topological functor. Prove that $\mathscr{A}$ has a generator and a cogenerator.

7.4.14 Let $U: \mathscr{A} \longrightarrow \mathsf{Set}$ be a topological functor. Prove that $\mathscr{A}$ is well-powered (resp., co-well-powered) if and only if U is fibre small.

8
Fibred categories

This chapter introduces the notion of fibration, which is a powerful tool in category theory, but also presents a deep reflexion on the bases of category theory themselves.

Very often in the previous chapters we have considered families of objects or morphisms in a category $\mathscr{C}$. This suggests the study of the category $\mathsf{Set}(\mathscr{C})$ of families of objects and morphisms in $\mathscr{C}$:

- the objects are the families $(C_i)_{i \in I}$ where I is a set and C_i is an object of $\mathscr{C}$;
- the morphisms $f: (C_i)_{i\in I} \longrightarrow (D_j)_{j \in J}$ are the pairs $(\alpha, (f_i)_{i\in I})$ with $\alpha: I \longrightarrow J$ a mapping in Set and each $f_i: C_i \longrightarrow D_{\alpha(i)}$ a morphism in $\mathscr{C}$.

Of special interest are the families $(C_i)_{i\in I}$ and $\left(1_I, (f_i: C_i \longrightarrow D_i)_{i \in I}\right)$ for a given set I, which constitute what will be called the "fibre" at I. The obvious projection functor $\mathsf{Set}(\mathscr{C}) \longrightarrow \mathsf{Set}$ will be the basic example of a fibration.

Working with set-indexed families as in the previous example very often leads to constructions or arguments which look innocent ... but are formally incorrect! For example given an I-indexed family of morphisms $(f_i: C_i \longrightarrow D_i)_{i\in I}$ in $\mathsf{Set}(\mathscr{C})$, few people really worry about considering the "set"

$$I_0 = \{i \in I \mid f_i \text{ is a monomorphism}\}.$$

But the sentence "f_i is a monomorphism" means

$$\forall C \in \mathscr{C} \ \forall u, v \in \mathscr{C}(C, C_i) \ \ f_i \circ u = f_i \circ v \Rightarrow u = v$$

... and no axiom of set theory will ever imply that I_0 is a set, since the formula contains a quantifier acting on a variable C which runs through something (namely, $\mathscr{C}$) which is not a set! Well, most of us will answer: "O.K., you are right, but who really cares? Leave this to the logicians, we are doing mathematics." And to some extent such an attitude is not really to blame: we "feel" that it should be possible to make things all right. For example by using the axiom of universes (see 1.1.4, volume 1), so that $\mathscr{C}$ becomes a set in a bigger universe and as a consequence I_0 does as well; and since $I_0 \subseteq I$, I_0 is a set in the same universe as I.

Now suppose – and this is the essence of the theory of fibrations – that we want to index things not just by the category of sets, but by an arbitrary category $\mathscr{E}$. Thus we consider a "fibration" $F: \mathscr{F} \longrightarrow \mathscr{E}$ and think of the objects and arrows of $\mathscr{F}$ as families indexed by the objects of $\mathscr{E}$. For example consider $I \in \mathscr{E}$ and $X, Y \in \mathscr{F}$ such that $F(X) = I = F(Y)$ together with a morphism $f: X \longrightarrow Y$ in $\mathscr{F}$ such that $F(f) = 1_I$. Even supposing that $\mathscr{E}$ is complete, cocomplete or whatever you want, you cannot possibly find a general construction in $\mathscr{E}$ producing a subobject $I_0 \subseteq I$ which represents intuitively "those indices for which f is a monomorphism", as in the set-indexed case. Even if vaguely overlooking some logical problems makes you feel free to consider that any formula φ on a given set I should define a subset I_0 of I, you will certainly not pretend that such a careless attitude can be carried over to a more "structured" category $\mathscr{E}$ (like groups, topological spaces, ...) in order to define a subobject $I_0 \subseteq I$ in $\mathscr{E}$; for example, the monomorphisms between two abelian groups certainly do not constitute a subgroup of the group of all homomorphisms! You will immediately point out that the formula φ should be somehow compatible with the "structure" of $\mathscr{E}$ in order to define a subobject in $\mathscr{E}$: this is the very fundamental notion of "definability" with respect to a fibration. And it is careless but common to overlook the fact that in the case of sets the property of being a set is not a structure at all.

8.1 Fibrations

We start with some definitions, fixing the terminology and the notation:

Definition 8.1.1 *Let $F: \mathscr{F} \longrightarrow \mathscr{E}$ be a functor. Given an object $I \in \mathscr{E}$, the fibre of F at I is the subcategory $\mathscr{F}_I$ of $\mathscr{F}$ defined in the following way:*

- *an object $X \in \mathscr{F}$ is in $\mathscr{F}_I$ when $F(X) = I$;*
- *if X, Y are objects in $\mathscr{F}_I$, a morphism $f: X \longrightarrow Y$ of $\mathscr{F}$ is in $\mathscr{F}_I$ when $F(f) = 1_I$.*

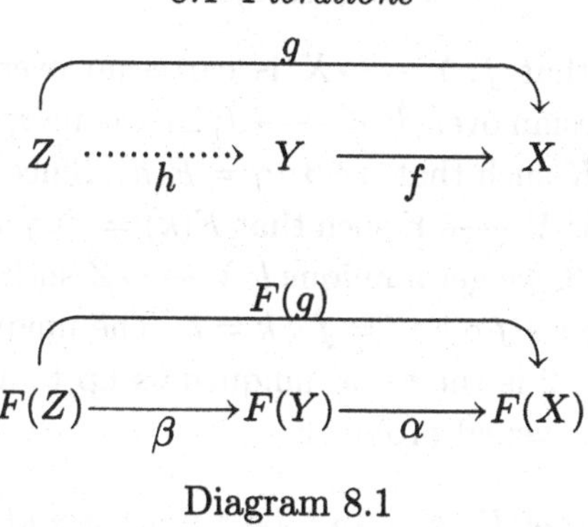

Diagram 8.1

Obviously, the fibres of F in 8.1.1 are by no means full subcategories of $\mathscr{F}$.

Definition 8.1.2 *Let $F\colon \mathscr{F} \longrightarrow \mathscr{E}$ be a functor and $\alpha\colon J \longrightarrow I$ a morphism of $\mathscr{E}$. An arrow $f\colon Y \longrightarrow X$ of $\mathscr{F}$ is cartesian over α if:*

(1) $F(f) = \alpha$;

(2) given $g\colon Z \longrightarrow X$ is a morphism of $\mathscr{F}$ such that $F(g)$ factors as $\alpha \circ \beta$, there exists a unique morphism $h\colon Z \longrightarrow Y$ in $\mathscr{F}$ such that $F(h) = \beta$ and $g = f \circ h$ (see diagram 8.1).

Definition 8.1.3 *A functor $F\colon \mathscr{F} \longrightarrow \mathscr{E}$ is a fibration when for every arrow $\alpha\colon J \longrightarrow I$ in $\mathscr{E}$ and every object X in the fibre over I, there exists in $\mathscr{F}$ a cartesian morphism $f\colon Y \longrightarrow X$ over α. We also call $\mathscr{F}$ a "category fibred over $\mathscr{E}$".*

There is a similar and equivalent way of presenting the definition of a fibred category: be careful, do not read it (see 8.1.7) without reading also the counterexample 8.1.8! There is also the notion of "indexed category" as in **Paré and Schumacher**: be careful again, it is not formally equivalent to the notion of fibred category; roughly speaking, it is a fibred category in which a distinguished choice of cartesian morphisms and coherent isomorphisms has been made once for all: it is something intermediate between a fibration (as in 8.1.3) and a split fibration (as in 8.3.3).

Lemma 8.1.4 *Let $F\colon \mathscr{F} \longrightarrow \mathscr{E}$ be a fibration.*

(1) The composite of two cartesian morphisms is again a cartesian morphism.

(2) If $f\colon Y \longrightarrow X$ and $g\colon Z \longrightarrow X$ are two cartesian morphisms over the same arrow $\alpha\colon J \longrightarrow I$ of $\mathscr{E}$, there exists a unique isomorphism h in the fibre $\mathscr{F}_J$ such that $g = f \circ h$.

Proof Assume that $f\colon Y \longrightarrow X$ is cartesian over $\alpha\colon J \longrightarrow I$ while $g\colon Z \longrightarrow Y$ is cartesian over $\beta\colon K \longrightarrow J$; choose morphisms $h\colon V \longrightarrow X$ and $\gamma\colon F(V) \longrightarrow K$ such that $\alpha \circ \beta \circ \gamma = F(h)$. Since f is cartesian over α we get a unique $k\colon V \longrightarrow Y$ such that $F(k) = \beta \circ \gamma$ and $f \circ k = h$. Since g is cartesian over β, we get a unique $l\colon V \longrightarrow Z$ such that $F(l) = \gamma$ and $g \circ l = k$. This implies $f \circ g \circ l = f \circ k = h$. The uniqueness condition is obvious. Statement 2 is the usual uniqueness up to an isomorphism for the solution of a universal problem. □

Definition 8.1.5 *Let $F\colon \mathscr{F} \longrightarrow \mathscr{E}$ be a functor and $\alpha\colon J \longrightarrow I$ a morphism of $\mathscr{E}$. An arrow $f\colon Y \longrightarrow X$ of $\mathscr{F}$ is precartesian over α if:*
(1) $F(f) = \alpha$;
(2) if $g\colon Z \longrightarrow X$ is a morphism of $\mathscr{F}$ such that $F(g) = \alpha$, there exists a unique morphism $h\colon Z \longrightarrow Y$ in the fibre $\mathscr{F}_J$ such that $g = f \circ h$.

Lemma 8.1.6 *Let $F\colon \mathscr{F} \longrightarrow \mathscr{E}$ be a fibration. Consider a morphism $f\colon Y \longrightarrow X$ in $\mathscr{E}$ mapped to $\alpha\colon J \longrightarrow I$ by F. The following conditions are equivalent:*
(1) f is cartesian over α;
(2) f is precartesian over α.

Proof Clearly, cartesianness implies precartesianness (make $\beta = 1_J$ in definition 8.1.2). Conversely if $f\colon Y \longrightarrow X$ is precartesian over $\alpha\colon J \longrightarrow I$, we can consider $g\colon Z \longrightarrow X$ as cartesian over α, since F is a fibration. Then f, g are both precartesian over α and therefore by a classical argument for universal problems, isomorphic via an isomorphism in the fibre over J. Since g is cartesian, f is cartesian as well. □

Proposition 8.1.7 *Let $F\colon \mathscr{F} \longrightarrow \mathscr{E}$ be a functor. The following conditions are equivalent.*
(1) F is a fibration;
(2) (a) for every arrow $\alpha\colon J \longrightarrow I$ in $\mathscr{E}$ and every object X in the fibre over I, there exists in $\mathscr{F}$ a precartesian morphism $f\colon Y \longrightarrow X$ over α;
(b) the composite of two precartesian morphisms is again precartesian.

Proof (1) implies (2) by lemmas 8.1.4 and 8.1.6. Conversely, let us prove that every precartesian morphism $f\colon Y \longrightarrow X$ over $\alpha\colon J \longrightarrow I$ is also cartesian. Consider therefore a morphism $g\colon Z \longrightarrow X$ such that

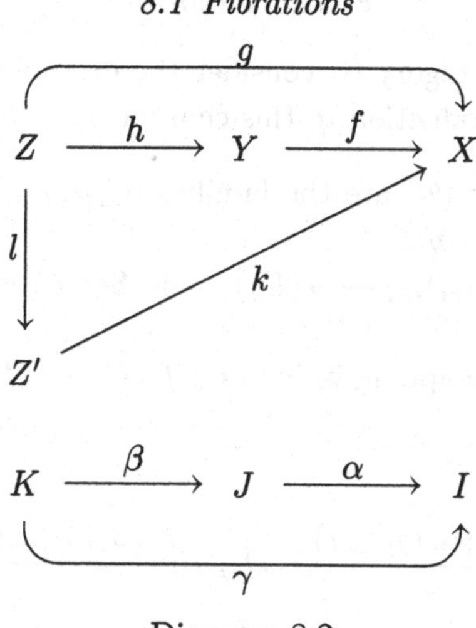

Diagram 8.2

$\gamma = F(g)$ factors as $\gamma = \alpha \circ \beta$ in $\mathscr{E}$, with $\beta: K \longrightarrow J$. Consider a precartesian morphism $h': Z' \longrightarrow Y$ over β. By assumption the composite $f \circ h': Z' \longrightarrow X$ is precartesian over $\alpha \circ \beta = \gamma$, thus there exists a unique morphism $h'': Z \longrightarrow Z'$ in the fibre over K such that $f \circ h' \circ h'' = g$. Putting $h = h' \circ h''$ yields $F(h) = \beta \circ 1_K = \beta$ and $f \circ h = g$. To prove the uniqueness of h, consider $\overline{h}: Z \longrightarrow Y$ such that $F(\overline{h}) = \beta$ and $f \circ \overline{h} = g$. Since h' is precartesian, there is a unique h''' in the fibre over K such that $h' \circ h''' = \overline{h}$; from $f \circ h' \circ h''' = f \circ \overline{h} = g$ we deduce $h''' = h''$ and thus $h = \overline{h}$. □

Counterexample 8.1.8
Here is an example of a functor $F: \mathscr{F} \longrightarrow \mathscr{E}$ satisfying condition (2)(a) of 8.1.7 and which is not a fibration.

Just consider the situation of diagram 8.2 where α, β, γ are the non-obvious morphisms of $\mathscr{E}$, f, g, h, k, l are the non-obvious morphisms of $\mathscr{F}$, $F(f) = \alpha$, $F(g) = \gamma$, $F(h) = \beta$, $F(k) = \gamma$, $F(l) = 1_K$ and all diagrams are commutative. It is obvious that in $\mathscr{F}$, the following morphisms are precartesian: f, h, k and of course, all identity morphisms. So condition (2)(a) is satisfied. But $g = f \circ h$ is not precartesian, since k does not factor through g. And f is not cartesian, since $F(k) = \gamma$ factors through $F(f) = \alpha$, while k does not factor through f. So there is no cartesian morphism over α.

Examples 8.1.9

8.1.9.a For every category $\mathscr{C}$, the identity functor on $\mathscr{C}$ is obviously a fibration: every morphism is cartesian.

8.1.9.b Given a category $\mathscr{C}$, consider the example $\varphi\colon \mathsf{Set}(\mathscr{C}) \longrightarrow \mathsf{Set}$ described in the introduction to this chapter:

- the objects of $\mathsf{Set}(\mathscr{C})$ are the families $(C_i)_{i\in I}$ where I is a set and C_i is an object of $\mathscr{C}$;
- a morphism $f\colon (C_i)_{i\in I} \longrightarrow (D_j)_{j\in J}$ in $\mathsf{Set}(\mathscr{C})$ is a pair $(\alpha, (f_i)_{i\in I})$ where
  $\alpha\colon I \longrightarrow J$ is a mapping in Set and $f_i\colon C_i \longrightarrow D_{\alpha(i)}$ is a morphism in $\mathscr{C}$;
- the composite
  $$(C_i)_{i\in I} \xrightarrow{(\alpha,(f_i)_{i\in I})} (D_j)_{j\in J} \xrightarrow{(\beta,(g_j)_{j\in J})} (E_k)_{k\in K}$$
  is the pair $\big(\beta\circ\alpha, (g_{\alpha(i)}\circ f_i)_{i\in I}\big)$;
- φ maps $(C_i)_{i\in I}$ to I and $\big(\alpha, (f_i)_{i\in I}\big)$ to α;

it is obvious that $\mathsf{Set}(\mathscr{C})$ is a category and φ is a functor. Now given a mapping $\alpha\colon J \longrightarrow I$ in Set and a family $(C_i)_{i\in I}$ of objects of $\mathscr{C}$, define $D_j = C_{\alpha(j)}$ and $f_j\colon D_j \longrightarrow C_{\alpha(j)}$ the identity morphism, for every index $j \in J$. This defines a morphism $\big(\alpha, (f_j)_{j\in J}\big)\colon (D_j)_{j\in J} \longrightarrow (C_i)_{i\in I}$ which is cartesian over α.

8.1.9.c Given a category $\mathscr{C}$ with finite limits, consider the category $\mathsf{Ar}(\mathscr{C})$ of arrows of $\mathscr{C}$ as in 1.2.7.c, volume 1:

- the objects of $\mathsf{Ar}(\mathscr{C})$ are the triples (A, f, B) where $f\colon A \longrightarrow B$ is an arrow of $\mathscr{C}$;
- a morphism $(u, v)\colon (A, f, B) \longrightarrow (C, g, D)$ is a pair of morphisms $u\colon A \longrightarrow C$, $v\colon B \longrightarrow D$ such that $g\circ u = v\circ f$.

The "codomain" functor

$$\partial_1\colon \mathsf{Ar}(\mathscr{C}) \longrightarrow \mathscr{C},\ \ \partial_1(A, f, B) = B,\ \ \partial_1(u, v) = v$$

is a fibration. Indeed, given a morphism $v\colon B \longrightarrow D$ in $\mathscr{C}$ and an object (C, g, D) in $\mathsf{Ar}(\mathscr{C})$, computing the pullback in diagram 8.3 yields a morphism $(u, v)\colon (A, f, B) \longrightarrow (C, g, D)$ in $\mathsf{Ar}(\mathscr{C})$ which, just by definition of a pullback, is cartesian over v. In this example, observe that the fibre over an object $B \in \mathscr{C}$ is exactly the category $\mathscr{C}/B$, as defined in 1.2.7.a, volume 1 – i.e. intuitively the category of B-indexed families in $\mathscr{C}$. This fibration is referred to as the canonical fibration of $\mathscr{C}$ over itself.

8.1.9.d Write **1** for the terminal category, with a single object and just the identity on it. For every category $\mathscr{C}$, the unique functor $\mathscr{C} \longrightarrow \mathbf{1}$ is

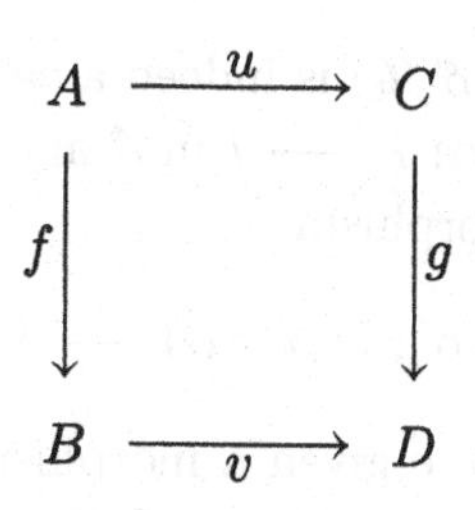

Diagram 8.3

a fibration since, given an object $C \in \mathscr{C}$, the identity on C is obviously a cartesian morphism.

Our next example is worth being singled out.

Example 8.1.10: The small fibrations.

Let us go back once more to the example of the introduction, also described in 8.1.9.b, supposing now that the category $\mathscr{C}$ is small. This ordinary small category $\mathscr{C}$ is thus a category internal to the category of sets (see 8.1.1, volume 1) and we shall write it as $\mathscr{C} = (C_0, C_1, d_0, d_1, i, c)$, where C_0 is the set of objects and C_1 the set of morphisms. Given a set I, an I-family $(M_i)_{i\in I}$ of objects of $\mathscr{C}$ is just a mapping $m: I \longrightarrow C_0$. Consider another set J and a mapping $n: J \longrightarrow C_0$ corresponding to the choice of a family $(N_j)_{j\in J}$ of objects. Given a mapping $\alpha: I \longrightarrow J$, a family $\left(f_i: M_i \longrightarrow N_{\alpha(i)}\right)_{i\in I}$ of morphisms of $\mathscr{C}$ corresponds to a mapping $f: I \longrightarrow C_1$ such that $d_0 \circ f = m$, $d_1 \circ f = n \circ \alpha$.

This construction generalizes immediately to the case of an internal category $\mathscr{C} = (C_0, C_1, d_0, d_1, i, c)$ in a category $\mathscr{E}$ with finite limits. One defines a fibration $\varphi: \mathscr{E}(\mathscr{C}) \longrightarrow \mathscr{E}$ in the following way:

- the objects of $\mathscr{E}(\mathscr{C})$ are the pairs (I, m) where I is an object of $\mathscr{E}$ and $m: I \longrightarrow C_0$ is a morphism of $\mathscr{E}$;
- a morphism $(I, m) \longrightarrow (J, n)$ of $\mathscr{E}(\mathscr{C})$ is a pair (α, f) with $\alpha: I \longrightarrow J$ and $f: I \longrightarrow C_1$ morphisms of $\mathscr{E}$ satisfying the conditions $d_0 \circ f = m$, $d_1 \circ f = n \circ \alpha$;
- the composite of two morphisms

$$(I, m) \xrightarrow{(\alpha, f)} (J, n) \xrightarrow{(\beta, g)} (K, l)$$

  is the pair $(\beta \circ \alpha, m \circ h)$, where $h: I \longrightarrow C_1 \times_{C_0} C_1$ is the unique factorization through the pullback obtained from the relation $d_1 \circ f = n \circ \alpha = d_0 \circ g \circ \alpha$;
- $\varphi(I, m) = I$ and $\varphi(\alpha, f) = \alpha$.

It is routine to check that $\mathscr{E}(\mathscr{C})$ is indeed a category and φ is a functor.

Now given a morphism $\alpha: I \longrightarrow J$ in $\mathscr{E}$ and an object (J, n) of $\mathscr{E}(\mathscr{C})$, one immediately gets a morphism

$$(\alpha, i \circ n \circ \alpha): (I, n \circ \alpha) \longrightarrow (J, n)$$

of $\mathscr{E}(\mathscr{C})$. It is cartesian since given a morphism $(\gamma, f): (K, l) \longrightarrow (J, n)$ such that $\gamma = \alpha \circ \beta$ for some $\beta: K \longrightarrow I$, the unique required factorization $(K, l) \longrightarrow (I, n \circ \alpha)$ is the pair (β, f).

Anticipating the considerations of the next section, we immediately make the following definition:

Definition 8.1.11 *A fibration $F: \mathscr{F} \longrightarrow \mathscr{E}$ over a category $\mathscr{E}$ with finite limits is called a small fibration when it is equivalent, in the 2-category* $\mathsf{Fib}(\mathscr{E})$ *of fibrations over $\mathcal{E}$, to the category $\varphi: \mathscr{E}(\mathscr{C}) \longrightarrow \mathscr{E}$ of 8.1.10, for some internal category $\mathscr{C}$ of $\mathscr{E}$.*

Let us conclude with some general processes for building new fibrations from given ones.

Proposition 8.1.12 *If $F: \mathscr{F} \longrightarrow \mathscr{E}$ and $G: \mathscr{E} \longrightarrow \mathscr{C}$ are fibrations, then so is the composite $G \circ F: \mathscr{F} \longrightarrow \mathscr{C}$.*

Proof Consider a morphism $a: A \longrightarrow B$ in $\mathscr{C}$ and an object $X \in \mathscr{F}$ such that $GF(X) = B$. Since G is a fibration, there is a cartesian morphism $\alpha: I \longrightarrow F(X)$ over a; since F is a fibration, there is a cartesian morphism $f: Y \longrightarrow X$ over α. It is then obvious that f is also cartesian over a. □

Proposition 8.1.13 *Given two categories $\mathscr{F}, \mathscr{E}$, the projection functor $p: \mathscr{F} \times \mathscr{E} \longrightarrow \mathscr{E}$ is a fibration.*

Proof Given a morphism $\alpha: J \longrightarrow I$ in $\mathscr{E}$ and an object (X, I) of $\mathscr{F} \times \mathscr{E}$, the morphism $(1_X, \alpha): (X, J) \longrightarrow (X, I)$ is obviously cartesian over α. □

Proposition 8.1.14 *If $F: \mathscr{F} \longrightarrow \mathscr{E}$ and $G: \mathscr{D} \longrightarrow \mathscr{C}$ are fibrations, then so is the functor $F \times G: \mathscr{F} \times \mathscr{D} \longrightarrow \mathscr{E} \times \mathscr{C}$.*

Proof Given a morphism $(\alpha, a): (J, B) \longrightarrow (I, A)$ in $\mathscr{E} \times \mathscr{C}$ and an object (X, M) in $\mathscr{F} \times \mathscr{D}$, choose $f: Y \longrightarrow X$ in $\mathscr{E}$ cartesian over α and $m: N \longrightarrow M$ in $\mathscr{D}$ cartesian over a; it is obvious that the morphism $(f, m): (Y, N) \longrightarrow (X, M)$ is cartesian over (α, a). □

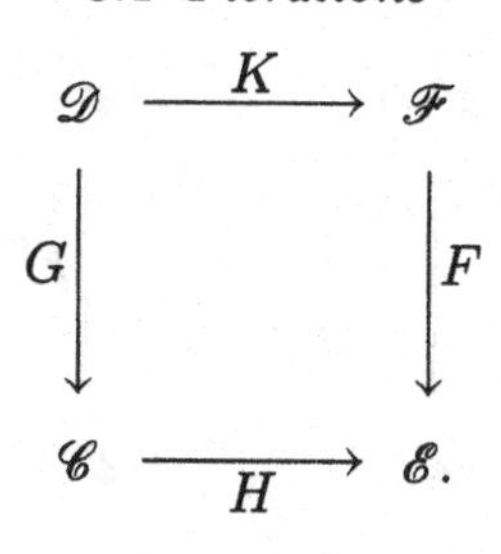

Diagram 8.4

Proposition 8.1.15 *Consider a pullback of categories and functors as in diagram 8.4. If F is a fibration, then so is G.*

Proof Consider a morphism $a: D \longrightarrow C$ in $\mathscr{C}$ and an object of $\mathscr{D}$ in the fibre over C, thus a pair (X, C) with $X \in \mathscr{F}$ and $H(C) = F(X)$. Choose a cartesian morphism $f: Y \longrightarrow X$ over the morphism $H(a): H(D) \longrightarrow H(C) = F(X)$ of $\mathscr{E}$. In particular the identities $F(Y) = H(D)$ and $F(f) = H(a)$ hold, thus $(f, a): (Y, D) \longrightarrow (X, C)$ is a morphism of $\mathscr{D}$. It is obviously cartesian over a. □

Proposition 8.1.16 *Let $F: \mathscr{F} \longrightarrow \mathscr{E}$ be a fibration and $\mathscr{D}$ a small category. The functor of composition with F,*

$$F^{\mathscr{D}}: \mathscr{F}^{\mathscr{D}} \longrightarrow \mathscr{E}^{\mathscr{D}},$$

is also a fibration.

Proof Consider a functor $M: \mathscr{D} \longrightarrow \mathscr{E}$ and a functor $P: \mathscr{D} \longrightarrow \mathscr{F}$ in the fibre over M; this means that $F \circ P = M$. Given another functor $N: \mathscr{D} \longrightarrow \mathscr{E}$ and a natural transformation $\mu: N \Rightarrow M$, we must construct a cartesian morphism $\eta: Q \Rightarrow P$ over μ.

For each object $D \in \mathscr{D}$, choose over $\mu_D: N(D) \longrightarrow M(D)$ a cartesian morphism $\eta_D: Q(D) \longrightarrow P(D)$. Given an arrow $d: D' \longrightarrow D$ in $\mathscr{D}$, the naturality of μ and the cartesianness of η_D imply that $P(d) \circ \eta_{D'}$ factors uniquely through η_D via a morphism $Q(d): Q(D') \longrightarrow Q(D)$. This defines the functor Q and the natural transformation η, which is cartesian over μ since each η_D is cartesian over μ_D. □

Corollary 8.1.17 *Let $F: \mathscr{F} \longrightarrow \mathscr{E}$ be a fibration and $\mathscr{D}$ a small category. The pullback of diagram 8.5, where $\Delta(E)$ is the constant functor on $E \in \mathscr{E}$, determines a fibration $\mathscr{F}^{(\mathscr{D})} \longrightarrow \mathscr{E}$.*

Proof Use 8.1.16 and 8.1.15. □

Observe that the fibre of $\mathscr{F}^{(\mathscr{D})}$ over an object $I \in \mathscr{E}$ has for objects those functors $M: \mathscr{D} \longrightarrow \mathscr{F}$ such that $F \circ M$ is the constant functor on

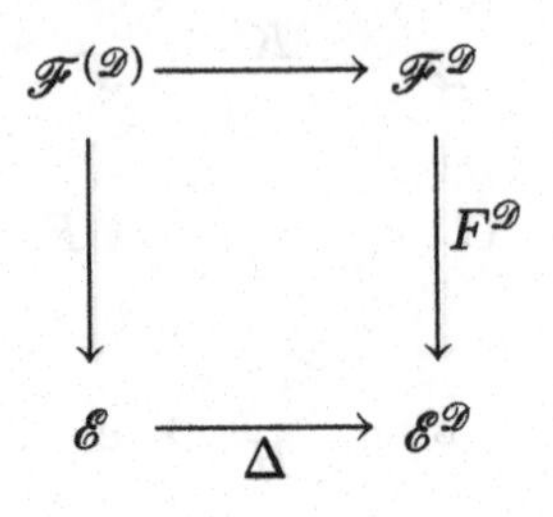

Diagram 8.5

I, thus exactly the functors $M: \mathscr{D} \longrightarrow \mathscr{F}_I$. The fibration $\mathscr{F}^{(\mathscr{D})}$ can thus be thought of as the fibration of diagrams of shape $\mathscr{D}$ in the fibres of $F: \mathscr{F} \longrightarrow \mathscr{E}$.

To avoid any ambiguity, let us make clear that the dual notion of a cofibration is obtained by dualizing both categories $\mathscr{F}$ and $\mathscr{E}$ in definition 8.1.3:

Definition 8.1.18 *A functor $F: \mathscr{F} \longrightarrow \mathscr{E}$ is a cofibration when the dual functor $F^*: \mathscr{F}^* \longrightarrow \mathscr{E}^*$ is a fibration.*

Notice that the notions of fibration or cofibration can also be "dualized" to the case of contravariant functors, by dualizing just one of the categories $\mathscr{F}$ or $\mathscr{E}$ in definition 8.1.3.

8.2 Cartesian functors

Cartesian functors are the "morphisms of fibred categories".

Definition 8.2.1 *Let $F: \mathscr{F} \longrightarrow \mathscr{E}$ and $G: \mathscr{G} \longrightarrow \mathscr{E}$ be two fibrations over the same base category $\mathscr{E}$. A cartesian functor $H: (\mathscr{F}, F) \longrightarrow (\mathscr{G}, G)$ is a functor $H: \mathscr{F} \longrightarrow \mathscr{G}$ such that:*
(1) $G \circ H = F$;
(2) G maps a cartesian morphism for F to a cartesian morphism for G.

Definition 8.2.2 *Let $F: \mathscr{F} \longrightarrow \mathscr{E}$ and $G: \mathscr{G} \longrightarrow \mathscr{E}$ be two fibrations over the same base category $\mathscr{E}$. If $H, K: (\mathscr{F}, F) \rightrightarrows (\mathscr{G}, G)$ are cartesian functors, a cartesian natural transformation $\theta: H \Rightarrow K$ is a natural transformation such that $G * \theta = F$.*

It is obvious that the fibrations over a base category $\mathscr{E}$, together with the cartesian functors and cartesian natural transformations between them, constitute a 2-category written $\mathsf{Fib}(\mathscr{E})$. We often write $\mathsf{Cart}(\mathscr{F}, \mathscr{G})$ for the category of cartesian functors and cartesian natural transformations from $(\mathscr{F}, F)$ to $(\mathscr{G}, G)$.

Examples 8.2.3

8.2.3.a Consider a functor $F: \mathscr{C} \longrightarrow \mathscr{D}$ between two categories and the corresponding fibrations φ: $\mathsf{Set}(\mathscr{C}) \longrightarrow \mathsf{Set}$, ψ: $\mathsf{Set}(\mathscr{D}) \longrightarrow \mathsf{Set}$ as in 8.1.9.b. The functor F induces a cartesian functor Φ: $\mathsf{Set}(\mathscr{C}) \longrightarrow \mathsf{Set}(\mathscr{D})$ over Set by putting

$$\Phi\big((C_i)_{i\in I}\big) = \big(F(C_i)\big)_{i\in I}, \quad \Phi\big(\alpha, (f_i)_{i\in I}\big) = \Big(\alpha, \big(F(f_i)\big)_{i\in I}\Big),$$

where the notation is that of 8.1.9.b.

8.2.3.b The previous example can easily be generalized to the context of 8.1.10. Consider a category $\mathscr{E}$ with finite limits, two internal categories $\boldsymbol{\mathscr{C}}, \boldsymbol{\mathscr{D}}$ and an internal functor $\boldsymbol{F}: \boldsymbol{\mathscr{C}} \longrightarrow \boldsymbol{\mathscr{D}}$, given by the two morphisms F_0: $C_0 \longrightarrow D_0$, F_1: $C_1 \longrightarrow D_1$. One immediately gets a cartesian functor $\mathscr{E}(\boldsymbol{F})$: $\mathscr{E}(\boldsymbol{\mathscr{C}}) \longrightarrow \mathscr{E}(\boldsymbol{\mathscr{D}})$ by putting, with the notation of 8.1.10, $\mathscr{E}(\boldsymbol{F})(I, m) = (I, F_0 \circ m)$ and $\mathscr{E}(\boldsymbol{F})(\alpha, f) = (\alpha, F_1 \circ f)$.

8.2.3.c In example 8.2.3.b, if $\boldsymbol{F}, \boldsymbol{G}$: $\boldsymbol{\mathscr{C}} \rightrightarrows \boldsymbol{\mathscr{D}}$ are two internal functors and $\boldsymbol{\theta}$: $\boldsymbol{F} \Rightarrow \boldsymbol{G}$ is an internal natural transformation, given thus by a morphism θ: $\mathscr{C}_0 \longrightarrow \mathscr{D}_1$, one gets a corresponding cartesian natural transformation $\mathscr{E}(\boldsymbol{\theta})$: $\mathscr{E}(\boldsymbol{F}) \Rightarrow \mathscr{E}(\boldsymbol{G})$ by putting $\mathscr{E}(\boldsymbol{\theta})_{(I,m)} = (1_I, \theta \circ m)$.

Proposition 8.2.4 *Consider a category $\mathscr{E}$ with finite limits and two internal categories $\boldsymbol{\mathscr{C}}, \boldsymbol{\mathscr{D}}$. The constructions of 8.1.10 and 8.2.3.b,c yield a full and faithful functor*

$$\mathsf{Int}(\boldsymbol{\mathscr{C}}, \boldsymbol{\mathscr{D}}) \longrightarrow \mathsf{Cart}\big(\mathscr{E}(\boldsymbol{\mathscr{C}}), \mathscr{E}(\boldsymbol{\mathscr{D}})\big)$$

where $\mathsf{Int}(\boldsymbol{\mathscr{C}}, \boldsymbol{\mathscr{D}})$ denotes the category of internal functors and internal natural transformations from $\boldsymbol{\mathscr{C}}$ to $\boldsymbol{\mathscr{D}}$.

Proof Checking the functoriality is just routine. The faithfulness is easy. Indeed given internal functors $\boldsymbol{F}, \boldsymbol{G}$: $\boldsymbol{\mathscr{C}} \rightrightarrows \boldsymbol{\mathscr{D}}$ and internal natural transformations $\boldsymbol{\theta}, \boldsymbol{\tau}$: $\boldsymbol{F} \Rightarrow \boldsymbol{G}$, one has by definition

$$\mathscr{E}(\boldsymbol{\theta})_{(C_0, 1_{C_0})} = (1_{C_0}, \theta), \quad \mathscr{E}(\boldsymbol{\tau})_{(C_0, 1_{C_0})} = (1_{C_0}, \tau),$$

from which the faithfulness follows.

The fullness of the construction results from the same relations. If Θ: $\mathscr{E}(\boldsymbol{F}) \Rightarrow \mathscr{E}(\boldsymbol{G})$ is a cartesian natural transformation, then for each object $(I, m) \in \mathscr{E}(\boldsymbol{\mathscr{C}})$ the relation $\mathscr{E}(\boldsymbol{\mathscr{D}}) * \Theta = \mathscr{E}(\boldsymbol{\mathscr{C}})$ implies that $\Theta_{(I,m)}$ is in the fibre over I. Therefore one has

$$\Theta_{(I,m)} = (1_I, f): (I, F_0 \circ m) \longrightarrow (I, G_0 \circ m)$$

with $d_0 \circ f = F_0 \circ m$ and $d_1 \circ f = G_0 \circ m$. In particular $\Theta_{(C_0, 1_{C_0})} = (1_{C_0}, \theta)$ for some θ: $C_0 \longrightarrow D_1$ satisfying $d_0 \circ \theta = F_0$, $d_1 \circ \theta = G_0$. We leave to the reader the straightforward verification that the usual naturality of Θ implies the condition on the morphism θ that it induces an internal natural transformation $\boldsymbol{\theta}: \boldsymbol{F} \Rightarrow \boldsymbol{G}$. □

Proposition 8.2.4 can be interpreted as the fact that the 2-category of internal categories in $\mathscr{E}$ is (equivalent to) a full sub-2-category of $\mathsf{Fib}(\mathscr{E})$. On the other hand $\mathscr{E}$ itself is a full subcategory of the category of internal categories in $\mathscr{E}$:

- each object $I \in \mathscr{E}$ determines the "discrete" internal category ∂I on I, namely $\partial I = (I, I, 1_I, 1_I, 1_I, 1_I)$ (see 8.1.6.a, volume 1);
- each morphism $\alpha: I \longrightarrow J$ in $\mathscr{E}$ determines a corresponding internal functor $\partial\alpha: \partial I \longrightarrow \partial J$ just via the pair $\partial\alpha = (\alpha, \alpha)$.

It is obvious from 8.1.2, volume 1, that this construction yields a full and faithful functor, because the axioms of internal functor $\boldsymbol{F}: \partial I \longrightarrow \partial J$ reduce immediately to the single relation $F_0 = F_1$. Composing this embedding with the functors of 8.2.4 thus yields a functor

$$\mathscr{E} \longrightarrow \mathsf{Fib}(\mathscr{E}), \;\; I \mapsto \mathscr{E}(\partial I), \;\; \alpha \mapsto \mathscr{E}(\partial\alpha).$$

Observe that since we are just working with discrete internal categories, the existence of finite limits in $\mathscr{E}$ is not necessary.

It is probably useful to compute explicitly the values of $\mathscr{E}(\partial I)$ and $\mathscr{E}(\partial\alpha)$.

Lemma 8.2.5 *Let $I \in \mathscr{E}$ be an object of an arbitrary category. With the previous notation, the fibration $\mathscr{E}(\partial I)$ over $\mathscr{E}$ is just the source functor*

$$\mathscr{E}/I \longrightarrow \mathscr{E}, \;\; (K, \beta) \mapsto K.$$

If $\alpha: I \longrightarrow J$ is a morphism of $\mathscr{E}$, the corresponding cartesian functor

$$\mathscr{E}/I \longrightarrow \mathscr{E}/J, \;\; (K, \beta) \longrightarrow (K, \alpha \circ \beta)$$

is just the composition with α.

Proof By 8.1.10, an object of $\mathscr{E}(\partial I)$ is indeed a pair (K, β) where $\beta: K \longrightarrow I$ is a morphism of $\mathscr{E}$. A morphism $(K, \beta) \longrightarrow (L, \gamma)$ in $\mathscr{E}(\partial I)$ is then a pair (ε, φ) where $\varepsilon: K \longrightarrow L$ and $\varphi: K \longrightarrow I$ are morphisms of $\mathscr{E}$ satisfying $\beta = \varphi$, $\gamma \circ \varepsilon = \varphi$; this is indeed equivalent to giving $\varepsilon: K \longrightarrow L$ such that $\gamma \circ \varepsilon = \beta$.

Now given $\alpha: I \longrightarrow J$, $\mathscr{E}(\partial\alpha)$ maps the object $(K, \beta) \in \mathscr{C}/I$ to the pair $(K, \alpha \circ \beta)$ and the morphism $(\varepsilon, \varphi): (K, \beta) \longrightarrow (L, \gamma)$ on $(\varepsilon, \alpha \circ \varphi)$; this is again composition with α. □

Proposition 8.2.6 *For an arbitrary category $\mathscr{E}$, the functor*

$$\mathscr{E} \longrightarrow \mathsf{Fib}(\mathscr{E}), \quad I \mapsto \mathscr{E}(\partial I), \quad \alpha \mapsto \mathscr{E}(\partial\alpha)$$

is full and faithful.

Proof Given a morphism $\alpha: I \longrightarrow J$ in $\mathscr{E}$, $\mathscr{E}(\partial\alpha)$ is just the composition with α (see 8.2.5) so that $\mathscr{E}(\partial\alpha)(I, 1_I) = (I, \alpha)$. This proves faithfulness.

Fullness follows from the same formula. If $H: \mathscr{E}/I \longrightarrow \mathscr{E}/J$ is a cartesian functor, from $H(I, 1_I) = (I, \alpha)$ we get a morphism $\alpha: I \longrightarrow J$. Since H is required to commute with the source functors, a morphism $\beta: (K, k) \longrightarrow (L, l)$ in $\mathscr{E}/I$ is mapped by H onto $\beta: (K, k') \longrightarrow (L, l')$ in $\mathscr{E}/J$. In particular $l \circ \beta = k$ and $l' \circ \beta = k'$. Considering a morphism $k: (K, k) \longrightarrow (I, 1_I)$ yields $k' = \alpha \circ k$, from which $H(K, k) = (K, \alpha \circ k)$. □

Proposition 8.2.6 yields bijections

$$\mathscr{E}(I, J) \cong \mathsf{Cart}\big(\mathscr{E}(\partial I), \mathscr{E}(\partial J)\big).$$

We shall generalize this somewhat by replacing $\mathscr{E}(\partial J)$ with an arbitrary fibration over $\mathscr{E}$. This yields the so-called "fibred Yoneda lemma"; the reason for this terminology will become more apparent in the next section. In the same spirit, the embedding in 8.2.6 will be called the "fibred Yoneda embedding".

Proposition 8.2.7 *Let $F: \mathscr{F} \longrightarrow \mathscr{E}$ be a fibration. For an object $I \in \mathscr{E}$, the fibre $\mathscr{F}_I$ is equivalent to the category $\mathsf{Cart}\big(\mathscr{E}(\partial I), \mathscr{F}\big)$ of cartesian functors and cartesian natural transformations from $\mathscr{E}(\partial I)$ to $(\mathscr{F}, F)$.*

Proof The proof requires the axiom of choice. Fix an object $X \in \mathscr{F}_I$ in the fibre over I. We shall construct a cartesian functor $H_X: \mathscr{E}/I \longrightarrow \mathscr{F}$ (see 8.2.5):

- for every object $(J, \alpha) \in \mathscr{E}/I$, we choose over α a cartesian morphism $\overline{\alpha}: X_\alpha \longrightarrow X$ in $\mathscr{F}$ and we put $H_X(J, \alpha) = X_\alpha$;
- for every morphism $\gamma: (J, \alpha) \longrightarrow (K, \beta)$ in $\mathscr{E}/I$ the relation $\beta \circ \gamma = \alpha$ implies the existence of a unique morphism $\widetilde{\gamma}: X_\alpha \longrightarrow X_\beta$ such that $\overline{\beta} \circ \widetilde{\gamma} = \overline{\alpha}$ and $F(\widetilde{\gamma}) = \gamma$; we put $H_X(\gamma) = \widetilde{\gamma}$.

The uniqueness condition in the definition of $\widetilde{\gamma}$ implies immediately that H_X is a functor. To prove that H_X is cartesian, using the previous notation, we choose a cartesian morphism $\gamma': X_\alpha \longrightarrow X_\beta$ over γ in $\mathscr{F}$; the composite $\overline{\beta} \circ \gamma'$ is cartesian over $\beta \circ \gamma = \alpha$ (see 8.1.4), thus there exists an isomorphism ε in the fibre $\mathscr{F}_J$ such that $\overline{\beta} \circ \gamma' \circ \varepsilon = \overline{\alpha}$. By

$$\begin{array}{ccc} X_\alpha & \xrightarrow{\overline{\alpha}_X} & X \\ {\scriptstyle (H_f)_\alpha}\downarrow & & \downarrow{\scriptstyle f} \\ Y_\alpha & \xrightarrow[\overline{\alpha}_Y]{} & Y \end{array}$$

Diagram 8.6

uniqueness of $\widetilde{\gamma}$, this yields $\widetilde{\gamma} = \gamma' \circ \varepsilon$, thus the cartesianness of $\widetilde{\gamma}$. So each morphism $H_X(\gamma)$ is cartesian and therefore H_X is a cartesian functor.

Fix now a morphism $f\colon X \longrightarrow Y$ in the fibre $\mathscr{F}_I$. To avoid any ambiguity, we put indices X, Y in the previous definitions of $\overline{\alpha}, \widetilde{\gamma}$. For every object $(J, \alpha) \in \mathscr{E}/I$, the composite $f \circ \overline{\alpha}_X$ factors uniquely through the cartesian morphism $\overline{\alpha}_Y$ in the fibre over J, yielding a morphism $(H_f)_\alpha\colon H_X(J,\alpha) \longrightarrow H_Y(J,\alpha)$; see diagram 8.6. The uniqueness condition in the definition of $(H_f)_\alpha$ implies immediately that those data define a natural transformation $H_f\colon H_X \Rightarrow H_Y$ which is of course cartesian and has all its components in the fibres of $\mathscr{F}$.

The functoriality of the previous construction is clear. Moreover, with the previous notation, observe that $(H_f)_{1_I} = f$, from which follows the faithfulness of the process. The same formula will imply the fullness. Indeed consider a cartesian natural transformation $\varphi\colon H_X \Rightarrow H_Y$. Its component on the object $(I, 1_I) \in \mathscr{E}/I$ is a morphism $f\colon X \longrightarrow Y$ which lies in the fibre $\mathscr{F}_I$, since φ is cartesian. Every object $(J, \alpha) \in \mathscr{E}/I$ yields a morphism $\alpha\colon (J, \alpha) \longrightarrow (I, 1_I)$ in $\mathscr{E}/I$ and thus, by naturality of φ, the relation $f \circ \widetilde{\alpha}_X = \widetilde{\alpha}_Y \circ \varphi_\alpha$. But for such a particular α one has $\widetilde{\alpha}_X = \overline{\alpha}_X$, so that $\varphi_\alpha = \big(H(f)\big)_\alpha$ by the uniqueness condition defining $\big(H(f)\big)_\alpha$. □

Putting $\mathscr{F} = \mathscr{E}(\partial J)$ for some object $J \in \mathscr{E}$, we observe that the fibre of $\mathscr{F}$ over $I \in \mathscr{E}$ has for objects the morphisms $\alpha\colon I \longrightarrow J$, a morphism $\gamma\colon \alpha \longrightarrow \beta$ being an endomorphism $\gamma\colon I \longrightarrow I$ such that $\beta \circ \gamma = \alpha$. Combining 8.2.6 and 8.2.7 thus yields an isomorphism of categories

$$\mathscr{E}(I, J) \cong \mathsf{Cart}\big(\mathscr{E}(\partial I), \mathscr{E}(\partial J)\big).$$

This justifies to some extent our claim that 8.2.7 can be thought of as some kind of generalization of 8.2.6.

8.3 Fibrations via pseudo-functors

Every category $\mathscr{E}$ can be seen as a 2-category where the only 2-cells are the identities. On the other hand, let us consider the 2-category Cat of (possibly large) categories, functors and natural transformations (see 7.1.4, volume 1). It thus makes sense to consider the 2-category $\mathsf{PsFun}(\mathscr{E}, \mathsf{Cat})$ of contravariant pseudo-functors, pseudo-natural transformations and modifications from $\mathscr{E}$ to Cat (see 7.5.1,2,3, volume 1). Clearly the situation is somewhat simpler than the general setting in section 7.5 of volume 1, since $\mathscr{E}$ has just trivial 2-cells. The reader who wants to handle more carefully the questions of size will be better off using a hierarchy of universes, but such precision is not really relevant for our purpose.

Theorem 8.3.1 *Let $\mathscr{E}$ be a category. There exists a 2-functor*

$$\varphi\colon \mathsf{PsFun}(\mathscr{E}, \mathsf{Cat}) \longrightarrow \mathsf{Fib}(\mathscr{E})$$

with the following properties:

(1) given two pseudo-functors $P, Q\colon \mathscr{E} \rightrightarrows \mathsf{Cat}$, φ induces an isomorphism of categories

$$\mathsf{PsFun}(P, Q) \longrightarrow \mathsf{Cart}\big(\varphi(P), \varphi(Q)\big);$$

(2) every fibration $F\colon \mathscr{F} \longrightarrow \mathscr{E}$ is isomorphic to a fibration $\varphi(P)$ arising from a pseudo-functor.

In short, the 2-categories $\mathsf{PsFun}(\mathscr{E}, \mathsf{Cat})$ and $\mathsf{Fib}(\mathscr{E})$ are 2-equivalent.

Proof To understand better the proof which will follow, it is sensible to start by constructing a pseudo-functor $P\colon \mathscr{E} \longrightarrow \mathsf{Cat}$ associated with a fibration $F\colon \mathscr{F} \longrightarrow \mathscr{E}$. This requires the axiom of choice.

- For an object $I \in \mathscr{E}$, $P(I)$ is the fibre $\mathscr{F}_I$.
- Given a morphism $\alpha\colon J \longrightarrow I$ in $\mathscr{E}$ we choose, for each object X in the fibre $\mathscr{F}_I$, a distinguished cartesian morphism $\alpha_X\colon X_\alpha \longrightarrow X$. We must define a functor $P\alpha\colon P(I) \longrightarrow P(J)$. We put $P(\alpha)(X) = X_\alpha$. Moreover if $f\colon X \longrightarrow Y$ is a morphism in the fibre $\mathscr{F}_I$, the composite $f \circ \alpha_X$ factors uniquely through the cartesian morphism α_Y via a morphism $X_\alpha \longrightarrow Y_\alpha$ in the fibre over J, which we choose as $P(\alpha)(f)$. Thus $P(\alpha)(f)$ is the unique morphism in the fibre over J such that $\alpha_Y \circ P(\alpha)(f) = f \circ \alpha_X$. The uniqueness condition in the definition of $P(\alpha)(f)$ implies immediately that $P(\alpha)$ is a functor.
- Given another morphism $\beta\colon K \longrightarrow J$ in $\mathscr{E}$, 8.1.4 implies immediately the existence of an isomorphism $P(\beta)P(\alpha)(X) \cong P(\alpha\beta)(X)$ in the fibre over K.

- Choosing in the same way $\alpha = 1_I$ in $\mathscr{E}$, we get immediately an isomorphism $P(1_I)(X) \cong X$; indeed the identity on $X \in \mathscr{F}_I$ is obviously a cartesian morphism, thus every cartesian morphism over 1_I is isomorphic to an identity.

Considering again lemma 8.1.4, it is straightforward to verify that P is indeed a pseudo-functor.

With the previous construction in mind, let us now construct a 2-functor

$$\varphi\colon \mathsf{PsFun}(\mathscr{E}, \mathsf{Cat}) \longrightarrow \mathsf{Fib}(\mathscr{E}).$$

Given a pseudo-functor $P\colon \mathscr{E} \longrightarrow \mathsf{Cat}$, we shall thus construct a fibration $G\colon \mathscr{G} \longrightarrow \mathscr{E}$ whose fibre at $I \in \mathscr{E}$ is precisely the category $P(I)$:

- an object of $\mathscr{G}$ is a pair (I, X) where $I \in \mathscr{E}$ and $X \in P(I)$ are respectively objects of $\mathscr{E}$ and $P(I)$;
- an arrow $(J, Y) \longrightarrow (I, X)$ in $\mathscr{G}$ is a pair (α, f) where $\alpha\colon J \longrightarrow I$ and $f\colon Y \longrightarrow P(\alpha)(X)$ are respectively arrows of $\mathscr{E}$ and $P(J)$;
- $G\colon \mathscr{G} \longrightarrow \mathscr{E}$ is just the first component functor, thus $G(I, X) = I$ and $G(\alpha, f) = \alpha$.

Clearly, we must still provide $\mathscr{G}$ with a composition which makes it a category, while G becomes a functor.

Consider arrows $(\alpha, f)\colon (J, Y) \longrightarrow (I, X)$, $(\beta, g)\colon (K, Z) \longrightarrow (J, Y)$ in $\mathscr{G}$. This yields the following composite in $P(K)$:

$$Z \xrightarrow{\ g\ } P(\beta)(Y) \xrightarrow{P(\beta)(f)} P(\beta)P(\alpha)(X) \xrightarrow{\ \cong\ } P(\alpha\beta)(X),$$

where the last isomorphism comes from the definition of pseudo-functor. Writing $f * g$ for this composite, the composition law is defined by the relation $(\alpha, f) \circ (\beta, g) = (\alpha \circ \beta, f * g)$. The associativity of this composition follows immediately from the first axiom for a pseudo-functor (see 7.5.1, volume 1). On the other hand the second axiom for a pseudo-functor indicates that the unit for the composition is equal to the isomorphism $X \xrightarrow{\cong} P(1_I)(X)$ given by the definition of a pseudo-functor. This proves that $\mathscr{G}$ is a category while the functoriality of G is obvious. Notice immediately that the fibre over an object $I \in \mathscr{E}$ is isomorphic to $P(I)$: this is again due to the isomorphism $X \cong P(1_I)(X)$ in the definition of a pseudo-functor.

To prove that $G\colon \mathscr{G} \longrightarrow \mathscr{E}$ is a fibration, let us consider a morphism $\alpha\colon J \longrightarrow I$ in $\mathscr{E}$ and an object $(I, X) \in \mathscr{G}_I$ in the fibre over I. The pair

$$(\alpha, 1_{P(\alpha)(X)})\colon (J, P(\alpha)(X)) \longrightarrow (I, X)$$

is thus a morphism of $\mathscr{G}$. This is a cartesian morphism because given another morphism in $\mathscr{G}$,

$$(X, f): (K, Z) \longrightarrow (I, X),$$

such that $\gamma = \alpha \circ \beta$ for some $\beta: K \longrightarrow J$, the composite h,

$$Z \xrightarrow{f} P(\gamma)(X) \xrightarrow{\cong} P(\beta)P(\alpha)(X),$$

yields the required unique factorization

$$(\beta, h): (K, Z) \longrightarrow (J, P(\alpha)(X)).$$

So $G: \mathscr{G} \longrightarrow \mathscr{E}$ is indeed a fibration.

Let us verify immediately that starting with a fibration $F: \mathscr{F} \longrightarrow \mathscr{E}$, constructing a corresponding pseudo-functor $P: \mathscr{E} \longrightarrow \mathsf{Cat}$ and the associated fibration $G: \mathscr{G} \longrightarrow \mathscr{E}$, the two fibrations are isomorphic. We freely use the previous notations, without redefining them. We construct a functor $H: \mathscr{F} \longrightarrow \mathscr{G}$.

- If $X \in \mathscr{F}$ is an object in the fibre over I, we define $H(X) = (I, X)$.
- If $Y \in \mathscr{F}$ is another object in the fibre over J and $f: Y \longrightarrow X$ is a morphism such that $F(f) = \alpha$, f factors uniquely as $f = \alpha_X \circ f_\alpha$ via a morphism $f_\alpha: Y \longrightarrow X_\alpha$ in the fibre over J; we define $H(f) = (\alpha, f_\alpha)$.

The functoriality of H follows immediately from the uniqueness condition in the definition of f_α.

Observe first that H is a cartesian functor. By construction $G \circ H = F$, since G is the first component functor. On the other hand if f is a cartesian functor, then both f and α_X are cartesian over α with the same codomain X; the factorization f_α is thus an isomorphism. Therefore $H(f) = (\alpha, f_\alpha)$ is isomorphic to the cartesian morphism $\big(\alpha, 1_{P(\alpha)(X)}\big)$ and so is cartesian.

It remains to prove that H is an isomorphism of categories. Obviously, H is bijective on the objects. To prove it is an equivalence, we must prove that given $X \in \mathscr{F}_I$, $Y \in \mathscr{F}_J$, the mapping

$$\mathscr{F}(Y, X) \longrightarrow \mathscr{G}\big((Y, J), (X, I)\big), \quad f \mapsto (\alpha, f_\alpha)$$

is a bijection, where for simplicity we have written $\alpha = F(f)$. This is obvious since α_X is a cartesian morphism.

We must now define

$$\varphi: \mathsf{PsFun}(\mathscr{E}, \mathsf{Cat}) \longrightarrow \mathsf{Fib}(\mathscr{E})$$

on the pseudo-natural transformations and the modifications. Let us consider two pseudo-functors $P, Q\colon \mathscr{E} \rightrightarrows \mathsf{Cat}$ and the corresponding fibrations $G\colon \mathscr{G} \longrightarrow \mathscr{E}$, $H\colon \mathscr{H} \longrightarrow \mathscr{E}$. Given a pseudo-natural transformation $\theta\colon P \Rightarrow Q$ is giving

- for each object $I \in \mathscr{E}$, a functor $\theta_I\colon P(I) \longrightarrow Q(I)$,
- for each arrow $\alpha\colon J \longrightarrow I$ of $\mathscr{E}$, an isomorphic natural transformation $\sigma_\alpha\colon \theta_J \circ P(\alpha) \Rightarrow Q(\alpha) \circ \theta_I$,

satisfying the compatibility conditions of 7.5.2, volume 1. From (θ, σ) we construct now a cartesian functor $L\colon \mathscr{G} \longrightarrow \mathscr{H}$.

- If $(I, X) \in \mathscr{G}_I$ is an object of $\mathscr{G}$, $L(I, X) = \big(I, \theta_I(X)\big)$.
- If $(\alpha, f)\colon (J, Y) \longrightarrow (I, X)$ is a morphism of $\mathscr{G}$, $L(\alpha, f)$ is the following composite:

$$\theta_J(Y) \xrightarrow{\theta_J(f)} \theta_J\big(P(\alpha)\big)(X) \xrightarrow{\sigma_{\alpha,X}} Q(\alpha)\big(\theta_I(X)\big).$$

It is routine to check the functoriality of L from the axioms for a pseudo-natural transformation. One has $K \circ L = G$ since K and G are just the first component functors. Moreover the cartesian morphism $\big(\alpha, 1_{P(\alpha)(X)}\big)$ of $\mathscr{G}$ is just by definition applied on $(\alpha, \sigma_{\alpha,X})$; since $\sigma_{\alpha,X}$ is an isomorphism in $Q(J)$, $(\alpha, \sigma_{\alpha,X})$ is isomorphic to the cartesian morphism $\big(\alpha, 1_{(Q\alpha)(X)}\big)$ via an isomorphism in the fibre $\mathscr{H}_J$ and is therefore an isomorphism.

Conversely if $N\colon \mathscr{G} \longrightarrow \mathscr{H}$ is a cartesian functor, let us prove it arises from a unique pseudo-natural transformation (θ, σ) via the previous construction. This construction imposes the relations $N(I, X) = \big(I, \theta_I(X)\big)$, $N\big(\alpha, 1_{P(\alpha)(X)}\big) = (\alpha, \sigma_{\alpha,X})$ and $N(\alpha, f) = \big(\alpha, \sigma_{\alpha,X} \circ \theta_J(f)\big)$ for each arrow $(\alpha, f)\colon (J, Y) \longrightarrow (I, X)$ in $\mathscr{G}$. This uniquely defines (θ, σ) in terms of N. Since N is cartesian, it maps the cartesian morphism $\big(\alpha, 1_{P(\alpha)(X)}\big)$ to another; therefore $(\alpha, \sigma_{\alpha,X})$ is isomorphic to the cartesian morphism $\big(\alpha, 1_{(Q\alpha)(X)}\big)$ via an isomorphism in the fibre $\mathscr{H}_J$; since the unique possible factorization is $(1_J, \sigma_{\alpha,X})$, $\sigma_{\alpha,X}$ is indeed an isomorphism in $Q(J)$. It is then straightforward to verify the axioms for a pseudo-natural transformation from the functoriality of F.

Finally consider another pseudo-natural transformation $(\tau, \varepsilon)\colon P \Rightarrow Q$ and the corresponding cartesian functor $M\colon \mathscr{G} \longrightarrow \mathscr{H}$. Giving a modification $\Xi\colon (\theta, \sigma) \Rightarrow (\tau, \varepsilon)$ is giving a family $\Xi_I\colon \theta_I \Rightarrow \tau_I$ of natural transformations, for $I \in \mathscr{E}$, with just the requirement that diagram 8.7 commutes since the only 2-cells in $\mathscr{E}$ are the identities. One defines $\xi\colon L \Rightarrow M$ by putting $\xi_{(I,X)} = \Xi_{I,X}$. Since each Ξ_I is natural, ξ is already natural

$$\begin{array}{ccc}
\theta_J P(\alpha)(X) & \xrightarrow{\sigma_{\alpha,X}} & Q(\alpha)(\theta_I X) \\
{\scriptstyle \Xi_{J,P(\alpha)(X)}}\downarrow & & \downarrow{\scriptstyle Q(\alpha)(\Xi_{I,X})} \\
\tau_J P(\alpha)(X) & \xrightarrow[\varepsilon_{\alpha,X}]{} & Q(\alpha)(\tau_I X)
\end{array}$$

Diagram 8.7

with respect to all the morphisms in a fibre; the additional axiom for Ξ to be a modification then yields the naturality of ξ, just by definition of L, M on an arbitrary morphism of $\mathscr{G}$. The cartesianness of ξ is just the fact that each morphism $\xi_{(I,X)}$ is in the fibre over I.

To conclude, it remains to verify that the previous construction mapping the modification Ξ to the cartesian natural transformation ξ is bijective. But giving ξ is giving a morphism $\xi_{(I,X)}$ in the fibre $\mathscr{H}_I$ over I for each object $X \in P(I)$. Putting $\Xi_{I,X} = \xi_{(J,X)}$ already yields a family of natural transformations $\Xi_I: \theta_I \Rightarrow \tau_I$, just by naturality of ξ. The additional requirement for Ξ to be a modification is just the naturality of ξ applied to the cartesian morphism $(\alpha, 1_{P(\alpha)(X)})$ in $\mathscr{G}$. □

Corollary 8.3.2 *Consider an object I of a category $\mathscr{E}$. Via the correspondence described in 8.3.1, the fibration $\mathscr{E}(\partial I) \longrightarrow \mathscr{E}$ of 8.2.5 corresponds with the "discrete representable functor"*

$$\mathscr{E}(-, I): \mathscr{E} \longrightarrow \mathsf{Cat}$$

mapping an object J to the discrete category having $\mathscr{E}(J, I)$ as set of objects.

Proof The fibration is the source functor $\mathscr{E}/I \longrightarrow \mathscr{E}$ (see 8.2.5), thus the fibre over J indeed has the arrows $J \longrightarrow I$ as objects and just the identities $J \longrightarrow J$ as morphisms. Given a morphism $\alpha: K \longrightarrow J$ and an object (J, j) in the fibre over J, the corresponding cartesian morphism is just $\alpha: (K, j \circ \alpha) \longrightarrow (J, j)$. Therefore "the" pseudo-functor corresponding with $\mathscr{E}/I \longrightarrow \mathscr{E}$ indeed acts by composition on the morphisms. Thus this pseudo-functor is exactly the usual representable functor $\mathscr{E}(-, I): \mathscr{E} \longrightarrow \mathsf{Set}$ ... where each set is identified with the corresponding discrete category. □

Corollary 8.3.2, together with theorem 8.3.1, now explains the terminology "Yoneda embedding" or "Yoneda lemma" used in 8.2.5 and 8.2.7. For example 8.2.7 indicates via those remarks an equivalence of categories

$$\mathsf{PsNat}(\mathscr{E}(-, I), P) \cong P(I)$$

where P is the pseudo-functor associated with the fibration $\mathscr{F} \longrightarrow \mathscr{E}$, $\mathscr{E}(-, I)$ is the (pseudo)-functor associated with the fibration $\mathscr{E}(\partial I) \longrightarrow \mathscr{E}$ and PsNat indicates the category of pseudo-natural transformations and modifications. This is indeed a Yoneda lemma for pseudo-functors.

The fibration $\mathscr{E}(\partial I) \longrightarrow \mathscr{E}$ presents the particular fact that it is the fibration associated with an actual functor $\mathscr{E} \longrightarrow \mathsf{Cat}$, not just with a pseudo-functor. A fibration $\mathscr{F} \longrightarrow \mathscr{E}$ associated with an actual functor $\mathscr{E} \longrightarrow \mathsf{Cat}$ thus has the particular property that a "compatible choice of cartesian morphisms" can be made: for each morphism $\alpha: J \longrightarrow I$ in $\mathscr{E}$ and each object $X \in \mathscr{F}_I$, applying $P(\alpha)$, one can choose a distinguished cartesian morphism $\alpha_X: P(\alpha)(X) \longrightarrow (X)$ in a way which is compatible with the identities and the composition, i.e. the distinguished cartesian morphisms α_X constitute a subcategory of $\mathscr{F}$. When this is the case, $\mathscr{F} \longrightarrow \mathscr{E}$ is called a "split fibration".

Definition 8.3.3 *A fibration $F: \mathscr{F} \longrightarrow \mathscr{E}$ is called a split fibration when, for each arrow $\alpha: J \longrightarrow I$ and each object $X \in \mathscr{F}_I$, it is possible to choose a distinguished cartesian morphism $\alpha_X: X_\alpha \longrightarrow X$ over α in such a way that all those morphisms α_X constitute a subcategory of $\mathscr{F}$.*

Proposition 8.3.4 *Each fibration $F: \mathscr{F} \longrightarrow \mathscr{E}$ is equivalent to a split fibration $G: \mathscr{G} \longrightarrow \mathscr{E}$.*

Proof Applying 8.3.1, the fibration $F: \mathscr{F} \longrightarrow \mathscr{E}$ can be described by a pseudo-functor

$$P: \mathscr{E} \longrightarrow \mathsf{Cat}, \quad I \mapsto P(I) = \mathscr{F}_I.$$

Applying our "fibred Yoneda lemma" (see 8.2.7), let us observe that P is equivalent in the 2-category $\mathsf{PsFun}(\mathscr{E}, \mathsf{Cat})$ to the actual functor

$$Q: \mathscr{E} \longrightarrow \mathsf{Cat}, \quad I \mapsto \mathsf{Cart}(\mathscr{E}(\partial I), \mathscr{F})$$

where the action of Q on a morphism $\alpha: J \longrightarrow I$ in $\mathscr{E}$ is just composition with the cartesian morphism $\mathscr{E}(\partial\alpha): \mathscr{E}(\partial J) \longrightarrow \mathscr{E}(\partial I)$ studied in 8.2.5. It remains by 8.3.1 to consider the fibration $G: \mathscr{G} \longrightarrow \mathscr{G}$ associated with the functor Q. It is equivalent to $F: \mathscr{F} \longrightarrow \mathscr{E}$ since Q is equivalent to P. To prove it is split, it suffices to consider the cartesian morphism $\big(\alpha, 1_{(Q\alpha)(H)}\big)$ for each arrow $\alpha: J \longrightarrow I$ in $\mathscr{E}$ and each cartesian functor $H \in Q(I)$: these morphisms constitute a subcategory of $\mathscr{G}$ because Q

is an actual functor, thus all the coherence isomorphisms are identities. □

We have seen in 8.1.9.b that every category $\mathscr{C}$ induces a fibration φ: Set($\mathscr{C}$)$\longrightarrow$Set. The "dual" of this fibration should be the fibration φ^*: Set($\mathscr{C}^*$)$\longrightarrow$Set associated with the dual category $\mathscr{C}^*$. Observe that Set($\mathscr{C}$) and Set($\mathscr{C}^*$) are by no means dual categories. Indeed consider a mapping $\alpha: I \longrightarrow J$ in Set and $\big(\alpha, (f_i)_{i\in I}\big): (C_i)_{i\in I} \longrightarrow (D_j)_{j\in J}$, a morphism in Set($\mathscr{C}$); this is a family $\big(f_i: C_i \longrightarrow D_{\alpha(i)}\big)_{i\in I}$ of morphisms in $\mathscr{C}$. On the other hand a morphism $\big(\alpha, (g_i)_{i\in I}\big): (C_i)_{i\in I} \longrightarrow (D_j)_{j\in J}$ in Set($\mathscr{C}^*$) is a family of morphisms $\big(g_i: C_i \longrightarrow D_{\alpha(i)}\big)_{i\in I}$ in the dual category $\mathscr{C}^*$, thus a family of morphisms $\big(g_i: D_{\alpha(i)} \longrightarrow C_i\big)_{i\in I}$ in $\mathscr{C}$. Such a family is by no means a morphism $(J_j)_{j\in J} \longrightarrow (C_i)_{i\in I}$ in Set($\mathscr{C}$) – except when α is a bijection! This last remark implies in particular that the fibres of Set($\mathscr{C}^*$) are the dual of the fibres of Set($\mathscr{C}$). Going back to 8.3.1 suggests the following definition.

Definition 8.3.5 *Consider a fibration $F: \mathscr{F} \longrightarrow \mathscr{E}$ and a corresponding pseudo-functor $P: \mathscr{E} \longrightarrow$ Cat generating $\mathscr{F}$ up to an isomorphism (see 8.3.1). The composite*

$$\mathscr{E} \xrightarrow{\ P\ } \mathsf{Cat} \xrightarrow{\ (-)^*\ } \mathsf{Cat}$$

where $(-)^*$ is the "dual category functor" corresponds by 8.3.1 with a fibred category $G: \mathscr{G} \longrightarrow \mathscr{E}$, called the dual fibration of $F: \mathscr{F} \longrightarrow \mathscr{E}$.

Clearly, the dual fibration is just defined up to isomorphism since the choice of the pseudo-functor P can only be made up to isomorphism.

Let us conclude this section with a point of terminology, related once more with the fact that the pseudo-functor P associated with a fibration is just defined up to isomorphism.

Definition 8.3.6 *Let $F: \mathscr{F} \longrightarrow \mathscr{E}$ be a fibration and $\alpha: J \longrightarrow I$ a morphism in $\mathscr{E}$. We call an "inverse image functor" $\alpha^*: \mathscr{F}_I \longrightarrow \mathscr{F}_J$ between the corresponding fibres any functor $P(\alpha): P(I) \longrightarrow P(J)$, where P is a pseudo-functor associated with the fibration $F: \mathscr{F} \longrightarrow \mathscr{E}$.*

Since two pseudo-functors corresponding with a same fibration are isomorphic, the inverse image functors α^* are just defined "up to isomorphism" (see 8.3.1). With the previous notations, given an object $X \in \mathscr{F}_J$ we write

$$\alpha_X: \alpha^*(X) \longrightarrow X$$

for a distinguished cartesian morphism. Again, α_X is just defined up to isomorphism in the fibre over I.

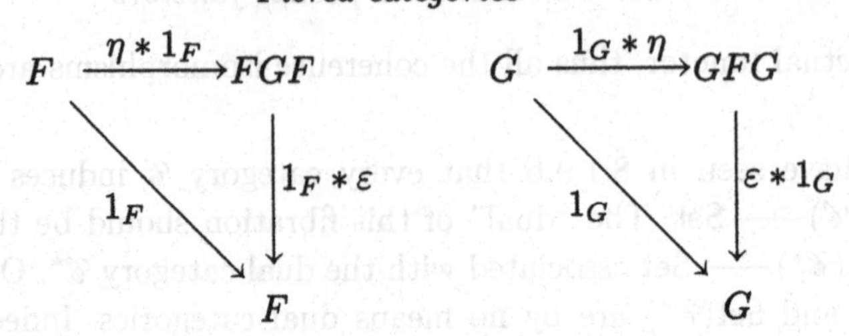

Diagram 8.8

8.4 Fibred adjunctions

The notion of "adjoint arrows" (see 7.1.2, volume 1) can be immediately applied to the 2-category $\mathsf{Fib}(\mathscr{E})$, for a base category $\mathscr{E}$:

Definition 8.4.1 *Consider two fibrations $F\colon \mathscr{F} \longrightarrow \mathscr{E}$, $G\colon \mathscr{G} \longrightarrow \mathscr{E}$ and two cartesian functors $K\colon \mathscr{F} \longrightarrow \mathscr{G}$, $L\colon \mathscr{G} \longrightarrow \mathscr{F}$ in the 2-category $\mathsf{Fib}(\mathscr{E})$. L is a fibred left adjoint to K when there exist two cartesian natural transformations*

$$\eta\colon 1_{\mathscr{G}} \Rightarrow K \circ L, \quad \varepsilon\colon L \circ K \Rightarrow 1_{\mathscr{F}}$$

such that the triangles of diagram 8.8 commute.

Proposition 8.4.2 *Consider two fibrations $F\colon \mathscr{F} \longrightarrow \mathscr{E}$, $G\colon \mathscr{G} \longrightarrow \mathscr{E}$ over $\mathscr{E}$ and a cartesian functor $K\colon \mathscr{F} \longrightarrow G$. The following conditions are equivalent:*

(1) K has a fibred left adjoint functor L;

(2) (a) for every object $I \in \mathscr{E}$, $K\colon \mathscr{F}_I \longrightarrow \mathscr{G}_I$ has a left adjoint functor L_I;

(b) for every morphism $\alpha\colon J \longrightarrow I$ and every object $X \in \mathscr{G}_I$, the morphism

$$\overline{\alpha}_X\colon L_J\alpha^*(X) \longrightarrow \alpha^* L_I(X),$$

which we shall describe now, is required to be an isomorphism.

To define the morphism $\overline{\alpha}_X$ of condition (2)(b), observe that the morphism

$$K(\alpha_{L_I X})\colon K\alpha^* L_I(X) \longrightarrow KL_I(X)$$

is cartesian since K is a cartesian functor; the composite

$$\alpha^*(X) \xrightarrow{\alpha_X} X \xrightarrow{\eta_X} KL_I(X),$$

where η is the unit of the adjunction $L_I \dashv K$, thus factors through $K(\alpha_{L_I X})$ via a morphism

$$\widetilde{\alpha}_X\colon \alpha^* X \longrightarrow K\alpha^* L_I X;$$

this yields by the adjunction $L_J \dashv K$ the required morphism

$$\overline{\alpha}_X \colon L_J\alpha^*(X) \longrightarrow \alpha^* L_I(X).$$

Proof The existence of a left adjoint L_I for each fibre means the existence of a functor $L_I \colon \mathscr{G}_I \longrightarrow \mathscr{F}_I$ for each fibre, together with morphisms

$$\eta_X \colon X \longrightarrow KL_I(X), \quad \varepsilon_Y \colon L_I K(Y) \longrightarrow Y$$

in the corresponding fibres, those morphisms satisfying both the naturality condition with respect to morphisms in the fibres and the triangular identities for adjunction.

Starting with a fibred adjoint L to K as in 8.4.1, it suffices to restrict L, η, ε to each fibre to get the required adjunctions $L_I \dashv K$. Since L is a cartesian functor, the morphism

$$L(\alpha_X) \colon L\alpha^*(X) \longrightarrow L(X)$$

is cartesian; by uniqueness (up to an isomorphism) of the cartesian morphism $\alpha_{L(X)}$, the unique factorization

$$\overline{\alpha}_X \colon L\alpha^*(X) \longrightarrow \alpha^* L(X)$$

is thus an isomorphism.

Conversely, suppose that conditions (2)(a), (2)(b) are satisfied and use the notation of the beginning of the proof. It remains to define the functor L on all the morphisms and prove the naturality of η, ε with respect to all the morphisms (not just the morphisms in a fibre). Since every morphism in a fibration factors as a morphism in a fibre followed by a cartesian morphism, it remains to handle the case of cartesian morphisms.

Considering the first part of the proof, we define $L(\alpha_X)$ as the composite

$$L_J\alpha^*(X) \xrightarrow{\ \overline{\alpha}_X\ } \alpha^* L_I(X) \xrightarrow{\ \alpha_{L_I X}\ } L_I(X);$$

the functoriality of L is immediate. The naturality of η with respect to the morphism α_X is just the commutativity of diagram 8.9. Notice that L is a cartesian functor since $L\alpha^* X$ is isomorphic to the cartesian morphism $\alpha_{L_I X}$, thus is cartesian.

To conclude the proof it remains to check the naturality of ε or, equivalently (see 3.1.5, volume 1), to prove that the pair $(L_I(X), \eta_X)$ is the reflection of $X \in \mathscr{G}$ along $K \colon \mathscr{F} \longrightarrow \mathscr{G}$. Indeed given $f \colon X \longrightarrow K(Y)$ in $\mathscr{G}$, just put $J = GK(Y) = F(Y)$, $\alpha = G(f) \colon I \longrightarrow J$ in $\mathscr{E}$. There is a

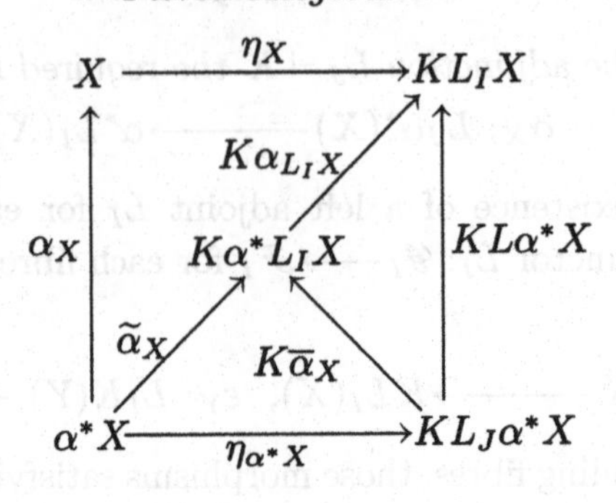

Diagram 8.9

unique factorisation of f as $f = \alpha_{K(Y)} \circ f'$, for $f'\colon X \longrightarrow \alpha^*K(Y)$ in the fibre $\mathscr{G}_I$. Since K is a cartesian functor, we get a composite

$$X \xrightarrow{f'} \alpha^*K(Y) \xrightarrow{\cong} K\alpha^*(Y)$$

which, by the adjunction $L_I \dashv K$, yields a unique factorization in the fibre $\mathscr{F}_I$, $f''\colon L_I(X) \longrightarrow \alpha^*(Y)$, and thus the required morphism

$$L_I(X) \xrightarrow{f''} \alpha^*(Y) \xrightarrow{\alpha_Y} Y. \qquad \square$$

In order to introduce a new useful construction on fibrations, we "prove" two lemmas.

Lemma 8.4.3 *Let $\mathscr{E}$ be a category. The category $\mathsf{Fib}(\mathscr{E})$ of fibrations over $\mathscr{E}$ has finite products.*

Proof The terminal object is the identity functor $\mathscr{E} = \mathscr{E}$. The product of two fibrations $F\colon \mathscr{F} \longrightarrow \mathscr{E}$, $G\colon \mathscr{G} \longrightarrow \mathscr{E}$ is just their pullback as functors (see 8.1.15 and 8.1.12). $\square$

The next result indicates that $\mathsf{Fib}(\mathscr{E})$ is "cartesian closed up to equivalences of categories".

Lemma 8.4.4 *Consider two fibrations $G\colon \mathscr{G} \longrightarrow \mathscr{E}$, $H\colon \mathscr{H} \longrightarrow \mathscr{E}$. There exists a fibration $H^G\colon \mathscr{H}^{\mathscr{G}} \longrightarrow \mathscr{E}$ such that for every fibration $F\colon \mathscr{F} \longrightarrow \mathscr{E}$, one has equivalences of categories*

$$\mathsf{Cart}(\mathscr{F} \times_{\mathscr{E}} \mathscr{G}, \mathscr{H}) \cong \mathsf{Cart}(\mathscr{F}, \mathscr{H}^{\mathscr{G}}).$$

Proof If the result holds, it must hold in particular when $\mathscr{F} = \mathscr{E}(\partial I)$ for some object $I \in \mathscr{E}$ (see 8.2.5). This means by 8.2.7 the existence of equivalences of categories

$$(\mathscr{H}^{\mathscr{G}})_I \cong \mathsf{Cart}(\mathscr{E}(\partial I), \mathscr{H}^{\mathscr{G}}) \cong \mathsf{Cart}(\mathscr{E}(\partial I) \times_{\mathscr{E}} \mathscr{G}, \mathscr{H}).$$

This indicates immediately how to construct $\mathscr{H}^{\mathscr{G}}$: it is the split fibration corresponding with the functor mapping I to $\mathsf{Cart}(\mathscr{E}(\partial I)\times_{\mathscr{E}}\mathscr{G},\mathscr{H})$ and acting by composition (see 8.2.5) on the arrows $\alpha: I \longrightarrow J$; see 8.3.1.

Now observe that for two fibrations $A: \mathscr{A} \longrightarrow \mathscr{E}$, $B: \mathscr{B} \longrightarrow \mathscr{E}$, replacing $(\mathscr{A}, A)$, $(\mathscr{B}, B)$ by equivalent fibrations (see 7.1.2, volume 1) will transform the category $\mathsf{Cart}(\mathscr{A},\mathscr{B})$ into an equivalent category. Since the final statement of the present proposition asserts the existence of an equivalence of categories, we can freely replace each fibration by an equivalent one. Applying 8.3.4, there is thus no restriction in supposing that all the fibrations involved in the problem are split fibrations. We recall that $\mathscr{H}^{\mathscr{G}}$, by construction, is already a split fibration.

Going back to the considerations of section 8.3, we shall work with functors $S, T: \mathscr{E} \rightrightarrows \mathsf{Cart}$ instead of split fibrations. Observe that a pseudo-natural transformation (see 7.5.2, volume 1) $S \Rightarrow T$ reduces to giving

- for every object $I \in \mathscr{E}$, a functor $\theta_I: S(I) \longrightarrow T(I)$,
- for every morphism $\alpha: J \longrightarrow I$ of $\mathscr{E}$, an isomorphic natural transformation $\tau_\alpha: T(\alpha) \circ \theta_I \Rightarrow \theta_J \circ S(\alpha)$,

in such a way that the following conditions hold:

- $\tau_{1_I} = 1_{\theta_I}$;
- if $\beta: K \longrightarrow J$ is another morphism of $\mathscr{E}$,

$$\tau_{\alpha\circ\beta} = \tau_\beta * 1_{S(\alpha)} = 1_{T(\beta)} * \tau_\alpha.$$

Observe finally that performing in $\mathsf{Fib}(\mathscr{E})$ the product of two fibrations $A: \mathscr{A} \longrightarrow \mathscr{E}$, $B: \mathscr{B} \longrightarrow \mathscr{E}$ is just performing their pullback as functors (see 8.4.3), thus the fibre $(\mathscr{A}\times_{\mathscr{E}}\mathscr{B})_I$ is just the product $\mathscr{A}_I \times \mathscr{B}_I$. So the product of fibrations in $\mathsf{Fib}(\mathscr{E})$ corresponds to the pointwise product of the corresponding pseudo-functors.

Given two functors $Q, R: \mathscr{E} \rightrightarrows \mathsf{Cat}$ one thus defines a new functor

$$R^Q: \mathscr{E} \longrightarrow \mathsf{Cat}, \quad I \mapsto \mathsf{PsNat}(\mathscr{E}(-,I) \times Q, R)$$

acting by composition on the arrows of $\mathscr{E}$ (see 8.3.2). Given a third functor $P: \mathscr{E} \longrightarrow \mathsf{Cat}$, we must prove the existence of an equivalence of categories

$$\mathsf{PsNat}(P \times Q, R) \cong \mathsf{PsNat}(P, R^Q).$$

Let us recall that, in terms of a pseudo-functor $S: \mathscr{E} \longrightarrow \mathsf{Cat}$, proposition 8.2.7 is just a pseudo-Yoneda lemma attesting the existence of an equivalence of categories

$$S(I) \cong \mathsf{PsNat}(\mathscr{E}(-,I),S)$$

for every object $I \in \mathscr{E}$. Putting $S = R^Q$, we have thus already, by definition of R^Q, the equivalences

$$\mathsf{PsNat}(\mathscr{E}(-,I)\times Q,R) \cong R^Q(I) \cong \mathsf{PsNat}(\mathscr{E}(-,I),R^Q),$$

which is the result in the case $P = \mathscr{E}(-,I)$.

Deducing the result for an arbitrary functor $P\colon \mathscr{E} \longrightarrow \mathsf{Cat}$ is rather straithforward, but a bit lengthy. Since we shall not need the result explicitly , but just the construction of the fibration $\mathscr{H}^{\mathscr{G}}$, we give the idea of the proof and leave the details as an exercise. Given a pseudo-natural transformation $(\theta,\tau)\colon P\times Q \Rightarrow R$, one defines a pseudo-natural transformation $(\overline{\theta},\overline{\tau})\colon P \Rightarrow R^Q$ by choosing $\overline{\theta}_I$ to be the following functor:

$$P(I) \longrightarrow \mathsf{PsNat}(\mathscr{E}(-,I)\times Q,R),\quad X \mapsto \overline{\theta}_I(X)$$

$$\overline{\theta}_I(X)_J\colon \mathscr{E}(J,I)\times Q(J) \longrightarrow R(J),\quad (\alpha,Y) \mapsto \theta_J(P(\alpha)(X),Y)$$

for objects $I,J \in \mathscr{E}$. Conversely, given a pseudo-natural transformation $(\overline{\theta},\overline{\tau})\colon P \Rightarrow R^Q$, one defines a corresponding pseudo-natural transformation $(\theta,\tau)\colon P\times Q \Rightarrow R$ by the formula $\theta_I(X,Y) = \overline{\theta}_I(X)_I$, where $X \in P(I)$ and $Y \in Q(I)$.

An alternative proof of the present proposition would have been to consider, for every pseudo-functor $S\colon \mathscr{E} \longrightarrow \mathsf{Cat}$, the equivalence of categories

$$\mathsf{PsNat}(P,\mathsf{PsNat}(Y-,S)) \cong \mathsf{PsNat}(P,S)$$

where Y is now the fibred Yoneda embedding mapping I to $\mathscr{E}(-,I)$ (see 8.3.2 and 8.2.6) and the equivalence follows immediately from the fibred Yoneda lemma (see 8.2.7). In view of 6.6.18, this equivalence presents P as a "pseudo-Cat-enriched limit" of Cat-representable functors. Developing the theory of those pseudo-Cat-enriched limits allows us then to mimic the proof of 6.1.9.d. □

From the previous "proof", let us at least extract the important definition of the exponentiation of two fibrations.

Definition 8.4.5 *Let $G\colon \mathscr{G} \longrightarrow \mathscr{E}$, $H\colon \mathscr{H} \longrightarrow \mathscr{E}$ be two fibrations. The fibration $H^G\colon \mathscr{H}^{\mathscr{G}} \longrightarrow \mathscr{E}$ is by definition the split fibration corresponding with the functor*

$$\mathscr{E} \longrightarrow \mathsf{Cat},\quad I \mapsto \mathsf{Cart}(\mathscr{E}/I \times_{\mathscr{E}} \mathscr{G}, \mathscr{H})$$

acting by composition on the morphisms of $\mathscr{E}$ (see 8.2.5).

With chapter 6 in mind $\mathscr{H}^{\mathscr{G}}$ can thus be thought as the fibration of cartesian functors from $\mathscr{G}$ to $\mathscr{H}$. When $\mathscr{C}$ is a category internal to $\mathscr{E}$, the fibration $\mathscr{H}^{\mathscr{E}(\mathscr{C})}$ will generally be written $\mathscr{H}^{\mathscr{C}}$ (see 8.2.4); this is intuitively the fibration of $\mathscr{H}$-valued presheaves on $\mathscr{C}$. When I is just an object of $\mathscr{E}$, the fibration $\mathscr{H}^{\mathscr{E}(\partial I)}$ will generally be written $\mathscr{H}^{I}$ (see 8.2.5); this is intuitively the fibration of I-indexed families in $\mathscr{H}$.

Let us conclude this section with a rather obvious "adjoint lifting theorem" for fibrations.

Proposition 8.4.6 *Consider fibrations $F\colon \mathscr{F} \longrightarrow \mathscr{E}$, $G\colon \mathscr{G} \longrightarrow \mathscr{E}$ and a cartesian functor $K\colon \mathscr{F} \longrightarrow \mathscr{G}$ having a fibred left adjoint functor. For every internal category $\mathscr{C}$ of $\mathscr{E}$, the cartesian functor $K^{\mathscr{C}}\colon \mathscr{F}^{\mathscr{C}} \longrightarrow \mathscr{G}^{\mathscr{C}}$ of composition with K has a fibred left adjoint functor.*

Proof For an object $I \in \mathscr{E}$, $K^{\mathscr{C}}_I$ is the functor

$$\mathsf{Cart}\big(\mathscr{E}/I \times_{\mathscr{E}} \mathscr{E}(\partial\mathscr{C}), \mathscr{F}\big) \longrightarrow \mathsf{Cart}\big(\mathscr{E}/I \times_{\mathscr{E}} \mathscr{E}(\partial\mathscr{C}), \mathscr{G}\big)$$

acting by composition with K. Its left adjoint is just the corresponding functor $L^{\mathscr{C}}_I$ acting by composition with L, where $L\colon \mathscr{G} \longrightarrow \mathscr{F}$ is a fibred left adjoint to K. The compatibility condition of 8.4.2 is obvious. Indeed, given a morphism $\alpha\colon J \longrightarrow I$, the functors α^* are just composition with

$$\mathscr{E}/\alpha \times_{\mathscr{E}} \mathsf{id}\colon \mathscr{E}/J \times_{\mathscr{E}} \mathscr{E}(\partial\mathscr{C}) \longrightarrow \mathscr{E}/I \times_{\mathscr{E}} \mathscr{E}(\partial\mathscr{C}),$$

while the functors $L^{\mathscr{C}}_I$ are just composition with $L\colon \mathscr{G} \longrightarrow \mathscr{F}$. Therefore, given a functor

$$X\colon \mathscr{E}/I \times_{\mathscr{E}} \mathscr{E}(\partial\mathscr{C}) \longrightarrow \mathscr{G},$$

the isomorphism $L^{\mathscr{C}}_J \alpha^* X \cong \alpha^* L^{\mathscr{C}}_I X$ is just the associativity relation

$$L \circ \big(X \circ (\mathscr{E}/\alpha \times_{\mathscr{E}} \mathsf{id})\big) = (L \circ X) \circ (\mathscr{E}/\alpha \times_{\mathscr{E}} \mathsf{id}). \qquad \square$$

8.5 Completeness of a fibration

Given a fibration $F\colon \mathscr{F} \longrightarrow \mathscr{E}$ and a small category $\mathscr{D}$, we have defined in 8.1.17 the fibration $\mathscr{F}^{(\mathscr{D})} \longrightarrow \mathscr{E}$ of diagrams of shape $\mathscr{D}$ in $\mathscr{F}$: the corresponding pseudo-functor (see 8.3.1) maps an object $I \in \mathscr{E}$ to the category $\mathsf{Fun}(\mathscr{D}, \mathscr{F}_I)$ of diagrams of shape $\mathscr{D}$ in the fibre $\mathscr{F}_I$; its action on an arrow $\alpha\colon J \longrightarrow I$ is just composition with the inverse image functor $\alpha^*\colon \mathscr{F}_I \longrightarrow \mathscr{F}_J$ (see 8.3.6).

Proposition 3.2.3, volume 1, suggests how to define limits of shape $\mathscr{D}$ in the fibration $\mathscr{F}$. It suffices to consider the diagram 8.10, which is commutative, and the corresponding factorization, still written Δ,

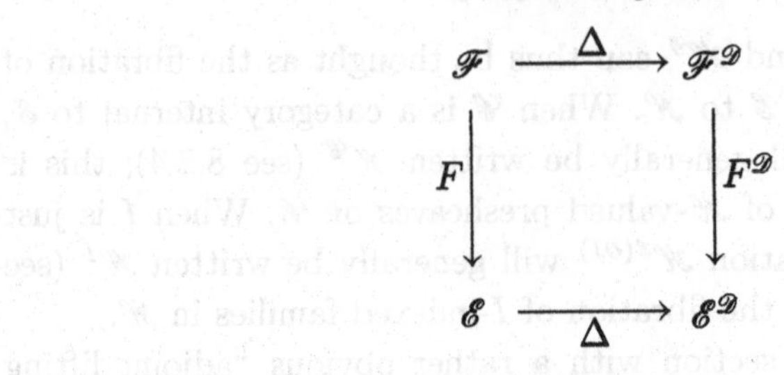

Diagram 8.10

$$\Delta: \mathscr{F} \longrightarrow \mathscr{F}^{(\mathscr{D})},$$

through the pullback in 8.1.17. Observe that $\Delta: \mathscr{F} \longrightarrow \mathscr{F}^{(\mathscr{D})}$ is the cartesian functor corresponding by 8.3.1 with the pseudo-natural transformation δ given by

$$\delta_I: \mathscr{F}_I \longrightarrow \mathsf{Fun}(\mathscr{D}, \mathscr{F}_I), \quad X \mapsto \Delta(X),$$

and having the identities as coherence isomorphisms. (We assume that a family of inverse image functors has been chosen once for all.)

Definition 8.5.1 *A fibration $F: \mathscr{F} \longrightarrow \mathscr{E}$ has all limits of shape $\mathscr{D}$, for a small category $\mathscr{D}$, when the cartesian functor*

$$\Delta: \mathscr{F} \longrightarrow \mathscr{F}^{(\mathscr{D})}$$

has a fibred right adjoint.

As expected, the existence of $\mathscr{D}$-limits in a fibration is related with the existence of $\mathscr{D}$-limits in the fibres.

Proposition 8.5.2 *Let $F: \mathscr{F} \longrightarrow \mathscr{E}$ be a fibration and $\mathscr{D}$ a small category. The following conditions are equivalent:*

(1) the fibration $(\mathscr{F}, F)$ has all limits of shape $\mathscr{D}$;

(2) for each fibre $\mathscr{F}_I$, $I \in \mathscr{E}$, and each functor $H: \mathscr{D} \longrightarrow \mathscr{F}_I$, the limit of H exists and is preserved by every inverse image functor α^*, for every morphism $\alpha: J \longrightarrow I$ in $\mathscr{E}$.

Proof This is an immediate consequence of 8.4.2 and 3.2.3, volume 1. □

It could appear natural to define the completeness of a fibration $F: \mathscr{F} \longrightarrow \mathscr{E}$ by requiring the existence of limits of shape $\mathscr{D}$, for every small category $\mathscr{D}$. But clearly such a definition of completeness would depend on the consideration of the internal categories $\mathcal{D}$ of the category of sets, while in the theory of fibrations over $\mathscr{E}$ one should rather consider the internal categories $\mathscr{D}$ of $\mathscr{E}$.

If $\mathscr{D}$ is an internal category of $\mathscr{E}$ and $F\colon \mathscr{F} \longrightarrow \mathscr{E}$ is a fibration, we have considered as a special case of 8.4.5 the fibration $F^{\mathscr{D}}\colon \mathscr{F}^{\mathscr{D}} \longrightarrow \mathscr{E}$ of "$\mathscr{F}$-valued presheaves on $\mathscr{D}$". We recall that $\mathscr{F}^{\mathscr{D}}$ is a split fibration whose fibre at $I \in \mathscr{E}$ is given by

$$(\mathscr{F}^{\mathscr{D}})_I \cong \mathsf{Cart}(\mathscr{E}/I \times_{\mathscr{E}} \mathscr{E}(\mathscr{D}), \mathscr{F}).$$

As in the case of ordinary categories, we want to consider a cartesian functor

$$\Delta\colon \mathscr{F} \longrightarrow \mathscr{F}^{\mathscr{D}}$$

corresponding with the choice of "constant diagrams of shape $\mathscr{D}$". This functor must clearly correspond via lemma 8.4.4 with the projection

$$\mathscr{F} \times_{\mathscr{E}} \mathscr{E}(\mathscr{D}) \longrightarrow \mathscr{F}.$$

Let us give an explicit description of it.

For every object $I \in \mathscr{E}$, we define a functor

$$\Delta I\colon \mathscr{F}_I \longrightarrow \mathsf{Cart}(\mathscr{E}/I \times_{\mathscr{E}} \mathscr{E}(\mathscr{D}), \mathscr{F})$$

in the following way. An object $X \in \mathscr{F}_I$ is mapped to the composite

$$\mathscr{E}/I \times_{\mathscr{E}} \mathscr{E}(\mathscr{D}) \longrightarrow \mathscr{E}/I \longrightarrow \mathscr{F},$$

where the first functor is just the projection and the second functor corresponds with $X \in \mathscr{F}_I$ by the pseudo-Yoneda-lemma (see 8.2.7), thus it maps $\alpha\colon J \longrightarrow I$ to $\alpha^*(X)$. This construction extends immediately to the arrows $f\colon Y \longrightarrow X$ of $\mathscr{F}/I$, just by considering the morphisms $\alpha^*(f)$. When $\mathscr{F}$ is a split fibration, the functors ΔI are natural in I, thus define by 8.3.1 the required cartesian functor Δ. When $\mathscr{F}$ is an arbitrary fibration, the functors $\Delta(I)$ are just pseudo-natural in I, but this again yields the required functor Δ.

Definition 8.5.3 *Let $F\colon \mathscr{F} \longrightarrow \mathscr{E}$ be a fibration and $\mathscr{D}$ an internal category of $\mathscr{E}$. The fibration $(\mathscr{F}, F)$ has all limits of shape $\mathscr{D}$ when the cartesian functor*

$$\Delta\colon \mathscr{F} \longrightarrow \mathscr{F}^{\mathscr{D}}$$

has a fibred left adjoint.

Given a fibration $F\colon \mathscr{F} \longrightarrow \mathscr{E}$, one might now want to define its completeness as the existence of all limits of shape $\mathscr{D}$, for all internal categories $\mathscr{D}$ of $\mathscr{E}$. This is not enough, for two reasons. First of all, without any further assumption on $\mathscr{E}$, there would not be any reason to find in $\mathscr{E}$ some particular internal categories $\mathscr{D}$ which would exhibit the existence of "basic" limit diagrams like pullbacks, equalizers, and so on. The second reason is more subtle and is related to the need for good stability conditions of the notion of completeness. We explain this by an example.

Suppose $\mathscr{C}$ is a complete category, in the usual sense. Obviously a good notion of completeness must imply that the canonical fibration $\mathsf{Set}(\mathscr{C}) \longrightarrow \mathsf{Set}$ of 8.1.9.b is complete, the limit of families being computed pointwise. But fix now a set I. One could consider an even more sophisticated category $\mathsf{Set}_I(\mathscr{C})$ where the indices are now the families $(J_i)_{i\in I}$ of sets and an object of $\mathscr{C}$ is thus a family of families of objects of $\mathscr{C}$: $\left((C_{ji_{j\in J_i}}\right)_{i\in I}$. It is easily observed (we shall do it immediately for an arbitrary fibration) that we get in this way a new fibration

$$\mathsf{Set}_I(\mathscr{C}) \longrightarrow \mathsf{Set}/I$$

where Set/I is the category of I-families of sets (see 1.2.7.a, volume 1). Clearly this new fibration should again be complete, with the limits of families of families still computed pointwise. Such a stability condition of the notion of completeness turns out to be important in practice and must be included in the definition.

Definition 8.5.4 *Let $F: \mathscr{F} \longrightarrow \mathscr{E}$ be a fibration and $I \in \mathscr{E}$ an object. The localization of F at I is the fibration*

$$F_{(I)}: \mathscr{F}_{(I)} \longrightarrow \mathscr{E}/I$$

whose fibre at $\alpha: J \longrightarrow I$ is just $\mathscr{F}_I$ and whose inverse image functor along $\gamma: (J,\alpha) \longrightarrow (K,\beta)$ is just γ^*.

It is obvious that $F_{(I)}$ is a fibration. We are now ready to define the completeness of a fibration.

Definition 8.5.5 *Let $F: \mathscr{F} \longrightarrow \mathscr{E}$ be a fibration, where $\mathscr{E}$ is a category with finite limits.*

(1) This fibration is finitely complete when it has all limits of shape $\mathscr{D}$, for every finite category $\mathscr{D}$;

(2) this fibration is internally complete when it has all limits of shape $\mathscr{D}$, for every internal category $\mathscr{D}$ in $\mathscr{E}$;

(3) this fibration is complete when, for every object $I \in \mathscr{E}$, the fibration $F_{(I)}: \mathscr{F}_{(I)} \longrightarrow \mathscr{E}/I$ is finitely and internally complete.

Proposition 8.5.6 *A complete fibration over a base category with finite limits is finitely and internally complete.*

Proof Put $I = \mathbf{1}$ (the terminal object) in 8.5.5. □

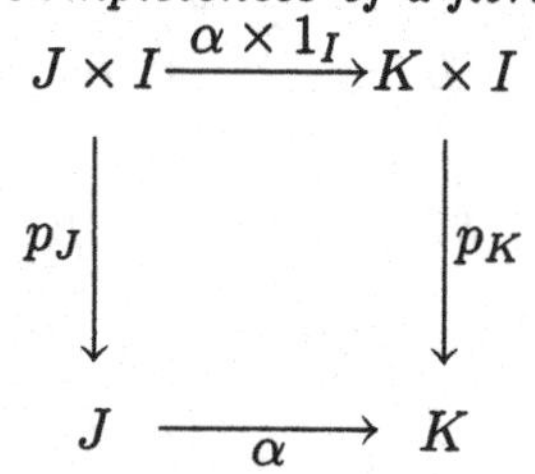

$$\begin{array}{ccc} J\times I & \xrightarrow{\alpha\times 1_I} & K\times I \\ {\scriptstyle p_J}\downarrow & & \downarrow{\scriptstyle p_K} \\ J & \xrightarrow[\alpha]{} & K \end{array}$$

Diagram 8.11

We shall now give a striking characterization of complete fibrations (see 8.5.9), which is also Bénabou's original definition of complete fibrations. We split the proof into several lemmas.

Lemma 8.5.7 *Let $F\colon \mathscr{F}\longrightarrow\mathscr{E}$ be a fibration, with $\mathscr{E}$ a finitely complete category. The following conditions are equivalent:*

(1) the fibration $(\mathscr{F},F)$ has all limits of shape ∂I, for every object $I\in\mathscr{E}$ (see 8.2.5);

(2) for every object $J\in\mathscr{E}$ the inverse image functor $p_J^*\colon \mathscr{F}_J\longrightarrow\mathscr{F}_{I\times J}$ has a right adjoint $\prod_J$ such that, for every morphism $\alpha\colon J\longrightarrow K$, the comparison morphism $\alpha^*\circ\prod_K\Rightarrow\prod_J\circ(\alpha\times 1_I)^*$ is an isomorphism (see diagram 8.11).

Proof Let us first make explicit the form of the "comparison morphism" in condition (2). Given an object $X\in\mathscr{F}_{I\times K}$, the counit of the adjunction $p_K^*\dashv\prod_K$ gives a morphism $\varepsilon_X\colon p_K^*\prod_K(X)\longrightarrow X$. This yields the composite

$$p_J^*\alpha^*\prod\nolimits_K(X)\xrightarrow{\ \cong\ }(\alpha\times 1_I)^*p_K^*\prod\nolimits_K(X)\xrightarrow{(\alpha\times 1_I)^*\varepsilon_X}(1_I\times\alpha)^*(X)$$

which corresponds via the adjunction $p_J^*\dashv\prod_J$ with the required morphism $\alpha^*\prod_K(X)\longrightarrow\prod_J(\alpha\times 1_I)^*(X)$. Moreover observe that we can replace the fibration $(\mathscr{F},F)$ in the proof by a split fibration; in view of 8.3.4, there is thus no restriction in supposing that $(\mathscr{F},F)$ is a split fibration. In this case $p_J^*\alpha^*=(\alpha\times 1_I)^*p_K^*$.

Let us now make explicit condition (1). It means the existence of a fibred right adjoint to the functor $\Delta\colon\mathscr{F}\longrightarrow\mathscr{F}^I$. The fibre at $J\in\mathscr{E}$ of the fibration $\mathscr{F}^I$ is (see 8.4.5, 8.2.5 and 8.2.7)

$$\begin{aligned}(\mathscr{F}^I)_J &\cong \mathsf{Cart}(\mathscr{E}/J\times_{\mathscr{E}}\mathscr{E}/I,\mathscr{F})\cong\mathsf{Cart}(\mathscr{E}/J\times I,\mathscr{F})\\ &\cong\mathscr{F}_{J\times I}.\end{aligned}$$

For a morphism $\alpha\colon K\longrightarrow J$, the inverse image functor α^* of $\mathscr{F}^I$ is just the inverse image functor $(\alpha\times 1_I)^*$ of $\mathscr{F}$. The cartesian functor

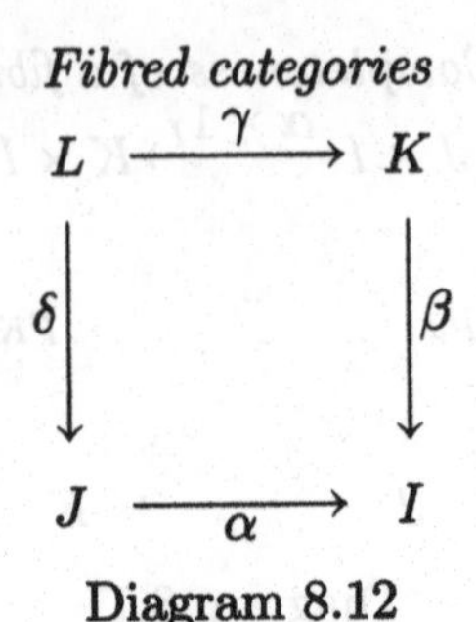

Diagram 8.12

$$\begin{array}{ccc}(L,\alpha\circ\delta)=(J,\alpha)\times(K,\beta) & \xrightarrow{\ \gamma\ } & (I,1_I)\times(K,\beta)=(K,\beta)\\ \big\downarrow{\scriptstyle\delta} & & \big\downarrow{\scriptstyle\beta}\\ (J,\alpha) & \xrightarrow[\ \alpha\ ]{} & (I,1_I)\end{array}$$

Diagram 8.13

Δ is defined by the fact that $\Delta_J\colon \mathscr{F}_J \longrightarrow \mathscr{F}_{J\times I}$ is the inverse image functor along the projection $p_J\colon J\times I \longrightarrow J$. The existence of a fibred right adjoint to Δ thus means the existence of a right adjoint $\prod_J\colon \mathscr{F}_{J\times I}\longrightarrow \mathscr{F}_J$ for each object $J\in\mathscr{E}$, together with a natural isomorphism $\alpha^*\circ\prod_K \cong \prod_J\circ(\alpha\times 1_I)^*$ for every morphism $\alpha\colon K\longrightarrow J$ in $\mathscr{E}$ (see 8.4.2), thus exactly condition (2). □

Lemma 8.5.8 *Let $F\colon \mathscr{F}\longrightarrow\mathscr{E}$ be a fibration, with $\mathscr{E}$ a finitely complete category. The following conditions are equivalent:*

(1) for every morphism $\alpha\colon J\longrightarrow I$ in $\mathscr{E}$, the fibration

$$F_{(I)}\colon \mathscr{F}_{(I)}\longrightarrow \mathscr{E}/I$$

has all limits of shape $\partial(\alpha)$;

(2) for every morphism $\alpha\colon J\longrightarrow I$ in $\mathscr{E}$ the functor $\alpha^*\colon \mathscr{F}_I\longrightarrow\mathscr{F}_J$ has a right adjoint $\prod_\alpha$ such that, for every pullback in $\mathscr{E}$ as in diagram 8.12, the comparison morphism $\alpha^*\circ\prod_\beta \Rightarrow \prod_\gamma\circ\delta^*$ is an isomorphism.

Proof This is an immediate consequence of 8.5.7, just observing that the pullback in condition (2) can equivalently be written as in diagram 8.13, in the category $\mathscr{E}/I$, where the product is just the pullback over I. □

Theorem 8.5.9 *Let $\mathscr{E}$ be a category with finite limits. A fibration $F\colon \mathscr{F}\longrightarrow\mathscr{E}$ is complete if and only if:*

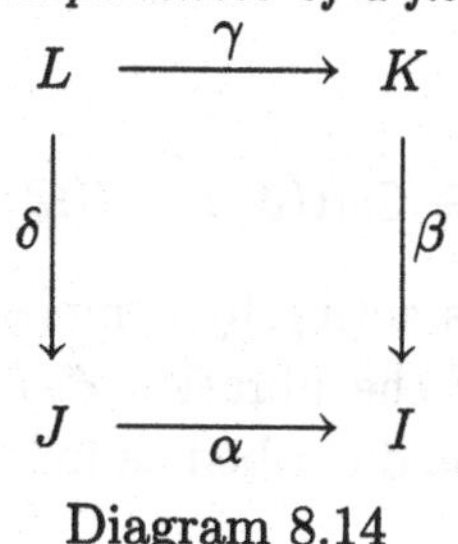

Diagram 8.14

(1) *the fibration* $(\mathscr{F}, F)$ *has all limits of shape* $\mathscr{D}$*, for every finite category* $\mathscr{D}$*;*

(2) *for every morphism* $\alpha\colon J \longrightarrow I$ *in* $\mathscr{E}$*, the corresponding inverse image functor* $\alpha^*\colon \mathscr{F}_I \longrightarrow \mathscr{F}_J$ *has a right adjoint* $\prod_\alpha$*;*

(3) *for every pullback in* $\mathscr{E}$*, as in diagram 8.14, the comparison morphism* $\alpha^* \circ \prod_\beta \Rightarrow \prod_\gamma \circ \delta^*$ *is an isomorphism.*

Proof For each object $I \in \mathscr{E}$, the fibration $F_{(I)}\colon \mathscr{F}_{(I)} \longrightarrow \mathscr{E}/I$ is built up from the fibres of $\mathscr{F}$ and the inverse image functors of $\mathscr{F}$ (see 8.5.4); thus if in $\mathscr{F}$ each fibre is finitely complete and each inverse image functor preserves finite limits, the same holds in $\mathscr{F}_{(I)}$. Applying 8.5.2, this is just saying that the finite completeness of the fibration $(\mathscr{F}, F)$ implies that of each fibration $(\mathscr{F}_{(I)}, F_{(I)})$. The converse is obvious, putting $I = 1$ in 8.5.5 (see 8.5.6).

Lemma 8.5.8 indicates that a complete fibration $(\mathscr{F}, F)$ satisfies conditions (2), (3) of the present theorem. So it remains to prove that conditions (1),(2), (3) in our statement imply the internal completion of each fibration $(\mathscr{F}_{(I)}, F_{(I)})$. This is the fibred version of theorem 2.8.1, volume 1. But observe immediately that the validity of conditions (1), (2), (3) of the statement for a fibration $F\colon \mathscr{F} \longrightarrow \mathscr{E}$ immediately implies the validity of the same conditions for each localized fibration $F_{(I)}\colon \mathscr{F}_{(I)} \longrightarrow \mathscr{E}/I$, with $I \in \mathscr{E}$; this is due again to the fact that $\mathscr{F}_{(I)}$ is built up from the fibres and the inverse image functors of $\mathscr{F}$. Thus it suffices to prove that conditions (1), (2), (3) of the statement imply that $\mathscr{F}$ has all limits of shape $\mathscr{D}$, for every internal category $\mathscr{D}$ of $\mathscr{E}$. This fact, applied to each fibration $F_{(I)}\colon \mathscr{F}_{(I)} \longrightarrow \mathscr{E}/I$, is precisely condition (2) in 8.5.5.

With the notation of 8.1.1, volume 1, let us fix an internal category $\mathscr{D} = (D_0, D_1, d_0, d_1, i, c)$ in $\mathscr{E}$. We must produce a fibred left adjoint to the cartesian functor $\Delta\colon \mathscr{F} \longrightarrow \mathscr{F}^{\mathscr{D}}$. Let us compute it explicitly. For every object $I \in \mathscr{E}$,

$$\mathscr{F}_I^{\mathscr{D}} \cong \mathsf{Cart}(\mathscr{E}/I \times_I \mathscr{E}(\mathscr{D}), \mathscr{F}),$$

the inverse image functors acting by composition (see 8.2.5). We must still compute the form of the fibration $\mathscr{E}/I \times_{\mathscr{E}} \mathscr{E}(D)$. Given an object $J \in \mathscr{E}$, its fibre at J can be described as follows (see 8.1.10):

- the objects are the pairs (j, m), where $j\colon J \longrightarrow I$ and $m\colon J \longrightarrow D_0$;
- a morphism $f\colon (j,m) \longrightarrow (k,n)$ in the fibre over J exists just when $j = k$, since the fibres of $\mathscr{E}/I$ are discrete categories (see 8.3.2); a morphism $f\colon (j,m) \longrightarrow (j,n)$ is then a morphism $f\colon J \longrightarrow K_1$ in $\mathscr{E}$ such that $d_0 \circ f = m$, $d_1 \circ f = n$.

This fibre of $\mathscr{E}/I \times_{\mathscr{E}} \mathscr{E}(\mathscr{D})$ at some object $J \in \mathscr{E}$ can thus be described in an equivalent way:

- the objects are the morphisms $\alpha\colon J \longrightarrow I \times D_0$ (put $\alpha = (j, m)$ in the previous description);
- a morphism $\gamma\colon \alpha \longrightarrow \beta$ is a morphism $\gamma\colon J \longrightarrow D_1$ such that
$$(1_I \times d_0) \circ \gamma = \alpha, \quad (1_I \times d_1) \circ \gamma = \beta;$$
put $\gamma = (j, f)$ in the previous description; composing the identities with the first projection yields $j = k$ and with the second projection, $d_0 \circ f = m$, $d_1 \circ f = n$.

Given the internal category $\mathscr{D}$ and the object I, multiplying by I all the data constituting $\mathscr{D}$,

$$(I \times D_0, I \times D_1, 1_I \times d_0, 1_I \times d_1, 1_I \times i, 1_I \times c)$$

yields a new internal category which we will denote $I \times \mathscr{D}$. Observe that considering the first projections $I \times D_0 \longrightarrow I$, $I \times D_1 \longrightarrow I$, ... those data can also be seen as an internal category in $\mathscr{E}/I$, which we shall denote $\mathscr{D}/I$. We have then the equivalences

$$\begin{aligned}\mathscr{F}_I^{\mathscr{D}} &\cong \mathsf{Cart}(\mathscr{E}/I \times_{\mathscr{E}} \mathscr{E}(\mathscr{D}), \mathscr{F}) \cong \mathsf{Cart}(\mathscr{E}(I \times \mathscr{D}), \mathscr{F}) \\ &\cong \mathsf{Cart}(\mathscr{E}/I(\mathscr{D}/I), \mathscr{F}_{(I)}) \cong \left(\mathscr{F}_{(I)}\right)_{1_I}^{\mathscr{D}/I},\end{aligned}$$

proving that the fibre of $\mathscr{F}^{\mathscr{D}}$ at I is just the fibre of $(\mathscr{F}_{(I)})^{\mathscr{D}_I}$ at the terminal object $1_I\colon I = I$ of $\mathscr{E}/I$. On the other hand the fibre of $\mathscr{F}$ at I is precisely the fibre of $\mathscr{F}_{(I)}$ at the terminal object 1_I of $\mathscr{E}/I$.

Constructing a fibred right adjoint $\nabla\colon \mathscr{F}^{\mathscr{D}} \longrightarrow \mathscr{F}$ to the cartesian functor Δ is constructing a right adjoint $\nabla_I\colon \mathscr{F}_I^{\mathscr{D}} \longrightarrow \mathscr{F}_I$ to Δ_I, for each object $I \in \mathscr{E}$, together with compatibility conditions (see 8.4.2). Our previous description of $\mathscr{F}_I^{\mathscr{D}}$ indicates that we must construct a right adjoint

$$\nabla_{1_I}: (\mathscr{F}/I)_{1_I}^{\mathscr{D}/I} \longrightarrow (\mathscr{F}/I)_{1_I}$$

to Δ_{1_I}, where 1_I is the terminal object of $\mathscr{E}/I$. Again since the conditions (1), (2), (3) of the statement are stable under localization at an object $I \in \mathscr{E}$, it suffices to prove the existence of ∇_1, for $\mathbf{1} \in \mathscr{E}$ the terminal object, and apply this result to $\mathscr{F}/I$ and $\mathscr{D}/I$ to get the existence of ∇_I.

Thus we must now consider the functor

$$\Delta_1: \mathscr{F}_1 \longrightarrow \mathsf{Cart}(\mathscr{E}(\mathscr{D}), \mathscr{F})$$

mapping an object $X \in \mathscr{F}_1$ to the "constant cartesian functor on X", i.e. the functor

$$\Delta_1(X): \mathscr{E}(\mathscr{D}) \longrightarrow \mathscr{F}$$

mapping $Y \in \mathscr{E}(\mathscr{D})_J$ to $\alpha_J^*(X)$, with $\alpha_J: J \longrightarrow \mathbf{1}$ in $\mathscr{E}$. We must construct a right adjoint ∇_1 to Δ_1. It suffices to mimic the proof of 8.1.1, volume 1. Let us consider a cartesian functor $H: \mathscr{E}(\mathscr{D}) \longrightarrow \mathscr{F}$; we must define an object $\nabla_1 H \in \mathscr{F}_1$.

The identity $1_{D_0}: D_0 = D_0$ is an object in the fibre $\mathscr{E}(\mathscr{D})_{D_0}$. Its inverse images in $\mathscr{E}(\mathscr{D})_{D_1}$, along the morphisms $d_0, d_1: D_1 \rightrightarrows D_0$, are just $d_0: D_1 \longrightarrow D_0$, $d_1: D_1 \longrightarrow D_0$ (see 8.1.10). On the other hand in the fibre $\mathscr{E}(\mathscr{D})_{D_1}$, the identity on D_1 yields a corresponding morphism $(D_1, d_0) \longrightarrow (D_1, d_1)$, since $d_0 \circ 1_{D_1} = d_0$ and $d_1 \circ 1_{D_1} = d_1$. Applying the cartesian functor H, we thus find an object $H_0 \in \mathscr{F}_{D_0}$, the image of $1_{D_0} \in \mathscr{E}(\mathscr{D})_{D_1}$. Since H is cartesian, $d_0^*(H_0) \cong H(D_1, d_0)$ and $d_1^*(H_0) \cong H(D_1, d_1)$. Therefore $H(1_{D_1})$ yields a morphism $\varphi: d_0^*(H_0) \longrightarrow d_1^*(H_0)$ in the fibre $\mathscr{F}_{D_1}$.

To understand the above intuitively consider the case where $\mathscr{F}$ is the fibration $\mathsf{Set}(\mathscr{C})$ for some complete category $\mathscr{C}$, as in example 8.1.9.b. Then $\mathscr{D}$ is a small category. Given a functor $H: \mathscr{D} \longrightarrow \mathscr{C}$,

$$H_0 = \big(H(U)\big)_{U \in D_0}, \; d_0^*(H_0) = \big(H(d_0 u)\big)_{u \in D_1}, \; d_1^*(H_0) = \big(H(d_1 u)\big)_{u \in D_1},$$

while $\varphi = \big(H(u)\big)_{u \in D_1}$. As will be observed in 8.5.9, given the mapping $\alpha_I: I \longrightarrow \mathbf{1}$ in Set, the corresponding adjoint functor

$$\prod_I: \mathsf{Set}(\mathscr{C})_I \longrightarrow \mathsf{Set}(\mathscr{C})_1$$

of α_I^* maps the family $(C_i)_{i \in I}$ to $\prod_{i \in I} C_i$.

Let us come back to our general construction. Given the morphism $\alpha_I\colon I \longrightarrow 1$ in $\mathscr{E}$, we write $\prod_I$ for the right adjoint to α_I^*. As in 2.8.1, volume 1, consider first the two objects $\prod_{D_0}(H_0)$ and $\prod_{D_0} d_1^*(H_0)$ in the fibre $\mathscr{F}_1$. We shall construct $a, b\colon \prod_{D_0}(H_0) \rightrightarrows \prod_{D_1} d_1^*(H_0)$ and define $\lambda_H\colon \nabla_1(H) \longrightarrow \prod_{D_0}(H_0)$ as the equalizer of a, b in $\mathscr{F}_1$. The morphism a corresponds via the adjunction $\alpha_{D_1}^* \dashv \prod_{D_1}$ with the composite $\overline{a}$:

$$\alpha_{D_1}^* \prod_{D_0}(H_0) = d_1^* \alpha_{D_0}^* \prod_{D_0}(H_0) \xrightarrow{d_1^* \varepsilon_{H_0}} d_1^*(H_0),$$

where ε is the counit of the adjunction $\alpha_{D_0}^* \dashv \prod_{D_0}$. On the other hand b corresponds via the adjunction $\alpha_{D_1}^* \dashv \prod_{D_1}$ with the composite $\overline{b}$:

$$\alpha_{D_1}^* \prod_{D_0}(H_0) = d_0^* \alpha_{D_0}^* \prod_{D_0}(H_0) \xrightarrow{d_0^* \varepsilon_{H_0}} d_0^*(H_0) \xrightarrow{\quad \varphi \quad} d_1^*(H_0).$$

To prove that $\nabla_1(H)$ is the coreflection of H along Δ_1, we must construct a cartesian natural transformation $\tau_H\colon \Delta_1 \nabla_1(H) \Rightarrow H$. Let us consider an object $(I, m) \in \mathscr{E}(\mathscr{D})$; we must define in the fibre $\mathscr{F}_I$ a morphism $\Delta_1 \nabla_1 H(I, m) \longrightarrow H(I, m)$, thus by definition of Δ_1 a morphism $\alpha_I^*(\nabla_1 H) \longrightarrow H(I, m)$. But in the fibration $\mathscr{E}(\mathscr{E})$, $(I, m) = m^*(D_0, 1_{D_0})$; therefore $H(I, m) \cong \alpha_m^* H_0$, since H is cartesian. The required morphism is then the composite

$$\alpha_I^* \nabla_1(H) \xrightarrow{\alpha_I^* \lambda_H} \alpha_I^* \prod_{D_0}(H_0) = \alpha_m^* \alpha_{D_0}^* \prod_{D_0}(H_0) \xrightarrow{\alpha_m^* \varepsilon_{H_0}} \alpha_m^*(H_0)$$

and it is routine to verify the naturality. Next, given an object $X \in \mathscr{F}_I$ and a cartesian natural transformation $\theta\colon \Delta_1(X) \Rightarrow H$, we must find a unique arrow $x\colon X \longrightarrow \nabla_1(H)$ in $\mathscr{F}_1$ such that $\tau_H \circ \Delta_1(x) = \theta$. Finding x is equivalent to finding $y\colon X \longrightarrow \prod_{D_0}(H_0)$ in $\mathscr{F}_1$ such that $a \circ y = b \circ y$. The morphism y corresponds, via the adjunction $\alpha_{D_0}^* \dashv \prod_{D_0}$, with the composite

$$\alpha_{D_0}^*(X) = \Delta_1 X(D_0, 1_{D_0}) \xrightarrow{\theta_{(D_0, 1_{D_0})}} H(D_0, 1_{D_0}) = H_0.$$

Applying the adjunction $\alpha_{D_1}^* \dashv \prod_{D_1}$, the equality $a \circ y = b \circ y$ follows from consideration of diagram 8.15. Each internal piece of the diagram is by definition commutative; thus the equality $\overline{a} \circ \alpha_{D_1}^* y = \overline{b} \circ \alpha_{D_1}^* y$ will follow from the commutativity of the outer diagram. This is the case by naturality of θ, since

$$d_0^* \theta_{(D_0, 1_{D_0})} = \theta_{d_0^*(D_0, 1_0)} = \theta_{(D_1, d_0)},$$
$$d_1^* \theta_{(D_0, 1_{D_0})} = \theta_{d_1^*(D_0, 1_0)} = \theta_{(D_1, d_1)},$$
$$\varphi = H(1_{D_1}),$$

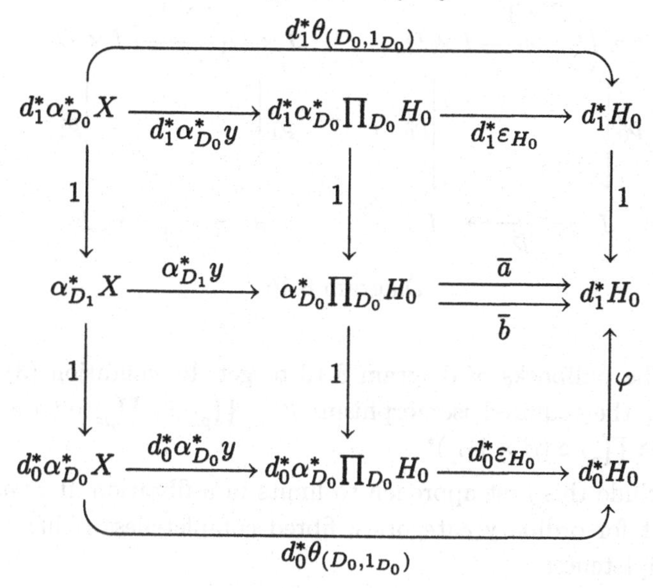

Diagram 8.15

$$1_{\alpha^*_{D_1}X} = \Delta_1(X)(1_{D_1}).$$

This concludes the proof that the functor $\Delta_1\colon \mathscr{F}_1 \longrightarrow \mathsf{Cart}(\mathscr{E}(\mathscr{D}), \mathscr{F})$ has a right adjoint ∇_1. We have already observed that, considering the fibration $F_{(I)}\colon \mathscr{F}_{(I)} \longrightarrow \mathscr{E}/I$ and the internal category $\mathscr{D}/I$ of $\mathscr{E}/I$, this implies that every functor $\Delta_I\colon \mathscr{F}_I \longrightarrow \mathsf{Cart}(\mathscr{E}/I \times_{\mathscr{E}} \mathscr{E}(\mathscr{D}), \mathscr{F})$ has a right adjoint ∇_I. It remains to prove that the family $(\nabla_I)_I$ of those right adjoints satisfies the compatibility condition of 8.4.2. This is not at all a "straightforward consequence of the naturality of the constructions": this fact depends on condition (3) of our statement. Let us first make explicit the construction of ∇_I. Let $H\colon \mathscr{E}/I \times_{\mathscr{E}} \mathscr{E}(\mathscr{D}) \longrightarrow \mathscr{F}$ be a cartesian functor. Just reorganizing the fibres over $\mathscr{E}/I$, this yields a cartesian functor $H\colon \mathscr{E}/I(\mathscr{D}/I) \longrightarrow \mathscr{F}/I$. Let us now write $p_0^I\colon I \times D_0 \longrightarrow I$ and $p_1^I\colon I \times D_1 \longrightarrow I$ for the first projections of the products defining $I \times \mathscr{D}$. The object $\nabla_I(H) \in \mathscr{F}_I$ is obtained via an equalizer

$$\nabla_I(H) \xrightarrow{\lambda^I_H} \prod\nolimits_{p_0^I}(H)(1_{I\times D_0}) \overset{a_I}{\underset{b_I}{\rightrightarrows}} \prod\nolimits_{p_1^I}(1_I \times d_1)^*(H)(1_{I\times D_0}).$$

Since the inverse image functors of $\mathscr{F}$ preserve equalizers (see 8.5.2), the compatibility condition required for the functor ∇_I will be implied by the analogous property for $\prod_{p_0}, \prod_{p_1}$, due to the naturality of the construction of a_I, b_I. Given a morphism $\beta\colon J \longrightarrow I$ in $\mathscr{E}$, it suffices to

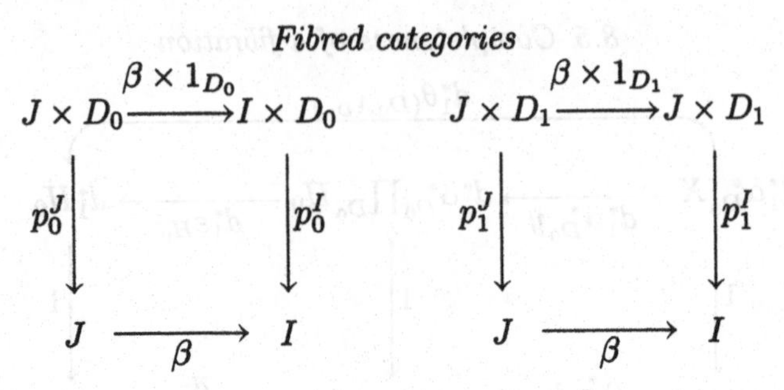

Diagram 8.16

consider the pullbacks of diagram 8.16 to get, by condition (3) of the statement, the required isomorphisms $\beta^* \circ \prod_{p_0^I} \cong \prod_{p_0^J} \circ (\beta \times 1_{D_0})^*$, $\beta^* \circ \prod_{p_1^I} \cong \prod_{p_1^J} \circ (\beta \times 1_{D_1})^*$. □

To conclude this first approach to limits in a fibration, it remains to verify that for ordinary categories, fibred completeness reduces to the usual completeness.

Proposition 8.5.10 *Let $\mathscr{C}$ be a category. The following conditions are equivalent:*
(1) $\mathscr{C}$ is complete;
(2) the fibration $\mathsf{Set}(\mathscr{C}) \longrightarrow \mathsf{Set}$ *is complete.*

Proof The completeness of the fibration $\mathsf{Set}(\mathscr{C}) \longrightarrow \mathsf{Set}$ implies the usual completeness of each fibre (see 8.5.2), thus in particular the completeness of the fibre over $\mathbf{1}$, which is $\mathscr{C}$.

Conversely if $\mathscr{C}$ is complete, for every set I the category $\mathscr{C}^I$ of I-indexed families of objects and arrows in $\mathscr{C}$ is complete and limits are computed pointwise. Given a mapping $\alpha\colon J \longrightarrow I$ between two sets, the corresponding inverse image functor $\alpha^*\colon \mathscr{C}^I \longrightarrow \mathscr{C}^J$ maps a family $(C_i)_{i\in I}$ to the family $\left(C_{\alpha(j)}\right)_{j\in J}$, thus preserves limits since these are computed pointwise.

Consider now a mapping $\alpha\colon J \longrightarrow I$ in Set and two families $(D_j)_{j\in J}$, $(C_i)_{i\in I}$ of objects of $\mathscr{C}$. Let us define

$$\prod_\alpha \colon \mathscr{C}^J \longrightarrow \mathscr{C}^I,\ (D_j)_{j\in J} \mapsto \left(\prod_{j\in\alpha^{-1}(i)} D_j\right)_{i\in I}$$

and analogously on the morphisms. A morphism

$$f\colon (C_i)_{i\in I} \longrightarrow \prod_\alpha (D_j)_{j\in J} \cong \mathscr{C}^I,\ (D_j)_{j\in J} \mapsto \left(\prod_{j\in\alpha^{-1}(i)} D_j\right)_{i\in I}$$

in $\mathscr{C}^J$ is an I-indexed family of morphisms $f_i\colon C_i \longrightarrow \prod_{j\in\alpha^{-1}(i)} D_j$, or equivalently a J-indexed family of morphisms $g_j\colon C_{\alpha(j)} \longrightarrow D_j$: just put

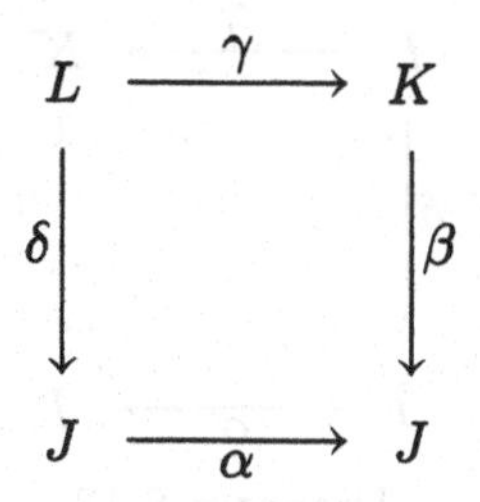

Diagram 8.17

$g_j = p_j \circ f_i$ if $\alpha(j) = i$. This proves that $\prod_\alpha$ is right adjoint to α^*.

Finally we must consider a pullback in Set, as in diagram 8.17, and a family $(X_j)_{j\in J}$ of objects of $\mathscr{C}$. We must prove the isomorphism $\beta^* \circ \prod_\alpha \cong \prod_\gamma \circ \delta^*$. Indeed

$$\Big(\beta^* \circ \prod\nolimits_\alpha\Big)(X_j)_{j\in J} = \beta^*\Big(\prod\nolimits_{j\in\alpha^{-1}(i)} X_j\Big)_{i\in I} = \Big(\prod\nolimits_{j\in\alpha^{-1}\beta(k)} X_j\Big)_{k\in K},$$

$$\Big(\prod\nolimits_\gamma \circ \delta^*\Big)(X_j)_{j\in J} = \prod\nolimits_\gamma (X_{\delta(l)})_{l\in L} = \Big(\prod\nolimits_{l\in\gamma^{-1}(k)} X_{\delta(l)}\Big)_{k\in K}.$$

To conclude the proof it remains to observe that

$$\begin{aligned}\{j \mid j\in J,\ j\in\alpha^{-1}\beta(k)\} &= \{j \mid j\in J,\ \alpha(j)=\beta(k)\}\\ &= \{\delta(l) \mid l\in L,\ \gamma(l)=k\}\\ &= \{\delta(l) \mid l\in L,\ l\in\gamma^{-1}(k)\}\end{aligned}$$

just because the pullback L is defined by

$$L = \{(j,k) \mid j\in J,\ k\in K,\ \alpha(j)=\beta(k)\}. \qquad \square$$

The construction of the functors $\prod_\alpha$ in 8.5.10 and the fact that their existence is equivalent to the existence of limits defined on discrete internal categories (see 8.5.8) justifies the following definition

Definition 8.5.11 *Let $F: \mathscr{F} \longrightarrow \mathscr{E}$ be a fibration, with $\mathscr{E}$ a finitely complete category. This fibration has $\mathscr{E}$-products when:*

(1) for every morphism $\alpha: J \longrightarrow I$ in $\mathscr{E}$ the functor $\alpha^*: \mathscr{F}_I \longrightarrow \mathscr{F}_J$ has a right adjoint $\prod_\alpha$;

(2) for every pullback as in diagram 8.18, in $\mathscr{E}$, the comparison morphism $\alpha^* \circ \prod_\beta \Rightarrow \prod_\gamma \circ \delta^*$ is an isomorphism.

Finally one defines

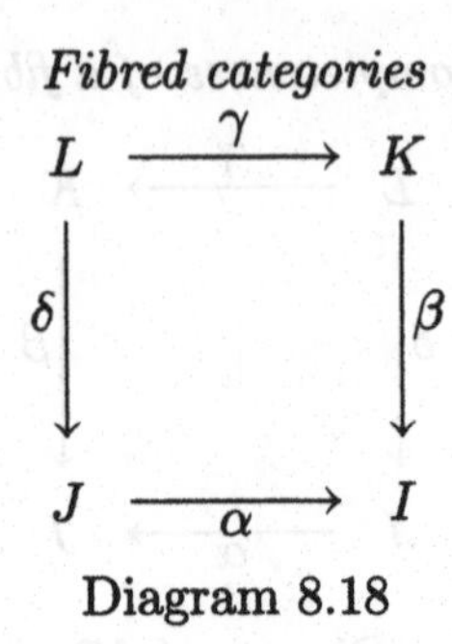

Diagram 8.18

Definition 8.5.12 *Let $\mathscr{C}$ be a category with finite limits.*

(1) A fibration $F: \mathscr{F} \longrightarrow \mathscr{E}$ is cocomplete when the corresponding dual fibration $F^*: \mathscr{F}^* \longrightarrow \mathscr{E}$ (see 8.3.5) is complete.

(2) A fibration $F: \mathscr{F} \longrightarrow \mathscr{E}$ has $\mathscr{E}$-coproducts when the dual fibration $F^*: \mathscr{F}^* \longrightarrow \mathscr{E}$ has $\mathscr{E}$-products.

8.6 Locally small fibrations

An essential feature in the definition of a category $\mathscr{C}$ is the requirement that the morphisms between two objects constitute a set. This is indeed crucial for proving, for example, the Yoneda lemma, on which many key results depend. When studying the fibrations over a base category $\mathscr{E}$, the "indexing objects" are now those of $\mathscr{E}$. In some sense, $\mathscr{E}$ now replaces the category of sets for many purposes. Therefore a special attention should be paid to those fibrations $F: \mathscr{F} \longrightarrow \mathscr{E}$ for which "the arrows between two objects can be represented by an object of $\mathscr{E}$".

To understand the definition better, let us consider again the basic example of the fibration $\mathsf{Set}(\mathscr{C}) \longrightarrow \mathsf{Set}$, for a small category $\mathscr{C}$ (see 8.1.9.b). Given two objects C, D in $\mathscr{C}$, i.e. in the fibre of $\mathsf{Set}(\mathscr{C})$ over $\mathbf{1}$, we are interested in considering all the arrows from C to D. This yields the set $\mathscr{C}(C, D)$. Now considering the mapping $\alpha: \mathscr{C}(C, D) \longrightarrow \mathbf{1}$, we can consider the two objects α^*C, α^*D in the fibre over $\mathscr{C}(C, D)$; they are just the constant families $(C)_{h\in\mathscr{C}(C,D)}$, $(D)_{h\in\mathscr{C}(C,D)}$. Between those two families, we have a very canonical morphism $f: \alpha^*(C) \longrightarrow \alpha^*(D)$, namely

$$(h: C \longrightarrow D)_{h\in\mathscr{C}(C,D)}$$

i.e. the family of all arrows from C to D. That this construction is canonical is attested by the following universal property. Consider a set J, the corresponding mapping $\beta: J \longrightarrow \mathbf{1}$ and a morphism $g: \beta^*C \longrightarrow \beta^*D$. This arrow g is a family of morphisms $(g_j: C \longrightarrow D)_{j\in J}$. This determines the unique mapping

$$\gamma: J \longrightarrow \mathscr{C}(C,D), \quad j \mapsto g_j$$

such that $\gamma^*(f) = g$. We have just expressed the representability of the morphisms between two objects C, D of the fibre over $\mathbf{1}$. It remains, by localization, to express the same property for two objects in an arbitrary fibre.

Definition 8.6.1 *Let $F: \mathscr{F} \longrightarrow \mathscr{E}$ be a fibration. This fibration is locally small when, for all objects $I \in \mathscr{E}$ and $X, Y \in \mathscr{F}_I$, it is possible to find*

(1) an object $\mathscr{F}(X,Y)$ in $\mathscr{E}$ and a morphism $\alpha_{XY}: \mathscr{F}(X,Y) \longrightarrow I$,
(2) a morphism $f_{XY}: \alpha^*_{XY}(X) \longrightarrow \alpha^*_{XY}(Y)$ in the fibre over $\mathscr{F}(X,Y)$,

these data being universal for these properties, which means that given

(1) an object J in $\mathscr{E}$ and a morphism $\beta: J \longrightarrow I$,
(2) a morphism $g: \beta^*(X) \longrightarrow \beta^*(Y)$ in the fibre over J,

there exists a unique morphism $\gamma: J \longrightarrow \mathscr{F}(X,Y)$ such that $\alpha_{XY} \circ \gamma = \beta$ and $\gamma^*(f_{XY}) = g$.

It is worth emphasizing here the fact that our axioms for a category already contain the requirement of "set-theoretical local smallness", i.e. the fact that the morphisms between two objects constitute a set. So in definition 8.6.1, the requirement is just that of "internal $\mathscr{E}$-smallness" (compare with definition 8.5.5).

Observe that giving $\alpha_{XY}: \mathscr{F}(X,Y) \longrightarrow I$ is precisely giving an object in $\mathscr{E}/I$, as suggested by the considerations preceding the proof.

Our first observation must obviously be

Proposition 8.6.2 *Let $\mathscr{C}$ be a category. The fibration* $\mathsf{Set}(\mathscr{C}) \longrightarrow \mathsf{Set}$ *is locally small.*

Proof Consider a set I and two families of objects $X = (X_i)_{i \in I}$, $Y = (Y_i)_{i \in I}$. The set $\mathscr{F}(X,Y)$ is not the set of morphisms from X to Y in $\mathsf{Set}(\mathscr{C})$ (how would it be possible to "fibre" it over I?); it is

$$\mathscr{F}(X,Y) = \coprod_{i \in I} \mathscr{C}(X_i, Y_i)$$

with $\alpha_{XY}: \mathscr{F}(X,Y) \longrightarrow I$ mapping a morphism $h \in \mathscr{C}(X_i, Y_i)$ to $i \in I$. The morphism $f_{XY}: \alpha^*_{XY}(X) \longrightarrow \alpha^*_{XY}(Y)$ is then the family of morphisms

$$(h: X_i \longrightarrow Y_i)_{h \in \coprod_{i \in I} \mathscr{C}(X_i, Y_i)}.$$

Next, given $\beta: J \longrightarrow I$ and $g: \beta^*(X) \longrightarrow \beta^*(Y)$, the required mapping $\gamma: J \longrightarrow \mathscr{F}(X,Y)$ maps the index $j \in J$ to $g_j: X_{\beta(j)} \longrightarrow Y_{\beta(j)}$. □

Now, a result which one should expect if the terminology is appropriate:

Proposition 8.6.3 *Let $\mathscr{E}$ be a finitely complete category. Every small fibration over $\mathscr{E}$ is locally small. In other words (see 8.1.11), for every internal category $\mathscr{C}$ of $\mathscr{E}$, the fibration $\mathscr{E}(\mathscr{C}) \longrightarrow \mathscr{E}$ is locally small.*

Proof Consider first two objects $(1,m)$, $(1,n)$ in the fibre of $\mathscr{E}(\mathscr{C})$ over 1. We must construct in $\mathscr{E}$ the object $\mathscr{C}(m,n)$ of "arrows of domain m and codomain n". Clearly the object $\mathscr{C}(m,\bullet)$ of all arrows of domain m should be the equalizer

$$\mathscr{C}(m,\bullet) \rightarrowtail C_1 \underset{m \circ \gamma}{\overset{d_0}{\rightrightarrows}} C_0,$$

where $\gamma: C_1 \longrightarrow 1$; in the same way the object $\mathscr{C}(\bullet,n)$ of all arrows of codomain n should be the equalizer

$$\mathscr{C}(\bullet,n) \rightarrowtail C_1 \underset{n \circ \gamma}{\overset{d_1}{\rightrightarrows}} C_0,$$

and finally $\mathscr{C}(m,n)$ should be the intersection

$$\mathscr{C}(m,n) = \mathscr{C}(m,\bullet) \cap \mathscr{C}(\bullet,n)$$

as subobjects of C_1.

We must now adapt this intuitive construction to the case of two objects (I,m), (I,n) in the fibre of $\mathscr{E}(\mathscr{C})$ over $I \in \mathscr{E}$. In other words, we must perform the previous construction over the terminal object in the localized fibration $F_{(I)}: \mathscr{F}_{(I)} \longrightarrow \mathscr{E}/I$. We thus consider the equalizers

$$\mathscr{C}(m,\bullet) \rightarrowtail I \times C_1 \underset{(1_I, m \circ p_I)}{\overset{1_I \times d_0}{\rightrightarrows}} I \times C_0,$$

$$\mathscr{C}(\bullet,n) \rightarrowtail I \times C_1 \underset{(1_I, n \circ p_I)}{\overset{1_I \times d_1}{\rightrightarrows}} I \times C_0,$$

where $p_I: I \times C_1 \longrightarrow I$ is the first projection. Then $\mathscr{C}(m,n)$ is defined as the intersection

$$\mathscr{C}(m,n) = \mathscr{C}(m,\bullet) \cap \mathscr{C}(\bullet,n)$$

as subobjects of $I \times C_1$. The morphism $\alpha_{mn}: \mathscr{C}(m,n) \longrightarrow I$ is the composite

$$\mathscr{C}(m,n) \overset{i}{\rightarrowtail} I \times C_1 \overset{p_I}{\longrightarrow} I.$$

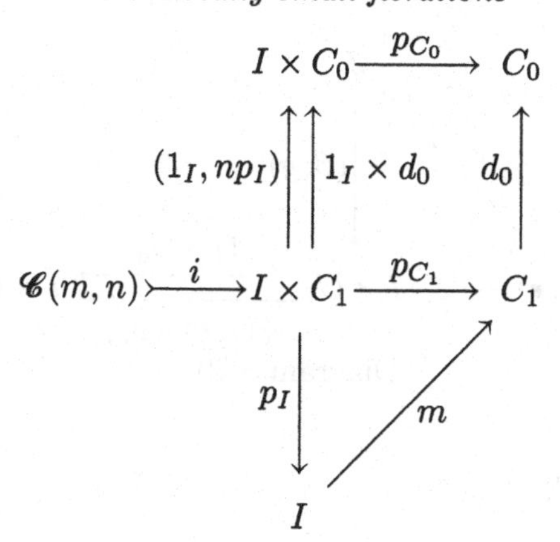

Diagram 8.19

We must still define the generic arrow $f_{mn}\colon \alpha^*_{mn}(I,m) \longrightarrow \alpha^*_{mn}(I,n)$. This is the composite

$$\mathscr{C}(m,n) \rightarrowtail^{i} I \times C_1 \xrightarrow{p_{C_1}} C_1.$$

Indeed,

$$\alpha^*_{mn}(I,m) = \big(\mathscr{C}(m,n), m \circ \alpha_{mn}\big), \quad \alpha^*_{mn}(I,n) = \big(\mathscr{C}(m,n), n \circ \alpha_{mn}\big).$$

Moreover, considering diagram 8.19, we get the relation

$$\begin{aligned} d_0 \circ f_{mn} &= d_0 \circ p_{C_1} \circ i = p_{C_0} \circ (1_I \times d_0) \circ i \\ &= p_{C_0} \circ (1_I, mp_I) \circ i = m \circ p_I \circ i \\ &= m \circ \alpha_{mn}. \end{aligned}$$

In the same way one gets $d_1 \circ f_{mn} = n \circ \alpha_{mn}$, proving that the morphism $f_{mn}\colon \alpha^*_{mn}(I,m) \longrightarrow \alpha^*_{mn}(I,n)$ is indeed in the fibre over $\mathscr{C}(m,n)$.

It remains to check the universality of our construction. Let us consider morphisms $\beta\colon J \longrightarrow I$ in $\mathscr{E}$ and $g\colon \beta^*(I,m) \longrightarrow \beta^*(I,n)$ in the fibre over J. The latter is thus a morphism $g\colon J \longrightarrow C_1$ in $\mathscr{E}$ such that $d_0 \circ g = m \circ \beta$, $d_1 \circ g = n \circ \beta$. Considering diagram 8.20, we find at once

$$(1_I \times d_0) \circ (\beta, g) = (\beta, d_0 \circ g) = (\beta, m \circ \beta) = (1_I, m \circ p_I) \circ (\beta, g),$$

from which we get a unique factorization of (β, g) through the equalizer $\mathscr{C}(m, \bullet)$. Analogously, the relation $d_1 \circ g = n \circ \beta$ implies that (β, g) factors through $\mathscr{C}(\bullet, n)$, from which finally (β, g) factors through $\mathscr{C}(m,n)$ via a unique morphism $\gamma\colon J \longrightarrow \mathscr{C}(m,n)$. One immediately has

$$\alpha_{mn} \circ \gamma = p_I \circ i \circ h = p_I \circ (\beta, g) = \beta.$$

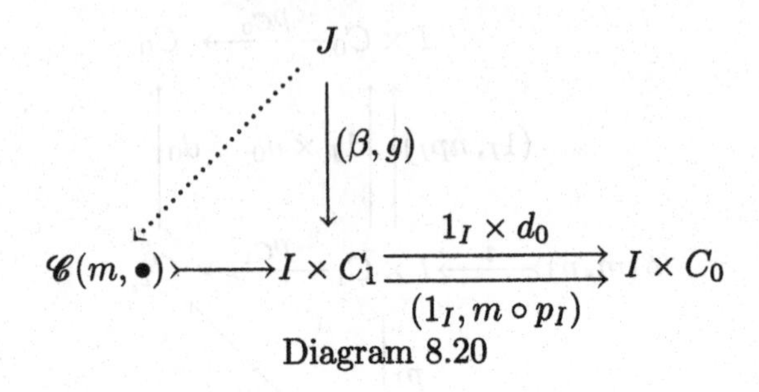

Diagram 8.20

On the other hand,

$$f_{mn} \circ \gamma = p_{C_1} \circ i \circ \gamma = p_{C_1} \circ (\beta, g) = g,$$

proving $\gamma^*(f_{mn}) = g$.

The uniqueness of such an arrow γ is attested by the relations

$$p_I \circ i \circ \gamma = \alpha_{m,n} \circ \gamma = \beta,$$
$$p_{C_1} \circ i \circ \gamma = f_{mn} \circ \gamma = g,$$

which imply $i \circ \gamma = (\beta, g)$, with i a monomorphism. □

In 1.3.2, volume 1, we observed that the functors and natural transformations from a category $\mathscr{C}$ to a category $\mathscr{D}$ again constitute a category – when $\mathscr{C}$ is small! This smallness condition was necessary to ensure that there is just a set of natural transformations between two functors from $\mathscr{C}$ to $\mathscr{D}$. This result carries over to the case of fibrations. We prove first two useful general lemmas:

Lemma 8.6.4 *If $F\colon \mathscr{F} \longrightarrow \mathscr{E}$ is a locally small fibration, then for every object $I \in \mathscr{E}$, the localized fibration $F_{(I)}\colon \mathscr{F}_{(I)} \longrightarrow \mathscr{E}/I$ is also locally small.*

Proof For an object $j\colon J \longrightarrow I$ of $\mathscr{E}/I$ and two objects X, Y in the fibre of $\mathscr{F}_{(I)}$ over (J, j), X, Y are in fact two objects in the fibre $\mathscr{F}_J$ of $\mathscr{F}$. Thus we get by assumption the morphisms

$$\alpha_{XY}\colon \mathscr{F}(X, Y) \longrightarrow J, \quad f_{XY}\colon \alpha^*_{XY}(X) \longrightarrow \alpha^*_{XY}(Y);$$

see 8.6.1. This immediately yields the morphism

$$\alpha_{XY}\colon (\mathscr{F}(X, Y), j \circ \alpha_{XY}) \longrightarrow (J, J)$$

in $\mathscr{E}/I$, from which the result follows. □

$$\begin{array}{ccc}
\mathscr{F}(X,Y)\times_I\mathscr{F}(Y,Z) & \xrightarrow{p_{YZ}} & \mathscr{F}(Y,Z) \\
{\scriptstyle p_{XY}}\downarrow & & \downarrow{\scriptstyle \alpha_{YZ}} \\
\mathscr{F}(X,Y) & \xrightarrow[\alpha_{XY}]{} & I
\end{array}$$

Diagram 8.21

Lemma 8.6.5 *Let $F\colon \mathscr{F} \longrightarrow \mathscr{E}$ be a locally small fibration on a category $\mathscr{E}$ with finite limits. For all objects $I \in \mathscr{E}$ and $X, Y, Z \in \mathscr{F}_I$, there exist a "composition morphism"*

$$\gamma_{XYZ}\colon \mathscr{F}(X,Y)\times_I\mathscr{F}(Y,Z) \longrightarrow \mathscr{F}(X,Z)$$

and a "unit morphism"

$$\varepsilon_X\colon I \longrightarrow \mathscr{F}(X,X)$$

presenting the category $\mathscr{F}_I$ as a category enriched over the monoidal category $\mathscr{E}/I$ provided with its cartesian product as a multiplication (see 6.2.1).

Proof Let us thus consider in $\mathscr{E}$ the pullback of diagram 8.21, i.e. the product in $\mathscr{E}/I$, and write $\alpha_{XYZ} = \alpha_{XY} \circ p_{XY} = \alpha_{YZ} \circ p_{YZ}$. By the local smallness of the fibration, the consideration of α_{XYZ} together with the composite

$$\alpha^*_{XYZ}(X) \xrightarrow{p^*_{XY}(f_{XY})} \alpha^*_{XYZ}(Y) \xrightarrow{p^*_{YZ}(f_{YZ})} \alpha^*_{XYZ}(Z)$$

yields the required composition morphism

$$\gamma_{XYZ}\colon \mathscr{F}(X,Y)\times_I\mathscr{F}(Y,Z) \longrightarrow \mathscr{F}(X,Z)$$

such that $\alpha_{XZ} \circ \gamma_{XYZ} = \alpha_{XYZ}$ and

$$\gamma^*_{XYZ}(f_{XZ}) = p^*_{YZ}(f_{YZ}) \circ p^*_{XY}(f_{XY}).$$

For the unit morphism ε_X, it suffices to consider the arrows

$$1_I\colon I \longrightarrow I,\ \ 1_X\colon X \longrightarrow X.$$

The universal property for local smallness implies the existence of a unique morphism $\varepsilon_X\colon I \longrightarrow \mathscr{F}(X,X)$ such that

$$\alpha_{XX} \circ \varepsilon_X = 1_I,\ \ \varepsilon^*_X(f_{XX}) = 1_X.$$

This already describes the properties of the composition and unit morphisms in terms of the fibration. From this it is routine to conclude the proof, an exercise which is left to the reader since the corresponding properties will not be needed in this book. □

Theorem 8.6.6 *Let $\mathscr{E}$ be a finitely complete category such that the canonical fibration of $\mathscr{E}$ over itself,*

$$\partial_1 \colon \mathsf{Ar}(\mathscr{E}) \longrightarrow \mathscr{E}$$

(see 8.1.9.c), is complete. If $F\colon \mathscr{F} \longrightarrow \mathscr{E}$ and $G\colon \mathscr{G} \longrightarrow \mathscr{E}$ are two fibrations over $\mathscr{E}$ with $(\mathscr{G}, G)$ small and $(\mathscr{F}, F)$ locally small, the fibration $F^G\colon \mathscr{F}^{\mathscr{G}} \longrightarrow \mathscr{E}$ is again locally small.

Proof Both the notions of completeness and of local smallness are stable under localization (see 8.5.5, 8.6.4). Therefore it suffices to prove the local smallness condition for two objects in the fibre over $\mathbf{1}$; the general statement for two objects in the fibre over $I \in \mathscr{E}$ then follows by working in the fibre over the terminal object 1_I in $\mathscr{E}/I$.

Since a fibration equivalent in $\mathsf{Fib}(\mathscr{E})$ to a locally small one is obviously locally small, there is no restriction in supposing that $F\colon \mathscr{F} \longrightarrow \mathscr{E}$ is a split fibration (see 8.3.4). We consider then an internal category $\mathscr{C}$ of $\mathscr{E}$ and take $(\mathscr{G}, G)$ to be the fibration $\mathscr{E}(\mathscr{C}) \longrightarrow \mathscr{E}$ of 8.1.10. We must prove that the fibration $\mathscr{F}^{\mathscr{C}} \longrightarrow \mathscr{E}$ (see 8.4.5) is locally small. As already observed, it suffices to prove the required condition for two objects in the fibre over $\mathbf{1}$. Recall that the fibre of $\mathscr{F}^{\mathscr{C}}$ over an object $I \in \mathscr{E}$ is

$$\mathscr{F}^{\mathscr{C}}_I \cong \mathsf{Cart}\big(\mathscr{E}(I \times \mathscr{C}), \mathscr{F}\big) \cong \mathsf{Cart}\big(\mathscr{E}/I(\mathscr{C}/I), \mathscr{F}_{(I)}\big);$$

see the proof of 8.5.9.

We consider two objects K, L in the fibre $\mathscr{F}^{\mathscr{C}}_1$, i.e. two cartesian functors $K, L\colon \mathscr{E}(\mathscr{C}) \rightrightarrows \mathscr{F}$. Again as observed in the proof of 8.5.9, K determines an object $K_0 \in \mathscr{F}_{C_0}$ and a morphism $\kappa\colon d_0^*(K_0) \longrightarrow d_1^*(K_0)$ in the fibre $\mathscr{F}_{C_1}$. In the same way L determines an object $L_0 \in \mathscr{F}_{C_0}$ and a morphism $\lambda\colon d_0^*(L_0) \longrightarrow d_1^*(L_0)$ in the fibre $\mathscr{F}_{C_1}$. The reader will again refer to the proof of 8.5.9 to remind himself of the intuition lying behind these data.

The assumption of completeness of the canonical fibration of $\mathscr{E}$ over itself implies, for every arrow $\alpha\colon J \longrightarrow I$ in $\mathscr{E}$, the existence of an adjoint pair of functors $\alpha^+ \dashv \prod_\alpha$,

$$\alpha^+\colon \mathscr{E}/I \longrightarrow \mathscr{E}/J, \quad \prod\nolimits_\alpha\colon \mathscr{E}/J \longrightarrow \mathscr{E}/I,$$

where α^+ is just pulling back along α; see 8.5.9. (We have written α^+ instead of α^* to avoid confusion with the inverse image functors in

$\mathscr{F}$.) When $I = \mathbf{1}$ is the terminal object, there is a unique morphism $\alpha: J \longrightarrow \mathbf{1}$ and we generally write $\prod_J$ instead of $\prod_\alpha$.

Intuitively, a morphism $\beta: K \longrightarrow J$ can be seen as a family $(K_j)_{j \in J}$; see 1.2.7.c, volume 1. The object $\prod_J (K, \beta) \in \mathscr{E}/\mathbf{1} \cong \mathscr{E}$ should then be thought of as the product $\prod_{j \in J} K_j$. Intuitively again, a natural transformation $\theta: K \Rightarrow L$ is a C_0-indexed family of morphisms

$$\big(\theta_U: K(U) \longrightarrow L(U)\big)_{U \in C_0},$$

i.e. an element of $\prod_{C_0} \mathscr{F}(K_0, L_0)$. The object of natural transformations $K \Rightarrow L$ should thus be a subobject of $\prod_{C_0} \mathscr{F}(K_0, L_0)$.

Let us consider first the objects $d_0^*(K_0)$, $d_1^*(K_0)$ in the fibre $\mathscr{F}_{C_1}$. By local smallness of $\mathscr{F}$, we get universal morphisms

$$\alpha_K: \mathscr{F}\big(d_0^*(K_0), d_1^*(K_0)\big) \longrightarrow C_1, \quad f_K: \alpha_K^* d_0^*(K_0) \longrightarrow \alpha_K^* d_1^*(K_0).$$

Considering then the morphisms

$$1_{C_1}: C_1 \longrightarrow C_1, \quad \kappa: d_0^*(K_0) \longrightarrow d_1^*(K_0),$$

we find a unique factorization $\beta_K: C_1 \longrightarrow \mathscr{F}\big(d_0^*(K_0), d_1^*(K_0)\big)$ in $\mathscr{E}$ such that $\alpha_K \circ \beta_K = 1_{C_1}$ and $\beta_K^*(f_K) = \kappa$. We can view all this as a morphism

$$\beta_K: (C_1, 1_{C_1}) \longrightarrow \Big(\mathscr{F}\big(d_0^*(K_0), d_1^*(K_0)\big), \alpha_K\Big)$$

in $\mathscr{E}/C_1$. Considering the unique morphism $\varepsilon_1: C_1 \longrightarrow \mathbf{1}$, the corresponding functor

$$\varepsilon_1^+: \mathscr{E} \cong \mathscr{E}/\mathbf{1} \longrightarrow \mathscr{E}/C_1$$

maps $\mathbf{1}$ to $\varepsilon_1^+(\mathbf{1}) \cong (C_1, 1_{C_1})$. Writing $\prod_{C_1}$ for the right adjoint of this functor, β_K corresponds by adunction with a morphism

$$\overline{\beta}_K: \mathbf{1} \longrightarrow \prod_{C_1} \Big(\mathscr{F}\big(d_0^*(K_0), d_1^*(K_0)\big), \alpha_K\Big).$$

In the same way one constructs

$$\overline{\beta}_L: \mathbf{1} \longrightarrow \prod_{C_1} \Big(\mathscr{F}\big(d_0^*(L_0), d_1^*(L_0)\big), \alpha_L\Big).$$

Intuitively, $\prod_{C_1} \big(\mathscr{F}(d_0^*(K_0), d_1^*(K_0)), \alpha_K\big)$ is the set of all families of morphisms indexed by C_1; $\overline{\beta}_K$ picks up the generic family $(u)_{u \in C_1}$.

Our next observation is quite general. Consider two objects X, Y in the fibre $\mathscr{F}_I$ and the corresponding universal data deduced from the local smallness of $(\mathscr{F}, F)$:

$$\alpha_{XY}: \mathscr{F}(X, Y) \longrightarrow I, \quad f_{XY}: \alpha_{XY}^*(X) \longrightarrow \alpha_{XY}^*(Y).$$

$$\begin{array}{ccc} \delta^+\mathscr{F}(X,Y) & \xrightarrow{\overline{\delta}} & \mathscr{F}(X,Y) \\ {\scriptstyle \delta^*\alpha_{XY}}\downarrow & & \downarrow{\scriptstyle \alpha_{XY}} \\ J & \xrightarrow[\delta]{} & I \end{array}$$

Diagram 8.22

Given an arrow $\delta: J \longrightarrow I$ in $\mathscr{E}$, we can consider the objects δ^*X, δ^*Y in the fibre $\mathscr{F}_J$, together with the universal data

$$\alpha_{\delta^*(X),\delta^*(Y)}: \mathscr{F}\big(\delta^*(X), \delta^*(Y)\big) \longrightarrow J,$$
$$f_\delta: \alpha^*_{\delta^*(X),\delta^*(Y)}\big(\delta^*(X)\big) \longrightarrow \alpha^*_{\delta^*(X),\delta^*(Y)}\big(\delta^*(Y)\big).$$

Considering in $\mathscr{E}$ the pullback of diagram 8.22, i.e. applying the inverse image functor δ^+ in the canonical fibration of $\mathscr{E}$ over itself, we can consider the morphisms

$$\delta^+(\alpha_{XY}): \delta^+\big(\mathscr{F}(X,Y)\big) \longrightarrow J,$$
$$\overline{\delta}^*(f_{XY}): \overline{\delta}^*\alpha^*_{XY}(X) \longrightarrow \overline{\delta}^*\alpha^*_{XY}(Y).$$

Since $\overline{\delta}^* \circ \alpha^*_{XY} = \big(\delta^+(\alpha_{XY})\big)^* \circ \delta^*$, this yields a unique factorization

$$\widetilde{\delta}: \delta^+\big(\mathscr{F}(X,Y)\big) \longrightarrow \mathscr{F}\big(\delta^*(X), \delta^*(Y)\big)$$

such that $\delta^*(\alpha_{XY}) = \alpha_{\delta^*(X),\delta^*(Y)} \circ \widetilde{\delta}$ and $\overline{\delta}^*(f_{XY}) = \widetilde{\delta}^*(f_\delta)$. In particular we have a morphism

$$\widetilde{\delta}: \delta^+\big(\mathscr{F}(X,Y), \alpha_{XY}\big) \longrightarrow \Big(\mathscr{F}\big(\delta^*(X), \delta^*(Y)\big), \alpha_{\delta^*(X),\delta^*(Y)}\Big)$$

in the fibre over J, which corresponds by the adjunction $\delta^+ \dashv \prod_\delta$ with a morphism

$$\widehat{\delta}: \big(\mathscr{F}(X,Y), \alpha_{XY}\big) \longrightarrow \prod\nolimits_\delta \Big(\mathscr{F}\big(\delta^*(X), \delta^*(Y)\big), \alpha_{\delta^*(X),\delta^*(Y)}\Big)$$

in $\mathscr{E}/I$. We can further apply $\prod_I$ to get

$$\prod\nolimits_I \widehat{\delta}: \prod\nolimits_I \big(\mathscr{F}(X,Y), \alpha_{XY}\big) \longrightarrow \prod\nolimits_J \Big(\mathscr{F}\big(\delta^*(X), \delta^*(Y)\big), \alpha_{\delta^*(X),\delta^*(Y)}\Big).$$

We shall in particular apply this construction with $\delta = d_0$ or $\delta = d_1$, choosing $X = K_0$ and $Y = L_0$. Intuitively, $\prod_{C_0} \widehat{d_0}$ maps a family $\big(t_U: K(U) \longrightarrow L(U)\big)_{U \in C_0}$ to the family $\big(t_U: K(U) \longrightarrow L(U)\big)_{u \in C_1}$, where $u\colon U \longrightarrow V$. In the same way $\prod_{C_0} \widehat{d_1}$ maps the same family of morphisms to $\big(t_V\colon K(V) \longrightarrow L(V)\big)_{u \in C_1}$.

We are now ready to define the object $\mathscr{F}^{\mathscr{C}}(K,L)$. For the sake of brevity, we shall now write $\prod_I \mathscr{F}(X,Y)$ instead of $\prod_I \big(\mathscr{F}(X,Y), \alpha_{XY}\big)$, when no confusion can occur. By lemma 8.6.5 we get a composition morphism

$$\gamma_K\colon \mathscr{F}\big(d_0^*(K_0), d_1^*(K_0)\big) \times_{C_1} \mathscr{F}\big(d_1^*(K_0), d_1^*(L_0)\big) \longrightarrow \mathscr{F}\big(d_0^*(K_0), d_1^*(L_0)\big).$$

Applying the functor $\prod_{C_1}\colon \mathscr{E}/C_1 \longrightarrow \mathscr{E}$, which preserves products since it has a left adjoint (see 3.2.2, volume 1), we get the composite γ_1:

$$\prod\nolimits_{C_1} (\gamma_K) \circ \Big(\overline{\beta}_K \times \prod\nolimits_{C_0} d_1\Big) : \prod\nolimits_{C_0} \mathscr{F}(K_0, L_0) \longrightarrow \prod\nolimits_{C_1} \mathscr{F}(d_0^* K_0, d_1^* L_0).$$

Intuitively, this composite maps a family $\big(\theta_U\colon K(U) \longrightarrow L(U)\big)_{U \in C_0}$ of morphisms to the family $\big(\theta_V \circ K(u)\big)_{u \in C_1}$, where $u\colon U \longrightarrow V$. In the same way starting with the composition morphism

$$\gamma_L\colon \mathscr{F}\big(d_0^*(K_0), d_0^*(L_0)\big) \times_{C_1} \mathscr{F}\big(d_0^*(L_0), d_1^*(L_0)\big) \longrightarrow \mathscr{F}\big(d_0^*(K_0), d_1^*(L_0)\big)$$

we get the composite γ_0:

$$\prod\nolimits_{C_1} (\gamma_L) \circ \Big(\prod\nolimits_{C_0} d_0 \times \overline{\beta}_L\Big) : \prod\nolimits_{C_0} \mathscr{F}(K_0, L_0) \longrightarrow \prod\nolimits_{C_1} \mathscr{F}(d_0^* K_0, d_1^* L_0).$$

Intuitively, this composite maps the family $\big(\theta_U\colon K(U) \longrightarrow L(U)\big)_{U \in C_0}$ to $\big(L(u) \circ \theta_U\big)_{u \in C_1}$. The object $\mathscr{F}^{\mathscr{C}}(K,L)$ of (pseudo)-natural transformations from K to L is thus defined as the equalizer of the two morphisms

$$\mathscr{F}^{\mathscr{C}}(K,L) \overset{i}{\rightarrowtail} \prod\nolimits_{C_0} \mathscr{F}(K_0, L_0) \underset{\gamma_0}{\overset{\gamma_1}{\rightrightarrows}} \prod\nolimits_{C_1} \mathscr{F}\big(d_0^*(K_0), d_1^*(L_0)\big).$$

Let us write $\alpha_{KL}\colon \mathscr{F}^{\mathscr{C}}(K,L) \longrightarrow \mathbf{1}$ for the unique morphism.

We must still define the "generic natural transformation"

$$f_{KL}\colon \alpha_{KL}^*(K) \Rightarrow \alpha_{KL}^*(L);$$

here $\alpha_{KL}^*(K)$ and $\alpha_{KL}^*(L)$ are just the composites

$$\mathscr{E}/\mathscr{F}^{\mathscr{C}}(K,L) \times_{\mathscr{E}} \mathscr{E}(\mathscr{C}) \overset{p}{\longrightarrow} \mathscr{E}(\mathscr{C}) \underset{L}{\overset{K}{\rightrightarrows}} \mathscr{F},$$

where the first functor p is just the second projection. For a pair (j,m), where

$$j\colon J \longrightarrow \mathscr{F}^{\mathscr{C}}(K,L), \quad m\colon J \longrightarrow C_0,$$

we must define in the fibre $\mathscr{F}_J$ a morphism

$$f_{(j,m)}: K(J,m) \longrightarrow L(J,m);$$

we have just dropped the indices K, L to avoid heavy notation. But $(J,m) = m^*(C_0, 1_{C_0})$ in $\mathscr{E}(\mathscr{C})$. Since K and L are cartesian, we must thus define

$$f_{(j,m)}: m^*K(C_0, 1_{C_0}) \longrightarrow m^*L(C_0, 1_{C_0}),$$

or in other words

$$f_{(j,m)}: m^*(K_0) \longrightarrow m^*(L_0).$$

To define it, consider the composite

$$J \xrightarrow{\ j\ } \mathscr{F}^{\mathscr{C}}(K,L) \rightarrowtail^{i} \textstyle\prod_{C_0} \mathscr{F}(K_0, L_0)$$

in $\mathscr{E} \cong \mathscr{E}/\mathbf{1}$. This corresponds by adjunction with a morphism

$$\overline{j}: J \times C_0 \longrightarrow \mathscr{F}(K_0, L_0)$$

in $\mathscr{E}/C_0$; thus $\alpha_{K_0L_0} \circ \overline{j} = p_{C_0}$. Consider now the composite

$$J \xrightarrow{(1_J, m)} J \times C_0 \xrightarrow{\ \overline{j}\ } \mathscr{F}(K_0, L_0) \xrightarrow{\alpha_{K_0L_0}} C_0$$

which is thus equal to m. It allows the definition

$$f_{(j,m)} = \big((1_J, m)^* \circ \overline{j}^*\big)(f_{K_0L_0}).$$

The naturality of f_{KL} is straightforward to check, just copying the definitions.

Finally we must check the universal property of the pair (α_{KL}, f_{KL}). We consider an object M in $\mathscr{E}$ and the unique morphism $\mu: M \longrightarrow \mathbf{1}$, together with a morphism $g: \mu^*(K) \longrightarrow \mu^*(L)$ in the fibre over M. The functors $\mu^*(K)$ and $\mu^*(L)$ are just the composites

$$\mathscr{E}/M \times_{\mathscr{E}} \mathscr{E}(\mathscr{C}) \xrightarrow{\ p\ } \mathscr{E}(\mathscr{C}) \underset{L}{\overset{K}{\rightrightarrows}} \mathscr{F}$$

and $g: K(p) \Rightarrow L(p)$ is a cartesian natural transformation. We must find a unique morphism $\nu: M \longrightarrow \mathscr{F}^{\mathscr{C}}(K,L)$ such that $\nu^*(f_{KL}) = g$.

Constructing ν is equivalent to constructing $\eta: M \longrightarrow \prod_{C_0} \mathscr{F}(K_0, L_0)$ such that $\gamma_1 \circ \eta = \gamma_0 \circ \eta$. By adjunction, constructing η reduces to constructing a suitable morphism $\overline{\eta}: M \times C_0 \longrightarrow \mathscr{F}(K_0, L_0)$ in $\mathscr{E}/C_0$, i.e. $\alpha_{K_0L_0} \circ \overline{\eta} = p_{C_0}$. To do this consider the product $M \times C_0$ and its two

$$\begin{array}{ccc}
(1_M \times d_0)^* p^*_{C_0} K_0 = p^*_{C_1} d^*_0 K_0 & \xrightarrow{p^*_{C_1}\kappa} & p^*_{C_1} d^*_1 K_0 = (1_M \times d^*_1) p^*_{C_0} K_0 \\
{\scriptstyle (1_M \times d_0)^* g_{(p_M, p_{C_0})}} \downarrow & & \downarrow {\scriptstyle (1_M \times d^*_1) g_{(p_M, p_{C_0})}} \\
(1_M \times d_0)^* p^*_{C_0} L_0 = p^*_{C_1} d_0 L_0 & \xrightarrow[p^*_{C_1}\lambda]{} & p^*_{C_1} d^*_1 L_0 = (1_M \times d^*_1) p^*_{C_0} L_0
\end{array}$$

Diagram 8.23

projections p_M, p_{C_0}. The pair (p_M, p_{C_0}) is an object of $\mathscr{E}/M \times_{\mathscr{E}} \mathscr{E}(\mathscr{C})$ in the fibre over $M \times C_0$; this yields a morphism

$$g_{(p_M, p_{C_0})} \colon K(M \times C_0, p_{C_0}) \longrightarrow L(M \times C_0, p_{C_0})$$

in the fibre over $M \times C_0$. But $(M \times C_0, p_{C_0}) = p^*_{C_0}(C_0, 1_{C_0})$ in the fibration $\mathscr{E}(\mathscr{C})$; since K and L are cartesian functors, this indicates that we have in fact obtained a morphism

$$g_{(p_M, p_{C_0})} \colon p^*_{C_0}(K_0) \longrightarrow p^*_{C_0}(L_0)$$

By universality of the pair $(\alpha_{K_0 L_0}, f_{K_0, L_0})$, there is a unique morphism $\overline{\eta} \colon M \times C_0 \longrightarrow \mathscr{F}(K_0, L_0)$ such that $\alpha_{K_0 L_0} \circ \overline{\eta} = p_{C_0}$ and $\overline{\eta}^*(f_{K_0 L_0}) = g_{(p_M, p_{C_0})}$. This defines the morphism $\eta \colon M \longrightarrow \prod_{C_0} \mathscr{F}(K_0, L_0)$.

The relation $\gamma_1 \circ \eta = \gamma_0 \circ \eta$ follows from the naturality of g. By adjunction, we have indeed to prove the equality of two morphisms

$$M \times C_1 \underset{\rho_0}{\overset{\rho_1}{\rightrightarrows}} \mathscr{F}(d^*_0(K_0), d^*_1(L_0))$$

over C_1. By the universal construction of the codomain, it suffices to check that

$$\rho^*_1\big(f_{d^*_0(K_0), d^*_1(L_0)}\big) = \rho^*_0\big(f_{d^*_0(K_0), d^*_1(L_0)}\big).$$

Making explicit all the definitions and considering the two morphisms

$$1_M \times d_0, 1_M \times d_1 \colon M \times C_1 \rightrightarrows M \times C_0$$

one concludes that the two morphisms in the fibre of $\mathscr{F}$ over $M \times C_1$ are the two paths in diagram 8.23. It suffices now to go back to the definitions of κ and λ, namely $\kappa = K(1_{C_1})$, $\lambda = L(1_{C_1})$, where the identity $1_{C_1} \colon (C_1, d_0) \longrightarrow (C_1, d_1)$ is a morphism of $\mathscr{E}(\mathscr{C})$. Considering now the product $M \times C_1$ in $\mathscr{E}$,

$$(1_M, 1_{C_1}) \colon (p_M, d_0) \longrightarrow (p_M, d_1)$$

is a morphism of $\mathscr{E}/M\times_{\mathscr{E}}\mathscr{E}(\mathscr{C})$ and the naturality of g applied to this morphism yields the commutativity of the previous diagram. □

8.7 Definability

When $F\colon \mathscr{F} \longrightarrow \mathscr{E}$ is a fibration and $I \in \mathscr{E}$ is an object, we think of the objects and arrows of the fibre $\mathscr{F}_I$ as I-indexed families, having in mind the canonical example $\mathsf{Set}(\mathscr{C}) \longrightarrow \mathsf{Set}$ of 8.1.9.b. As already indicated in the introduction, there is *a priori* no reason for us to be able to speak of the "subfamilies satisfying some given property". For example if $f\colon X \longrightarrow Y$ is a morphism in the fibre $\mathscr{F}_I$, how can we indicate the subobject $J \rightarrowtail I$ in $\mathscr{E}$ "where f is monomorphic"? Being able to do it means that the notion of monomorphism is "definable" in the fibration $F\colon \mathscr{F} \longrightarrow \mathscr{E}$.

Definition 8.7.1 *Let $F\colon \mathscr{F} \longrightarrow \mathscr{E}$ be a fibration. A class $\mathscr{C}$ of objects of $\mathscr{F}$ is definable when:*

(1) if $\alpha\colon J \longrightarrow I$ in $\mathscr{E}$ and $X \in \mathscr{F}_I$ are such that $X \in \mathscr{C}$, then $\alpha^* X \in \mathscr{C}$;
(2) given objects $I \in \mathscr{E}$, $X \in \mathscr{F}_I$, there exists a subobject $\alpha\colon J \rightarrowtail I$ such that $\alpha^*(X) \in \mathscr{C}$ and α is universal for this property, i.e. given $\beta\colon K \longrightarrow I$ such that $\beta^*(X) \in \mathscr{C}$, there exists a (unique) factorization $\gamma\colon K \longrightarrow J$ such that $\beta = \alpha \circ \gamma$.

Observe that the uniqueness of γ is implied by the requirement that α has to be a monomorphism.

Definition 8.7.2 *Let $F\colon \mathscr{F} \longrightarrow \mathscr{C}$ be a fibration. A class $\mathscr{C}$ of arrows in the fibres of $\mathscr{F}$ is definable when:*

(1) if $\alpha\colon J \longrightarrow I$ in $\mathscr{E}$ and $f\colon X \longrightarrow Y$ in $\mathscr{F}_I$ are such that $f \in \mathscr{C}$, then $\alpha^*(f) \in \mathscr{C}$;
(2) given an object $I \in \mathscr{E}$ and an arrow $f\colon X \longrightarrow Y$ in $\mathscr{F}_I$, there exists a subobject $\alpha\colon J \rightarrowtail I$ such that $\alpha^*(f) \in \mathscr{C}$ and α is universal for this property, i.e. given $\beta\colon K \longrightarrow I$ such that $\beta^*(f) \in \mathscr{C}$, there exists a (unique) factorization $\gamma\colon K \longrightarrow J$ such that $\beta = \alpha \circ \gamma$.

Definition 8.7.3 *Let $F\colon \mathscr{F} \longrightarrow \mathscr{E}$ be a fibration. By a definable subfibration of F we mean a subcategory $\mathscr{C} \subseteq \mathscr{F}$ together with the restriction $F\colon \mathscr{C} \longrightarrow \mathscr{E}$, such that:*

(1) $F\colon \mathscr{C} \longrightarrow \mathscr{E}$ is still a fibration;
(2) the class of objects of $\mathscr{C}$ is definable;
(3) the class of those morphisms of $\mathscr{C}$ lying in the fibres is definable.

Proposition 8.7.4 *Let $\mathscr{A}$ be a small category. The definable subfibrations of the canonical fibration* $\mathsf{Set}(\mathscr{A})\longrightarrow\mathsf{Set}$ *(see 8.1.9.b) are equivalent to the canonical fibrations* $\mathsf{Set}(\mathscr{B})\longrightarrow\mathsf{Set}$*, for $\mathscr{B}$ a subcategory of $\mathscr{A}$.*

Proof If $\mathscr{B}\subseteq\mathscr{A}$ is a subcategory, $\mathsf{Set}(\mathscr{B})$ is a subfibration of $\mathsf{Set}(\mathscr{A})$. Given a set I and a family $(C_i)_{i\in I}$ of objects of $\mathscr{C}$, it suffices to put

$$J=\{i\in I\,|\,B_i\in\mathscr{B}\}$$

to satisfy condition (2) of definition 8.7.1. Indeed, if $\beta\colon K\longrightarrow I$ is such that $\beta^*(C_i)_{i\in I}\in\mathscr{B}$, one has $C_{\beta(k)}\in\mathscr{B}$ for each index $k\in K$, from which β factors (uniquely) through the subobject J. In the same way if $(f_i\colon C_i\longrightarrow D_i)_{i\in I}$ is a family of arrows of $\mathscr{C}$, one defines

$$J=\{i\in I\,|\,f_i\in\mathscr{B}\}$$

to satisfy condition (2) of 8.7.2.

Conversely, fix a definable subfibration $\mathscr{C}\longrightarrow\mathsf{Set}$ of $\mathsf{Set}(\mathscr{A})\longrightarrow\mathsf{Set}$. The fibre of $\mathsf{Set}(\mathscr{A})$ over the singleton $\mathbf{1}$ is precisely $\mathscr{A}$. We define $\mathscr{B}$ to be the fibre of $\mathscr{C}$ at $\mathbf{1}$. By construction of $\mathsf{Set}(\mathscr{A})$, this defines $\mathscr{B}$ up to equivalence (depending on the choice of a distinguished singleton). We must prove that the two fibrations $\mathscr{C}\longrightarrow\mathsf{Set}$ and $\mathsf{Set}(\mathscr{B})\longrightarrow\mathsf{Set}$ are equivalent over Set. By 8.3.1, it suffices to prove they have equivalent fibres.

If $(A_i)_{i\in I}$ is an object of $\mathsf{Set}(\mathscr{A})$ in the fibre over I, consider the corresponding subset $J\subseteq I$ given by definability of the objects of $\mathscr{C}$. For every index $i\in I$, we can consider the inclusion $\beta\colon\{i\}\longrightarrow I$. By 8.7.1, β factors through J if and only if A_i is in the fibre of $\mathscr{C}$ over $\{i\}$, i.e. $i\in J$ if and only if $A_i\in\mathscr{B}$. Thus $(A_i)_{i\in I}\in\mathscr{C}$ if and only if for each index i, $A_i\in\mathscr{B}$. The argument for the arrows is perfectly analogous. □

Counterexample 8.7.5

We consider again a category $\mathscr{A}$ and the corresponding canonical fibration $\mathsf{Set}(\mathscr{A})\longrightarrow\mathsf{Set}$. Not all subfibrations of this fibration are definable. For an elementary counterexample, choose $\mathscr{A}$ to be the discrete category with just two objects X, Y. The full subcategory of $\mathsf{Set}(\mathscr{A})$ whose objects are the families $(A_i)_{i\in I}$ with all the A_i's equal is obviously a subfibration of $\mathsf{Set}(\mathscr{A})$. But it does not correspond with any subcategory of $\mathscr{A}$.

Let us now prove two important stability properties of definable subfibrations.

Proposition 8.7.6 *If a subfibration is definable, all its localizations are again definable subfibrations.*

Proof Let $F: \mathscr{F} \longrightarrow \mathscr{E}$ be a fibration and $F: \mathscr{C} \longrightarrow \mathscr{E}$ a definable subfibration. The localized fibration $F_{(I)}: \mathscr{C}_{(I)} \longrightarrow \mathscr{E}/I$ is just obtained by restricting $F_{(I)}: \mathscr{F}_{(I)} \longrightarrow \mathscr{E}/I$ to $\mathscr{C}$. Giving an object or an arrow in the fibre of $\mathscr{C}_{(I)}$ over $\beta: K \longrightarrow I$ is just giving it in the fibre of $\mathscr{F}$ over K (see 8.5.4). The corresponding subobject $\alpha: J \rightarrowtail K$ in $\mathscr{E}$ obtained by definability of $\mathscr{C}$ immediately yields the subobject $\alpha: (J, \beta\alpha) \longrightarrow (K, \beta)$ of $\mathscr{E}/I$ exhibiting the definability of $\mathscr{C}_{(I)}$. □

Proposition 8.7.7 *If a fibration is locally small, all its definable subfibrations are again locally small.*

Proof Consider a locally small fibration $F: \mathscr{F} \longrightarrow \mathscr{E}$ and a definable subfibration $F: \mathscr{C} \longrightarrow \mathscr{E}$. Consider also $I \in \mathscr{E}$ and two objects X, Y in the fibre of $\mathscr{C}$ over I. This yields by 8.6.1 two universal morphisms $\alpha_{XY}: \mathscr{F}(X,Y) \longrightarrow I$ in $\mathscr{E}$ and $f_{XY}: \alpha^*_{XY}(X) \longrightarrow \alpha^*_{XY}(Y)$ in the fibre of $\mathscr{F}$ over $\mathscr{F}(X,Y)$. But since $\mathscr{C}$ is definable, by 8.7.2 we can find a subobject $\beta_{XY}: \mathscr{C}(X,Y) \rightarrowtail \mathscr{F}(X,Y)$ such that $\beta^*_{XY}(f_{XY})$ is in the fibre of $\mathscr{C}$ over $\mathscr{C}(X,Y)$ and is universal for this property. The composite $\alpha_{XY} \circ \beta_{XY}$ together with $\beta^*_{XY}(f_{XY})$ exhibits the local smallness of $\mathscr{C}$. Indeed given $\gamma: J \longrightarrow I$ in $\mathscr{E}$ and $g: \gamma^*(X) \longrightarrow \gamma^*(Y)$ in $\mathscr{C}$, we get first a unique $\delta: J \longrightarrow \mathscr{F}(X,Y)$ such that $\alpha_{XY} \circ \delta = \gamma$ and $\delta^*(f_{XY}) = g$. But since g is in $\mathscr{C}$, this morphism δ factors through β_{XY} via some ε, thus $\beta_{XY} \circ \varepsilon = \delta$. This yields finally a unique $\varepsilon: J \longrightarrow \mathscr{C}(X,Y)$ such that $\alpha_{XY} \circ \beta_{XY} \circ \varepsilon = \alpha_{XY} \circ \delta = \gamma$ and $\varepsilon^* \beta^*_{XY}(f_{XY}) = \delta^*(f_{XY}) = g$. □

We conclude with some existence theorems of definable classes.

Theorem 8.7.8 *Let $\mathscr{E}$ be a category with finite limits. If $F: \mathscr{F} \longrightarrow \mathscr{E}$ is a locally small fibration, the class of isomorphisms is definable in $\mathscr{F}$.*

Proof Consider objects $I \in \mathscr{E}$ and $X, Y \in \mathscr{F}_I$ and the pullback of diagram 8.24, where $\varepsilon_X, \varepsilon_Y, \gamma_{XYX}, \gamma_{YXY}$ are defined in 8.6.5 and τ is the twisting isomorphism on the pullback of α_{XY}, α_{YX} (notation of 8.1.1).

We shall prove first that the composite

$$\mathsf{Iso}(X,Y) \xrightarrow{\delta_{XY}} \mathscr{F}(X,Y) \times_I \mathscr{F}(Y,X) \xrightarrow{p_{XY}} \mathscr{F}(X,Y)$$

is a monomorphism, which we shall write σ_{XY}. By definition of pullbacks, giving an arrow $\rho: J \longrightarrow \mathsf{Iso}(X,Y)$ in $\mathscr{E}$ is equivalent to giving

$$\begin{array}{ccc} \mathsf{Iso}(X,Y) & \xrightarrow{\beta_{XY}} & I \\ \downarrow{\scriptstyle \delta_{XY}} & & \downarrow{\scriptstyle (\varepsilon_X, \varepsilon_Y)} \\ \mathscr{F}(X,Y)\times_I \mathscr{F}(Y,X) & \xrightarrow[(\gamma_{XYX}, \gamma_{YXY}\circ\tau)]{} & \mathscr{F}(X,X)\times_I \mathscr{F}(Y,Y) \end{array}$$

Diagram 8.24

arrows

$$u: J \longrightarrow \mathscr{F}(X,Y), \quad v: J \longrightarrow \mathscr{F}(Y,X), \quad \mu: J \longrightarrow I,$$

such that

$$\gamma_{XYZ} \circ \begin{pmatrix} u \\ v \end{pmatrix} = \varepsilon_X \circ \mu, \quad \gamma_{YXY} \circ \begin{pmatrix} v \\ u \end{pmatrix} = \varepsilon_Y \circ \mu.$$

Since $\sigma_{XY} \circ \rho = u$, σ_{XY} will be a monomorphism as long as u determines ρ, thus as long as u determines v and μ.

By 8.1.1, u is uniquely determined by the composite $\beta = \alpha_{XY} \circ u$ and the morphism $g = u^*(f_{XY})$; in the same way v is uniquely determined by the composite $\beta' = \alpha_{YX} \circ v$ and the morphism $g' = v^*(f_{YX})$. By assumption on u, v, μ, $\beta = \beta'$ and moreover, with the notations of 8.6.5

$$\begin{aligned} \mu &= \alpha_{XX} \circ \varepsilon_X \circ \mu = \alpha_{XX} \circ \gamma_{XYX} \circ \begin{pmatrix} u \\ v \end{pmatrix} \\ &= \alpha_{XYX} \circ \begin{pmatrix} u \\ v \end{pmatrix} = \alpha_{XY} \circ p_{XY} \circ \begin{pmatrix} u \\ v \end{pmatrix} = \alpha_{XY} \circ u = \beta; \end{aligned}$$

thus $\beta' = \beta = \mu$. In particular the following arrows are composable in $\mathscr{F}_J$:

$$u^*(f_{XY}): \mu^*(X) \longrightarrow \mu^*(Y), \quad v^*(f_{YX}): \mu^*(Y) \longrightarrow \mu^*(X).$$

Let us prove they are mutually inverse.

By assumption diagram 8.25 is commutative. By definition of γ_{XYX} (see 8.6.5)

$$\gamma^*_{XYX}(f_{XX}) = p^*_{YX}(f_{YX}) \circ p^*_{XY}(f_{XY}),$$

thus

$$\left(\begin{pmatrix} u \\ v \end{pmatrix}^* \circ \gamma^*_{XYX}\right)(f_{XX})$$

$$\begin{array}{ccc} J & \xrightarrow{\binom{u}{v}} & \mathscr{F}(X,Y)\times_I\mathscr{F}(Y,X) \\ \mu\downarrow & & \downarrow\gamma_{XYX} \\ I & \xrightarrow[\varepsilon_X]{} & \mathscr{F}(X,X) \end{array}$$

Diagram 8.25

$$= \left(\binom{u}{v}^* \circ p^*_{YX}\right)(f_{YX}) \circ \left(\binom{u}{v}^* \circ p^*_{XY}\right)(f_{XY})$$

$$= \left(p_{YX} \circ \binom{u}{v}\right)^*(f_{YX}) \circ \left(p_{XY} \circ \binom{u}{v}\right)^*(f_{XY})$$

$$= v^*(f_{YX}) \circ u^*(f_{XY}).$$

But by definition of ε_X (see 8.6.5)

$$(\mu^* \circ \varepsilon^*_X)(f_{XX}) = \mu^*(1_X) = 1_{\mu^*(X)}$$

from which $v^*(f_{YX})\circ\mu^*(f_{XY}) = 1_{\mu^*(X)}$. The converse equality is proved in an analogous way.

As a first consequence, knowledge of the morphism u determines $\beta' = \mu = \alpha_{XY} \circ u$, but also $g = v^*(f_{YX})$ which is just the inverse of $u^*(f_{XY})$. So u determines both μ and v and σ_{XY} is a monomorphism.

We are now ready to prove that isomorphisms are definable in $\mathscr{F}$. Obviously, isomorphisms are mapped onto isomorphisms by the inverse image functors. Next fix a morphism $f\colon X \longrightarrow Y$ in the fibre $\mathscr{F}_I$, for some object $I \in \mathscr{E}$. The identity $1_I\colon I \longrightarrow I$ together with the morphism $f\colon 1^*_I(I) \longrightarrow 1^*_I(I)$ yields a unique morphism $\nu\colon I \longrightarrow \mathscr{F}(X,Y)$ such that $\alpha_{XY} \circ \nu = 1_I$ and $\nu^*(f_{XY}) = f$. Computing the pullback of diagram 8.26 yields a monomorphism τ_f which will be proved to have the required properties.

We have to consider a morphism $\mu\colon J \longrightarrow I$ in $\mathscr{E}$ such that $\mu^*(f)$ is an isomorphism in the fibre $\mathscr{F}_J$. Let us put $u = \nu\circ\mu$, thus $u\colon J \longrightarrow \mathscr{F}(X,Y)$ and $u^*(f_{XY}) = \mu^*\nu^*(f_{XY}) = \mu^*(f)$. Considering next the morphisms

$$\mu\colon J \longrightarrow I, \quad \mu^*(f)^{-1}\colon \mu^*(Y) \longrightarrow \mu^*(X)$$

yields a unique factorization $v\colon J \longrightarrow \mathscr{F}(Y,X)$ in $\mathscr{E}$ such that $v^*(f_{XY}) = \mu^*(f)^{-1}$. Finally u, v, μ constitute a triple as at the beginning of the proof

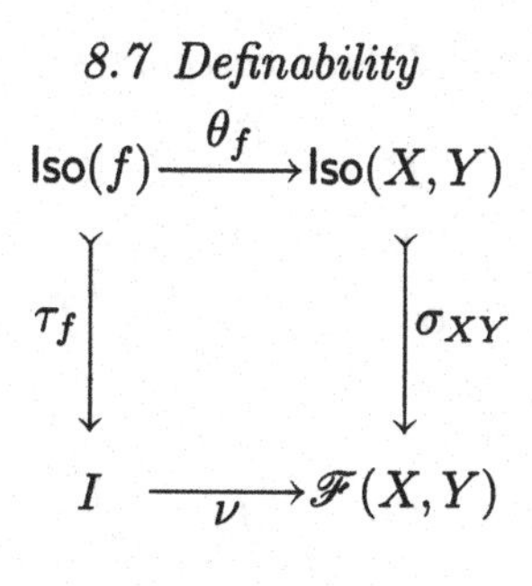

Diagram 8.26

and determine a morphism $w\colon J \longrightarrow \mathsf{Iso}(X,Y)$ such that $\sigma_{XY} \circ w = \nu \circ \mu$. This yields the required factorization $J \longrightarrow \mathsf{Iso}(f)$. □

Theorem 8.7.8 and its consequences are worth a comment. A morphism $f\colon X \longrightarrow Y$ in a category $\mathscr{C}$ is an isomorphism when

$$\exists g \in \mathscr{C}(Y,X) \quad f \circ g = 1_Y, \quad g \circ f = 1_X.$$

In this formula, the existential quantifier acts on a variable g running through a set $\mathscr{C}(Y,X)$, thus by the comprehension scheme the formula determines a subset $\mathsf{Iso}(X,Y) \subseteq \mathscr{C}(X,Y)$ of those arrows $f\colon X \longrightarrow Y$ satisfying the required property. The local smallness of $\mathscr{C}$ has been used twice, for $\mathscr{C}(Y,X)$ and for $\mathscr{C}(X,Y)$, for us to be allowed to apply the comprehension scheme.

Replacing "isomorphism" by "monomorphism" would lead us to consider the formula

$$\forall Z \in \mathscr{C} \ \forall u,v \in \mathscr{C}(Z,X) \quad f \circ u = f \circ v \Rightarrow u \circ v.$$

This formula contains a quantifier acting on a variable Z which runs through a class $\mathscr{C}$, not just a set. Thus in most set theories, the formula will not determine a subset $\mathrm{Mono}(X,Y) \subseteq \mathscr{C}(X,Y)$ of those arrows $f\colon X \longrightarrow Y$ which satisfy the required property, because the comprehension scheme cannot be applied.

But now suppose the category $\mathscr{C}$ has finite limits. In that case we can consider the kernel pair of a morphism $f\colon X \longrightarrow Y$, as in diagram 8.27, (see 2.5.6, volume 1). We know that f is a monomorphism if and only if $\alpha = \beta$, thus if and only if the equalizer $k\colon K \longrightarrow P$ of α, β is an isomorphism. This reduces the property "f is a monomorphism" to the properties "α is equal to β" or "k is an isomorphism", which can now be handled in any set theory.

These considerations underline the fact that working with finitely complete fibrations, in the following results, is not just a convenient technical

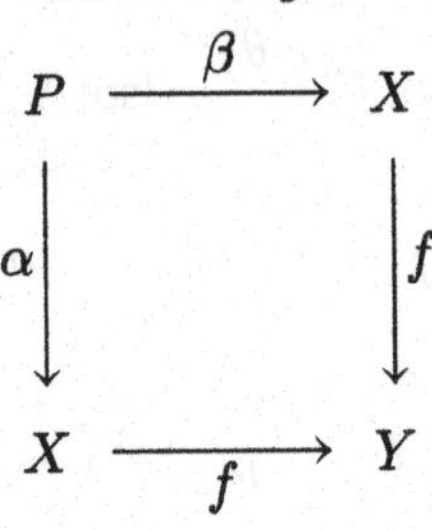

Diagram 8.27

requirement, but is really a crucial assumption. The reader could argue that defining a finite limit requires also a quantification on the class of all objects of the category, to express the universal property of the limit. This is true, but this is also an observation of a different nature. Proving a property for every object of a category does not require any comprehension scheme at all; it is constructing the set of those objects satisfying the property which requires the comprehension scheme.

Proposition 8.7.9 *Let $F: \mathscr{F} \longrightarrow \mathscr{E}$ be a locally small and finitely complete fibration over a category $\mathscr{E}$ with finite limits. The class of monomorphisms in the fibres of $\mathscr{F}$ is definable.*

Proof Given an arrow $f: X \longrightarrow Y$ in the fibre $\mathscr{F}_I$ over $I \in \mathscr{E}$, we compute the equalizer k_f of the kernel pair (α_f, β_f) of f (see 2.5.6, volume 1) in the fibre $\mathscr{F}_I$. We know that f is a monomorphism iff $\alpha_f = \beta_f$, thus iff k_f is an isomorphism.

Since inverse image functors preserve kernel pairs and equalizers, they preserve monomorphisms. Next if $f: X \longrightarrow Y$ is an arrow as above, consider by 8.7.8 and 8.7.2 the subobject $\alpha: J \rightarrowtail I$ universal for the fact that $\alpha^*(k_f)$ is an isomorphism. Again since the inverse image functors preserve kernel pairs and equalizers, $k_{\alpha^*(f)} = \alpha^*(k_f)$ and thus $\alpha: J \rightarrowtail I$ is universal for the fact that $k_{\alpha^*(f)}$ is an isomorphism, i.e. $\alpha^*(f)$ is a monomorphism. □

Proposition 8.7.10 *Let $F: \mathscr{F} \longrightarrow \mathscr{E}$ be a locally small and finitely complete fibration over a category $\mathscr{E}$ with finite limits. The class of terminal objects in the fibres of $\mathscr{F}$ is definable.*

Proof Given an object $X \in \mathscr{F}_I$ in the fibre over I, let us consider the unique arrow $\varepsilon_X: X \longrightarrow 1_I$ in $\mathscr{F}_I$, where 1_I denotes the terminal object of the fibre. Since inverse image functors preserve terminal objects, given a morphism $\alpha: J \longrightarrow I$ in $\mathscr{E}$, $\alpha^*(X)$ is a terminal object in

$\mathscr{F}_J$ precisely when $\alpha^*(\varepsilon_X)$ is an isomorphism. Thus the universal monomorphism $\alpha: J \rightarrowtail I$ for $\alpha^*(X)$ terminal is just the universal morphism $\alpha: J \rightarrowtail I$ for $\alpha^*(\varepsilon_X)$ an isomorphism (see 8.7.8). □

Definitions 8.7.1 and 8.7.2 explain the notion of definability for a class of objects or a class of arrows. In fact the notion of definability can be immediately extended to a class of diagrams of arbitrary shape $\mathscr{D}$.

Definition 8.7.11 *Let $F: \mathscr{F} \longrightarrow \mathscr{E}$ be a fibration and $\mathscr{D}$ a small category. A class $\mathscr{C}$ of diagrams of shape $\mathscr{D}$ in the fibres of $\mathscr{F}$ is definable when the class $\mathscr{C}$ of corresponding objects in the fibration $\mathscr{F}^{(\mathscr{D})} \longrightarrow \mathscr{E}$ is definable (see 8.1.7).*

We recall that the fibre of $\mathscr{F}^{(\mathscr{D})}$ over an object $I \in \mathscr{E}$ is precisely the category of diagrams of shape $\mathscr{D}$ in the fibre $\mathscr{F}_I$. One observes immediately that definition 8.7.2 is just a special case of 8.7.11, taking for $\mathscr{D}$ the category $\bullet \longrightarrow \bullet$ with two distinct objects and just one non-trivial arrow between them. Here is a special case of interest:

Definition 8.7.12 *Let $F: \mathscr{F} \longrightarrow \mathscr{E}$ be a fibration. Equality is definable in $(\mathscr{F}, F)$ when the class of parallel pairs of arrows $f, g: X \rightrightarrows Y$ such that $f = g$, in the fibres of $\mathscr{F}$, is definable.*

Proposition 8.7.13 *Let $F: \mathscr{F} \longrightarrow \mathscr{E}$ be a locally small and finitely complete fibration over a category $\mathscr{E}$ with finite limits. Then equality is definable in $(\mathscr{F}, F)$.*

Proof Given two arrows $f, g: X \rightrightarrows Y$ in the fibre $\mathscr{F}_I$ over $I \in \mathscr{E}$, $f = g$ precisely when the equalizer $k_{fg}: K_{fg} \longrightarrow X$ of (f, g) is an isomorphism. Therefore the universal monomorphism $\alpha: J \longrightarrow I$ such that $\alpha^*(f) = \alpha^*(g)$ is the universal monomorphism such that $\alpha^*(k_{fg})$ is an isomorphism (see 8.7.8), just because α^* preserves equalizers and thus $\alpha^*(k_{fg}) = k_{\alpha^*(f)\alpha^*(g)}$. □

Proposition 8.7.14 *Let $F: \mathscr{F} \longrightarrow \mathscr{E}$ be a locally small and finitely complete fibration over a category $\mathscr{E}$ with finite limits. Binary products are definable in $(\mathscr{F}, F)$.*

Proof We must consider the class of those pairs of arrows

$$X \xleftarrow{p_X} Z \xrightarrow{p_Y} Y$$

which are the projections of the product $X \times Y$ in a fibre. Another such diagram $X \xleftarrow{f} T \xrightarrow{g} Y$, is the product of X, Y when the unique factorization $t: T \longrightarrow Z$ is an isomorphism.

Thus given a diagram $X \xleftarrow{f} T \xrightarrow{g} Y$ in the fibre over I and the factorization $t\colon T \longrightarrow X \times Y$ as above, the universal subobject $\alpha\colon J \rightarrowtail I$ such that $(\alpha^*(f), \alpha^*(g))$ is a product, is just the universal subobject such that $\alpha^*(t)$ is an isomorphism, since α^* preserves binary products. □

More generally, one has

Proposition 8.7.15 *Let $F\colon \mathscr{F} \longrightarrow \mathscr{E}$ be a locally small and finitely complete fibration over a category $\mathscr{E}$ with finite limits. Given a finite category $\mathscr{D}$, the limits of diagrams of shape $\mathscr{D}$ are definable in $(\mathscr{F}, F)$.*

Proof From $\mathscr{D}$, we construct the category $\overline{\mathscr{D}}$ obtained by adding one object M to $\mathscr{D}$, an identity arrow on M and one arrow $q_D\colon M \longrightarrow D$ for each object of $\mathscr{D}$; we impose, for each arrow $d\colon D \longrightarrow D'$ in $\mathscr{D}$, the relation $d \circ q_D = q_{D'}$, which defines the composition law in $\overline{\mathscr{D}}$. We are interested in the class of those diagrams $H\colon \overline{\mathscr{D}} \longrightarrow \mathscr{F}_I$ of shape $\overline{\mathscr{D}}$ in the fibres of $\mathscr{F}$, for which $\big(H(M), (H(q_D))_{D \in \mathscr{D}}\big)$ is the limit of the diagram $H\colon \mathscr{D} \longrightarrow \mathscr{F}_I$.

For an arbitrary diagram $H\colon \overline{\mathscr{D}} \longrightarrow \mathscr{F}_I$ in the fibre over I, consider its limit $\big(L, (p_D)_{D \in \mathscr{D}}\big)$ and the unique factorization $h\colon H(M) \longrightarrow L$ such that $p_D \circ h = H(q_D)$ for each object $D \in \mathscr{D}$. Now $\big(H(M), (H(q_D))_{D \in \mathscr{D}}\big)$ is a limit diagram precisely when h is an isomorphism. Therefore the universal subobject $\alpha\colon J \rightarrowtail I$ such that $\big(\alpha^* H(M), (\alpha^* H(q_D))_{D \in \mathscr{D}}\big)$ is a limit diagram is just the universal subobject such that $\alpha^*(h)$ is an isomorphism, because α^* preserves finite limits. □

8.8 Exercises

8.8.1 Given a base category $\mathscr{E}$, construct the fibration of fibred categories over $\mathscr{E}$: the fibre at an object $I \in \mathscr{E}$ is the category $\mathsf{Fib}(\mathscr{E}/I)$ of fibred categories and cartesian functors over $\mathscr{E}/I$.

8.8.2 Define (up to equivalence) the dual of a fibration $F\colon \mathscr{F} \longrightarrow \mathscr{E}$ directly from $(\mathscr{F}, F)$, without using the axiom of choice and thus the correspondence with pseudo-functors.

8.8.3 Let $F\colon \mathscr{F} \longrightarrow \mathscr{E}$ be a fibration over a finitely complete category $\mathscr{E}$. One gets a new fibration $\mathsf{Fam}(\mathscr{F}) \longrightarrow \mathscr{E}$ in the following way:

- an object of $\mathsf{Fam}(\mathscr{F})$ is a pair (α, X), where $\alpha\colon J \longrightarrow I$ is a morphism of $\mathscr{E}$ and $X \in \mathscr{F}_J$;

- with analogous notation, a morphism $(\alpha, X) \longrightarrow (\alpha', X')$ is a triple (j, i, f) where $j: J \longrightarrow J'$, $i: J \longrightarrow J'$, $f: X \longrightarrow X'$ are morphisms and $i \circ \alpha = \alpha' \circ j$, $F(f) = i$;
- the projection functor $\mathsf{Fam}(\mathscr{F}) \longrightarrow \mathscr{E}$ maps (α, X) to I and (i, j, f) to i.

Intuitively, an object of $\mathsf{Fam}(\mathscr{F})$ in the fibre over I is an I-indexed family of families. Prove that this construction extends to a monad $\mathsf{Fam}\colon \mathsf{Fib}(\mathscr{E}) \longrightarrow \mathsf{Fib}(\mathscr{E})$. Prove that the fibration $F: \mathscr{F} \longrightarrow \mathscr{E}$ has $\mathscr{E}$-coproducts iff the unit of the monad $\mathscr{F} \longrightarrow \mathsf{Fam}(\mathscr{F})$ admits a fibred left adjoint functor.

8.8.4 Choose $\mathscr{E}$ to be the category of commutative rings with units. Define a fibration $\mathscr{F} \longrightarrow \mathscr{E}$ by choosing as fibre over the ring R the category of modules on R. Prove that this fibration is cocomplete.

8.8.5 Let $\mathscr{E}$ be a finitely complete and cocomplete category. Prove that the fibration $d_1\colon \mathsf{Ar}(\mathscr{E}) \longrightarrow \mathscr{E}$ is finitely cocomplete iff finite colimits are universal in $\mathscr{E}$. When this is the case, prove that the fibration is cocomplete.

8.8.6 Consider the canonical fibration $d_1\colon \mathsf{Ar}(\mathsf{Cat}) \longrightarrow \mathsf{Cat}$ of the category Cat of small categories and functors over itself. Show that for every category $I \in \mathsf{Cat}$, the "diagonal" cartesian functor $\mathsf{Cat} \longrightarrow \mathsf{Ar}(\mathsf{Cat})^I$ has a fibred right adjoint, but that nevertheless the fibration $\mathsf{Ar}(\mathsf{Cat}) \longrightarrow \mathsf{Cat}$ is not complete, because the inverse image functors do not have a right adjoint.

8.8.7 Let $\mathscr{E}$ be a category with finite limits. Prove that the fibration of fibred categories over $\mathscr{E}$ is complete (see 8.8.1) and has $\mathscr{E}$-indexed coproducts.

8.8.8 Let $\mathscr{E}$ be a category with finite limits, $F: \mathscr{F} \longrightarrow \mathscr{E}$ a fibration and $\mathscr{D}$ an internal category of $\mathscr{E}$. If the fibration $(\mathscr{F}, F)$ is complete or cocomplete, show that the same holds for the fibration $\mathscr{F}^{\mathscr{D}} \longrightarrow \mathscr{E}$.

8.8.9 Let $\mathscr{E}$ be a category with finite limits and $F: \mathscr{F} \longrightarrow \mathscr{E}$ a locally small fibration. Given objects $X \in \mathscr{F}_I$, $Y \in \mathscr{F}_J$ in the fibres over I, J, define an object $\mathrm{Hom}(X, Y) \in \mathscr{E}$, morphisms

$$\partial_0\colon \mathrm{Hom}(X,Y) \longrightarrow I, \quad \partial_1\colon \mathrm{Hom}(X,Y) \longrightarrow J$$

in $\mathscr{E}$ and a morphism of $\mathscr{F}$,

$$\varphi_{XY}\colon \partial_0^* X \longrightarrow \partial_1^* Y,$$

universal for these data. [Hint: consider the product $I \times J$ and the object $\mathscr{F}(p_X^*(X), p_Y^*(Y))$.]

8.8.10 Let $\mathscr{E}$ be a category with finite limits. A fibration $F\colon \mathscr{F} \longrightarrow \mathscr{E}$ is locally small iff given objects $X, Y \in \mathscr{F}$, there exist an object $Z \in \mathscr{F}$, a cartesian arrow $f\colon Z \longrightarrow X$ and a morphism $g\colon Z \longrightarrow Y$, universal for these data. [Hint: with the notation of 8.8.9, $Z = \partial_0^* X$ in the fibre over $\mathrm{Hom}(X, Y)$.]

8.8.11 Let $\mathscr{C}$ be a category. In the case of the fibration $\mathsf{Set}(\mathscr{C}) \longrightarrow \mathsf{Set}$, compute explicitly the objects $\mathscr{F}(X, Y)$, $\mathrm{Hom}(X, Y)$, Z of 8.8.9 and 8.8.10 in the case of two objects X, Y in the same fibre.

8.8.12 Let $\mathscr{E}$ be a category with finite limits. Prove that when the canonical fibration $d_1\colon \mathsf{Ar}(\mathscr{E}) \longrightarrow \mathscr{E}$ of $\mathscr{E}$ over itself has $\mathscr{E}$-indexed products, it is locally small. [Hint: given two arrows $u\colon J \longrightarrow I$, $v\colon K \longrightarrow I$ compute their pullback (L, u', v') and consider $\prod_u (v')$.]

8.8.13 Let $F\colon \mathscr{F} \longrightarrow \mathscr{E}$ be a fibration with $\mathscr{F}$ non-empty. If the fibration is locally small and $\mathscr{E}$ has an initial object $\mathbf{0}$, prove that the fibre $\mathscr{F}_0$ is equivalent to the terminal category.

8.8.14 Let $F\colon \mathscr{F} \longrightarrow \mathscr{E}$ be a locally small fibration. Show that for every epimorphism $\alpha\colon J \longrightarrow I$ in $\mathscr{E}$, the inverse image functor $\alpha^*\colon \mathscr{F}_I \longrightarrow \mathscr{F}_J$ is faithful.

8.8.15 Let $\mathscr{E}$ be the category of R-modules, for a ring R. Prove that in the canonical fibration $d_1\colon \mathsf{Ar}(\mathscr{E}) \longrightarrow \mathscr{E}$ of $\mathscr{E}$ over itself, the equality is not definable.

8.8.16 Let $F\colon \mathscr{F} \longrightarrow \mathscr{E}$ be a fibration and $\mathscr{C}$ a definable class of objects of $\mathscr{F}$. Given a strong epimorphism $\alpha\colon J \longrightarrow I$ and an object $X \in \mathscr{F}_I$, prove that $X \in \mathscr{C}$ iff $\alpha^*(X) \in \mathscr{C}$. Generalize this result by replacing α with a strongly epimorphic family.

8.8.17 Let $F\colon \mathscr{F} \longrightarrow \mathscr{E}$ be a fibration, with $\mathscr{E}$ finitely complete. If $\mathscr{C}, \mathscr{D}$ are definable classes of objects in $\mathscr{F}$, prove that their intersection $\mathscr{C} \cap \mathscr{D}$ is again definable.

8.8.18 Let $F\colon \mathscr{F} \longrightarrow \mathscr{E}$ be a fibration for which the class of isomorphisms in the fibres of $\mathscr{F}$ is definable. Show that when $\mathscr{E}$ has an initial object $\mathbf{0}$, the fibre $\mathscr{F}_0$ is a groupoid (i.e. all its arrows are invertible).

8.8.19 Let $F\colon \mathscr{F} \longrightarrow \mathscr{E}$, $G\colon \mathscr{G} \longrightarrow \mathscr{E}$, $H\colon \mathscr{H} \longrightarrow \mathscr{E}$ be fibrations; let also $U\colon \mathscr{F} \longrightarrow \mathscr{H}$, $V\colon \mathscr{G} \longrightarrow \mathscr{H}$ be cartesian functors over $\mathscr{E}$. Define a fibration $(U, V) \longrightarrow \mathscr{E}$, obtained by computing fibrewise the comma categories. If $\mathscr{F}, \mathscr{G}$ are locally small and the equality is definable in $\mathscr{H}$, prove that (U, V) is locally small.

8.8.20 Suppose $\mathscr{E}$ has finite limits. With the notation of 8.8.19, prove that $\mathscr{H}$ is locally small iff for every small fibration $\mathscr{F}, \mathscr{G}$ and all cartesian functors U, V, the fibration (U, V) is small.

8.8.21 Consider fibrations $F: \mathscr{F} \longrightarrow \mathscr{E}$, $G: \mathscr{G} \longrightarrow \mathscr{E}$, $H: \mathscr{H} \longrightarrow \mathscr{E}$ and a cartesian functor $U: \mathscr{F} \longrightarrow \mathscr{G}$ over $\mathscr{E}$. Prove that when $\mathscr{F}$ is small, $\mathscr{G}$ is locally small and $\mathscr{H}$ is cocomplete, the cartesian functor $\mathscr{H}^F: \mathscr{H}^{\mathscr{G}} \longrightarrow \mathscr{H}^{\mathscr{F}}$ induced by the composition with F has a fibred left adjoint functor. [Hint: use the fibred comma categories of 8.8.19 to mimic the proof of 3.7.2, volume 1.]

Bibliography

J. Adamek, H. Herrlich and G. Strecker, *Abstract and concrete categories,* Wiley, 1990

J. Adamek, and J. Rosicky, Locally presentable and accessible categories, *London Math Soc. Lecture Notes,* **189**, 1994, Cambridge University Press

F.W. Anderson and K.R. Fuller, *Rings and categories of modules,* Springer, 1973

R.F. Arens and J.L. Kelley, Characterizations of the space of continuous functions over a compact Hausdorff space, *Trans. of the Amer. Math. Soc.,* **62**, 1947, 499–508

M. Barr, Coequalizers and free triples, *Math. Zeit.,* **116**, 1970, 307–322

M. Barr, Exact categories, *Springer LNM,* **236**, 1971, 1–120

M. Barr and J. Beck, Homology and standard constructions, *Springer LNM,* **80**, 1969, 245–335

M. Barr and C. Wells, *Toposes, triples and theories,* Springer, 1985

A. Bastiani and C. Ehresmann, Categories of sketched structures, *Cahiers de Top. et Géom. Diff.,* **13**, 1973, 1–105

J. Bénabou, Catégories avec multiplication, *Comp. Rend. Acad. Sc. Paris,* **256**, 1963, 1887–1890

J. Bénabou, Algèbre élémentaire dans les catégories avec multiplication, *Comp. Rend. Acad. Sc. Paris,* **258**, 1964, 771–774

J. Bénabou, Categories relatives, *Comp. Rend. Acad. Sc. Paris,* **260**, 1965, 3824–3827

J. Bénabou, Catégories algébriques et théorie de la descente, *Sém. Dubreil,* **21**, 1967-68

J. Bénabou, Structures algébriques dans les catégories, *Cah. de Top. et Géom. Diff.,* **10**, 1968, 1–126

J. Bénabou, Fibrations petites et localement petites, *Comp. Rend. Acad. Sc. Paris,* **281**, 1975, A897–900

J. Bénabou, Leçons sur les catégories fibrées, *unpublished notes by J. R. Roisin* , 1980, Louvain-la-Neuve

J. Bénabou, Fibered categories and the foundations of naive category theory, *J. of Symb. Logic,* **50**, 1985, 10–37

D. Buchsbaum, Exact categories and duality, *Trans. of the Amer. Math. Soc.,* **80**, 1955, 1–34

I. Bucur and A. Deleanu, *Introduction to the theory of categories and functors,* Wiley, 1968

J. Cigler, V. Losert and P. Michor, *Banach modules and functors on categories of Banach spaces,* Marcel Dekker, 1979

B. Day and G.M. Kelly, On topological product maps preserved by pullbacks or products, *Proc. Cambridge Phil. Soc.,* **67**, 1970, 553–558

B. Day and G.M. Kelly, Enriched functor categories, *Springer LNM,* **137**, 1970, 1–38

E. Dubuc, Kan extensions in enriched category theory, *Springer LNM,* **145**, 1970

J. Duskin, Variations on Beck's tripleability criterion, *Springer LNM,* **106**, 1969, 74–129

S. Eilenberg and G.M. Kelly, Closed categories, *Proc. Conf. on Cat. Alg. La Jolla 1965, Springer* , 1966, 421-562

S. Eilenberg and J. Moore, Adjoint functors and triples, *Illinois J. of Math.,* **9**, 1965, 381–398

P. Freyd, *Abelian categories,* Harper and Row, 1964

P. Freyd, Algebra valued functors in general categories and tensor products in particular, *Colloq. Math.,* **14**, 1966, 89–106

P. Freyd and A. Scedrov, *Categories - Allegories,* North-Holland, 1990

P. Gabriel, Des catégories abéliennes, *Bull. Soc. Math. France,* **90**, 1962, 323–448

P. Gabriel and F. Ulmer, Lokal präsentierbare Kategorien, *Springer LNM,* **221**, 1971

J. Gray, Fibred and cofibred categories, *Proc. Conf. on Cat. Alg. La Jolla 1965, Springer,* 1966, 21-83

P. Grillet, Regular categories, *Springer LNM,* **236**, 1971, 121-22

A. Grothendieck, Sur quelques points d'algèbre homologique, *Tohôku Math. J.,* **9**, 1957, 119–221

A. Grothendieck and J.L. Verdier, SGA4, 1963-1964, exposé 1: Préfaisceaux, *Springer LNM,* **269**, 1972, 1–218

H. Herrlich, Categorical topology, *Gen. Top. and its Appl.,* **1**, 1971, 1–15

J. Isbell, Subobjects, adequacy, completeness and categories of algebras, *Razprawy Mat.,* **36**, 1964, 1–32

T. Jech, *Set theory,* Academic Press, 1978

P.T. Johnstone, Adjoint lifting theorems for categories of algebras, *Bull. London Math. Soc.,* **7**, 1975, 294–297

J. Kelley, *General topology,* Van Nostrand, 1955

G.M. Kelly, Monomorphisms, epimorphisms and pullbacks, *J. of the Austr. Math. Soc.,* **9**, 1969, 124–142

G.M Kelly, Adjunction for enriched categories, *Springer LNM,* **106**, 1969, 166–177

G.M. Kelly, *Basic concepts of enriched category theory,* London Math. Soc. Lecture Notes, **64**, 1982, Cambridge University Press

G.M. Kelly and S. Mac Lane, Coherence in closed categories, *J. of Pure and Appl. Alg.,* **1**, 1971, 97–140

J. Kennisson, On limit preserving functors, *Illinois J. of Math.,* **12**, 1968, 616–619

H. Kleisly, Every standard construction is induced by a pair of adjoint functors, *Proc. Amer. Math. Soc.,* **16**, 1965, 544–546

A. Kock, Closed categories generated by commutative monads, *Aarhus Univ. Prep. Series,* **13**, 1968–69

C. Lair, Catégories modelables et catégories esquissables, *Diagrammes,* **6**, 1981

J. Lambeck, *Lectures on rings and modules,* Blaisdell, 1966

J. Lambeck, Torsion theories, additive semantics and rings of quotients, *Springer LNM,* **177**, 1971

F.W. Lawvere, Functorial semantics of algebraic theories, *Proc. Nat. Acad. Sc.,* **50**, 1963, 869–872

F. Linton, Autonomous categories and duality of functors, *J. Alg.,* **2**, 1965, 315–349

F. Linton, Some aspects of equational categories, *Proc. of the Conf. on Cat. Alg. La Jolla 1965, Springer* , 1966, 84–94

S. Lubkin, Embedding of abelian categories, *Trans. of the Amer. Math. Soc.,* **97**, 1960, 410-417

S. Mac Lane, *Categories for the working mathematician,* Springer, 1971

E. Manes, *Algebraic theories,* Springer, 1976

J. Mersch, Chasses sur les diagrammes des catégories abéliennes, *Bull. Soc. Roy. Scient. Liège,* **33**, 1964, 411–450

B. Mitchell, The full embedding theorem, *Amer. J. of Math,* **86**, 1964, 619–637

B. Mitchell, *Theory of categories,* Academic Press, 1965

K. Morita, Duality for modules and its applications to the theory of rings with minimum conditions, *Sci. Rep. Tokyo, Kyoiku Daigaku,* **6**, 1958, 83–142

R. Paré and D. Schumacher, Abstract families and the adjoint functor theorem, *Springer LNM,* **661**, 1978, 1–125

B. Pareigis, *Categories and functors,* Academic Press, 1970

M.C. Pedicchio and S. Solomini, On a "good" dense class of topological spaces, *J. of Pure and Appl. Algebra,* **42**, 1986, 287–295

N. Popescu, *Abelian categories with applications to rings and modules,* Academic Press, 1973

N. Popescu and L. Popescu, *Theory of categories,* Stijhoff & Noordhoff, 1979

H. Schubert, *Categories,* Springer, 1972

B. Stenström, *Rings of quotients,* Springer, 1975

G. Wraith, Algebraic theories, *Aarhus Univ. Lec. Notes Series,* **22**, 1970

Index

The page number I.123 indicates page 123 in volume I